An Introduction to the Practical Aspects of Clinical Hyperthermia

An Introduction to the Practical Aspects of Clinical Hyperthermia

Edited by
S. B. Field and J. W. Hand

MRC Cyclotron Unit, Hammersmith Hospital,
Ducane Road, London W12 0NS

Taylor & Francis
London • New York • Philadelphia
1990

UK	Taylor & Francis Ltd, 4 John St., London WC1N 2ET
USA	Taylor & Francis Inc., 1900 Frost Rd., Suite 101, Bristol, PA 19007

British Library Cataloguing in Publication Data

An introduction to the practical aspects of clinical hyperthermia.
  1. Man. Cancer. Hyperthermia therapy
  I. Field, S. B. II. Hand, J. W.
616.99'40632

ISBN 0-85066-788-7

Library of Congress Cataloging-in-Publication Data

An Introduction to the practical aspects of clinical hyperthermia/edited by S. B. Field, J. W. Hand
      p. cm.
  Includes bibliographical references.
  ISBN 0-85066-788-7
  1. Cancer—Thermotherapy. 2. Clinical trials. I. Field, Stanley B. II. Hand, Jeffrey, W.,
  [DNLM: 1. Hyperthermia, Induced.   WB 469 I61]
RC271.T5158 1990
616.99'40632—dc20
DNLM/DLC   90–10719
for Library of Congress   CIP

Cover design by Caroline Archer

Typeset in 10/12 Century School Book by Chapterhouse, The Cloisters, Formby

Printed in Great Britain by Burgess Science Press, Basingstoke on paper which has a specified pH value on final paper manufacture of not less than 7.5 and is therefore 'acid free'.

Contents

Contents

Contributors

J. B. ANDERSEN is Professor at the Institute of Electronic Systems, Aalborg University, Fredrik Bajers Vej 7, 9220 Aalborg, Denmark.

T. C. CETAS is Professor of Radiation Oncology and Head of Division of Physics, Department of Radiation Oncology, University of Arizona Health Sciences Center, Tucson, AZ 85724, USA. He is also Professor in the Departments of Electrical and Computer Engineering and Aerospace and Mechanical Engineering, University of Arizona.

C–K CHOU is a Research Scientist and Associate Director at the Department of Radiation Research, City of Hope National Medical Center, Duarte, CA 91010, USA.

C. T. COUGHLIN is Chief of the Section of Radiation Oncology and Professor of Clinical Medicine at Dartmouth-Hitchcock Medical Center, Hanover, NH 03756, USA.

O. DAHL is Professor and Head of Department of Therapeutic Oncology and Radiophysics, University of Bergen, N–5021 Haukeland Hospital, Bergen, Norway.

C. H. DURNEY is Professor of Electrical Engineering at the Department of Electrical Engineering, University of Utah, Salt Lake City, UT 84112, USA. He is also Professor of Bioengineering at the Department of Bioengineering, University of Utah.

N. S. FAITHFULL is Director of Medical Research at Delta Biotechnology Limited, Castle Court, Castle Boulevard, Nottingham NG7 1FD, UK. He is also Consultant Anaesthetist in Queens Medical Centre, Nottingham, and a Special Lecturer in Anaesthesia at the University of Nottingham.

S. B. FIELD is Head of the Hyperthermia Section, MRC Cyclotron Unit, Hammersmith Hospital, Ducane Road, London W12 0HS, UK.

J.W. HAND is a Senior Scientist in the Hyperthermia Section, MRC Cyclotron Unit, Hammersmith Hospital, Ducane Road, London W12 0HS, UK.

L. HEINZL is an Associate Professor at the Institute of Electronic Systems, Aalborg University, Fredrik Bajers Vej 7, 9220 Aalborg, Denmark.

S.N. HORNSLETH is a PhD candidate at the Institute of Electronic Systems, Aalborg University, Fredrik Bajers Vej 7, 9220 Aalborg, Denmark.

J.W. HUNT is a Senior Staff Scientist at the Physics Division, The Ontario Cancer Institute and Department of Medical Biophysics, University of Toronto, 500 Sherbourne Street, Toronto, Ontario, Canada M4X 1K9.

D.S. KAPP is Professor and Director of Clinical Hyperthermia at the Department of Radiation Oncology, Stanford University School of Medicine, Stanford, CA 94395, USA.

J.J.W. LAGENDIJK is a Clinical Physicist at the Department of Radiotherapy, Academic Hospital, Heidelberglaan 100, 3584 CX Utrecht, The Netherlands.

O. MELLA is a Consultant in Medical Oncology and Radiotherapy at the Department of Oncology, University of Bergen, N–5021 Haukeland Hospital, Bergen, Norway.

J.L. MEYER is a Radiation Oncologist and Chairman, Radiation Oncology Department, St. Francis Memorial Hospital, 900 Hyde Street, San Francisco, CA 94109, USA.

J. OVERGAARD is Head of the Danish Cancer Society Department of Experimental Clinical Oncology, Norrebrogade 44, DK 8000, Aarhus C, Denmark.

K.D. PAULSEN is an Assistant Professor at the Thayer School of Engineering, Dartmouth College, Hanover, NH 03755, USA.

G.P. RAAPHORST is Head of Medical Physics at the Ottawa Regional Cancer Center, 190 Melrose Avenue, Ottawa, Ontario, Canada K1C 4K7.

P. RASKMARK is an Associate Professor at the Institute of Electronic Systems, Aalborg University, Fredrik Bajers Vej 7, 9220 Aalborg, Denmark.

H.S. REINHOLD is Professor at the Department of Experimental Radiotherapy, Erasmus University, c/o Daniel den Hoed Kliniek, Rotterdam and the Radiobiological Institute TNO, Lange Kleiweg 151, Rijswijk, The Netherlands.

P. R. STAUFFER is an Assistant Adjunct Professor at the Department of Radiation Oncology, University of California, San Francisco, CA 94143, USA.

G. C. van RHOON is a Physicist at the Department of Hyperthermia, Dr Daniel den Hoed Cancer Center, Groene Hilledijk 301, 3075 EA Rotterdam, The Netherlands.

A. P. van den BERG is a Biochemist at the Department of Experimental Radiotherapy, Erasmus University, PO Box 1738, 3000 DR Rotterdam, The Netherlands.

J. van der ZEE is a Hyperthermia Oncologist at the Department of Hyperthemia, Dr Daniel den Hoed Cancer Center, Groene Hilledijk 301, 3075 EA Rotterdam, The Netherlands.

Introduction

S. B. Field and J. W. Hand

The role of hyperthermia in human cancer therapy is currently undergoing evaluation in many institutions throughout the developed world. Hyperthermia may be defined as therapy in which tissue temperature is raised to 41 °C or higher by external means in opposition to the body's control processes whose purpose is to maintain temperature around the normal set-point. Although a rationale for using hyperthermia to treat malignant disease began to be developed in the 1960s, the task of inducing and monitoring hyperthermia in patients has proven to be technically difficult. Physicists and engineers have developed techniques which are currently being evaluated by oncologists in the clinic but there is still a need for improved methods, particularly for the treatment of deep-seated tumours. Nevertheless, clinical results achieved so far appear promising and some controlled, randomized clinical trials are underway.

In this book, a group of biologists, engineers, physicists and oncologists review aspects of clinical hyperthermia ranging from the biological rationale for its use in the treatment of malignant disease, through various means of clinical implementation, to projections of the future role and practice of this form of therapy. The purpose is to assimilate the considerable expertise within the group of contributors into a volume which will be a useful and up-to-date source of practical information for those entering or working in the field of hyperthermic oncology.

1. Historical perspectives of hyperthermic therapy

The application of heat in the treatment of disease has been practised for centuries. In India, the Ayurvedic system which involves a month-long programme of feeding with oils and rice, heating by steam baths and the administration of purgatives, has been used since around 3000 BC. It is still practised today for a variety of illnesses. The Edwin Smith papyrus, also dating from around 3000 BC, describes the use in ancient Egypt of a 'fire drill' to burn away a tumour in the breast. Therapeutic use of more-moderate temperature has also been known for a long time. In the fourth century, Rufus of Ephesus observed that the occurrence of fever could be beneficial in many diseases, including cancer. More recently, in 1866, W. Busch, a German physician, described the disappearance of a sarcoma on the neck of a patient who developed a high fever associated with erysipelas. Similar reports were made by others some 20 years later (Fehleisen, 1883; Bruns, 1888). Reports such as these led W.C. Coley, a surgeon in New York, to administer bacterial pyrogens to cancer patients (Coley, 1893). In 1898, F. Westermark, a Swedish gynaecologist, used a coil containing hot water as a controlled localized source of heat in the treatment of uterine tumours.

Such early studies and observations were followed by many reports of tumours responding to both whole-body and localized hyperthermic treatments induced by a variety of techniques. Shortly after the discovery of X-rays by Roentgen in 1895, the use of hyperthermia as a radiosensitizing agent was suggested by Schmidt (1909). In the 1920s and 1930s, diathermy was often used in the treatment of cancers in a wide range of anatomical sites (Rohdenburg and Prime, 1921; Arons and Sokolov, 1937). Many of the early reports describing the uses of various forms of diathermy claimed frequency-specific effects for the electromagnetic energy involved and it was Mittleman *et al.* (1941) who, recognizing the need for careful dosimetry, measured temperatures and related them to absorbed energy. As technology developed, higher-frequency electromagnetic fields were used. An early report of the therapeutic use of microwaves (375 MHz) was that by Denier in 1936. Krusen *et al.* (1947) and Leden *et al.* (1947) pioneered microwave diathermy shortly after powerful sources of higher frequency fields (≈3000 MHz) had become available for non-military applications. The work of Schwan (1957, 1959) and Lehmann *et al.* (1962, 1965) led to a deeper understanding of electromagnetic heating, highlighting the problems associated with the very limited penetration of 2450 MHz fields in tissues.

The use of ultrasound as a method of heating was suggested by Freundlich *et al.* (1932), and early reports of its clinical use were given by Pohlman *et al.* (1939) and Horvath (1944). A number of clinical studies using ultrasound alone followed but results were discouraging. In 1949 a statement to the effect that ultrasound was not suitable for cancer therapy and that its clinical use should be discontinued was made at Erlangen during the Ultrasound in Medicine Conference. During the following two decades, however, encouraging results were obtained from studies in which ultrasound was combined with ionizing radiation (Woeber, 1965). Since ultrasound had been used for heating normal tissues during physiotherapy (Lehmann, 1965), there was a resurgence of interest in developing ultrasound techniques in the 1970s as part of the increased activity in hyperthermia at that time (Lele, 1975).

Despite technical developments throughout the first half of the present century, considerable problems were encountered with heat delivery and temperature measurement. Early clinical reports were mostly anecdotal, containing little information on the temperatures actually achieved. Although details of many experimental studies were published during this time, interest in the clinical use of hyperthermia declined, principally because of the technical difficulties encountered and the rapid development of surgery, radiotherapy and chemotherapy as the major methods of treating malignant disease.

A revival of interest in hyperthermia was provided by a group in Rome (Cavaliere *et al.*, 1967) who carried out a series of biochemical studies into the effects of elevated temperature on normal and malignant cells from rodents. They observed that heat caused a greater inhibition of respiration in tumour cells than in normal ones and concluded that neoplastic cells were more sensitive to heat than their normal counterparts. Whether or not this is true remains a contentious issue, but these studies did provide a stimulus to further research in the field. The group also reported encouraging responses in 22 cancer patients treated by hyperthermic perfusion. Results were particularly good in the case of melanomas.

In recent years there has been a systematic investigation of the possible anti-cancer effects of hyperthermia with a variety of experimental studies being reported and critical analyses of a vast literature of clinical reports. The results confirm that temperature rises of only a few degrees have profound effects on cells and tissues and that hyperthermia undoubtedly has an anti-tumour effect. Numerous biological studies, mostly involving temperatures in the range 41–46 °C, have demonstrated a clear rationale for expecting hyperthermia to have a greater effect on tumours than normal tissues.

2. Rationale for the clinical use of hyperthermia

The reasons for using hyperthermia in the treatment of malignant disease rest principally on differences between the vasculature and blood supply in tumours and normal tissues, as discussed in depth in Chapter 4. Briefly, the blood supply in tumours is frequently disorganized and heterogeneous, which can result in some areas being poorly perfused, even though the average blood flow in tumours is not necessarily less than that in normal tissues. Some consequences are:

(a) Since blood flow is an important part of the cooling mechanism, some tumours or regions within tumours will have a reduced capacity for cooling and may therefore become hotter than surrounding normal tissues when subjected to a local hyperthermic-treatment field. Moreover, there may be a substantial increase in blood flow in normal tissues in response to an increase in temperature. There will be a smaller response in tumour blood flow to such an insult; indeed, there is evidence from experimental tumours that hyperthermia results in a reduction in blood flow. This would increase further any temperature differential between tumour and normal tissue. There is some clinical evidence of a temperature differential being achieved between heated tumours and adjacent normal tissues but the magnitude of this is highly variable.

(b) It is well known that deficiencies in tumour vasculature result in the development of hypoxic cells. Such cells are resistant to ionizing radiation. Various regimens have been devised in radiotherapy with the aim of destroying hypoxic focii, but the clinical results, in terms of improving tumour control, have not been particularly encouraging, although research in this area is on-going. In contrast, hypoxia has relatively little effect on the response of cells to hyperthermia. However, those cells which are short of oxygen are likely to switch to anaerobic metabolism, the end product of which is lactic acid, and so hypoxic regions are likely to be at low pH. This is particularly relevant for tumours which appear to have a natural tendency towards anaerobic metabolism. In addition, such regions are deprived of nutrients and it is known that both low pH and nutrient deficiency sensitize cells to killing by hyperthermia.

Combined treatment with hyperthermia and radiotherapy might also be expected to have therapeutic advantages. Cells in the DNA-synthetic phase of the cell cycle are particularly sensitive to hyperthermia but relatively resistant to X-rays so that a treatment involving both agents may, under some circumstances, be advantageous. Also, as outlined in (b), those tumour cells which are most resistant to X-rays are likely to be the most sensitive to hyperthermia. Thus the use of both agents should result in increased killing of tumour cells. The combination of hyperthermia and chemotherapeutic agents may also lead to an enhanced therapeutic effect since uptake of a drug and its cytotoxicity may be increased.

3. Technical aspects of clinical hyperthermia

Techniques in clinical hyperthermia may be classified into three broad categories: whole body, regional and localized. Whole-body hyperthermia, in which core temperature is maintained at up to 42°C for a few hours, is technically the least demanding. Thermal energy is introduced into the body either non-invasively by one of a variety of methods, including hot air, hot wax, infra-red or radio-frequency radiation, or invasively (e.g. by heating the blood extra-corporeally) whilst thermal losses are minimized (e.g. using insulating blankets). Although a relatively uniform temperature can be achieved within deep-seated tumours, the maximum permissible temperature is relatively low.

In the two other categories, the aim is to achieve temperatures in the range 42–45°C. With regional hyperthermia, large portions of the trunk or whole limbs are heated and often a difference in perfusion of the tumour and normal tissue is relied upon to produce a temperature differential. In localized hyperthermia a smaller volume of tissue local to the tumour is heated. Most methods used for this purpose involve non-ionizing electromagnetic waves (radiofrequencies and microwaves) or ultrasound, although in some cases extracorporeal heating of the blood (e.g. for regional hyperthermia of limbs) and direct heat sources (e.g. interstitial implants heated by magnetically induced currents, DC electric currents or hot water) may be used. The techniques may be either non-invasive or invasive, the latter using interstitial implants or intracavitary devices inserted into body cavities.

To date, most experience has been gained with non-invasive local treatments for superficial tumours. In the case of electromagnetic techniques, the maximum depth at which effective heating can be achieved is approximately 2 cm. In situations where ultrasound can be used this is increased to around 4 cm. To achieve electromagnetic energy deposition in deep tissues, a relatively low frequency (below ≈ 100 MHz) must be used. Since the minimum dimensions of the 'focus' are about half of the wavelength in the medium (≈ 14 cm at 100 MHz in tissue with high water content) and heat transport mechanisms acting within the body distribute this energy to an even larger volume, localization of heating is then lost. In the case of ultrasound, highly focused beams can be used since the wavelengths involved are of the order of 1 mm. Thus, local heating at depth is possible in sites where this modality can be used.

The second major technical problem in clinical hyperthermia is that of thermometry. Ideally, the temperature distribution produced within the treated region would be measured non-invasively with adequate spatial and temporal resolutions. Although several techniques are under investigation in the laboratory (e.g. microwave or ultrasound radiometry, applied potential tomography, magnetic resonance imaging), the possible clinical use of these methods (perhaps with the exception of microwave radiometry) is unlikely in the near future. On the other hand, invasive thermometry has been developed into a useful clinical procedure, although progress is needed in areas such as determining the locations of sensors relative to applicators and of estimating temperatures at locations other than those of the sensors.

4. Overview of the book

The response of cells and tissues to elevated temperature is frequently compared with that following ionizing radiation since much of the research into thermobiology follows closely the principles laid down in radiobiology. The first chapter, by G. P. Raaphorst, is concerned with the effects of hyperthermia on mammalian cells. Here the possible mechanisms by which elevated temperature can lead to cell death are discussed. Factors which affect thermal sensitivity and substances which may be used to modify it are considered, as is the question regarding the relative sensitivities of normal and neoplastic cells. Chapter 2 by S. B. Field is a discussion of *in vivo* responses with emphasis on normal tissues. Tissue studies on thermotolerance, interactions between hyperthermia and other modalities, particularly radiation, and time–temperature relationships are fundamental for planning hyperthermic treatment on a rational basis.

The question of how to define a thermal dose is dealt with by S. B. Field and G. P. Raaphorst in Chapter 3. This is a difficult area in which there is currently no general agreement. The concept of dose and how it relates to thermobiology is discussed and the most commonly accepted formulation is described.

Central to the rationale for the use of hyperthermia in cancer therapy are the differences between blood flow and metabolism in tumours and normal tissues and how these differences may change as a result of hyperthermia. The role of the vasculature, including its modification by heat or other modalities, and changes in microcirculation and energy metabolism are topics discussed by H. S. Reinhold and A. P. van den Berg in Chapter 4.

The potential for combining hyperthermia with chemotherapy is reviewed in Chapter 5 by O. Dahl and O. Mella. The chapter includes a discussion of the mechanisms of interaction between hyperthermia and anticancer drugs, a review of the results of experimental studies indicating possibilities for enhancing the therapeutic effects and a review of current clinical experience.

Non-invasive heating in clinical practice is the topic of Chapter 6 by D. S. Kapp and J. L. Meyer. The techniques used to treat superficial or deep-seated tumours and the possibilities for thermometry are discussed from a clinical view point and are related to the factors which determine clinical outcome and toxicities. In Chapter 7, C. T. Coughlin presents an overview, also from a clinical point of view, of the three main techniques for interstitial hyperthermia: radio-frequency currents, microwave antennas and ferromagnetic seeds. The relative advantages and drawbacks of the methods are discussed, as are likely future developments in interstitial methods. The topic of whole-body hyperthermia (WBHT) is reviewed by J. van der Zee and her colleagues in Chapter 8. Following a discussion of the principles of WBHT, the technical performance of a system is described in detail. An important consideration in WBHT is the induced physiological disturbance. The chapter ends with a review of the clinical efficacy and toxicities of this form of hyperthermic therapy.

The progress of clinical hyperthermia depends on the outcome of carefully designed clinical trials (which in turn depend on the availability of suitable techniques for heating and measuring temperatures). The topic of clinical trials is the subject of Chapter 9 by J. Overgaard. Following a discussion of protocol design, the clinical indications obtained to date are reviewed with emphasis on the potential for the combination of hyperthermia and radiotherapy. The chapter also includes a discussion on the selection of tumours for controlled trials. This is particularly important in the case of hyperthermia, bearing in mind the technical limitations of heat delivery.

Technical aspects of clinical hyperthermia are addressed in Chapters 10 through 18. In Chapter 10, C. H. Durney describes the basic principles of electromagnetic fields, how these fields interact with tissues and the ways in which they are generated and transmitted. In Chapter 11, L. Heinzl and his colleagues continue the theme of electromagnetic aspects of hyperthermia by discussing the various types of applicator which are used to transfer electromagnetic energy to the patient. The general categories into which these applicators can be grouped are described and illustrated by specific examples. There is also a discussion of the practical aspects of feeding applicators with radio-frequency power. In Chapter 12, the discussion of electromagnetically induced hyperthermia is extended to the realm of computer models of various applicators. K. D. Paulsen provides a comprehensive survey of solution methods. The chapter outlines current capabilities in this important field and concludes with the author's insight into the likely future directions in this area. A common trend in these chapters is that, despite the inherent mathematical nature of the topics covered, the authors, mindful of the varied backgrounds of their readers, have avoided detailed mathematical accounts.

In Chapter 13, P. R. Stauffer describes the basic principles of the techniques available for interstitial hyperthermia. Details of methods employing 'hot sources' as well as the more commonly used radio-frequency and microwave techniques are given. Consideration is also given to the requirements for temperature measurement, and the chapter concludes with general guidelines on choosing a method of heating. In Chapter 14 the discussion moves to the use of ultrasound in clinical hyperthermia. In a comprehensive review of the topic, J. W. Hunt discusses the basic physical aspects underlying these techniques through to the design and performance of various ultrasound systems.

An area of major importance in clinical hyperthermia is that of thermometry. In a review of this topic in Chapter 15, T. C. Cetas begins with a discussion of the fundamentals of thermometry. The basic principles of the various types of thermometer used in hyperthermia are then described. The methods available for calibration are described and followed by a detailed discussion of the various sources of error which may occur in clinical thermometry and techniques for minimizing these. Finally, the problem of locating precisely the position of probes implanted in patients is addressed.

Heat transport mechanisms in tissues and the development of thermal models are topics of great importance in treatment planning. In Chapter

16, J. J. W. Lagendijk begins a review of heat transfer mechanisms by considering fundamental aspects. Subsequently, heat transfer in vascularized tissues is addressed, leading to a discussion of the role of the conventional bioheat transfer equation in thermal modelling.

J. W. Hand addresses the topic of quality assurance in hyperthermia in Chapter 17. The discussion covers our current capabilities in treatment planning and guidelines regarding the choice of applicator for a particular treatment. Various aspects of a quality assurance programme are outlined including means of characterizing heating devices and the performance and calibration requirements for thermometry. The chapter ends with a discussion of treatment practices, including the need for adequate documentation of data. In the final chapter, C. K. Chou provides a comprehensive review of the biological effects of electromagnetic radiation waves and ultrasound together with a discussion of safety standards and criteria relating to exposure to these non-ionizing radiations. The chapter concludes with practical recommendations to ensure that hyperthermic treatments are both safe and effective.

References

Arons, I. and Sokolov, B. (1937) Combined roentgenotherapy and ultra-shortwave, *American Journal of Surgery*, **36**, 533–543.

Bruns, P. (1888) Die Heilwirkung des Erysipels auf Geschwulste, in Bruns, P. (Ed.), *Bietrage zur klinischen Chirurgie. Mittheilungen aus der chirurgischen Klinik zu Tubingen*, Vol. 3, Tubingen, Verlag der H. Laupp'schen Buchhandlung, pp. 443–466.

Busch, W. (1866) Uber den Einfluss welche heftigere Erysipelen Zuweilig auf organisierte Neuildungenausuben, *Verhandlugen Naturhist. Preuss. Rhein. Wesphal.*, **23**, 28–30.

Cavaliere, R., Ciocatto, E. C., Giovanella, B. C., Heidelberger, C., Johnson, R. O., Margottini, M., Mondovi, B., Morrica, G. and Rossi-Fanelli, A. (1967) Selective heat sensitivity of cancer cells, *Cancer*, **20**, 1351–1381.

Coley, W. C. (1893) The treatment of malignant tumors by reported inoculations of erysipelas: with a report of ten original cases, *American Journal of Medical Science*, **105**, 486–511.

Denier, A. (1936) Les ondes herziennes ultracourtes de 80 cm, *Journal de Radiologie et d'Electricite*, **20**, 193–197.

Fehleisen, R. (1883) *Die Atiologie des Erysipels*, Berlin, Fischer.

Freundlich, H., Sollher, K. and Fogowski, F. (1932) Clinige biologische wircungen von ultraschallwellen, *Klinische Wochenschrift*, **11**, 1512–1513.

Horvath, J. (1944) Ultraschallwirkung beim menschlichen Sarkom, *Strahlentherapie*, **75**, 119.

Krusen, F. H., Herrick, J. F., Leden, U. N. and Wakim, K. G. (1947) Preliminary report of experimental studies of the heating effect of microwaves (radar) in living tissue, *Proceedings of Staff Meeting, Mayo Clinic*, **22**, 209–224.

Leden, U. N., Herrick, J. F., Wakim, K. G. and Krusen, F. H. (1947) Preliminary studies on the heating and circulating effects of microwaves (radar), *British Journal of Physical Medicine*, **10**, 177–184.

Lehmann, J. F. (1965) Ultrasound therapy, in Licht, S. (Ed.), *Therapeutic Heat and Cold*, New Haven, Conn., Elizabeth Licht, pp. 321–386.

Lehmann, J. F., McMillan, J. A., Brunner, G. D. and Johnston, V. C. (1962) Heating patterns produced in specimens by microwaves of the frequency 2456 MHz when applied by A, B, and C directors, *Archives of Physical Medicine*, **43**, 538–546.

Lehmann, J. F., Johnston, V. C., McMillan, J. C., Silverman, D. R., Brunner, G. D. and Rathbun, L. A. (1965) Comparison of deep heating by microwaves at frequencies 2456 and 900 megacycles. *Archives of Physical Medicine*, **46**, 307–314.

Lele, P. P. (1975) Hyperthermia by ultrasound, in Wizenberg, M. J. and Robinson, J. E. (Eds.), *Proceedings of the International Symposium on Cancer Therapy by Hyperthermia and Radiation*, Philadelphia, The American College of Radiology, pp. 168–175.

Mittlemann, E., Osborne, S. L. and Coulter, J. S. (1941) Short wave diathermy power absorption and deep tissue temperature, *Archives of Physical Therapy*, **22**, 133–139.

Pohlman, R., Richter, R. and Parow, E. (1939) Uber die Austbreitand und Absorption des ultraschalls im mensleichen Gewebo und Seintherapeutisch Wirkung an Ischias Plexusneuralgie, *Deutsch Medische Wochenschrift*, **65**, 251–256.

Rohdenberg, G. L. and Prime, F. (1921) Effect of combined radiation and heat on neoplasms, *Archives of Surgery*, **2**, 116–129.

Schmidt, W. E. (1909) Zur Rontgenbehandlung tiefliegender Tumoren, *Fortscr. Roentgenstr.*, **14**, 134.

Schwan, H. P. (1957) Electrical properties of tissues and cells, *Advances in Biology, Medicine and Physics*, **5**, 147–209.

Schwan, H. P. (1959) Alternating current spectroscopy of biological substances, *Proceedings of IRE*, **47**, 1841–1855.

Westermark, F. (1898) Uber die Behandlung des ulcerirenden Cervixcarcinoms mittels konstanter Warme. *Zentralblatt für Gynaekologie*, **22**, 1335–1339.

Woeber, K. (1965) The effect of ultrasound in the treatment of cancer, in Kelly, E. (Ed.), *Ultrasonic Energy*, Urbana, University of Illinois Press, pp. 135–149.

Disclaimer

1 Fundamental aspects of hyperthermic biology

G. P. Raaphorst

1.1. Introduction

Temperature plays a major role in many biological processes. The structure and function of cells, their constituent organelles and macromolecular function, the balance between stability and dynamics, have all been shown to tolerate a relatively narrow temperature range (for reviews see Pain, 1987; Lee and Chapman, 1987).

The advent of the use of hyperthermia in cancer therapy has led to an increase in research into cellular and basic mechanisms underlying the response of biological systems to raised temperatures (Dethlefsen and Dewey, 1982; Overgaard, 1984). Even though the mechanisms are still not well understood there is a growing wealth of data that can be used to aid design of hyperthermia treatment and to avoid major pitfalls. In this chapter some aspects of cellular thermal biology and some of the possible mechanisms underlying these responses will be reviewed. Since the body of literature concerning cellular hyperthermia responses is very large it is not possible to cover all areas and, therefore, only selected topics will be discussed. Because much of the research on thermal biology has been done in relation to its properties as a radiosensitizer, and many of the techniques in radiobiology have been employed in these investigations, comparisons to radiation effects and some discussion of radiosensitization will be presented.

1.2. Heat sensitivity of mammalian cells

1.2.1. General features

Sensitivity of cells in culture to therapeutic regimens can provide a guidance to therapy, and differential sensitivity to mammalian cell lines bears on the potential for therapeutic gain.

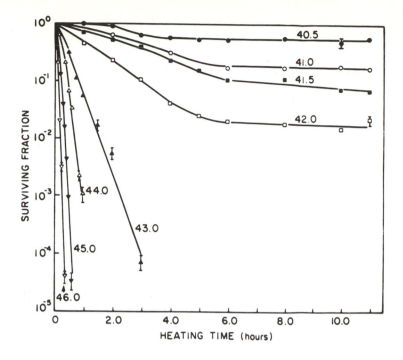

Figure 1.1 Survival of Chinese hamster V79 cells after heating in a water bath for the times and temperatures shown. (From Raaphorst and Szekely (1988) with permission.)

The response of Chinese hamster cells to hyperthermia shows that it is both time and temperature dependent (Figure 1.1). These types of responses have been noted for many different animal and human cell lines studied in culture. At low hyperthermic temperatures a plateau of resistance called thermotolerance is reached while at higher temperatures this does not develop during heating but can develop after heating (Henle and Dethlefsen, 1987). The combination of time and temperature determine the degree of cell killing and comprise a thermal dose (see Chapter 3). The data in Figure 1.1 which are consistent with results *in vivo* (Chapter 2) show that relatively small changes in temperature can have a large effect on cell killing. Therefore, careful temperature monitoring during clinical hyperthermia is important.

1.2.2. Arrhenius analysis

The kinetics of cell killing by hyperthermia can be analysed in terms of Arrhenius plots. Such analysis can be used to relate time and temperature for thermally induced end-points, such as cell killing, and can form the basis for development of a thermal dose as discussed in Chapter 3.

Dewey *et al.* (1977) presented an analysis of cellular heat inactivation based on the thermodynamics of heat inactivation described by Johnson *et al.* (1954) and Snell *et al.* (1965). From this analysis the equations for Arrhenius presentation were developed, i.e.

$$1/D_0 = A\,e^{-\Delta H/2T} \tag{1.1}$$

or

$$\ln 1/D_0 = \ln A - (\Delta H/2)(1/T), \tag{1.2}$$

where $1/D_0$ represents the reaction rate constant, D_0 the slope of the heat survival curve, ΔH the activation energy, T the absolute temperature and A are effectively constant over a narrow temperature range of 43–46°C. Subsequent discussions by Henle (1983) indicated that caution should be exercised in the use of these analyses because irreversibility of cell death cannot be approximated by near-equilibrium conditions inherent in the development and the use of this model. Caution should also be exercised in using this model for indicating mechanisms because instead of one rate-limiting step in a series of steps, the molecular events may be coupled involving feedback and co-operativity. However, keeping these potential pitfalls in mind the model can be used to compare cell killing by hyperthermia and to develop the concept of thermal dose.

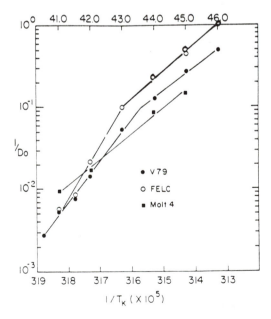

Figure 1.2 Arrhenius plots of the inverse of the slope of the heat survival curves versus the inverse absolute temperature of three cell lines heated over a temperature range from 41 to 46°C. (Data from Raaphorst *et al.* (1983b) with permission.)

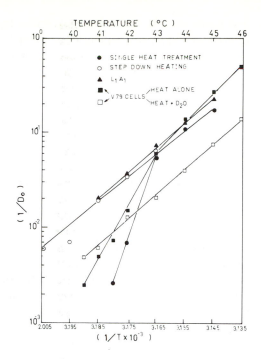

Figure 1.3 Arrhenius plots of Chinese hamster ovary (CHO) cells given single heat treatment and step-down heat treatment (data redrawn after Henle, 1980). The response of L_1A_1 cells (redrawn after Nielson and Overgaard, 1979) and Chinese hamster lung (V79) cells heated in the presence or absence of D_2O (redrawn from Azzam *et al.*, 1982) is also shown.

The data shown in Figure 1.1 can be presented in the Arrhenius form and plotted using equation (1.2) (shown in Figure 1.2). The slope of the curve gives the activation energy, and for the data for hamster cells and erythroleukaemia cells, and from the results presented by others (Henle, 1983; Bauer and Henle, 1979; Henle and Dethlefsen, 1980), it is clear that the Arrhenius plot can be defined by at least two different slopes over the temperature range 41–46°C. In fact, the work of Landry and Marceau (1978) showed that the Arrhenius plot of cell survival data ranging over a broad temperature range of 41–55°C had a continuously changing slope. The Arrhenius plot can be used to establish the relationship between time and temperature for isoeffect in cell killing, and is often defined by two slopes. The factor (R) relating time and temperature is different above ($R=2$) and below ($R=4–8$) the transition point in the Arrhenius plot (Dewey *et al.*, 1977; Sapareto *et al.*, 1978; Gerner, 1984) (see Chapter 3).

It is well known that at these lower temperatures thermotolerance develops during heating (see Figure 1.1) and that this could influence the slope (D_0) calculated from the survival curve and thus affect the value of R. The three sets of data presented in Figure 1.3 provide evidence that the value of R at temperatures less than 43 °C was probably greater due to the onset of thermotolerance. First, the results of Azzam *et al.* (1982) show that treatment with D_2O can result in the elimination of the development of thermotolerance and also the transition in the Arrhenius plot. Secondly, when L_1A_2 cells (Figure 1.3) and human T-leukaemic cells (MOLT-4) (Figure 1.2) are heated at temperatures of 41–43 °C they do not develop thermotolerance at lower temperatures and the data plotted in the Arrhenius form does not show a transition. Thirdly, when cells are heated at higher temperatures before heating at temperatures less than 43 °C (step-down heating), thermosensitization occurs, thermotolerance development at low temperatures is inhibited and the transition in the Arrhenius plot is eliminated (Henle, 1983). Thus, intrinsically the R value may not change but the process of thermotolerance causes an apparent change. Hence R could be represented by a single value and $R' = R + TF$ should be a modified value to account for thermotolerance, where TF is a thermotolerance factor.

Inactivation energies ranging from 500 to 660 kJ/mol have been calculated in the temperature range 41–46.0 °C when thermotolerance is absent. This possibly represents a protein inactivation process, because this is the reported energy range of protein denaturation (Johnson *et al.*, 1954; Brown and Crozier, 1927). At higher temperatures, the work of Landry and Marceau (1978) indicate lower inactivation energies which may imply a different inactivation process. However, since many biological processes are co-operative and controlled by feedback, caution should be exercised in using only inactivation energy as proof for, or identification of, a specific target in thermal inactivation. As seen in the following sections, inactivation may be a highly interactive process of many aspects of cell structure and function.

1.2.3. Normal versus transformed cells

Early studies indicated that the transformed cell may be more sensitive than the normal cell and this was thought to be one of the factors contributing to therapeutic advantage in clinical hyperthermia (Cavalieri *et al.*, 1967; Gionvanella *et al.*, 1973, 1976).

Table 1.1 summarizes some *in vitro* studies comparing the thermal sensitivities of normal and malignant cells. Many of the studies indicate that transformed cells are more thermally sensitive than normal cells although some studies indicate no differences. Some of the documented differences may be related to the enhanced glycolysis observed in transformed cells, which results in lactic acid production and acidification of the culture medium, thereby causing thermal sensitization. Our studies with the C3H–10T1/2 cell system show that cells transformed by X-rays or by transfection with oncogenes display a range of thermal sensitivities,

Table 1.1 Thermal sensitivity of normal and malignant cells.

Reference	Cell types	Temperature (°C)	Relative sensitivity[1]
Giovanella et al. (1976)	Melanocytes: melanoma	43	T > N
	Intestinal: colon carcinoma	43	T > N
	Fibroblasts: fibrosarcoma	43	T > N
	Neuroepithelial: terotocarcinoma human cells	42.5	T > N
Giovanella et al. (1973)	Mouse embryonal: MCA-induced sarcomas	42.5	T > N
Hahn (1980)	C3H–10T1/2 normal: transformed	41–43.5	T > N[2]
	Not correlated with the degree of malignancy[2]		
Symonds et al. (1981)	Mouse haemopoietic stem cells: leukaemic L1210	43	T ≈ N
Okumura et al. (1979)	Human normal: malignant	39–44	T > N
Chen and Heidel-berger (1969)	Mouse prostate normal: transformed	43	T > N
Levine and Robbins (1970)	Human normal: cancer	42	T > N[2] Plateau phase[2]
Kase and Hahn (1976)	Human normal: SV 40T transformed	43	T > N log, T ≈ N plateau
Auersperg (1966)	Human fibroblasts: carcinoma (epithelial)	46	T < N
Kachani and Sabin (1969)	Hamster normal: X-ray transformed	41.6	T ≈ N
	Normal: drug transformed		T ≈ N
	Normal: virus transformed		T ≈ N
Robbins et al. (1983)	AKR mouse bone marrow:	41.8	
	Leukaemia	42.5	T > N
Raaphorst et al. (1985, 1987b)	C3H–10T1/2	42–45	
	Normal: X-ray transformed		T < N, T = N
	Normal: H-Ras transformed		T > N, T = N

[1] T=tumour or transformed cell; N=normal cell.
[2] From Raaphorst and Szekely (1988) with permission.
See this review for individual references.

some of which are greater than the normal parental cell line and some which are less (Raaphorst *et al.*, 1985b, 1987). Although large differences exist in cellular and tissue thermal sensitivity, one cannot generalize that such differences relate to transformed versus normal cells. The *in vivo* response of tumours may be more strongly modulated by pH, oxygen status, nutrient status and other micro-environmental factors (Gerweck, 1988).

1.2.4. Differential cell sensitivity

Differences in thermal sensitivity of mammalian cells have been studied for seven different cell lines cultured under identical conditions. Since the nature of the cell culture medium can play a role in thermal sensitivity, comparisons should be made with caution (Raaphorst *et al.*, 1979b; Raaphorst and Azzam, 1980a). The data presented in Figure 1.4 show the following:

(a) Cellular heat sensitivity can vary extensively even when the cells are cultured under identical conditions. For example, after 5 h of heating at 42.5 °C there was greater than four orders of magnitude difference in cell killing between mouse LP59 and pig kidney cells. Other studies have also shown leukaemic cells to be thermally sensitive (Raaphorst *et al.*, 1983a, b; Robbins *et al.*, 1983).

(b) Thermotolerance did not develop in the same way in all cell lines. In five of the cell lines it developed after 4–5 h of heating, and was characterized by a resistant plateau in the survival curves. A plateau developed earlier and transiently for rat kangaroo cells and not at all for mouse LP59 cells. Such transient thermal tolerance plateaux were also observed for crypt survival of mouse jejunum (Hume and Marigold, 1980). In addition, a much lower maximum temperature threshold for development of thermotolerance during heating was observed in thermally sensitive leukaemia cell lines (Raaphorst *et al.*, 1983b).

(c) The sequence of relative thermal sensitivity among the cell lines changed with altering the temperature from 42.5 to 45.5 °C.

(d) Cell lines from the same animal have different sensitivity, as seen for the Chinese hamster lines heated at 42.5 °C.

(e) A shoulder on survival curve (indicating accumulation of sublethal heat damage) was present for most cell lines but not for the HeLa cell curve as has also been shown by others (Leith *et al.*, 1977).

(f) The spread in thermal sensitivity was much greater than the spread in radiosensitivity indicating a need for more caution in thermal therapy because of the potential for greater differences in tissue sensitivity.

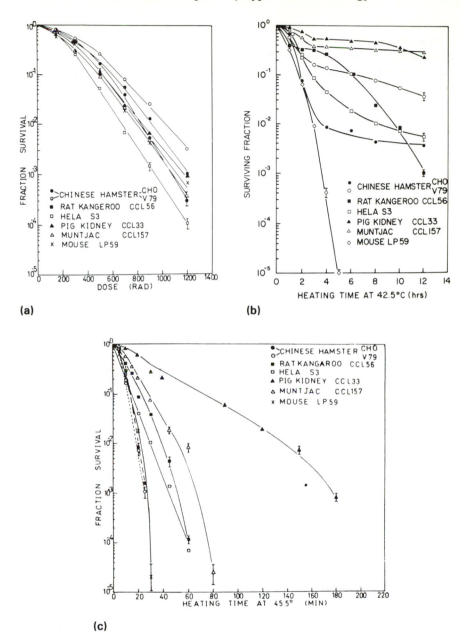

Figure 1.4 Survival of seven cell lines cultured under identical conditions: (a) survival after irradiation; (b) survival after heating to 42.5 °C; (c) survival after heating to 45.5 °C. (From Raaphorst *et al.* (1979b) with permission.)

(g) There was no correlation between thermal sensitivity and radiosensitivity, as has also been shown in other studies (Gerweck and Burlett, 1978; Raaphorst *et al.*, 1989; Kampinga *et al.*, 1986). It was also shown that thermal sensitivity could not be correlated with chromosome number, DNA content, cell volume and cell population doubling time.

Cells derived from similar tissues can also have extensive differences in thermal sensitivity as shown in Figure 1.5 for three glioma lines heated to 41 or 42°C (Raaphorst *et al.*, 1989). Such differences were also observed in human melanoma xenografts derived from different individuals (Rofstad and Brustad, 1986). This study also found a poor correlation between *in vitro* and *in vivo* thermal sensitivity of human melanoma xenografts and this may be related to micro-environmental factors, which play a crucial role in tissue thermal sensitivity (see Chapter 4).

Differences in thermal sensitivity are further emphasized in Figure 1.6 for cells derived from tissues of different origins. Endothelial cells showed a relatively high level of thermal resistance compared to the cell data presented in Figure 1.4. This might suggest that endothelial cells and,

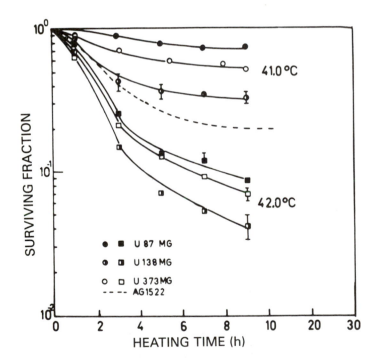

Figure 1.5 The survival of three human glioma cell lines after heating to 41 and 42°C. The dashed curve represents the survival of a normal human fibroblast line after heating to 42°C and has been included for comparison. (From Raaphorst *et al.* (1989) with permission.)

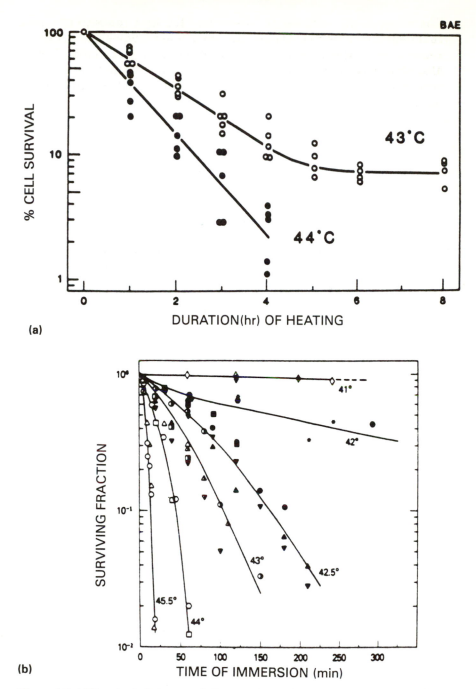

Figure 1.6 (a) Survival of exponentially growing bovine aorta endothelial cells after heating to 43 and 44 °C (from Rhee and Song (1984) with permission). (b) Survival of human bone marrow cells heated at temperatures ranging from 41 to 45.5 °C (data from Bromer *et al.* (1982) with permission).

consequently, capillary damage *in vivo* may not be a limiting factor for hyperthermia. However, care must be taken in drawing such conclusions because *in vitro* and *in vivo* results do not always agree, as was seen in the studies of Rofstad and Brustad (1986). In addition, Fajardo *et al.* (1985) using cellular attachment capacity as an end-point rather than colony formation showed human endothelial cells to be more sensitive than human fibroblasts. Further studies showed that angiogenesis was a much more heat-sensitive end-point than the attachment assay, and this may play an important role in tumour therapy. Figure 1.6 also shows the response of human bone marrow cells. These cells were very heat sensitive and did not show the development of thermotolerance.

Thus, cellular studies have shown that heat sensitivity can vary greatly depending on species, tissue of origin, as well as for tissues of similar origin. Some very general conclusions can tentatively be drawn, such as heightened sensitivity of cells derived from haematopoietic tissues compared to fibroblasts or glioma cells, but in general the variation is considerable.

1.3. Factors influencing thermal sensitivity

1.3.1. The oxygen effect

Hypoxia is known to make cells radioresistant and such radioresistance can lead to failure in cancer therapy for tumours containing hypoxic cells (Kallman, 1972). Figure 1.7, however, illustrates that aerobic and hypoxic cells show no difference in thermal sensitivity. Table 1.2 summarizes the results of reports on the effect of hypoxia on heat sensitivity and thermal enhancement of radiosensitivity. In the majority of these studies, hypoxic cells were equally heat sensitive or more sensitive than aerobic cells (i.e. the oxygen enhancement ratio (OER) is less than or equal to 1). The study by Gerweck *et al.* (1979) showed that the OER was 1 for acute hypoxia but less than 1 for chronic hypoxia, which may more closely model the clinical situation. However, many of the other studies presented in Table 1.2 also indicate an OER of less than 1 for acute hypoxia in relation to thermal cell killing. It has also been shown that for combined heat and X-ray treatments, hyperthermia caused a reduction in the radiation OER, although most results indicate that the OER remains unchanged. Thus, hypoxia in tumours may offer some advantage in therapy involving hyperthermia, or hyperthermia and radiation, compared to radiotherapy, where it clearly protects.

Table 1.2 Effect of hypoxia combined with heat and radiation.

Reference[2]	Cells	Temperature (°C)	Hypoxia	OER heat alone	OER X alone	OER heat+X[1]
Badanidiyor and Hopwood (1985)	CHO	42.5	Acute		3.1	2.5 ▽X / X▽
Bass et al. (1978)	Hela / CHO	42, 43	Acute	>1 / 1	3.1 / 3.1	
Durand (1978)	V79 small / Spheroids large	42–46	Acute	1 / <1	Normal	
Gerweck et al. (1981)	CHO	43	Acute	1	3.1	
Gerweck et al. (1979)	CHO	42	Acute / Chronic	1 / <1		
Gerweck et al. (1974)	CHO	45.5	Acute	<1	2.5	—
Kim et al. (1983)	HeLa	40.5	Acute	—	Normal	3.2 X▽
Mivechi et al. (1981)	BP-8	37–45	Acute	1	Normal	Normal X▽
Nielsen (1981)	L_1A_2	42	Acute 0–24 h	1	Normal	
Overgaard and Bichel (1977)	JB-1-E	42.5	Acute	<1		
Power and Harris (1977)	V79 / EMT6	43	Acute	1 / 1	3.0 / 3.2	3.8 ▽X / 2.9 ▽X
Rajaratnam et al. (1981)	V79	43 / 44	Acute / 16 h	>1 / >1		
Robinson et al. (1974)	Bone / Marrow	42.5 / 43	Acute	—	2.47	1.69 ▽,X / 1.38

[1] ▽ = heat treatment, X = X-rays, ▽,X = simultaneous treatment.
[2] From Raaphorst and Szekely (1988). See this review for individual references.

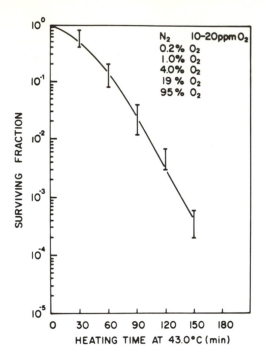

Figure 1.7 Survival of CHO cells heated under hypoxic or aerobic conditions.
(Redrawn from Gerweck et al, 1981.)

1.3.2. Nutrient dependence

It is well known that cells in solid tumours can have altered metabolic status due to the deprivation of nutrients caused by inadequate vascular development. Studies by Hahn (1974) showed that cells held in balanced salt solutions or in culture medium which was depleted or lacking in serum became thermally sensitized. This sensitization was lost in about 2 h after the culture medium was refreshed. Reduced nutrient status could cause even greater thermal sensitivity during fractionated treatments, and the recovery of cells from heat damage was also inhibited under conditions of nutrient deprivation (Li et al., 1980a; Gerweck et al., 1984). Further studies showed that the thermal sensitivity achieved by nutrient deprivation was not dependent on the ionic composition of the medium solution or on the presence of glucose in the absence of other culture media components (Raaphorst and Azzam, 1985b).

Variation in the oxygen effect may be related to variation in the nutrient level of the culture medium. Results presented in studies by Kim et al. (1978) and Gerweck et al. (1984) showed that while the OER for X-rays was not affected by glucose content in the medium the OER for thermal cell killing was strongly influenced by the presence or absence of glucose, i.e. cells became more sensitive to hyperthermia under conditions combining hypoxia with the absence of glucose. These results indicate that hypoxia and nutrient factors together may play an important role in cellular thermal sensitivity.

1.3.3. The effect of pH

Numerous studies (reviewed by Wike–Hooley *et al.*, 1984) have shown that tumour tissues can have a lower pH than normal tissues. Such differences in pH can have an impact on cancer therapy and may be exploited to provide therapeutic gain.

Thermal cell killing exhibits the opposite effect of pH to that of radiation cell killing. While many studies have indicated increased radioresistance at low pH (Freeman *et al.*, 1981; Rottinger *et al.*, 1980), cells held at acidic pH are more heat sensitive than cells at normal pH of 7.4 (Figure 1.8). These data show that the effect is especially large for low-temperature heating and thus may provide a therapeutic advantage in therapy, especially when high temperatures are difficult to achieve. The enhanced thermal sensitivity at low pH has been observed in many cell lines,

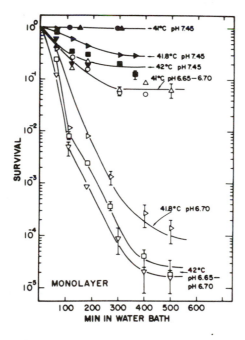

Figure 1.8 Survival of Chinese hamster ovary (CHO) cells heated in alkaline and acidic media. The pH was changed 2h before heating. (Data from Freeman *et al.* (1981) with permission.)

especially for low-temperature protracted heating (Wike–Hooley *et al.*, 1984; Meyer *et al.*, 1979). In addition, several studies showed that intracellular pH plays the major role in thermal sensitization. However, adaptation to acidic pH can result in a reduction in the degree of heat sensitization caused by low extracellular pH (Chu and Dewey, 1988; Hofer and Mivechi, 1980).

1.3.4. Cell cycle effects

It is well known that cellular radiosensitivity depends on the stage of cell cycle, cells generally being more resistant during S phase. Variation of cellular sensitivity to heat also occurs for cells in various stages of the cell cycle (Westra and Dewey, 1971). Figure 1.9 shows that Chinese hamster cells are more sensitive to hyperthermia during S phase of the cell cycle compared to G1, the opposite to X-ray sensitivity. When cells were given combined treatment of heat and radiation the variation through the cell cycle was greatly reduced. The results of Mackey and Dewey (1988) show that the G1- and S-phase cells had the same inactivation rate (Arrhenius analysis) but with a temperature 'offset' of 1.5 °C. Earlier data (Dewey *et al.*, 1971) showed that the enhanced S-phase sensitivity may be related to

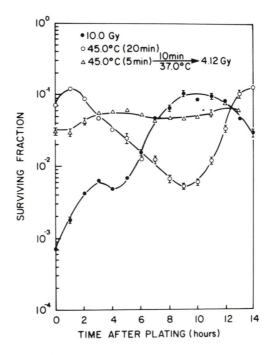

Figure 1.9 Survival of V79 cells under heat and radiation regimens at various stages of the cell cycle. Cells were synchronized by mitotic selection. (Data from Raaphorst *et al.* (1985a) with permission.)

the induction of chromosomal aberrations, which does not occur in the other phases of the cell cycle. This may also be related to the sensitivity difference between the two phases (Mackey and Dewey, 1988).

1.4. Hyperthermia and radiation

1.4.1. Thermoradiosensitization

The induction of increased radiation sensitivity by hyperthermia treatment has been observed in many mammalian cell systems studied (for reviews see Raaphorst, 1989; Konnings, 1987; Dewey, 1989).

Thermoradiosensitization depends on heating time (Figure 1.10) and temperature (Figure 1.11). The continual increase in radiosensitization by heating was also observed for heating after irradiation. At lower temperatures, such as 41 °C or 42 °C, thermoradiosensitization did not

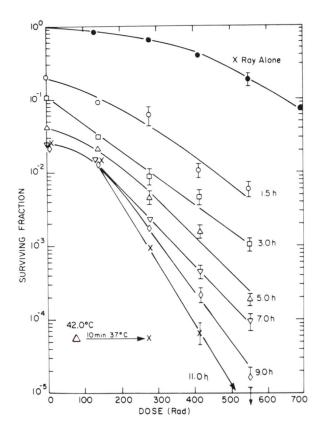

Figure 1.10 The effect of heating at 42.0 °C on the radiosensitivity of V79 cells. Heating was completed 10 min before acute X-irradiation.

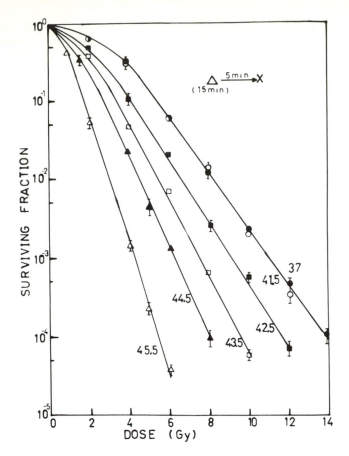

Figure 1.11 Response of V79 cells heated for 15 min at the temperatures indicated. Heating was completed 10 min before acute irradiation.

increase at long heating times, which may be due to development of thermotolerance (Spiro *et al.*, 1982a). In general, a similar approach to thermal dose may also apply to thermoradiosensitization with reasonable accuracy (Sapareto *et al.*, 1979, 1984), but the relationship depends on temperature as well as heat doses (Sapareto *et al.*, 1979) and may be modified by many factors, such as thermotolerance, treatment sequencing or culture conditions.

Thermotolerance development by heating causes both thermal resistance and decreased thermoradiosensitization (Henle, 1987b). A detailed study by Dewey and Holahan (1987) indicated that thermotolerance for heat resistance and radiosensitization occurred with roughly the same kinetics of development and decay.

Table 1.3 shows a summary of studies evaluating the development of thermotolerance and thermoradiosensitization. It is clear that not all

Table 1.3 Heat radiosensitization and thermotolerance.

Study	Cell line	Temperature (°C)	Type	Effect of thermotolerance
Freeman *et al.* (1979)	CHO	42	Chronic	+
Haveman (1983a,b)	M80/13	43/42	Acute	+,+
Harton-Eaton *et al.* (1984)	CHO	43	Acute	−
Henle *et al.* (1979)	CHO	45	Acute	+
Holahan *et al.* (1984)	CHO	42	Chronic	+
Holahan *et al.* (1986)	CHO	45.5	Acute	+
Jorristma *et al.* (1986)	HeLa-S3	42	Chronic	+
Jorristma *et al.* (1985)	HeLa-Se	44	Acute	−
		42–44		−
Majima *et al.* (1985)	CHO	44	Acute	−
Mivechi and Li (1986)	CFU	42/44	Acute	−
Nielson (1983)	L1A2	42	Acute	−
Raaphorst and Azzam (1983)	V79	41	Chronic	+
	V79	42	Chronic	−
	V79	45	Acute	+
	C3H/10T$_{1/2}$	42	Chronic	−
Streffer *et al.* (1984)	MeWo	42	Chronic	−
Van Rijn *et al.* (1984)	H35	41–42.5	Chronic	+
		42.5	Acute	+

From Raaphorst (1989). See this review for individual references.

studies show heat-induced resistance to radiosensitization. This may be in part due to the use of different heating, times, temperatures, sequencing and cell type. Two studies with Chinese hamster V79 and CHO cells demonstrated cell line dependence, tolerance in radiosensitization occurring during chronic heating at 42 °C in CHO cells but occurring at a lower temperature (41 °C) in V79 cells (Raaphorst and Azzam, 1983; Freeman *et al.*, 1979). This difference seems to be related to different temperature thresholds for development of chronic thermotolerance.

The sequencing of heat and radiation play an important role in thermo-radiosensitization and the interaction of the two types of damage. Figure 1.12 shows the sequence dependence of three different cell lines (Raaphorst *et al.*, 1983b). The data vary for different lines and indicate that maximum interaction occurs (synergistic cell killing) for the simultaneous treatments for FELC and V79 cells, but to varying degrees. This cell line dependence has also been observed in other studies (Raaphorst and Azzam, 1985a; Li and Kal, 1977; Cohen *et al.*, 1988) and may be related to the capacity of the cell to repair heat or radiation damage and/or to the conditions during incubation. Results with the radiosensitive repair-deficient AT cells show no sequence dependence, indicating possible involvement of repair (Raaphorst and Azzam, 1985a). The lack of sequence dependence in the T-leukaemia Molt-4 line and other T-leukaemia cell lines may also relate to its reduced capacity to repair damage (Szekely and Lobreau, 1985; Cohen *et al.*, 1988). Conditions of low pH and hypoxia during incubation between heat and radiation treatment can result in reduction or inhibition of recovery (Dewey *et al.*, 1980; Badanidiyoor and

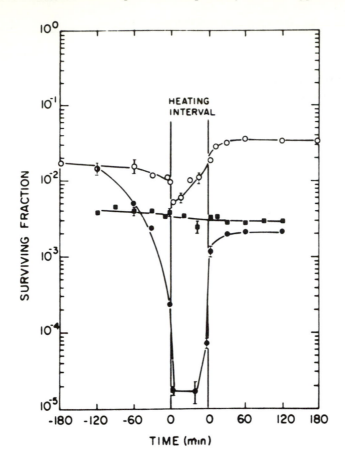

Figure 1.12 Survival of cells exposed to different sequences of heat treatment and irradiation. The negative abscissa represents incubation at 37°C between heating and irradiation. The interval between the vertical bars represents the heating time and symbols between the vertical bars indicate the results of simultaneous heat and radiation treatments. The heating intervals were 90 min at 42.0°C for V79 cells (●——●), 60 min at 42.0°C for FELC (○——○), and 60 min at 42.0°C for MOLT-4 cells (■——■). The V79, FELC, and MOLT-4 cells received X-ray doses of 6.0, 3.0, and 2.0 Gy, respectively. (From Raaphorst *et al.* (1983b) with permission.)

Hopwood, 1985; Figure 1.13). Factors such as reduced pH or reduced oxygen levels, which occur in some tumours, may lead to a therapeutic advantage if heat and radiation are separated in time, allowing normal tissue to recover while tumour tissue remains sensitized (see also Chapter 2).

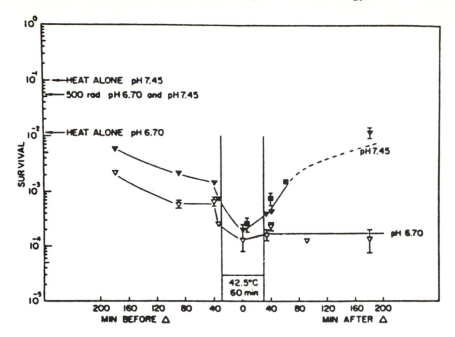

Figure 1.13 The effect of varying the sequence between heat and radiation on survival of CHO cells. Radiation (500 rad) was administered either during hyperthermia (points between the vertical bars) before heating (points to the left of the vertical bars) or after heating (points to the right of the vertical bars). The pH was maintained with carbon dioxide, at either 7.45 or 6.70, before and during the treatments. (From Dewey *et al.* (1980) with permission.)

1.4.2. Damage repair

It is well known that mammalian cells can repair radiation damage, which is operationally defined as potentially lethal damage (PLD) and sublethal damage (SLD) (Elkind and Sutton, 1960; Philips and Tolmach, 1966). The repair and/or fixation of radiation damage can lead to large changes in response of mammalian cells. The observation of recovery from radiation damage led to the question of the existence of SLD and PLD resulting from heating and whether such damage could be repaired or modified, by analogy with radiation damage.

Split-dose treatment with radiation results in restoration of the survival curve shoulder, characteristic of SLD repair, whereas for heating, split-dose treatments can result in large increases in thermal resistance characterized by an increase in the survival curve slope (D_0) and no restoration of the shoulder (Figure 1.14). This phenomenon is termed thermotolerance (see Henle and Dethlefsen, 1978) and *in vitro*, it develops in a similar time to that of SLD recovery and thus makes SLD repair

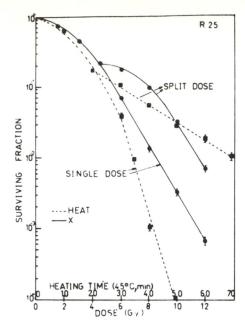

Figure 1.14 Survival of C3H/10T$_1$ transformed mouse embryo cells (R25) exposed to single or split-dose irradiation, and cells given single or split-dose hyperthermia treatment at 45.0°C. (Data from Raaphorst (1989) with permission.)

measurements difficult. Nielsen and Overgaard (1979) showed that incubation between heat fractions at 42°C first resulted in restoration of the shoulder of the survival curve at 4 h (indicating repair of SLD), and this was followed by an increase in D_0 for longer times indicating development of thermotolerance. Studies on the effect of D$_2$O treatment during split-dose recovery showed inhibition of development of thermotolerance and possibly SLD repair, but the two effects cannot be identified separately (Raaphorst and Azzam, 1982b). Recent work by Rastogi *et al.* (1988) showed that pretreatment with carbonyl-cyanide *m*-chlorophenylydrazone (an uncoupler of oxidative phosphorylation) can induce a heat response characterized by a larger survival curve shoulder, possibly indicating a larger capacity for SLD accumulation during heating.

Repair of PLD for cells held in plateau phase after treatment is shown in Figure 1.15 for mouse cells after irradiation or after heating (Raaphorst and Feeley, 1990). PLD recovery occurred after heat and radiation treatment at two survival levels tested. The half-times for R-PLD (repair of radiation-induced PLD) ranged from 1.5 to 2.5 h while the half-time for H-PLD (repair of heat-induced PLD) was much longer (7–10 h). Post-treatment by anisotonic saline caused fixation of radiation damage but not heat damage. These results show that PLD after irradiation and heating are different. Other studies showed that low temperatures

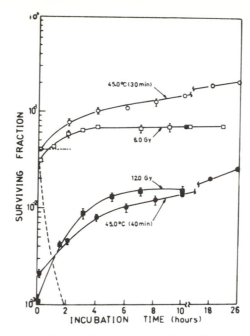

Figure 1.15 **Repair of radiation- and heat-induced (45.0°C) potentially lethal damage in C3H–10T1/2 cells incubated in plateau phase after treatment. The dotted and dashed curves show the survival of cells exposed to 0.05 M sodium chloride after heat treatment (45°C, 30 min) and irradiation (6 Gy), respectively. (Data from Raaphorst and Feeley (1989) with permission.)**

(0–4°C) inhibited R-PLD but not H-PLD, low pH inhibited H-PLD and increased R-PLD and DDC treatment inhibited H-PLD but increased R-PLD (Evans *et al.*, 1983, 1982; Li *et al.*, 1980a; Leeper and Henle, 1975). These results provide further evidence indicating that mechanisms of H-PLD and R-PLD are different. In addition, radiation damage and repair can be correlated with recovery of DNA strand breaks and chromosomal aberrations, while heat damage and its repair does not appear to be correlated with these end-points. Temporal correlations of heat PLD can be made with recovery of the normal protein-to-DNA ratio (Clark *et al.*, 1981), recovery of normal DNA synthesis (Wong and Dewey, 1982), β-polyermase activity (Spiro *et al.*, 1982b), membrane blebling (Kapiszewsak and Hopwood, 1986), RNA and protein synthesis (Warters and Stone, 1982) and cytoskeletal systems (Wachsberger and Coss, 1989), but none of these end-points have been directly implicated in cellular heat recovery. Thus, the mechanisms of hyperthermic damage and its repair remain poorly understood and need further investigation.

Several studies have shown that hyperthermia can inhibit repair of radiation-induced SLD and PLD (Raaphorst *et al.*, 1979a; Li *et al.*, 1976; Murthy *et al.*, 1977). The inhibition of radiation damage by hyperthermia may be an important factor in thermoradiosensitization. Several studies show that thermal inhibition of radiation damage is associated with inhibition of repair of both DNA single- and double-strand breaks. This subject has been reviewed elsewhere (Raaphorst, 1989; Dewey, 1989; Konings, 1987).

1.5. Mechanisms of thermal damage

Hyperthermia induces changes in mammalian cells at many levels and each of these must be examined as a possible target for thermally induced cell inactivation. The understanding of the mechanisms of thermal cellular inactivation may lead to improved strategies for its use in cancer therapy. This section will explore thermal effects on various cellular components to give some insight into mechanisms of thermal damage.

1.5.1. Membranes

Membranes of mammalian cells are highly complex and range in biological function from acting as a simple insulator in myelin to complex functional activity in other cells and to respiratory function in mitochondria. The molecular composition of membranes is highly complex having many different types of lipids (over 1000 having been reported in erythrocyte membranes) and having protein-to-lipid ratios ranging from 1.5 to 4 (Zahler, 1969; Korn, 1969). In addition, interactions between lipids and proteins are important in providing the correct environment for biological activity of membrane-active proteins (Cossins and Shinitzky, 1984; Hesket *et al.*, 1976; Griffith and Jost, 1972). Hyperthermic damage in the mammalian-cell membrane may affect one or more molecules in the membrane and relate to single or multiple molecular interactions and functions.

Since homeoviscous regulation in cells is well known (Cossins *et al.*, 1981) many studies have focused on changes in cellular plasma-membrane fluidity as a possible mechanism in thermal cellular inactivation. In a mutant of *Escherichia coli* requiring fatty acid, Yatvin and coworkers (Yatvin, 1977; Dennis and Yatvin, 1981) showed good correlation between membrane fluidity and thermal sensitivity. In mammalian cells, however, the correlation is less clear, although interpretation may be difficult since the concept of membrane fluidity could refer to many different types of organizational changes which may also be related to measurement technique (for a review see Anghileri, 1986). Lepock and coworkers (1983) failed to show any correlation between hyperthermia-induced cell death and phase transitions in bulk cell membrane lipids. Furthermore, no correlation was demonstrated between heat sensitivity of cells and increased membrane fluidity induced by butylated hydroxytoluene

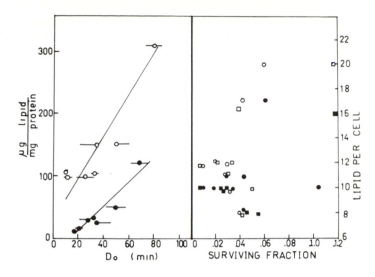

Figure 1.16 Relation between thermal response and membrane lipid composition.
The left panel shows cholesterol (●) and phospholipid (○) concentrations versus the
D_0 of cell survival curves for several different cell lines (data redrawn after Cress *et
al.*, 1982). The right panel shows the cholesterol content (●, ■) and phospholipid
content (○, □) of cells versus cell survival after heating at 43°C (120 min ○, ●) and
45°C (40 min □, ■) of seven transformed and one normal C3H/10T$_i$ cell lines (data
redrawn after Raaphorst *et al.*, 1985b).

(Lepock *et al.*, 1981) or increased microviscosity using cholesteryl
hemisuccinate (Yatvin *et al.*, 1983).

Membrane-lipid composition plays a role in membrane fluidity and
cholesterol is known to reduce lipid order when added below the phase
transition (gel phase) and to increase order when added in the liquid-
crystalline phase (Anghileri, 1986). The work of Cress and coworkers
(1980, 1982), summarized in Figure 1.16(a), shows a correlation between
cellular membrane cholesterol or phospholipid content and thermal
sensitivity for several cell types tested. However, a study by Raaphorst *et
al.* (1985b) shown in Figure 1.16(b) shows no correlation between thermal
sensitivity and lipid composition of seven transformed cell lines derived
from the same parental line. A lack of correlation was also found by
Konings and Ruifrok (1985).

Better correlations between thermal sensitivity and membrane
composition and fluidity were found for cells grown at different
temperatures to allow homeoviscous adaptation. Anderson *et al.* (1981)
showed that mammalian cells grown at lower temperatures displayed
increased membrane fluidity, correlating with cholesterol content and
inversely with thermal sensitivity. Such correlations were also observed
in fish FHM cells and CHO cells and to a lesser extent in cells of rat
embryos (Bowler *et al.*, 1981; Bates *et al.*, 1985; Mendez, 1982; Culver and

Gerner, 1982). Yatvin and coworkers showed a good correlation between thermal sensitivity of mammalian cells and the relative degree of enrichment of unsaturated fatty acids which leads to increased membrane fluidity (Hidvegi *et al.*, 1980; Mulcahy *et al.*, 1981), a result supported by several other reports (for a review see Bowler, 1987).

Further support for membrane involvement in the thermal cell-killing process comes from studies using membrane-modifying agents. Local anaesthetics and alcohols, which are membrane-active agents and partition into membrane lipids, are thought to potentiate the heat fluidization of membranes and are shown to increase thermal cell killing (Yatvin *et al.*, 1987; Seeman, 1972; Paterson *et al.*, 1972). Local anaesthetics such as lidocaine and procaine enhanced thermal cell killing in cells in culture and *in vivo* (Paterson *et al.*, 1972; Yau, 1979; Yatvin *et al.*, 1979). Alcohols were also shown to increase thermal sensitivity (Li and Hahn, 1978). Other workers showed that the fluidizing effect of a series of monohydric alcohols did not correlate with the degree of thermal sensitization although the results did support protein denaturation (Massicotte–Nolan *et al.*, 1981). These data suggest that perhaps the fluidizing effect on thermal sensitivity may saturate and involve proteins.

As was suggested earlier, changes in lipid phase may affect structure and function of membrane proteins. This has been evaluated by many investigators and has been reviewed by Bowler (1987) and Anghileri (1986). At 45 °C, changes were found in Na^+/K^+ ATPase activity and Rb^+ ion transport, which were variable among the studies, but Bates and Mackillop (1985) showed that ouabain-sensitive Rb^+ ion influx increased reversibly at temperatures below 45 °C and was irreversibly inhibited above this temperature. Hyperthermia was also shown to inhibit protein-mediated amino acid transport, reduce the number of protein-binding sites for growth factors and cause irreversible phase transitions in both mitochondrial and plasma-membrane proteins (Kwock *et al.*, 1985; Calderwood and Hahn, 1983; Lepock *et al.*, 1983).

Although not all results agree, there is a large body of evidence suggesting plasma membrane involvement in thermally induced cell death. The evidence points to a model in which modification of the lipid environment can result in further changes in membrance protein structure and function which can affect the cell. Whether such changes are solely involved in cell demise or whether they precipitate further damage in other components of the cell is not clear. However, the work of Roti Roti and Wilson (1984) provides an interesting clue, showing that membrane-active agents which increase thermal sensitivity also increase the protein content of the nucleus. An interpretation suggests a cascade of events leading to cell death. If this is the case, the application of Arrhenius analysis to determine mechanism could be misleading because it indicates only one rate-limiting step in the process.

1.5.2. Morphology and cytoskeleton

Morphological changes resulting from hyperthermia have been studied extensively. Erythrocyte fragmentation occurs rapidly (within 1 s) at 50°C, which is the denaturation temperature for the cytoskeletal protein spectrin (Brandts *et al.*, 1977; Coakley, 1987). In cultured mammalian cells the process of changing cell morphology, such as bleb formation, in response to hyperthermia occurs more slowly (Borrelli *et al.*, 1986) which may be related to the complexity of the cellular cytoskeleton (Coakley, 1987). Studies showed that elevated temperatures disrupt cytoplasmic microtubule organization *in vivo* (Rieder and Beyer, 1977) and cause disassembly *in vitro* (Coss *et al.*, 1982). The use of immunostaining techniques showed that hyperthermia at 43 and 45°C caused an irreversible disruption of microtubular organising centres (Lobreau *et al.*, 1988). Three cytoskeleton subsystems (microtubules, actin stress fibres and vimentin intermediate filaments) have different heat sensitivities, with the latter collapsing about the nucleus after 5 min heating at 45°C. All three systems collapsed after 15 min heating and this occurred concomitantly with cell rounding (Coss and Wachsberger, 1987; Welsh and Suhan, 1985). Coakley (1987) has reviewed the composition and structure of the cytoskeleton in eukaryotic cells and has shown its association with the nuclear membrane and the plasma membrane. Such an association may be involved in membrane-induced effects being transmitted to the nucleus and involving nuclear structure and function.

Hyperthermia-induced membrane damage in the form of bleb formation has been observed in many studies and correlated with the loss of cell viability (Borrelli *et al.*, 1986; Bass *et al.*, 1982; Zielke–Jemme and Hopwood, 1982). This correlation held in G1-phase but not in S-phase cells, where additional mechanisms associated with DNA processing may be impaired by hyperthermia to result in cell death (Borrelli *et al.*, 1986). The type of membrane damage observed after heating (blebing, cell rounding) was similar to effects of agents such as trypsin, local anaesthetics, actinomycin D and cytochalasin which are known to affect membrane-associated microtubular apparatus and may be influenced by calcium levels (Coakley, 1987). Thus, the blebing of the cell membrane resulting from hyperthermia may be associated with the heat-induced breakdown of the cytoskeleton. Such a disruption of cellular structure could have an impact on nuclear structure and function as discussed in the next section.

1.5.3. Nucleus

A wide range of studies using many different techniques have shown that hyperthermia can disrupt nuclear structure and function. Morphological studies have shown vessiculation of the nuclear membrane, decreased heterochromatin and increased perinuclear granules after heating (Heine *et al.*, 1971; Warters *et al.*, 1986). Changes in the nucleolar structure have also been observed for temperatures ranging from 41 to 45 °C (Simard and Bernhard, 1967).

Hyperthermia has been shown to inhibit DNA synthesis. This process has been studied in terms of initiation of replicons, DNA chain elongation and assembly into chromosomes. Reduced incorporation of ^{3}H thymidine into DNA after hyperthermia is also indicative of inhibition of DNA synthesis (Wong and Dewey, 1982). Using sucrose gradient and alkaline elution techniques it was shown that hyperthermia inhibited the replicon initiation step of DNA synthesis (Wong and Dewey, 1982; Davis *et al.*, 1983). Furthermore, the degree of inhibition was dependent on the time and temperature of the hyperthermia treatment and correlated with the differences in thermal sensitivity of HeLa and CHO cells (Wong and Dewey, 1986).

The elongation of DNA molecules into replicons, clusters of replicons and into complete parental-sized molecules was also affected by hyperthermia. Wong and Dewey (1982) and Warters and Stone (1983a, b) showed that the processing time of small DNA molecules into larger units was increased two-fold (60 min to 120–135 min) when cells were heated at 45 °C for 15 min. A three-fold reduction of single-stranded nascent DNA into duplex molecules was also observed. Recovery of the elongation process was completed in 6 h in HeLa cells but was longer in CHO cells (Warters and Stone, 1984; Wong and Dewey, 1982). The activation energy for the inhibition of elongation was found to be 512kJ/mol, which is consistent with protein denaturation but not DNA denaturation, for which the activation energy was observed to be less than 63kJ/mol (Warters and Stone, 1983b).

The chromatin assembly process is also sensitive to hyperthermia. The results of Warters and Roti Roti (1981) using resistance to M nuclease digestion as an indicator of chromatin assembly showed that heating at 43.5–48 °C extensively lengthened this process. The step of ligation into chromosome-sized pieces was also affected by hyperthermia, as was shown by the increase in smaller units up to 5–15 h post-hyperthermia (45.5 °C, 15 min) (Wong and Dewey, 1982).

The induction of DNA strand breaks by hyperthermia has also been examined. Most studies indicate that hyperthermia alone does not cause DNA strand breaks although not all results are in agreement (for a review see Raaphorst, 1989). In some studies, DNA strand breakage has been observed, but this usually developed upon incubation at 37 °C after heating and was dependent on heating time and temperature (Wong and Dewey, 1982; Corry *et al.*, 1977; Mills and Meyn, 1981). The post-hyperthermia development of DNA strand breakage has been reported to result primarily from cell necrosis (Warters and Henle, 1982). There is little or no evidence that hyperthermia induces DNA double-strand breaks (Corry *et al.*, 1977; Radford, 1983); however, there is evidence that hyperthermia causes chromosomal aberrations in S-phase cells (Dewey *et al.*, 1978). These are thought to result from damaged and discontinued replication sites during S phase. When hyperthermia is combined with radiation, inhibition of repair of radiation-induced DNA-strand break damage is observed (Raaphorst, 1989; Dewey, 1989).

Some of the enzymes involved in DNA processing have also been evaluated for their response to hyperthermic damage. Polymerase α (primarily involved in replication) and polymerase β (primarily involved in DNA repair) have been studied and the latter was found to be more heat sensitive than the former (Dube *et al.*, 1977; Spiro *et al.*, 1983). Inactivation energies of 580–675 kJ/mol were observed for polymerases (Denman *et al.*, 1982). In fact, short heat treatments up to 45°C increased the activity of polymerase β (Wawra and Dolejs, 1979), which may explain the initial increased level of strand-break repair observed in some experiments (Dikomey, 1982). Polymerase β inactivation by hyperthermia was protected by polyols, sensitized by local anaesthetics, and became resistant during thermotolerance in whole cells but did not follow this correlation for the isolated enzymes (for a review see Jorritsma, 1989). Although hyperthermia inactivated polymerases, this did not closely correlate with cell killing, especially in the shoulder region of the survival curve. This lack of correlation and the fact that modification of enzyme response (protection and sensitization) is related to the cellular environment indicates the importance of the nuclear environment, where such enzymes may be associated with the nuclear matrix.

Several studies indicated that hyperthermia can cause increases in the nuclear protein content (Tomasovic *et al.*, 1978; Roti Roti *et al.*, 1982) and one such result is shown in Figure 1.17(a). Further, it was shown that a large proportion of the excess protein was associated with the nuclear matrix (Warters *et al.*, 1986). Although the protein localization process is well established, little is known about these proteins except that they are not histones and that one is heat shock protein 70. Further studies (reviewed by Roti Roti, 1982) showed changes in chromatin structure resulted in changes in enzymatic access to nuclear DNA. These data indicate that the damage leading to cell death may be related to disruption of nuclear function due to changes in nuclear structure. Figure 1.17(b) shows a comparison between cell survival and nuclear protein content for many different experiments with good correlation with cell survival modified by sensitizers, protectors or thermotolerance (Roti Roti and Lazlo, 1988). Thus, the model of thermal cell killing of a cascading effect including disruption of the plasma membrane and cytoskeletal attachment (see section 1.5.2) resulting in cytoskeletal collapse towards the nucleus (Coss and Wachsberger, 1987), absorption of protein into the nuclear matrix and disruption of nuclear function fits much of the data (Roti Roti and Lazlo, 1988; Dewey *et al.*, 1980; Roti Roti, 1982).

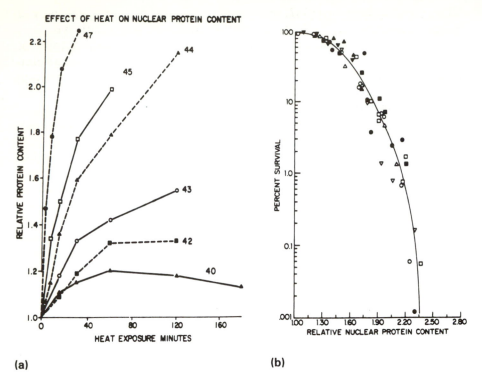

Figure 1.17 (a) The effect of heating at various temperatures on the nuclear chromatin content of HeLa cells. (b) Survival as a function of protein content in HeLa cells measured by flow cytometry immediately after heating (45°C). Thermal response was modified using ethanol (○, ●), procaine (□, ■), glycerol (△, ▲) and through thermotolerance development (▽, ▼; first log survival only). In each experiment the open symbols represent heat alone, and solid symbols heat plus modifier. From Roti Roti *et al.* (1982) with permission.

1.5.4. Metabolism and energy

Several studies have shown that energy deprivation through nutrient depletion or inhibition of metabolic energy production can cause increased thermal sensitivity (Calderwood *et al.*, 1985; Nagle *et al.*, 1980; Haveman and Hahn, 1981; Hahn, 1982; Calderwood and Dickson, 1983). Furthermore, the results of Kim *et al.* (1980) show that enhanced thermal sensitization can be achieved when combining hypoxia with glucose deprivation. The results of Gerweck *et al.* (1984, 1988) show a good correlation between cellular ATP levels (controlled by varying nutrient medium, oxygen and glucose) and cell killing induced by hyperthermia. These data suggest that energy status may play a role in thermally induced cell death.

Recently, Calderwood (1987) reviewed the role of energy and transport of energy substrates in cellular responses to heat. It was shown that hyperthermia could inhibit normal and insulin-stimulated uptake of energy substrates in cells. The inhibition of stimulated uptake paralleled the kinetics of cell killing and had an inactivation energy within the range of those reported for cell killing (625 kJ/mol) (Calderwood, 1987; Calderwood and Hahn, 1983).

Some changes in mitochondria, such as assumption of dense morphology immediately after heating and subsequent swelling, have been observed (Calderwood, 1987; Overgaard and Overgaard, 1976; Welsh and Suhan, 1985). The mitochondrial condensation has been linked to increased phosphorylation (Hackenbrock *et al.*, 1971) which may be stimulated by ATP demand resulting from heat stress. Recent results show that Lonidamine, a drug being tested in cancer therapy, which acts directly on cellular condensed mitochondria (Szekely *et al.*, 1989), is known to inhibit glycolysis and can cause increased thermal sensitization (Raaphorst *et al.*, 1990; Natali *et al.*, 1984). These data indicate a possible role for inhibition of glycolysis in thermal inactivation.

The role of respiration has been examined through studies of oxygen utilization. Many normal and malignant cells show inhibition of oxygen utilization resulting from hyperthermia, but these data have not been compared with cell survival (Mondovi, 1976). Landry *et al.* (1985) showed that cells selected for respiration deficiency by ethidium bromide treatment exhibited similar thermal sensitivity to cells with full respiration function, whilst Bowler (1981) showed that inhibition of respiration induced in blowfly muscle by hyperthermia correlated with viability. The results on respiration are therefore inconclusive and warrant further study.

Direct evaluation of modification of ATP and energy charge by hyperthermia in cells under normal nutrient conditions was also made by Calderwood and coworkers (1985, 1987). These data show a lack of correlation between ATP depletion (reduction of energy charge) and loss of viability after heating at 45°C (Figure 1.18). When cells were deprived of nutrients (Figure 1.18) there was a correlation between cellular ATP levels and thermal sensitivity. Thus, it appears that reduced energy levels may increase thermal sensitivity in cells in which energy production is reduced or compromised. Such conditions may be found in tumours (Gerweck, 1988) and could make these tissues more sensitive to heat than normal tissue. Recent studies using ^{31}P magnetic resonance spectroscopy show changes in cell and tissue metabolites after heating, which supports this concept (Sijens *et al.*, 1987, 1989; Gerweck, 1988).

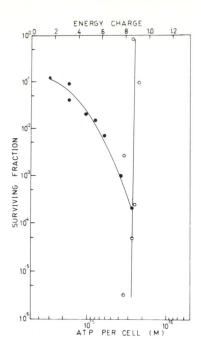

Figure 1.18 Survival of CHO cells plotted as a function of cellular ATP concentration (●). ATP levels were changed by varying the oxygen and glucose concentration. At 37°C there was little to no loss in survival. (Data redrawn from Gerweck *et al.* (1984).)
Comparison between energy charge (=[ATP]+½[ADP]/[ATP]+[ATP]+[AMP]) and survival in HA-1 cells (○). (Data redrawn from Calderwood *et al.* (1985).)

1.6. Modifiers of thermal sensitivity

1.6.1. Membrane-active agents

As indicated previously, the cell membrane is thought to be involved in the early stages of thermal damage leading to cell death. A host of membrane partitioning agents have been shown to increase thermal sensitivity.

Local anaesthetics are known to interact with the cell membrane (Seeman, 1972) and Yau (1979) showed that thermal sensitivity in cultured mammalian cells was enhanced by procaine. Thermal sensitization occurred at clinically achievable concentrations of procaine, increased with drug concentration and depended on temperature. Of the agents tested, dibucaine appeared to be the most active, consistent with its activity as a nerve blocker and other membrane-associated effects (Yau and Kim, 1980; Seeman, 1972). Thermal sensitization occurred in both G1 and S phases of the cell cycle (Coss and Dewey, 1982). In prokaryotic cells, a good correlation between membrane-fluidizing effect and thermal sensitization was found (Yatvin, 1977; Yatvin *et al.*, 1987), although this was not correlated in mammalian cells (Lepock, 1982). However, in mammalian cells, bulk measurement of membrane phase changes may not be sensitive enough to detect such changes in microdomains (Anghileri, 1986) which could in turn affect membrane protein structure and function (Chan and Wang, 1984).

Alcohols have also been shown to be thermal sensitizers (Li and Hahn, 1978). The action of alcohol on mammalian cells is very similar to that of heat, in that it also induces thermotolerance and resistance to ethanol or adriamycin, and is affected by pH and by D_2O (Li *et al.*, 1980b, 1982). Ethanol has been labelled as a heat analogue. The alcohols are known to partition into membranes (Shinitzky, 1984) and increase membrane fluidity (Yatvin, 1987). These agents were also shown to have an effect on proteins (McElhaney, 1985). While the membrane-fluidizing effect of a series of monohydric alcohols did not correlate with thermal sensitization, a correlation with protein denaturation occurred, indicating this to be a critical step in the sensitization process (Massicotte–Nolan *et al.*, 1981). These findings do not exclude the possibility of modification of membrane microdomains affecting protein structure and function.

Other membrane-active agents such as amphotericin B, a polyene antibiotic which binds to sterols of plasma membranes, and some organic solvents increase thermal sensitivity (Hahn and Li, 1977; Li and Hahn, 1978) while butylated hydroxytoluene, a membrane active agent that alters membrane fluidity, does not cause thermal sensitization (Lepock *et al.*, 1981). Further studies involving membrane subcomponents and microstructure are needed to elucidate fully the role of the membrane in thermal damage.

1.6.2. Polyamines

Polyamines are small polycationic peptides and are involved in stabilization of DNA and methylation of RNA, and can cause increased rigidity of membranes. Gerner *et al.* (1980) showed that exposure of cells to exogenous concentrations of polyamines causes thermal sensitization. Sensitization by these compounds depended on concentration and chain length, and maximum sensitization developed when the two terminal amino groups were separated by 10–11 Å. This distance may be related to a critical dimension of target structures. In addition, it was shown that low levels of exogenous polyamines could sensitize cells expressing thermotolerance. Gerner *et al.* (1983) and Fuller and Gerner (1982) showed that inhibition of polyamine biosynthesis increased thermal sensitivity which had a complex time dependence. They proposed that intracellular polyamines may be active in the mediation of thermal sensitivity and thermotolerance. Recently, others have proposed mechanisms whereby oxidation products of polyamines may mediate thermal sensitization (Henle *et al.*, 1986). Inhibitors of polyamine biosyntheses are being tested in tumour models as potentiators of thermal responsiveness.

1.6.3. Thiols

Cysteamine and cysteine are thiol-containing compounds which cause thermal sensitization which increases rapidly in the region of 42–43°C (Kapp and Hahn, 1979). Further studies by Mitchell and Russo (1983) using cysteine and n-acetyl-cysteine indicate that blocking the amino group of cysteine with an acetyl group reduced the thermal sensitization which therefore may be related to the production of free radicals through thiol auto-oxidation. Further studies by Issels *et al.* (1985) suggest the involvement of activated oxygen species which are generated through auto-oxidation of thiols in the presence of oxygen. The reduction of cellular glutathione through enzymatic inhibition of its production by buthionine sulphoxamine (BSO) or oxidation through diethylmaleate (DEM) reduces levels to less than 5% in V79 cells and increased cellular heat sensitivity. The effect of BSO and DEM on GSH reduction and thermal sensitization occurred to a lesser extent at higher temperatures (43–45°C) and is cell line dependent (Konings and Penninga, 1985). Results on human glioma cells show maximum reduction of GSH by BSO was 20% with no concomitant thermal sensitization. Heating itself caused an increase in cellular GSH levels, possibly indicating a protective response to heating. Further studies by Freeman *et al.* (1985) showed the process of thermal sensitization through GSH depletion to depend on oxygen tension. Thus, manipulation of these levels in cells and tissues may be a way of achieving a greater thermal response.

1.6.4. Thermal protectors

1.6.4(a)   Glycerols, polyols and monosaccharides

Glycerol and polyhydroxy compounds such as polyols and sugars were shown to protect against cell killing and catalytic and structural properties of proteins (Bach *et al.*, 1979; Henle *et al.*, 1983, 1984; Henle, 1987a). Protection by glycerol occurred only when cells were heated in its presence and as it reached full equilibrium in cells. The protection also increased with concentration (Henle, 1987a). The mechanism is thought to be through molecular stabilization which occurs due to increased hydrophobic interactions between solvent and solute macromolecules (Bach *et al.*, 1979).

The larger polyhydroxy glycerol analogues (polyols) also showed the ability to confer thermal protection when present during heating (Henle,

1987a). Initial exposure resulted in thermal sensitization, possibly due to osmotic shock imparted to cells by the addition of polyols, but as these compounds equilibrated in the cell during about 1–3 h, thermal protection occurred. As with glycerol, such protection may be related to stabilization of cellular macromolecules and structures.

Physiological sugars such as glucose, galactose and mannose were also shown to result in thermal protection of cells but after relatively long equilibration times (0.5–5 h) (Henle, 1987a). Heat protection by galactose was marked by an intracellular accumulation of free galactose and its corresponding polyols, dulcitol and sorbitol, in contrast to protection by glucose which showed no significant intracellular accumulations of free sugars or polyols. This correlated with the much more rapid and greater degree of thermal protection by galactose, compared to glucose, and suggests stabilization of cellular macromolecules by these compounds.

1.6.4(b) Deuterium oxide

Thermal protection by D_2O has been shown in several investigations (Azzam *et al.*, 1982; Raaphorst and Azzam, 1982b; Fisher *et al.*, 1982; Li *et al.*, 1982). Protection occurred only when D_2O was present during heating, and increased with concentration but did not change the inactivation energy for cell killing although it did eliminate the transition in the Arrhenius curve. These data suggest that protection resulted not from a change in mechanism of cell killing but through stabilization of cellular macromolecules and structures resulting in the observed shift in the temperature threshold by 1.6–1.8°C. Further studies by Lin *et al.* (1984) showed D_2O protection of microtubular structures, indicating stabilization of the cytoskeleton and supporting the concept of protection through molecular and structural stabilization.

D_2O was shown to inhibit the development of thermotolerance and protein synthesis (Raaphorst and Azzam, 1982b). As was pointed out by Henle (1987a), heat protectors can become inhibitors of cellular functions through their excess stabilization of proteins thus inhibiting function. This may also be the mechanism of inhibition of the development of thermotolerance by D_2O.

1.7. Thermotolerance

Thermotolerance (the development of resistance to hyperthermia) resulting from prior exposure to heat or chemicals is an induced trait and not the selection of genetically resistant subpopulations (Raaphorst and Azzam, 1980b). Chronic heating at low temperatures (40–42°C; see Figure 1.1) allows the development of thermotolerance as characterized by a resistance plateau in the survival curve. This phenomenon occurs in almost all mammalian cells although there are a few exceptions (see Section 1.3). Thermotolerance can also develop during incubation at $37\,^{\circ}$C after acute heating and the extent of its development as well as development time and decay time depend on many factors, such as time and temperature of the inducing heat treatment, the environmental conditions of the cell (pH, nutrient media, oxygen status), the cell cycle phase, cell growth state and cell type. In addition, a host of other agents, some of which are inhibitors of protein synthesis, such as anisotonicity, D_2O, glycerol, amino acid analogues, modifiers of cellular redox state and local anaesthetics, also modify thermotolerance (Henle and Dethlefsen, 1978; Li and Mivechi, 1986; Gerner, 1983; Burdon, 1987; Henle, 1987b).

The mechanisms involved in the induction of thermotolerance are not fully understood. Correlations of thermotolerance have been made with production of some classes of heat shock proteins (Li and Laszlo, 1985; Li and Hahn, 1984), but evidence against such a causal relationship also exists, such as thermotolerance development without production of heat-shock proteins. This topic has been reviewed by Landry (1987). Other mechanisms, such as production of endogenous heat protectors (polyols, sugars, etc.), cellular damage repair mechanisms, reduced polyamine leakage from the cell, membrane factors, thiol factors and ionic effects have been reviewed by Henle (1987a) and remain unclear as mechanisms in cellular thermotolerance development.

Thermotolerance has also been shown to modify the degree of thermosensitization to other agents such as radiation (see previous section on hyperthermia and radiation) and chemotherapeutic agents. This phenomenon has been extensively reviewed (Dewey and Holahan, 1987; Henle *et al.*, 1989; Hahn and Li, 1982; Hahn, 1983; Hahn, 1987).

The modification of response to hyperthermia, radiation and chemotherapeutic agents by thermotolerance can have extensive clinical implications. Such effects on normal and tumour cells and tissues may present opportunities for improvement of therapeutic gain. However, at present the phenomenon is not clearly understood and the conservative approach is to avoid the development of thermotolerance in the clinic through prolonged fractionation intervals. More research and understanding of this phenomenon are urgently needed.

1.8. Summary

In order to exploit hyperthermia optimally in cancer therapy, characterization of the various responses in different mammalian cell and tissue systems is necessary. Studies with mammalian cells in culture show a much larger variation in thermal response than in radiation response and there appears to be no correlation between them. Although some studies show tumour cells to be more sensitive than normal cells this is not consistently the case for all cell systems tested. A host of factors, such as pH, hypoxia, nutrient depletion, cell cycle distribution and repair, make hyperthermia and radiation complementary, as well as synergistic, agents for cancer therapy.

The numerous studies of various cellular substructures support a model of cell inactivation involving a cascade of events. Membrane-active agents have been shown to modify thermal sensitivity and also cause changes in chromatin within the nucleus. Compounds or conditions (D_2O, polyols, pressure) which stabilize cellular macromolecules render thermal protection. These data and many others support a model in which the first event of heat damage is at the membrane, altering: (1) fluidity, (2) protein structure, (3) membrane transport, (4) cell communication and (5) anchorage of the cytoskeleton. Subsequent events could be changes in the cytoplasmic milieu through changes in ionic composition and other molecules leaking into or out of the cells. These changes and loss of anchorage result in collapse of the cytoskeleton towards the nucleus. These events result in changes at the periphery of the nucleus as well as within it, including changes in chromatin structure and function. The final result is the impairment of nuclear processes such as DNA replication, transcription, and repair, resulting in cellular demise unless the process is reversed or repaired.

References

Anderson, R. L., Minton, K. W., Li, G. C. and Hahn, G. M. (1981) Temperature induced homeoviscous adaptation in Chinese hamster ovary cells, *Biochem. Biophys. Acta*, **641**, 334–348.

Anghileri, L. J. (1986) Role of tumour cell membrane in hyperthermia, in Anghileri, L. J. and Robert, J. (Eds), *Hyperthermia in Cancer Treatment*, Vol. 1, Boca Raton, CRC Press, pp. 1–36.

Azzam, E. I., George, I. and Raaphorst, G. P. (1982) Alteration of thermal sensitivity of Chinese hamster cells by D_2O treatment, *Radiat. Res.*, **90**, 644–648.

Bach, J. F., Oakenfull, D. and Smith, M. B. (1979) Increased thermal stability of proteins in the presence of sugars and polyols, *Biochem.*, **18**, 5191–5196.

Badanidiyoor, S. R. and Hopwood, L. E. (1985) Effect of hypoxia on recovery from damage induced by heat and radiation in plateau-phase CHO cells, *Radiat. Res.*, **101**, 312–325.

Bass, H., Coakley, W. T., Moore, J. L. and Tilley, D. (1982) Hyperthermia-induced changes in the morphology of CHO-K1 and their refractile inclusions, *J. Therm. Biol.*, **7**, 231–242.

Bates, D. A., Legrimellic, D., Bates, J. H. T., Loufti, A. and Mackillop, W. J. (1985) Effects of thermal adaptation at 40°C on membrane viscosity and the sodium pump in Chinese hamster ovary cells, *Cancer Res.*, **45**, 3895–4899.

Bates, D. A. and Mackillop, W. J. (1985) The effect of hyperthermia on sodium–potassium pump in Chinese hamster ovary cells, *Radiat. Res.*, **103**, 441–451.

Bauer, K. D. and Henle, K. J. (1979) Arrhenius analysis of heat survival curves from normal and thermotolerant CHO cells, *Radiat. Res.*, **78**, 251–263.

Borrelli, M. J., Wong, R. S. L. and Dewey, W. C. (1986) A direct correlation between hyperthermia-induced membrane blebbing and survival in synchronous G1 CHO cells, *J. Cell Physiol.*, **126**, 181–190.

Bowler, K. (1987) Cellular heat injury, are membranes involved? in Bowler, K. and Fuller, B. J. (Eds), *Temperature and Animal Cells*, Cambridge, Company of Biologists, pp. 157–185.

Bowler, K. (1981) Heat death and cellular heat injury, *J. Therm. Biol.*, **6**, 171–179.

Bowler, K., Laudien, H. and Laudien, I. (1981) Cellular heat injury, *J. Thermal. Biol.*, **8**, 426–430.

Brandts, J. F., Erickson, L., Lysko, K., Schwartz, A. T. and Toverna, R. D. (1977) Calorimetric studies of the structural transitions of the human erythrocyte membrane. The involvement of spectrum in the A transition, *Biochem.*, **16**, 3450–3454.

Bromer, R. H., Mitchell, J. B. and Soares, N. (1982) Response of human hematopoietic precurser cells (CFUC) to hyperthermia and radiation, *Cancer Res.*, **42**, 1261–1265.

Brown, L. A. and Crozier, W. J. (1927) Rate of killing of cladocerans at higher temperatures, *J. Gen. Physiol.*, **11**, 25–36.

Burdon, R. H. (1987) Thermotolerance and heat shock proteins, in Bowler, K. and Fuller, B. J. (Eds), *Temperature and Animal Cells*, Cambridge, Company of Biologists, pp. 269–283.

Calderwood, S. K. (1987) Role of energy in cellular responses to heat, in Bowler, K. and Fuller, B. J. (Eds) *Temperature and Animal Cells*, Cambridge, Company of Biologists, pp. 213–233.

Calderwood, S. K., Bump, E. A., Stevenson, M. A., Van Kersen, I. and Hahn, G. M. (1985) Investigation of adenylate energy charge, phosphorylation potential and ATP concentration in cells stressed with starvation and heat, *J. Cell Physiol.*, **124**, 261–268.

Calderwood, S. K. and Dickson, J. A. (1983) pH and tumour response to hyperthermia, *Adv. Radiat. Biol.*, **19**, 135–186.

Calderwood, S. K. and Hahn, G. M. (1983) Thermal sensitivity and resistance of insulin receptor binding, *Biochem. Biophys. Acta*, **76**, 1–8.

Cavaliere, R., Ciocatto, E. C., Giovanella, B. C., Heidlerger, C., Johnson, R. O., Margottini, M., Mondovi, B., Moricca, B. and Rossi–Fanelli, A. (1967) Selective heat sensitivity of cancer cells. Biochemical and clinical studies, *Cancer*, **20**, 1351–1381.

Chan, D. S. and Wang, H. H. (1984) Local anesthetics can interact electrostatically with membrane proteins, *Biochem. Biophys. Acta*, **770**, 55–64.

Chu, G. L. and Dewey, W. C. (1988) The role of low intracellular or extracellular pH in sensitization to hyperthermia, *Radiat. Res.*, **114**, 154–167.

Clark, E. P., Dewey, W. C. and Lett, J. J. (1981) Recovery of CHO cells from hyperthermic potentiation to X-rays: repair of DNA and chromatin, *Radiat. Res.*, **85**, 302–313.

Coakley, W. T. (1987) Hyperthermia effects on the cytoskeleton and on cell morphology, in Bowler, K. and Fuller, B. J. (Eds), *Temperature and Animal Cells*, Cambridge, Company of Biologists, pp. 187–211.

Cohen, J. D., Robins, H. I., Mulcahy, T. R., Gipp, J. J. and Bouck, N. (1988) Interaction between hyperthermia and irradiation in two human thymoblastic leukemia cell lines *in vitro*, *Cancer Res.*, **48**, 3576–3580.

Corry, P. M., Robinson, S. and Getz, S. (1977) Hyperthermic effects on DNA repair mechanisms, *Radiol.*, **123**, 475–482.

Coss, R. A. and Dewey, W. C. (1982) Heat sensitization of G1 and S phase cells by procaine hydrochloride, *Radiat. Res.*, **92**, 615–617.

Coss, R. A. and Wachsberger, P. R. (1987) Role of cytoskeletal damage in heat killing of G1 populations of CHO cells, in Fielden, E. M., Fowler, J. F., Hendry, J. H. and Scott, D. (Eds), *Proceedings of the Eighth International Congress on Radiation Research*, London, Taylor and Francis, p. 340.

Coss, R. A., Dewey, W. C. and Bamburg, J. R. (1982) Effects of hyperthermia on dividing Chinese hamster ovary cells and on microtubules *in vitro*, *Cancer Res.*, **42**, 1059–1071.

Cossins, A. R. and Sinitsky, M. (1984) Adaption of membranes to temperature pressure and exogenous lipids, in Shinitzky, M. (Ed.) *Physiology of Membrane Fluidity*, Vol. II, Boca Raton, CRC Press, pp. 1–20.

Cossins, A. R., Bowler, K. and Prosser, C. L. (1981) Homeoviscous adaptation and its effect on membrane band-bound proteins, *J. Thermal Biol.*, **6**, 183–187.

Cress, A. E. and Gerner, E. W. (1980) Cholestorol levels inversely reflect the thermal sensitivity of mammalian cells in culture, *Nature, Lond.*, **283**, 677–679.

Cress, A. E., Culver, P. S., Moon, T. E. and Gerner, E. W. (1982) Correlation between amount of cellular membrane components and sensitivity to hyperthermia in a variety of mammalian cell lines in culture, *Cancer Res.*, **42**, 1716–1721.

Culver, P.S. and Gerner, E.W. (1982) Temperature acclimation and specific cellular components in regulation of thermal sensitivity of mammalian cells, in Dethlefson, L. (Ed.) *Cancer Therapy by Hyperthermia, Drugs and Radiation*, Bethesda, Maryland, US Department of Health ahd Human Services, pp. 99–101.

Davis, R.C., Bowden, G.T. and Cress, A.E. (1983) The effect of heat and radiation on the initiation and elongation process of DNA synthesis, *Int. J. Radiat. Biol.*, **43**, 379–390.

Denman, D.L., Spiro, I.J. and Dewey, W.C. (1982) Effects of hyperthermia on DNA polymerases alpha and beta: relationship to thermal cell killing and radiosensitization, *J. Natl Cancer Inst.*, **61**, 37–39.

Dennis, W.H. and Yatvin, M.B. (1981) Correlation of hyperthermic sensitivity and membrane microviscosity in *E. coli* K1060, *Int. J. Radiat. Biol.*, **39**, 265–271.

Dethlefsen, L. and Dewey, W.C. (Eds) (1982) Proceedings of the Third International Symposium on Cancer Therapy by Hyperthermia, Drugs and Radiation, *J. Natl Cancer Inst. Monogr.*, **61**.

Dewey, W.C. (1989) Mechanisms of thermal radiosensitization, in Urano, M. and Dauple, E. (Eds), *Hyperthermia and Oncology*, Vol. 2, Utrecht, VSP.

Dewey, W.C. and Holahan, P.K. (1987) Thermotolerance as a modifier of radiation toxicity, in Henle, K.J. (Ed.), *Thermotolerance and Thermophily*, Vol. 1, Boca Raton, CRC Press, pp. 113–125.

Dewey, W.C., Westra, A., Miller, H.H. and Nagasawa, H. (1971) Heat induced lethality and chromosomal damage in synchronized Chinese hamster cells treated with 5-bromodeoxyuridine, *Int. J. Radiat. Biol.*, **20**, 505–520.

Dewey, W.C., Hopwood, L.E., Sapareto, S.A. and Gerweck, L.E. (1977) Cellular responses to combinations of hyperthermia and radiation, *Radiol.*, **123**, 464–474.

Dewey, W.C., Sapareto, S.A. and Betten, D.A. (1978) Hyperthermic radiosensitization of synchronous Chinese hamster cells: relationships between lethality and chromosome aberrations, *Radiat. Res.*, **76**, 48–59.

Dewey, W.C., Freeman, M.L., Raaphorst, G.P., Clark, E.P., Wong, R.S.L., Highfield, D.P., Spiro, I.J., Tomasovic, S.F., Denmer, D.L. and Coss, R.P. (1980) Cell biology of hyperthermia and radiation, in Meyn, R.E. and Withers, H.R. (Eds), *Radiation Biology in Cancer Research*, New York, Raven Press, pp. 589–621.

Dube, D.K., Seal, G. and Loeb, L.A. (1977) Differential heat sensitivity of mammalian DNA polymerases, *Biochem. Biophys. Res. Commun.*, **76**, 483–487.

Dikomey, E. (1982) Effect of hyperthermia at 42 and 45°C on repair of radiation-induced DNA strand breaks in CHO cells, *Int. J. Radiat. Biol.*, **41**, 603–614.

Elkind, M.M. and Sutton, H. (1960) Radiation response of mammalian cells grown in culture. 1. Repair of X-ray damage in surviving Chinese hamster cells, *Radiat. Res.*, **13**, 556–593.

Evans, R.G., Engle, C., Wheatly, C. and Nielsen, J. (1982) Modification of the sensitivity and repair of potentially lethal damage by DDC during and following exposure of plateau phase cultures of mammalian cells to radiation and *cis*-diamminedichloroplatinum II, *Cancer Res.*, **42**, 3074–3078.

Evans, R.G., Nielsen, J., Engel, C. and Wheathley, C. (1983) Enhancement of heat sensitivity and modification of repair of potentially lethal heat damage in plateau phase cultures of mammalian cells by DDC, *Radiat. Res.*, **93**, 319–325.

Fajardo, L.F., Schreiber, A.B., Kelley, N.I. and Hahn, G.M. (1985) Thermal sensitivity of endothelial cells, *Radiat. Res.*, **103**, 276–285.

Fischer, G.A., Li, G.C. and Hahn, G.M. (1982) Modification of thermal response by D_2O, 1 cell survival and the temperature shift, *Radiat. Res.*, **92**, 530–540.

Freeman, M.L., Raaphorst, G.P. and Dewey, W.C. (1979) The relationship of heat killing and thermal radiosensitization to the duration of heating at 42°C, *Radiat. Res.*, **78**, 172–175.

Freeman, M.L., Holahan, E.V., Highfield, D.P., Raaphorst, G.P., Spiro, I.J. and Dewey, W.C. (1981) The effect of pH on hyperthermic and X-ray induced cell killing, *Int. J. Radiat. Oncol.*, **7**, 211–216.

Freeman, M.L., Malcolm, A.W. and Meredith, A.W. (1985) Role of glutathione in cell survival after hyperthermic treatment of Chinese hamster ovary cells, *Cancer Res.*, **45**, 6308–6313.

Fuller, D.J.M. and Gerner, E.W. (1982) Polyamines: a dual role in modulation of cellular sensitivity to heat, *Radiat. Res.*, **92**, 439–444.

Gerner, E.W. (1984) Definition of thermal dose, biological isoeffect relationships and dose for temperature-induced cytotoxicity, in Overgaard, J. (Ed.), *Hyperthermic Oncology*, Vol. 2, London, Taylor and Francis, pp. 245–251.

Gerner, E. W. (1983) Thermotolerance, in Storm, F. K. (Ed.), *Hyperthermia in Cancer Therapy*, Boston, GK Hall, pp. 141–162.

Gerner, E. W., Holmes, D. K., Stickney, D. G., Noterman, J. A. and Fuller, D. J. M. (1980) Enhancement of hyperthermia inducted cytotoxicity by polyamines, *Cancer Res.*, **40**, 432–438.

Gerner, E. W., Stickney, D. G., Herman, T. S. and Fuller, D. J. M. (1983) Polyamines and polyamine biosynthesis in cells exposed to hyperthermia, *Radiat. Res.*, **93**, 340–352.

Gerweck, L. E. (1988) Modifiers of thermal effects: environmental factors, in Urano, M. and Douple, E. (Eds), *Thermal Effects on Cells and Tissues*, Vol. 1, VSP, pp. 83–98.

Gerweck, L. E. and Burlett, P. (1978) The lack of correlation between heat and radiation sensitivity in mammalian cells, *Int. J. Radiat. Oncol.*, **4**, 283–285.

Gerweck, L. E. and Suzuki, N. (1988) NMR spectroscopy prediction and detection of response to therapy, in Sugahara, T. and Saito, M. (Eds), *Hyperthermic Oncology*, Vol. 2, Kyoto, Japan, Taylor and Francis, pp. 278–281.

Gerweck, L. E., Nygaard, T. G. and Burlett, M. (1979) Response of cells to hyperthermia under acute and chronic hypoxic conditions, *Cancer Res.*, **39**, 966–972.

Gerweck, L. E., Richards, B. and Jennings, M. (1981) The influence of variable oxygen concentration on the response of cells to heat and/or x-irradiation, *Radiat. Res.*, **85**, 314–320.

Gerweck, L. E., Dahlberg, W. K., Epstein, L. F. and Shimm, D. S. (1984) Influence of nutrient and energy deprivation on cellular response to single and fractionated heat treatments, *Radiat. Res.*, **99**, 573–581.

Giovanella, B. C., Morgan, A. C., Stehlin, J. S., Williams, L. J. (1973) Selective lethal effect of supranormal temperatures on mouse sarcoma cells, *Cancer Res.*, **33**, 2568–2578.

Giovanella, B. C., Stehlin, J. S., Morgan, A. C. (1976) Selective lethal effect of supranormal temperatures on human neoplastic cells, *Cancer Res.*, **36**, 3944–3950.

Griffith, O. H. and Jost, P. C. (1972) in Solomon, A. and Karnovsky, M. (Eds), *Lipid Protein Associations in Molecular Specialization and Symmetry in Membrane Function*, Cambridge, Harvard University Press, Chapt. 2.

Hackenbrock, C. R., Rehn, T. G., Wienback, E. C. and Lemasters, T. L. (1971) Oxidative phosphorylation and ultrastructural transformation in mitochondria in the intact ascites tumour cell, *J. Cell. Biol.*, **51**, 123–137.

Hahn, G. M. (1974) Metabolic aspects of the role of hyperthermia in mammalian cell inactivation and their possible relevance to cancer treatment, *Cancer Res.*, **34**, 3117–3123.

Hahn, G. M. (1982) *Hyperthermia in Cancer*, New York, Plenum Press.

Hahn, G. M. (1983) Hyperthermia to enhance drug delivery, *Rational Basis for Chemotherapy*, New York, Alan R. Liss, p. 427.

Hahn, G. M. (1987) Thermotolerance and drugs, in Henle, K. J. (Ed.), *Thermotolerance*, Vol. 1, *Thermotolerance and Thermophily*, Boca Raton, CRC Press, pp. 97–112.

Hahn, G. M. and Li, G. C. (1977) Interaction of amphotericin β and 43°C hyperthermia, *Cancer Res.*, **37**, 761–764.

Hahn, G. M and Li, G. C. (1982) Interaction of hyperthermia and drugs; treatments and probes, *J. Natl Cancer Inst. Monogr.*, **61**, 317–323.

Haveman, J. and Hahn, G. M. (1981) The role of energy in hyperthermia induced mammalian cell inactivation. A study of the effects of glucose starvation and an uncoupler of oxidative phosphorylation, *J. Cell. Physiol.*, **108**, 237–241.

Heine, U., Severak, L., Kondratick, J. and Bonar, R. A. (1971) The behaviour of HeLa-S3 cells under influence of supranormal temperatures, *J. Ultrastruct. Res.*, **34**, 375–396.

Henle, K. J. (1987a) Heat protectors of mammalian cells, in Henle, K. J. (Ed.), *Thermotolerance, Mechanisms of Heat Resistance*, Vol. 2, Boca Raton, CRC Press, pp. 127–143.

Henle, K. J. (1987b) Thermotolerance in cultured mammalian cells, in Henle, K. J. (Ed.), *Thermotolerance and Thermophily*, Vol. 1, Boca Raton, CRC Press, pp. 13–71.

Henle, K. J. (1980) Sensitization to hyperthermia below 43°C induced by step-down heating, *J. Natl Cancer Inst.*, **64**, 1479–1483.

Henle, K. J. (1983) Arrhenius analysis of thermal responses, in Storm, F. K. (Ed.), *Hyperthermia in Cancer Therapy*, Boston, Hall Medical Publishers, pp. 47–53.

Henle, K. J. and Dethlefsen, L. A. (1978) Heat fractionation and thermotolerance: a review, *Cancer Res.*, **38**, 1843–1851.

Henle, K. J. and Dethlefsen, L. A. (1980) Time temperature relationships for heat-induced killing of mammalian cells, *Ann. New York Acad. Sci.*, **335**, 234–253.

Henle, K. J., Peck, J. W. and Higashikubo, R. (1983) Protection against heat-induced cell killing by polyols *in vitro*, *Cancer Res.*, **43**, 1624–1627.

Henle, K. J., Monson, T. P., Moss, A. J. and Nagle, W. A. (1984) Protection against thermal cell death in Chinese hamster ovary cells by glucose galactose or mannose, *Cancer Res.*, **44**, 5499–5504.

Henle, K. J., Moss, A. J. and Nagle, W. A. (1986) Mechanism of spermidine cytotoxicity at 37°C and 43°C in Chinese hamster ovary cells, *Cancer Res.*, **46**, 175–182.

Henle, K. J., Nagle, W. A. and Moss, A. J. (1989) Effect of thermotolerance on the cellular heat-radiation response, in Urano, M. (Ed.), *Hyperthermia and Oncology*, Vol. 2, Utrecht, VSP, Chapt. 3.

Hesket, T. R., Smith, G. A., Houslay, M. D., McGill, K. A., Birdsall, N. J. M., Metcalfe, J. C. and Warren, G. B. (1976) Annular lipids determine the ATPase activity of a calcium transport protein complexed with dipalmitoyllecithin, *Biochem.*, **15**, 4145–4150.

Hidvegi, E. J., Yatvin, M. B., Dennis, W. H. and Hidvegi, E. (1980) Effect of altered membrane lipid composition and procaine on hyperthermic killing of ascites tumour cells, *Oncol.*, **37**, 360–363.

Hume, S. P. and Marigold, J. C. L. (1980) Transient heat induced thermal resistance in the small intestine of mouse, *Radiat. Res.*, **82**, 526–535.

Hofer, K. G. and Mivechi, N. F. (1980) Tumour cell sensitivity to hyperthermia as a function of extracellular and intracellular pH, *J. Natl Cancer Inst.*, **65**, 621–625.

Issels, R. D., Bourier, S. and Li, G. (1985) Influence of cysteamine on thermotolerance in Chinese hamster ovary cells, *33rd Proc. Radiat. Res.*, **CC9**, 20.

Johnson, F. H., Eyring, H. and Palissar, M. (1954) *The Kinetic Basis of Molecular Biology*, New York, Wiley, pp. 3–460.

Jorritsma, J. B. M. (1989) Hyperthermic radiosensitization; role of DNA repair, in Konings, A. W. T. *et al.* (Eds), *Annual Reviews on Hyperthermia*, Vol. 1, in press.

Kallman, R. J. (1972) The phenomenon of reoxygenation and its complications for fractionated radiotherapy, *Radiol.*, **105**, 135–142.

Kampinga, H. H., Jorritsma, J. B. M., Burgman, P. and Konings, A. W. T. (1986) Differences in heat induced cell killing as determined in three mammalian cell lines do not correspond with the extent of heat radiosensitization, *Int. J. Radiat. Biol.*, **50**, 675–684.

Kapiszewsak, M. and Hopwood, L. E. (1986) Changes in bleb formation following hyperthermia treatment of Chinese hamster ovary cells, *Radiat. Res.*, **105**, 405–412.

Kapp, D. S. and Hahn, G. M. (1979) Thermosensitization by sulphydryl compounds of exponentially growing Chinese hamster cells, *Cancer Res.*, **39**, 4630–4635.

Kim, J. H., Kim, S. H. and Hahn E. W. (1978) Killing of glucose-deprived hypoxic cells with moderate hyperthermia, *Radiat. Res.*, **75**, 448–451.

Kim, S. H., Kim, J. H., Hahn, E. W. and Ensign, N. A. (1980) Selective killing of glucose and oxygen deprived HeLa cells by hyperthermia, *Cancer Res.*, **40**, 3459–3462.

Konings, A. W. T., (1987) Effects of heat and radiation on mammalian cells, *Radiat. Phys. Chem.*, **30**, 339–349.

Konings, A. W. T. and Penninga, P. (1985) On the importance of the level of glulathione and the activity of the pentose phosphate pathway in heat sensitivity and thermotolerance, *Int. J. Radiat. Biol.*, **48**, 409–422.

Konings, A. W. T. and Ruifrok, A. C. C. (1985) Role of membrane lipids and membrane fluidity in thermosensitivity and thermotolerance of mammalian cells, *Radiat. Res.*, **102**, 86–98.

Korn, E. D. (1969) Cell membranes. Structure and synthesis, *Annual Rev. Biochem.*, **38**, 263–281.

Kwock, L., Lin, P. S. and Hefter, K. (1985) A comparison of the effects of hyperthermia on cell growth in human T and B lymphoid cells: relationship to alterations in plasma membrane transport properties, *Radiat. Res.*, **101**, 197–206.

Landry, J. (1987) Heat shock proteins and thermotolerance, in Anghileri, L. J. and Robert, J. (Eds), *Hyperthermia in Cancer Treatment*, Vol. 1, Boca Raton, CRC Press, pp. 37–58.

Landry, J. and Marceau, N. (1978) Rate limiting events in hyperthermic cell killing, *Radiat. Res.*, **75**, 573–585.

Landry, J., Chretien, P., DeMuys, T. N. and Morais, R. (1985) Induction of thermotolerance and heat shock protein-synthesis in normal and respiration deficient chick embryo fibroblasts, *Cancer Res.*, **45**, 2240–2247.

Lee, D. C. and Chapman, D. (1987) The effect of temperature on biological membranes and their models, in Bowler, K. and Fuller, B. J. (Eds), *Temperature and Animal Cells*, Cambridge, Company of Biologists, pp. 35–52.

Leeper, D. B. and Henle, K. J. (1975) Hyperthermia: effects of different temperatures on normal and tumour cells, *Proceedings of the International Symposium on Cancer Therapy by Hyperthermia and Radiation, Washington, DC*, pp. 47–60.

Leith, J. T., Miller, R. C., Gerner, E. W. and Boone, M. L. M. (1977) Hyperthermic potentiation. Biological aspects and application to radiation therapy, *Cancer Res.*, **39**, 766–779.

Lepock, J. R. (1982) Involvement of membranes in cellular response to hyperthermia, *Radiat. Res.*, **92**, 433–438.

Lepock, J. R., Nolan, P. M., Rule, G. S. and Kruuv, J. (1981) Lack of correlation between hyperthermic cell killing, thermotolerance and membrane fluidity, *Radiat. Res.*, **87**, 300–313.

Lepock, J. R., Cheng, H. K., Al-Qysi, H. and Kruuv, J. (1983) Thermotropic lipid and protein transitions in Chinese hamster lung cell membranes: Relationship to hyperthermic killing, *Can. J. Biochem. Cell Biol.*, **61**, 421–427.

Li, G. C. and Hahn, G. M. (1984) Mechanisms of thermotolerance, in Overgaard, J. (Ed.), *Hyperthermic Oncology*, Vol. 2, London, Taylor and Francis, p. 231.

Li, G. C. L. and Hahn, G. M. (1978) Ethanol-induced tolerance to heat and adriamycin, *Nature, Lond.*, **274**, 694–701.

Li, G. C. and Kal, H. B. (1977) Effect of hyperthermia on radiation response of two mammalian cell lines, *Europ. J. Cancer*, **13**, 65–69.

Li, G. C. and Laszlo, A. (1985) Thermotolerance in mammalian cells: a possible role for heat shock proteins, in Atkinson, B. G. and Walden, D. B. (Eds), *Changes in Gene Expression in Response to Environmental Stress*, New York, Academic Press, p. 227.

Li, G. C. and Mivechi, N. F. (1986) Thermotolerance in mammalian systems: a review, in Anghileri, L. J. and Robert, J. (Eds), *Hyperthermia in Cancer Treatment*, Vol. 1, Boca Raton, CRC Press, pp. 59–77.

Li, G. C., Evans, R. G. and Hahn, G. M. (1976) Modification and inhibition of repair of potentially lethal X-ray damage by hyperthermia, *Radiat. Res.*, **67**, 491–501.

Li, G. C., Shiu, B. S. and Hahn, G. M. (1980a) Recovery of cells from heat-induced potentially lethal damage: effect of pH and nutrient environment, *Int. J. Radiat. Oncol. Biol. Phys.*, **6**, 577–582.

Li, G. C., Shiu, E. C. and Hahn, G. M. (1980b) Similarities in cellular inactivation by hyperthermia or by ethanol, *Radiat. Res.*, **82**, 257–268.

Li, G. C., Fischer, G. A. and Hahn, G. M. (1982) Modification of thermal response by D_2O. II. Thermotolerance and the specific inhibition of development, *Radiat. Res.*, **92**, 541–551.

Lin, P. S., Hefter, K. and Hok, C. (1984) Modification of membrane function, protein synthesis and heat killing effect in cultured Chinese hamster cells by glycerol and D_2O, *Cancer Res.*, **44**, 5776–5784.

Lobreau, A. U., Raaphorst, G. P. and Szekely, J. G. (1988) The use of immunostaining to quantify X-ray and heat-induced multiple microtubule organising centers in normal and transformed cells, *Bas. Appl. Histochem.*, **32**, 263–278.

Mackey, M. A. and Dewey, W. C. (1988) Time–temperature analysis of cell killing of synchronous G1 and S phase Chinese hamster cells *in vitro*, *Radiat. Res.*, **113**, 318–333.

Massicotte-Nolan, P., Glofcheski, D. J., Kruuv, J. and Lepock, J. R. (1981) Relationship between hyperthermic cell killing and protein denaturation by alcohols, *Radiat. Res.*, **87**, 284–299.

McElhaney, R. N. (1985) Membrane fluidity, phase state and membrane function in prokaryotic micro-organisms, in Aloia, R. C. (Ed.), *Membrane Fluidity in Biology*, Vol. 4, New York, Academic Press, pp. 147–208.

Mendez, R. G., Minton, K. W. and Hahn, G. M. (1982) Lack of correlation between membrane lipid composition and thermotolerance in Chinese hamster ovary cells, *Biochem. Biophys. Acta*, **692**, 168–170.

Meyer, K. R., Hopwood, L. E. and Gillette, E. J. (1979) The thermal response of mouse adenocarcinoma cells at low pH, *Europ. J. Cancer*, **15**, 1219–1222.

Mills, M. D. and Meyn, R. E. (1981) Effects of hyperthermia on repair of radiation-induced DNA strand breaks, *Radiat. Res.*, **87**, 314–328.

Mitchell, J. B. and Russo, A. (1983) Thiols, thiol depletion and thermosensitivity, *Radiat. Res.*, **95**, 471–485.

Mondovi, B. (1976) Biochemical and ultrastructural lesions, in Wizenberg, M. J. and Robinson, J. E. (Eds), *Proceedings of the First International Symposium on Cancer Therapy by Hyperthermia and Radiation*, Chicago, American College of Radiology, pp. 3–15.

Mulcahy, R. T., Gould, M. N., Hidvergi, E., Elson, C. E. and Yatvin, M. B. (1981) Hyperthermia and surface morphology of P388 ascites tumour cells: Effects of membrane modification, *Int. J. Radiat. Biol.*, **39**, 95–106.

Murthy, A. K., Harris, J. R. and Belli, B. A. (1977) Hyperthermia and radiation response of plateau phase cells. Potentiation and radiation damage repair, *Radiat. Res.*, **70**, 241–247.

Nagle, W. A., Moss, A. J., Roberts, H. G. and Baker, M. D. (1980) Effects of 5-thio-D-glucose on cellular ATP levels and DNA rejoining in hypoxic and aerobic Chinese hamster cells, *Radiol.*, **137**, 203–211.

Natali, P. G., Salsano, F., Viora, M., Nista, A., Malorni, W., Marolli, A. and De Martino, C. (1984) Inhibition of aerobic glycolysis in normal and neoplastic lymphoid cells induced by lonidamine, *Oncology*, **41**, 7–14.

Nielsen, O. S. and Overgaard, J. (1979) Effect of extracellular pH on thermotolerance and recovery of hyperthermic damage *in vitro*, *Cancer Res.*, **39**, 2772–2778.

Overgaard, J. (Ed.) (1984) *Proceedings of the Fourth International Symposium of Hyperthermia Oncology*, Vols 1 and 2, Taylor and Francis, London.

Overgaard, K. and Overgaard, J. (1976) Pathology of heat damage studies on the histopathology in tumour tissue exposed to *in vivo* hyperthermia, in Wizenberg, M. J. and Robinson, J. E. (Eds), *Proceedings of the International Symposium on Cancer Therapy by Hyperthermia and Radiation*, Chicago, American College of Radiology, pp. 115–127.

Pain, R. H. (1987) Temperature and macromolecular structure and function, in Bowler, K. and Fuller, B. J. (Eds), *Temperature and Animal Cells*, Cambridge, Company of Biologists, pp. 21–34.

Paterson, S. J., Butler, K., Huana, P., Labelle, J., Smith, I. C. P. and Schneider, H. (1972) Effect of alcohols on lipid bilayers: a spin label study, *Biochem. Biophys. Acta*, **226**, 597–602.

Philips, R. A. and Tomach, L. J. (1966) Repair of potentially lethal damage in X-irradiated HeLa cells, *Radiat. Res.*, **29**, 413–432.

Raaphorst, G. P. (1989a) Thermal radiosensitization *in vitro*, in Urano, M. and Douple, E. (Eds), *Hyperthermia and Oncology*, Vol. 2, Utrecht, VSP Utrecht, in press.

Raaphorst, G. P. and Azzam, E. I. (1980a) Dependence on heat and X-ray sensitivity of V79 cells on growth media and various serum combinations, *Int. J. Radiat. Biol.*, **38**, 677–693.

Raaphorst, G. P. and Azzam, E. I. (1980b) Heat and radiation sensitivity of Chinese hamster V79 cells and nine clones selected from survivors of a thermal tolerant cell population, *Int. J. Radiat. Oncol.*, **6**, 1577–1581.

Raaphorst, G. P. and Azzam, E. I. (1982a) The thermal sensitivity of normal and ataxia telangiectasia human fibroblasts, *Int. J. Radiat. Oncol.*, **8**, 1947–1950.

Raaphorst, G. P. and Azzam, E. I. (1982b) The effect of D_2O on thermal sensitivity and thermotolerance in cultured Chinese hamster V79 cells, *J. Thermal. Biol.*, **7**, 147–153.

Raaphorst, G. P. and Azzam, E. I. (1983) Thermal radiosensitization in Chinese hamster V79 and mouse C3H-10T½ cells. The thermotolerance effect, *Br. J. Cancer*, **48**, 45–54.

Raaphorst, G. P. and Azzam, E. I. (1985a) Hyperthermia enhances X-ray cell killing in normal and homozygous and heterozygous ataxia telangiectasia human cells, *Int. J. Radiat. Oncol. Biol. Phys.*, **11**, 855–859.

Raaphorst, G. P. and Azzam, E. I. (1985b) Response of hamster cells to hyperthermia in the presence of solutions of monovalent and divalent salts, *Int. J. Hypertherm.*, **4**, 383–385.

Raaphorst, G. P. and Feeley, M. M. (1989) Hyperthermic damage and its repair or fixation measured at the cellular level, in Sugahara, T. and Saito, M. (Eds), *Hyperthermic Oncology*, Vol. 2, Kyoto, Japan, Taylor and Francis, pp. 181–182.

Raaphorst, G. P. and Feeley, M. M. (1990) Comparison of hyperthermia and radiation PLD, *Radiat. Res.*, in press.

Raaphorst, G. P. and Szekely, J. G. (1988) Thermal enhancement of cellular radiation damage: a review of complementary and synergistic effects, *J. Scan. Elect. Microscopy*, 2, 513–535.

Raaphorst, G. P., Freeman, M. L. and Dewey, W. C. (1979a) Radiosensitivity and recovery from radiation damage in cultured CHO cells exposed to hyperthermia at 42.5 or 45.5°C, *Radiat. Res.*, 79, 390–402, 1979.

Raaphorst, G. P., Romano, S. L., Mitchell, J. B., Bedford, J. S. and Dewey, W. C. (1979b) Intrinsic differences in heat and/or X-ray sensitivity of seven mammalian cell lines cultured and treated under identical conditions, *Cancer Res.*, 39, 396–401.

Raaphorst, G. P., Azzam, E. I., Borsa, J. and Einspenner, M. (1983a) Hyperthermia in a differentiating murine erythroleukemia cell line. I. cell killing by heat and radiation, *Int. J. Radiat. Biol.*, 44, 275–283.

Raaphorst, G. P., Szekely, J., Lobreau, A. and Azzam, E. I. (1983b) A comparison of cell killing by heat and/or X-rays in Chinese hamster V79 cells, friend erythroleukemia mouse cells and human thymocyte MOLT-4 cells, *Radiat. Res.*, 94, 340–349.

Raaphorst, G. P., Broski, A. P. and Azzam, E. I. (1985a) Sensitivity to heat, radiation and heat plus radiation of Chinese hamster cells synchronized by mitotic selection, thymidine block or hydroxyurea, *J. Therm. Biol.*, 10, 177–181.

Raaphorst, G. P., Vadasz, J., Azzam, E. I., Sargent, M., Borsa, J. and Einspenner, M. (1985b) Comparison of heat and/or radiation sensitivity and membrane composition of seven X-ray transformed C3H-10T½ cells, *Cancer Res.*, 45, 5452–5456.

Raaphorst, G. P., Spiro, I. J., Azzam, E. I. and Sargent, M. (1987) Normal cells and malignant cells transfected with the *HRas* oncogene have the same heat sensitivity in culture, *Int. J. Therm. Biol.*, 3, 209–216.

Raaphorst, G. P., Feeley, M. M., Danjoux, C. E., Martin, L., Maroun, J. and DeSanctis, A. J. (1990) The effect of lonidamine on radiation and thermal responses of human and rodent cell lines, *Int. J. Radiat. Oncol.*, submitted.

Raaphorst, G. P., Feeley, M. M., DaSilva, V. F., Danjoux, C. E. and Gerig, L. H. (1989) Comparison of heat and radiation sensitivity of three human glioma cell lines, *Biol. Phys.*, 17, 615–621.

Radford, I. R. (1983) Effects of hyperthermia on the repair of X-ray induced DNA double strand breaks in mouse L cells, *Int. J. Radiat. Biol.*, 43, 551–557.

Rastogi, D., Nagle, W. A., Henle, K. J., Moss, A. J. and Rastogi, S. P. (1988) Uncoupling of oxidative phosphorylation does not induce thermotolerance in cultured Chinese hamster cells, *Int. J. Hypertherm.*, 4, 333–344.

Rhee, J. G. and Song, C. W. (1984) Thermosensitivity of bovine aortic endotheial cells in culture; *in vitro* clonogenicity study, in Overgaard, J. (Ed.), *Hyperthermic Oncology*, London, Taylor and Francis, pp. 157–160.

Rieder, C. and Bajer, A. S. (1977) Heat-induced reversible hexagonal packing of spindle microtubules, *J. Cell. Biol.*, 74, 717–725.

Robbins, H. I., Steeves, R. A., Clark, A. W., Martin, P. A., Miller, K., Dennis, W. A. (1983) Differential sensitivity of AKR murine leukemia and normal bone marrow cells to hyperthermia, *Cancer Res.*, 43, 4951–4955.

Rofstad, E. K. and Brustad, T. (1986) Differences in thermoradiosensitization among cloned cell lines isolated from a single human melanoma xenograft, *Radiat. Res.*, 106, 147–155.

Roti Roti, J. L. (1982) Heat-induced cell death and radiosensitization; molecular mechanisms, *J. Natl Cancer Inst. Monogr.*, 61, 3–10.

Roti Roti, J. L. and Laszlo, A. (1988) The effects of hyperthermia on cellular macromolecules, in Urano, M. and Douple, E. (Eds), *Hyperthermia and Oncology*, Vol. 1, VSP, Utrecht, pp. 13–56.

Roti Roti, J. L. and Wilson, C. F. (1984) The effects of alcohols, procaine and hyperthermia on the protein content of nuclei and chromatin, *Int. J. Radiat. Biol.*, 46, 25·33.

Roti Roti, J. L., Higashikubo, R., Blair, O. C. and Uygler, N. (1982) Cell-cycle position and nuclear protein content, *Cytometry*, 3, 91–96.

Rottinger, E. M., Mendonca, M., Gerweck, L. E. (1980) Modification of pH induced cellular inactivation by irradiation-glial cells, *Int. J. Radiat. Oncol.*, 6, 1659–1662.

Sapareto, S. A. and Dewey, W. C. (1984) Thermal dose determination in cancer therapy, *Int. J. Radiat. Oncol.*, 10, 787–800.

Sapareto, S. A., Hopwood, L. E., Dewey, W. C., Raju, M. R. and Gray, J. W. (1978) Effects of hyperthermia on survival and progression of Chinese hamster ovary cells, *Cancer Res.*, **38**, 393–400.

Sapareto, S. A., Raaphorst, G. P. and Dewey, W. C. (1979) Cell killing and the sequencing of hyperthermia and radiation, *Int. J. Radiat. Oncol. Biol. Phys.*, **5**, 343–347.

Seeman, P. (1972) The membrane actions of anesthetics and tranquilizers, *Pharmacol. Rev.*, **24**, 583–655.

Shinitzky, M. (1984) *Physiology of Membrane Fluidity*, Vol. 2, Boca Raton, CRC Press.

Sijens, P. E., Bovee, W. M. M., Seijkens, D., Koole, P., Los, G. and VanRijssel, R. H. (1987) Murine mammary tumour response to hyperthermia and radiotherapy evaluated by *in vivo* ^{31}P-nuclear magnetic resonance spectroscopy, *Cancer Res.*, **47**, 6467–6473.

Sijens, P. E., Bovee, W. M. M., Koole, P. and Schipper, J. (1989) Phosphorous NMR study of the response of a murine tumour to hyperthermia as a function of treatment time and temperature, *Int. J. Hyperthermia*, **5**, 351–357.

Simard, R. and Bernhard, W. (1967) Heat sensitive cellular function located in the nucleolus, *J. Cell. Biol.*, **34**, 61–76.

Snell, F. M., Shulman, S., Spencer, R. P. and Moos, C. (1965) *Biophysical Principles of Structure and Function*, Reading, Mass., Addison–Wesley, pp. 342–346.

Spiro, I. J., Denman, D. L. and Dewey, W. C. (1983) Effect of hyperthermia on isolated DNA B polymerase, *Radiat. Res.*, **95**, 68–77.

Spiro, I. J., Sapareto, S. A., Raaphorst, G. P. and Dewey, W. C. (1982a) Effect of chronic and acute heat conditioning on development of thermal tolerance, *Int. J. Radiat. Biol.*, **8**, 53–58.

Spiro, I. J., Denman, D. L. and Dewey, W. C. (1982b) Effect of hyperthermia on CHO, DNA polymerase α and β, *Radiat. Res.*, **89**, 134–149.

Szekely, J. and Lobreau, A. (1985) High radiosensitivity of the Molt-4 leukemic cell lines, *Int. J. Radiat. Biol.*, **48**, 277–284.

Szekely, J. G., Lobreau, A. U., Delaney, S., Raaphorst, G. P. and Feeley, M. M. (1990) Morphological effects of lonidamine on human tumour cells, *J. Scanning Elect. Micros.*, in press.

Tomasovic, S. P., Turner, G. N. and Dewey, W. C. (1978) Effects of hyperthermia on nonhistone proteins isolated with DNA, *Radiat. Res.*, **73**, 535–552.

Wachsberger, P. R. and Coss, R. A. (1990) Acrylamide sensitization of heat response of cytoskeleton and cytotoxicity in attaching and well spread CHO cells, *Cell Motility and the Cytoskeleton*, in press.

Warters, R. L. and Henle, K. J. (1982) DNA degradation in Chinese hamster ovary cells after exposure to hyperthermia, *Cancer Res.*, **42**, 4427–4432.

Warters, R. L. and Roti Roti, J. L. (1981) The effect of hyperthermia on replicating chromatin, *Radiat. Res.*, **88**, 69–78.

Warters, R. L. and Stone, O. L. (1982) Inhibition of DNA and histone protein synthesis at hyperthermia temperatures, *Radiat. Res.*, **91**, 280–288.

Warters, R. L. and Stone, O. L. (1983a) The effects of hyperthermia on DNA replication in HeLa cells, *Radiat. Res.*, **93**, 71–84.

Warters, R. L. and Stone, O. L. (1983b) Macromolecular synthesis in HeLa cells after thermal shock, *Radiat. Res.*, **96**, 646–652.

Warters, R. L. and Stone, O. L. (1984) Histone protein and DNA synthesis in HeLa cells after thermal shock, *J. Cell. Physiol.*, **118**, 153–160.

Warters, R. L., Yasui, L. S., Sharma, R. and Roti Roti, J. L. (1986) Heat shock (45 °C) results in an increase of nuclear matrix protein mass in HeLa cells, *Int. J. Radiat. Biol.*, **50**, 253–268.

Wawra, E. and Dolejs, I. (1979) Evidences for the function of DNA polymerase-B in unscheduled DNA synthesis, *Nucl. Acids Res.*, **7**, 1675–1686.

Welsh, W. J. and Suhan, J. P. (1985) Morphological study of mammalian stress response characterization of changes in cytoplasmic organelle, cytoskeleton and nucleoli and appearance of intranuclear actin filaments in rat fibroblasts after heat shock treatment, *J. Cell. Biol.*, **101**, 1198–1211.

Westra, A. and Dewey, W. C. (1971) Variation in sensitivity to heat shock during the cell-cycle of Chinese hamster cells *in vitro*, *Int. J. Radiat. Biol.*, **19**, 467–477.

Wike-Hooley, J. L., Haveman, J. and Reinhold, H. S. (1984) The relevance of tumour pH to the treatment of malignant disease, *Radiotherapy and Oncology*, **2**, 343–366.

Wong, R. S. L. and Dewey, W. C. (1982) Molecular studies on hyperthermic inhibition of DNA synthesis in Chinese hamster ovary cells, *Radiat. Res.*, **92**, 370–395.

Wong, R. S. L. and Dewey, W. C. (1986) Effect of hyperthermia on DNA synthesis, in Anghileri, L. J. and Robert, J. (Eds), *Hyperthermia in Cancer Treatment*, Vol. 1, Boca Raton, CRC Press, pp. 80–89.

Yatvin, M. B. (1977) The influence of membrane lipid composition and procaine on hyperthermic death of cells, *Int. J. Radiat. Biol.*, **32**, 513–522.

Yatvin, M. B., Clifton, K. H. and Dennis, W. H. (1979) Hyperthermia and local anesthetics. Potentiation of survival in tumour bearing mice, *Science*, **205**, 195–196.

Yatvin, M. B., Vorphal, J. W., Gould, M. N. and Lyte, M. (1983) The effect of membrane modification and hyperthermia on the survival of P-388 and V79 cells, *Eur. J. Cancer Clin. Oncol.*, **19**, 1247–1253.

Yatvin, M. B., Dennis, W. H., Elegbede, J. and Elson, C. E. (1987) Sensitivity of tumour cells to heat and ways of modifying the response, in Bowler, K. and Fuller, B. J. (Eds), *Temperature and Animal Cells*, Cambridge, Company of Biologists, pp. 235–267.

Yau, T. M. (1979) Procaine-mediated modification of membranes and of the response to X-irradiation and hyperthermia in mammalian cells, *Radiat. Res.*, **80**, 523–541.

Yau, T. M. and Kim, S. C. (1980) Local anesthetics as hypoxic radiosensitizers, oxic radioprotectors and potentiators of hyperthermic killing in mammalian cells, *Br. J. Radiol.*, **53**, 687–692.

Zahler, P. (1969) The structure of the erythrocyte membrane, *Experientia*, **25**, 449.

Zielke–Jemme, B. and Hopwood, L. (1982) Time lapse cinemicrographic observations of heated G1 phase Chinese hamster ovary cells, *Radiat. Res.*, **92**, 332–342.

2 *In vivo* aspects of hyperthermic oncology

S. B. Field

2.1. Introduction

The reasons why hyperthermia might be useful in treating cancer are given in the introduction to this book. The possible mechanisms by which cells respond to heat damage are discussed in Chapter 1, which also includes a detailed discussion of cellular response to hyperthermia alone or in combination with radiation. In order to formulate a rationale for the efficient clinical use of hyperthermia, it is essential also to have some understanding of the response of normal tissue and tumours *in situ*. Most of the available information on this topic comes from experiments on rodents. The majority of these studies are phenomenological rather than basic research, but they do provide valuable insights into tissue response to hyperthermia. In particular, it is possible to investigate aspects of long-term injury, which is not possible using *in vitro* systems. Nevertheless, it must be remembered that since the majority of these studies have used rodent tissues, which in most cases have been heated by immersion in hot baths, only general patterns of response can be determined and quantitative findings should be treated with considerable caution.

2.2. The expression of thermal damage

Temperatures of around 41 °C or higher can be lethal to mammalian cells, as described in Chapter 1. The same is true for damage to tissues. It is clear from *in vitro* studies that membrane injury is a major component of heat damage to cells. As a result they die and lyse rapidly, i.e. far sooner than after equally lethal doses of ionizing radiation. The delay in tissue response following irradiation is generally accepted to be related to cellular proliferation because in nearly all tissues cell death following ionizing radiation occurs only when cells attempt to divide. As a result, slowly proliferating tissues may not exhibit radiation injury until weeks

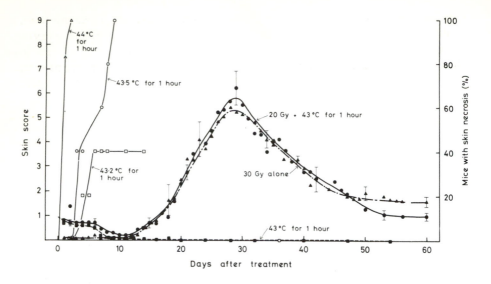

**Figure 2.1 The response of the mouse ear to hyperthermia or ionizing radiation.
Open symbols give the percentage of ears showing necrosis after heat alone (right-
hand axis): (△) 44°C for 1 h; (○) 43.5°C for 1 h; (□) 43.2°C for 1 h. Closed symbols give
skin reaction using an arbitrary numerical score (left-hand axis). Scores of 1 and 2
indicate degrees of erythema. Scores of 3–9 indicate increasing areas of moist
desquamation: (■) hyperthermia alone 43°C for 1 h; (▲) X-rays alone; (●) X-rays
followed by hyperthermia at 43°C for 1 h. (From Law *et al.* (1978).)**

or months after irradiation. In contrast, because hyperthermia kills cells
through membrane damage, they die in interphase so that cell death, lysis
and, hence, tissue manifestation of heat damage occurs very rapidly, even
in slowly proliferating tissues. This difference in tissue reaction to heat or
ionizing radiation is illustrated in Figure 2.1 for the skin of the mouse.
Following a dose of ionizing radiation, the principle response of the skin is
a wave of erythema and desquamation reaching a peak at 25–30 days, the
timing being related to proliferation of the epithelial cells. The reaction
then heals within 6 or 8 weeks, depending upon dose. In contrast,
hyperthermia rapidly causes a deep reddening and swelling, and,
depending on the treatment, even necrosis within the first day or two
following treatment. If, however, a mild hyperthermia treatment, which
itself causes no visible reaction, is combined with a dose of ionizing
radiation, the reaction is qualitatively similar to that following
irradiation alone. Thus, a given radiation response can be achieved with a
smaller dose when it is combined with hyperthermia, i.e. there is thermal
enhancement of the radiation response. This aspect will be dealt with in
more detail in Section 2.6.

The difference in time course of the reaction following either
hyperthermia or ionizing radiation is a very general phenomenon and has
been demonstrated in the skin at various sites, cartilage, testis, small

intestine, kidney and spinal cord. Studies on mouse small intestine have shown that heat not only causes death of the proliferative cells in the crypt but also kills post-mitotic epithelial cells of the villi. Hume *et al.* (1983) showed that the time course of expression of thermal injury was similar in both crypts and villi, but that the non-proliferating epithelial cells of the villi were somewhat more susceptible to thermal injury than the proliferating cells in the crypts. This result is in direct contrast to that following ionizing radiation which only sterilizes proliferating cells in the crypts and inflicts virtually no damage to the post-mitotic cells of the villi. Scanning electron microscopy has demonstrated changes in villus morphology as early as 5 min following hyperthermia with maximal damage appearing at approximately 2 h (Kamel *et al.*, 1985). Villus damage is probably related, at least in part, to injury resulting from damaged capillaries. Vascular damage is an extremely important component of tissue injury and is discussed in detail in Chapter 4.

Thus, it is clear that damage due to heat is of a different nature and follows a different time course from that resulting from ionizing radiation. However, with cells *in vitro* or with tumours *in vivo*, the differences may not be readily apparent, because in both cases the end-points result only from the numbers of cells killed, with the mode and timing of cell death hardly influencing the result. With normal tissues, however, it is possible to separate the effects of two modalities as illustrated in Figure 2.1.

In general, the pathology following hyperthermia is similar to but less severe than that of a thermal burn. Depending on the magnitude of the treatment, hyperthermia may be followed by oedema, focal haemorrhage, granulocytic infiltration and possibly even necrosis, which may all occur within the first day or two. There may also be later changes which include loss of blood vessels, monocyte infiltration and fibrosis. There is, of course, considerable interest in the question of late reactions to hyperthermia, but most observations indicate that once the acute response has healed there appear to be no late effects. However, we cannot be quite certain of this and there may be reactions in some cases which have not yet been discovered. As discussed in Chapter 4, damage to supporting tissues, particularly the vasculature, plays a crucial role in the pathology of heat damage, both to tumours and to normal tissues. The topic of the pathological effects of hyperthermia in normal tissues has been reviewed by Fajardo (1984).

2.3. *Quantitation of heat injury* in vivo

It is possible to obtain survival curves following hyperthermia *in situ* and their characteristics are similar to those described in Chapter 1 for cells *in vitro*. The majority of *in vivo* studies, however, rely on an assessment of gross tissue response. There are numerous examples in the literature, one of which is illustrated in Figure 2.2. In this experiment, heat was used to

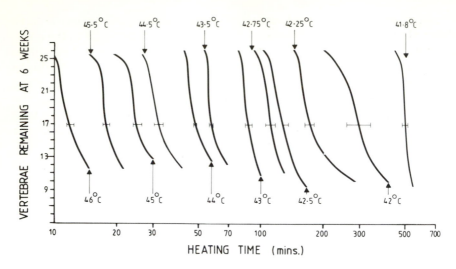

Figure 2.2 Dose-response curves for loss of vertebrae in the tails of young rats. (From Morris and Field (1985).)

cause necrosis in the tail of a young rat, assessed radiographically at 6 weeks following hyperthermia, so that the number of lost vertebrae could easily be assessed. Figure 2.2 shows the number of remaining vertebrae following hyperthermia treatments at different times and different temperatures. The figure shows clearly that at any given temperature, once the time of heating is sufficiently long to reach the threshold of injury, a small further increase in heating time amounting to approximately 20%, or an increase in temperature of only 0.5°C, will increase the probability of necrosis to 100%. This finding illustrates clearly the danger of causing tissue necrosis when a hyperthermia treatment is given which approaches the limit of normal tissue tolerance. Small differences in tissue temperature are very likely to occur in clinical treatment and therefore may lead to marked variation in biological response. Very great care must be taken to monitor properly temperature at as many points as possible. However, if tumours are more sensitive than normal tissues or if they are at a slightly higher temperature than the surrounding normal tissues, then hyperthermia should result in a clear therapeutic advantage.

It is very important for a variety of reasons, some of which have already been discussed in Chapter 1, to have a clear understanding of the relationship between temperature and time of heating to produce a given level of reaction. Such isoeffect relationships may be derived from dose-response curves such as those illustrated in Figure 2.2. In this particular example, the reduction in the number of intact vertebrae from the normal of 25 to 17 is taken as the index of response. The time of heating, at this

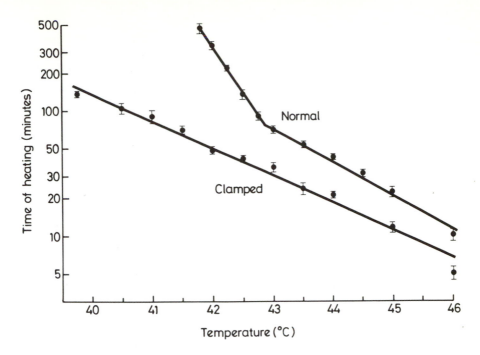

Figure 2.3 Relationships between time of heating under normal conditions and temperature to produce a given loss of vertebrae in the tail of a baby rat (upper curve). When the blood supply was clamped during hyperthermia (lower curve), the tissue became substantially more sensitive. (From Morris and Field (1985).)

chosen level of response, is then plotted on a logarithmic scale as a function of temperature, as illustrated in Figure 2.3. The upper curve in Figure 2.3 was obtained for tails heated under normal conditions. It is seen that at approximately 42.5 °C there is a transition in the curve, such that for temperatures above it a 1 °C temperature change was approximately equivalent to a factor of two in heating time, whereas at temperatures below the transition, a temperature change of 1 °C was approximately equivalent to a factor of six in heating time. As discussed in Chapter 1, such a transition in the time-temperature relationship or in an Arrhenius plot (which is fairly similar) is extremely common. It may result from a different mechanism of cell killing or, more likely, the reduced effect at lower temperatures may result from the induction of thermotolerance during the relatively long heating times. (The topic of thermotolerance was reviewed for cells *in vitro* in Chapter 1 and will be discussed for tissues below.) In contrast to the upper curve shown in Figure 2.3, when the blood supply to the tails was occluded by a clamp, there were two major changes in the isoeffect relationship. Firstly, the tissue became considerably more sensitive, equivalent to a factor of approximately two in heating time, which is most likely due to a reduction in pH. Secondly, the transition at

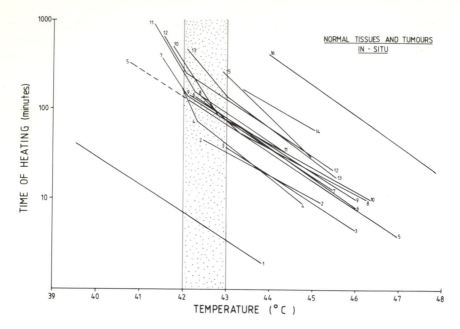

Figure 2.4 Relationship between time of heating and temperature for a given level of damage in a range of normal tissues and tumours *in situ*: (1) mouse testis weight loss; (2) rat tumour 9L, heated *in vivo*, assayed *in vitro*; (3) mouse jejunum LD_{50};; (4) mouse jejunum, 50% loss of crypts; (5) mouse sarcoma 180, majority cure; (6) baby rat tail, 50% necrosis; (7) mouse ear skin, 50% necrosis; (8) rat skin epilation; (9) baby rat tail, 5% stunting; (10) baby rat tail, whole-tail necrosis; (11) mouse tumour C3H/Tif regrowth; (12) mouse tumour F(Sal) (TCD_{50}); (13) mouse foot skin epilation; (14) mouse skin, feet and leg reactions; (15) mouse tumour C3H, 50% cures; (16) pig and human skin necrosis and cutaneous burns. (From Field and Morris (1983).)

42.5°C was clearly eliminated. It is known that a reduction in pH can reduce or abolish thermotolerance and this is considered to be the most likely cause of this difference between the isoeffect curves.

Since tumours (or at least some regions of tumours) might exist in conditions of hypoxia and hence low pH, these regions may be represented by the lower curve in Figure 2.3, whereas normal tissues are more likely to be represented by the upper curve. As a result, the maximum differential between tumours and normal tissues might be achieved by heating at relatively low temperatures for long periods. Unfortunately, however, this may not be the most practical way of administering clinical hyperthermia.

A summary of *in vivo* data of the type shown in Figure 2.3 is given in Figure 2.4. In general, the features of the upper isoeffect curve in Figure 2.3 are retained throughout, the transition generally occurring between 42 and 43°C. Absolute tissue sensitivity is represented by the vertical position of the isoeffect lines, and is seen to be extremely tissue dependent. The variation in sensitivity between the most-sensitive tissue represented in Figure 2.4, i.e. mouse testis, and most resistant tissue, i.e. human skin, is more than a factor of 100 in heating time. Despite this massive difference in absolute tissue sensitivity, the time–temperature relationships are extremely consistent. In general, a change in temperature of 1 °C is equivalent to a change in heating time by a factor of two at temperatures above the transition and by a factor of approximately six at temperatures below (Field and Morris, 1983). These results are discussed further in the discussion of heat dose (see Chapter 3).

2.4. Thermotolerance

By analogy with radiation therapy, clinical hyperthermia is usually given in several fractions so that a knowledge of the biological response to fractionated hyperthermia becomes extremely important in the rational design of a clinical treatment. A major consideration is the induction of thermotolerance, i.e. cells and tissues become resistant to heating as a result of prior hyperthermia. At the phenomenological level, two types of thermotolerance have been identified, i.e. that which develops during prolonged heating at temperatures below a critical value, normally about 42°C, and that which develops between individual, more acute, hyperthermia treatments providing the tissue is allowed to return to physiological temperature during the interval. Thermotolerance is both a very general and a very large effect. It may result in an increase in slope of a survival curve by as much as a factor of 15 or an increase in duration of heating to produce a given effect by a factor of four or more. The effect is transient, and the time course appears to be strongly cell or tissue-type dependent.

A typical example of thermotolerance *in vivo* is shown in Figure 2.5, again using the end-point of necrosis in the rat tail. In this example, a test treatment at a temperature of 44.5°C was given. Under normal circumstances it took approximately 30 min at this temperature to produce the chosen injury end-point. When a more gentle hyperthermia treatment was given (which alone appears to cause no change) the tail becomes increasingly resistant to the test treatment at 44.5 °C. In the case shown in Figure 2.5 the maximum thermotolerance occurred at approximately 10 h after the mild priming treatment, at which time the duration of heating at 44.5 °C needed to cause the end-point increased by a factor of more than four. At longer times, thermotolerance decays, the tissues slowly returning towards normal sensitivity; however, this may take as long as a week or two.

Almost all cells and tissues show this type of response, the maximum extent varying between an increase in heating time by a factor of between 2 and 4.5 (Field, 1985).

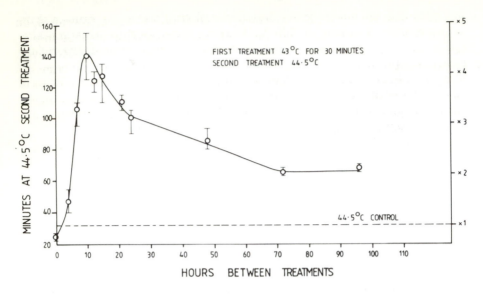

Figure 2.5 Thermotolerance in the rat tail following 43°C for 30 min, tested by measuring the time required to cause a given level of necrosis at 44.5°C at various intervals after the initial treatment. The fine line shows the effect in the absence of prior treatment. (From Field and Morris (1985).)

The extent of thermotolerance and also its timing have been shown to be related to the effectiveness of the first treatment. The greater the effectiveness of the first primary heat dose, the greater will be the extent of thermotolerance and the longer it will take before this maximum is reached (Nielsen and Overgaard, 1982; Field and Morris, 1985). The decay of thermotolerance, however, does not appear to relate to the priming treatment, although it is markedly dependent on the type of tissue. It has been suggested that thermotolerance decays more quickly in rapidly dividing cells and tissues, but overall the available data do not support the hypothesis that thermotolerance kinetics are dependent primarily on cell proliferation kinetics (Field and Majima, 1989).

A crucial question is how the extent and kinetics of thermotolerance alters when a series of fractionated hyperthermia treatments is given. It appears from the available data that with multiple fractions it is not possible to exceed the maximum extent of thermotolerance which can be achieved following a single treatment. However, if multiple treatments are given such that, for each, thermotolerance would be less than the maximum possible, then as fractionation continues, the degree of thermotolerance may steadily increase up to this maximum, but not above. After a multifraction treatment the decay is simply that which would occur if only the last treatment had been given (Law, 1988). Thermotolerance *in vivo* and *in vitro* follow similar patterns, which strongly suggest that it is a cellular rather than a physiological phenomenon. Reduced pH may cause a reduction in thermotolerance, on which basis it has been suggested that thermotolerance may be less in some tumours (or regions of tumours) than in normal tissues. However, experimental results so far have not given a clear indication that tumours have a reduced capacity to develop thermotolerance, and there is no basis to be optimistic that differences in thermotolerance between tumours and normal tissues may lead to a therapeutic advantage (Field, 1985; Field and Majima, 1989).

The cellular aspects of thermotolerance are discussed in Chapter 1 and the topic of thermotolerance in general has been reviewed in detail by Henle (1987).

2.5. Step-down sensitization

It has been shown that when cells or tissues are heated at temperatures of approximately 43 °C or higher, followed by temperatures of 42 °C or lower, then the response to the lower temperature is greater than it would have been in the absence of the prior higher-temperature treatment (Henle, 1980). This phenomenon has been termed step-down sensitization, and it can be important clinically if there are marked temperature variations during treatment. In fact, variations in temperature are difficult to avoid in any clinical treatment and whether or not step-down sensitization will occur depends on the degree of variation. However, it has also been proposed that use may be made of the 'step-down' effect by giving a fairly short high-temperature treatment, followed by the bulk of treatment at a lower temperature. This could avoid much of the discomfort and pain of a lengthy treatment at a relatively high temperature. This aspect of step-down sensitization has been discussed recently by Lindegaard and Overgaard (1990).

It has been suggested that step-down sensitization may be the result of abolishing the effect of thermotolerance. However, it is more likely to be an independent phenomenon, since its time course is different from that of thermotolerance, being present immediately after the first heating. Also, it decays substantially more rapidly than thermotolerance.

2.6. Interaction between hyperthermia and radiation

Hyperthermia can cause an increase in the effectiveness of ionizing radiation as well as resulting in direct thermal injury. In terms of radiation survival curves, hyperthermia causes an increase in intrinsic radiation sensitivity (i.e. a decrease in D_0) and it sometimes causes a reduction in the 'shoulder' or extrapolation number giving rise to reduced potential for repair.

The response of skin to heat alone or heat in combination with ionizing radiation is illustrated in Figure 2.1 and is typical of many tissues. As already explained, direct thermal injury is observed earlier than the expression of radiation damage. A mild heat treatment which causes no measurable effect may enhance the response to ionizing radiation, it being qualitatively similar to that following radiation alone. The enhancing effect is normally expressed in terms of a thermal enhancement ratio (TER) defined as the ratio of doses of ionizing radiation to cause a given level of response in the presence or absence of hyperthermia, respectively. The threshold for TER values becoming greater than 1 is approximately 60 min at 41 °C, or equivalent. As the magnitude of the heat treatment is then increased the TER increases and maximum values in the region of 3–4 have been reported. At still higher heat doses, hyperthermia normally begins to cause direct thermal injury and it becomes difficult or impossible to measure a TER.

Whether or not the TER for tumours is likely to be greater than that for normal tissues is of course a crucial question. The rationale for expecting tumours to be more sensitive than normal tissues is, however, different when heat is used to enhance the effect of ionizing radiation compared to that for heat alone, e.g.:

1. The oxygen enhancement ratio for ionizing radiation, which is approximately 3 for low LET (linear energy transfer) radiation does not change significantly for the combination of radiation and hyperthermia. Expressed differently, the thermal enhancement ratio is similar for hypoxic and oxygenated cells.
2. *In vitro* studies indicate that, in contrast to the effects of heat alone, reduced pH does not sensitize tissue to the combined effects of heat and radiation.
3. Only when the tumour becomes hotter than the surrounding normal tissues will the combined treatment give a therapeutic advantage.

The clinical rationale for using heat to enhance radiation damage is therefore weaker than that for heat alone. However, from the available clinical studies (see Chapter 9) it is clear that hyperthermia as a single modality is unlikely to be of great value. Hyperthermia will almost always be combined with a conventional anticancer treatment, normally radiation therapy. Therefore, ways must be sought to combine hyperthermia and radiation to maximum advantage.

2.6.1. Temporal relationships between heat and ionizing radiation

It is well known that the TER is maximal when the two modalities are given simultaneously. As heat and radiation are separated in time, the enhancing effect decreases, as illustrated for various normal tissues in Figure 2.6. The data appear to be more consistent when heat is given

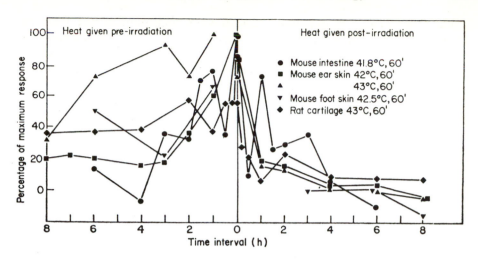

Figure 2.6 The time course of decay of heat potentiation of X-ray damage in normal tissues for hyperthermia given either before or after irradiation. Values have been normalized to the percentage of the maximum response for each curve (minimum time interval). When heat is applied 3–5 h after irradiation, thermal enhancement is reduced to zero. (From Hume and Field (1978).)

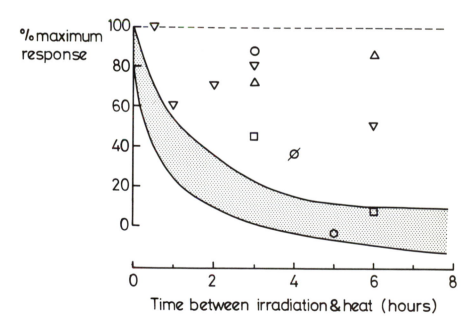

Figure 2.7 The decay of thermal enhancement of X-ray damage by heat given following irradiation. The shaded area denotes responses of normal tissues (derived from Figure 2.6). The dashed line and symbols denote values obtained from various sources for different normal tissues. (Based on a figure from Field and Bleehen (1979).)

following X-rays (rather than the reverse order) in which case the interaction is reduced to zero when the time interval is greater than approximately 3–4 h. In tumours, however, the effects of direct-heat injury may be greater so that separating heat and radiation may give a therapeutic advantage whilst still not increasing the damage to normal tissues (see Figure 2.7). In this situation, the two treatment modalities thus act independently and the clinical advantages for using heat to damage a tumour would be expected to apply. Heat given 3–4 h after a radiotherapy treatment would be expected to add to the radiation effect, selectively destroying those cells which are resistant to X-rays. There should be no enhancement of radiation damage, thus eliminating the need to alter the radiation dose, which is an important clinical advantage. This treatment technique has already been demonstrated as being clinically effective (Overgaard and Overgaard, 1987). However, some hyperthermists still prefer to treat with heat and radiation separated by as small a time interval as is possible.

2.6.2. Time–Temperature relationships

Time–Temperature relationships for the combination of heat and radiation have been reviewed by Hume and Marigold (1985). Although there are limited data, it is clear that there is a clear relationship between time and temperature for the thermal enhancement of radiation damage *in vivo* as there is for direct heat damage. The transition temperature is, however, slightly lower, at approximately 41.5°C, for thermal enhancement, and the relationship between time and temperature for a given TER appears to be such that 1°C is equivalent to a factor of 2–4 in heating time for temperatures above this transition and a factor of 5–6 in heating time for temperatures below it. These factors are clearly not significantly different from those for heat alone, except for the difference in transition temperature of approximately 1°C.

2.6.3. Fractionation of heat and X-rays

Radiotherapy is normally given in fractionated schedules and it is common also to give several hyperthermia treatments. It is, therefore, important to know whether or not thermotolerance influences the enhancement of radiation damage. The results of a number of experimental studies indicate that prior hyperthermia does reduce the radiosensitizing effect of a subsequent heat treatment, i.e. the TER is reduced. However, the effect is less than that for direct heat damage. Thermotolerance will, therefore, influence treatment by heat and radiation, but to a lesser extent than for the direct effects of hyperthermia. (See reviews by Field (1985) and Field and Majima (1989).)

Clearly, there is not sufficient information to utilise thermotolerance for therapeutic advantage, and thus it is reasonable to avoid its influence as far as possible. A practical therapeutic approach is therefore to combine conventional radiotherapy fractionation with hyperthermia given approximately once a week to limit the effect of thermotolerance and separated from the previous radiation treatment by at least a few hours to avoid interaction between the two modalities.

2.7. Re-treatment of previously irradiated tissue

Tumours which have failed treatment by conventional therapy are clearly excellent candidates for re-treatment with hyperthermia. Such treatment is normally without complications although Douglas *et al.* (1981) did report hyperthermia-induced myelopathy in some patients who had received neurotoxic drugs or local irradiation of the spinal cord. However, this only occurred when hyperthermia was given within 2 months of previous treatment.

Experimental studies on rodent skin and intestine have shown an increased sensitivity to hyperthermia when heat is given following irradiation treatment. The effect is transient and the time course of the increase in heat sensitivity is variable and also dependent on the prior radiation treatment (see review by Law, 1988). Clearly, these results show that there may be a residual effect of prior radiotherapy, and it is plausible that when hyperthermia becomes more efficient, by improvements in heating technology, complications following a more severe heat treatment might result when hyperthermia is given following a previous course of radical radiotherapy or chemotherapy.

2.8. Conclusions

This chapter has concentrated on effects of normal tissue since these limit the treatment which may be given. Sufficient knowledge is now available of heat dose-response curves *in vivo*, isoeffect relationships, thermotolerance, step-down sensitization and the various interactions between hyperthermia and radiotherapy to plan treatment on a reasonably rational basis. However, we do not yet have available sufficient information on differences in response between tumours and normal tissue to plan treatments on an optimal basis.

References

Douglas, M. A., Parks, L. C. and Bebin, J., 1981, Sudden myelopathy secondary to therapeutic total body hyperthermia after spinal-cord irradiation, *New Eng. J. Med.*, **10**, 583–585.

Fajardo, L. F., 1984, Pathological effects of hyperthermia in normal tissues, *Cancer Res.*, **44**(suppl.), 4826s–4835s.

Field, S. B. (1985) Clinical implications of thermotolerance, in Overgaard, J. (Ed.), *Hyperthermic Oncology, 1984*, Vol. 2, London, Taylor and Francis, pp. 235–244.

Field, S. B. and Bleehen, N. M. (1979) Hyperthermia in the treatment of cancer, *Cancer Treatment Reviews*, **6**, 63–94.

Field, S. B. and Majima, H. (1989) Sugahara, T. (Ed.), *Kinetics of Thermotolerance. Hyperthermic Oncology*, Vol. 2, *Summary Papers*, London, Taylor and Francis, pp. 205–210.

Field, S. B. and Morris, C. C. (1983) The relationship between heating time and temperature; its relevance to clinical hyperthermia, *Radiotherapy and Oncology*, **1**, 179–186.

Field, S. B. and Morris, C. C. (1985) Experimental studies of thermotolerance *in vivo*. 1. The baby rat tail model. *Int. J. Hyperthermia*, **1**, 235–246.

Henle, K. J. (1980) Sensitization to hyperthermia below 43 °C induced in Chinese hamster ovary cells by step down heating, *J. Nat. Cancer Inst.*, **68**, 1033–1036.

Henle, K. J. (1987) *Thermotolerance*, Vols I and II, Boca Raton, CRC Press.

Hume, S. P. and Field, S. B. (1978) Hyperthermic sensitization of mouse intestine to damage by X-rays: the effect of sequence and temporal separation of the two treatments, *Brit. J. Radiology*, **51**, 302–307.

Hume, S. P. and Marigold, J. C. L. (1985) Time temperature relationships for hyperthermal radiosensitisation in mouse intestine. Influence of thermotolerance, *Radiotherapy and Oncology*, **3**, 165–171.

Hume, S. P., Marigold, J. C. L. and Michalowski, A. (1983) The effect of local hyperthermia on nonproliferative, compared with proliferative, epithelial cells of the mouse intestinal mucosa, *Radiat. Res.*, **94**, 252–262.

Kamel, H. M. H., Hume, S. P., Carr, K. E. and Marigold, J. C. L. (1985) Structural changes in mouse small intestinal villi following lower body hyperthermia, *Scanning Electron Microscopy*, **2**, 849–858.

Law, M. P. (1988) The response of normal tissues to hyperthermia, in Urano, M. and Douple, E. (Eds), *Hyperthermic Oncology*, Vol. 1, VSP, pp. 121–159.

Law, M. P., Ahier, R. G. and Field, S. B. (1978) The response of the mouse ear to heat applied alone or combined with X-rays, *Brit. J. Radiology*, **51**, 132–138.

Lindegaard, J. C. and Overgaard, J. (1990) Step down heating in a C3H mammary carcinoma *in vivo*. Effects of varying the time and the temperature of the sensitizing treatment, *Int. J. Hyperthermia*, in press.

Morris, C. C. and Field, S. B. (1985) The relationship between heating time and temperature for rat tail necrosis with or without occlusion of the blood supply, *Int. J. Radiat. Biol.*, **47**, 41–48.

Nielsen, O. S. and Overgaard, J. (1982) Influence of time and temperature on the kinetics of thermotolerance in L_1A_2 cells *in vitro*, *Cancer Res.*, **42**, 4190–4196.

Overgaard, J. and Overgaard, M. (1987) Hyperthermia as an adjuvant to radiotherapy in the treatment of malignant melanoma, *Int. J. Hyperthermia*, **3**, 483–502.

3   Thermal dose

S. B. Field and G. P. Raaphorst

3.1. Introduction

The concept of dose is far from simple. It is defined in the Oxford English Dictionary as 'a definite quantity of a medicine, or something regarded as analogous to a medicine, given or prescribed to be given at one time'. It is, in addition, 'a definite amount of some ingredient added to wine to give it a special character'. Perhaps for the present purpose the first definition is the more important! The purpose of the dose is to provide a number which relates to a specific biological response. A knowledge of the dose, therefore, provides a means of predicting the biological response to a given treatment and hence also a means of communicating when one wishes to compare biological responses given in different places or at different times. It follows that the principal requirements of a dose are that it relates to the biological response in a relevant manner, it should be a well-defined and measurable physical quantity and there should be a proper means of intercomparison.

3.2. Dose of ionizing radiation or drugs

It is useful in considering the concept of thermal dose to examine the use of dose in other aspects of medicine, e.g. ionizing radiation or drug therapy. With radiation the accepted unit of dose is based on the deposition of energy into tissue. The unit 1 gray (Gy) is defined as 1 joule (J) of energy deposited in 1 kg of tissue. However, the biological response to a given dose of ionizing radiation depends also on many other factors, as shown in Table 3.1. For example, the effectiveness depends on the size of the dose itself, large doses being more effective per Gy than small doses. As stated above, dose is defined for single doses only and, if multiple treatments are given, fairly complex expressions are used in attempts to describe the effects of fractionated treatment. Such expressions are certainly not identical for all

Table 3.1   Ionizing radiation: factors other than dose which affect biological response.

Intrinsic cell or tissue susceptibility
Dose rate
Overall treatment time
Quality of the radiation
Environmental factors, e.g. oxygenation
Presence of other modifying factors
Cellular factors, e.g. cell cycle distribution
Combination with other modalities
Previous history

Table 3.2   Administration of drugs: factors other than mass or concentration which affect biological response.

Intrinsic cell or tissue susceptibility
Route of administration
Period of administration
Drug metabolism
Period of action
Distribution of drug in the body
Environmental conditions
Cellular factors
Modifying factors (cellular, environmental)
Combination treatments
Previous history

types of cell tissue response and there is by no means universal agreement on the best approach to solving the problem of relating different fractionated radiation treatments.

Dose in the application of a medicine is perhaps even more difficult. It is normally described in terms of the mass and concentration of a drug. But, again, this definition applies only to a single application. As with ionizing radiation the response to a given dose of drug will depend on many other factors than the drug dose itself. Some of these are illustrated in Table 3.2. The effects of multiple treatment with drugs are probably even more complex than with ionizing radiation and the means of computing the effects of such treatments are, in general, not available. Despite these difficulties, however, there is a universal acceptance in the use of dose to describe the effects of both ionizing radiation and drug therapy. The difficulties are noted but are far from solved. We have simply become accustomed to using dose with these modalities.

3.3. *The concept of dose in hyperthermia*

In contrast to ionizing radiation, absorbed energy cannot be used satisfactorily to predict the biological response to hyperthermia. This has been illustrated by Hahn (1982), who considered the situation of a thermally insulated culture dish initially at 37°C and heated quickly to 43°C. This action will not result in cell killing. If, however, the system is maintained at 43°C with no energy allowed to leave or enter, cell killing will begin simply as the result of cells being at the elevated temperature. As illustrated in Chapters 1 and 2 the effects of hyperthermia depend on the temperature and on the duration of heating, but not on the energy required to produce a temperature rise. On this basis it is clear that if hyperthermia could be defined as a fixed and constant temperature, the duration of heating would be a reasonable way of expressing thermal dose, the unit being time. The use of this parameter adequately fulfils the conditions as stated above and, certainly, is not less appropriate than energy deposition for ionizing radiation or mass and concentration for drugs.

The difficulty with this approach is that it should apply to all temperatures and, in any case, in clinical practice it is not possible to maintain a uniform temperature. With both temperature and duration of heating being variables in any treatment, it is necessary to find some means of relating a treatment to an *equivalent* time at a chosen reference temperature. The solution to this problem, however, is far from trivial.

One of the first proposals for a heat unit was by Atkinson (1977), who suggested relating heat dose to a standard heat survival curve. Unfortunately, this procedure is far from satisfactory because heat survival curves are extremely variable. An alternative approach, based on the thermodynamics of heat inactivation, was made by Dewey *et al.* (1977), and discussed more recently by Gerner (1987). Based on the Arrhenius equation, the reaction rate is described by

$$k = A \, e^{-\Delta H/RT} \tag{3.1}$$

where k is the reaction rate constant, ΔH is the activation energy, T is the temperature and R is the gas constant. This is a valid approach providing measurements are made under quasi-steady-state conditions and the same chemical step remains limiting over the whole temperature range, because the process could involve more than one step, each of which might become rate limiting under different conditions. In practice, both R and A are effectively constant over the clinically useful temperature range. ΔH is found to be approximately 600 kJ/mol at temperatures above about 42.5°C, consistent with values for heat damage to proteins. However, great caution should be exercised in the use of such analysis in the description of cell killing.

As early as 1931, Pinchus and Fisher reported that inhibition of cell growth fitted the relationship that for a given level of growth inhibition the time of heating decreased by a factor of two for every degree increase above 44 °C. Dewey and colleagues (1977), presented the formula based on *in vitro* observations, relating two treatments at different temperatures,

$$\frac{t_2}{t_1} = R^{T_1 - T_2} \tag{3.2}$$

where t is the time of heating, T is the temperature and R, in this case, is a constant. Hahn (1982), derived a similar relationship between time of heating and temperature based on the classical target theory approach, i.e. a cell is inactivated if its instantaneous kinetic energy exceeds a specific critical value at any time during the heating. Equation (3.2) provides a means of relating treatments with different temperatures and times of heating, but takes no account of differential tissue sensitivity. Sapareto and Dewey (1984) proposed that 43°C should be used as the reference temperature and that all treatments be described as equivalent minutes of heating at 43 °C. By making this transformation from 'time' to 'equivalent time' a biological factor has been built into an otherwise pure physical unit. However, despite it being a hybrid between a biological and physical unit it does not invalidate the use of the term dose. It is normallay referred to as the thermal isoeffect dose (TID).

In order to calculate the TID from the above formula the value, or values, of R must be found. A wide range of studies both with cells *in vitro* and tissues *in vivo* have shown that the time–temperature relationships can be defined by at least two different slopes within the temperature range 41–46°C. However, in a review of available data from studies *in vivo* and *in vitro* by Field and Morris (1983), it was found that for almost all tissues there was a single transition in the time–temperature relationship occurring between 42 and 43°C. Above the transition the slopes of all curves were fairly similar giving a value for R of approximately 2. Below the transition, although there were far fewer data, the value for R was found to be approximately 6 (Table 3.3, see also Chapter 2).

It is well known that in the lower temperatures range below the transition, thermotolerance develops during heating, which could explain the transition itself and hence the altered value for R at lower temperatures and longer times of heating. Therefore, it might be preferable to represent R by a single value, i.e. that pertaining to

Table 3.3   Values for R in the formula $\frac{t_2}{t_1} = R^{T_1 - T_2}$.

	Above transition	Below transition
Summary of *in vitro* results	2.0±0.08	5.0±0.06
Summary of *in vivo* results	2.1±0.07	6.4±0.04

The transition temperature was between 42 and 43°C (from Field and Morris (1983)).

temperatures above the transition where there is no development of thermotolerance, and to apply a modifying factor to account for thermotolerance where appropriate.

In either case, these considerations apply to hyperthermia given at a constant temperature. In practice, it takes a finite time for the desired temperature to be reached and it is then impossible to keep it constant. There will be temperature fluctuations during treatment followed by a finite time of cooling when the power is switched off. This is dealt with by integration of equation (3.2) during the whole treatment procedure. However, variations in temperature may have a marked effect in determining the biological response, for example, by either the development of thermotolerance or the occurrence of step-down sensitization as described in Chapters 1 and 2. Clearly, these effects have the potential to invalidate equation (3.2). The extent to which this is likely to happen in clinical practice was tested experimentally by Field and Morris (1985). In these studies a series of temperature variations were simulated experimentally. It was observed that, in general, despite the fact that both thermotolerance and step-down sensitization were known to play an important role in the experimental system used, the maximum difference observed between the measured and predicted effects in any of the studies was approximately 20% in heating time, equivalent to an error in temperature of 0.3 °C. Such a change is relatively trivial compared with other uncertainties in a typical hyperthermia treatment.

3.4. *Combination of heat with other modalities*

It must not be assumed that the parameters which apply to the effects of heat alone are the same as those for the combination of hyperthermia with other modalities. For heat combined with radiotherapy a few data are available for the isoeffect relationships. The parameters appear to be reasonably similar to those for heat alone, values for R being approximately 2.5 for temperatures above the transition and 5.9 for those below. However, transition appears to occur at approximately 1 °C lower than for heat alone (Hume and Marigold, 1985; Field, 1987 (by courtesy of Dr M. P. Law)). Clinically, heat and radiotherapy are rarely, if ever, given simultaneously and as the interval between the two modalities is increased they become increasingly independent, the interaction between them fading to zero by about 3 or 4 h (see Chapters 1 and 2). As this occurs, the parameters appropriate for heat alone become increasingly relevant in the calculations of heat dose.

Unfortunately, there appear to be no relevant data for estimating the thermal dose parameters for the combination of hyperthermia and chemotherapy.

3.5. *Problems with thermal dose equivalent (TID)*

A number of difficulties are involved in the definition of thermal dose as follows:

1. The transition temperature which is normally given as being between 42 and 43°C was derived from various cell lines *in vitro* and tissues *in vivo* as discussed above. However, very recent results obtained from human cell lines studied *in vitro* (Hahn, 1989) show transition temperatures in the region of 44°C or higher. Clearly, the choice of transition temperature will have a marked effect on the numerical value of the TID.

2. Although it is fairly well established that the value for R is close to 2 for temperatures above the transition there is considerable uncertainty about the value in the lower temperature range. The choice of this parameter will have an important effect on the TID for treatments which include temperatures below the transition.

3. Whilst it seems unlikely that moderate fluctuations in temperature will invalidate the TID equation, it is likely that long heating-up times or duration of treatment will give rise to the induction of thermotolerance and hence a smaller biological effect than predicted by the TID formula. In contrast, any major reductions in temperature during treatment will result in step-down sensitization and cause the lower temperature treatment to be more effective than predicted.

4. It is difficult or even impossible to heat tissue uniformly. As a result, the TID will vary spatially throughout the heated volume. It is generally considered that the most important values of the TID will be the minimum within the tumour, since this will determine the least-efficiently heated region, and the maximum value in normal tissue, since this is likely to relate to normal tissue tolerance.

5. The time–temperature relationships for the interactions between heat and radiotherapy or heat and chemotherapy are likely to be different from those for heat alone. In addition, the degree of interaction between the modalities will depend on their temporal relationships.

6. There is no known way to account for the effects of fractionated exposures. Transient thermotolerance is a widely universal phenomenon and unless the separation between hyperthermal treatments is fairly long, i.e. at least a week and probably more, the subsequent treatment will be affected. In these circumstances TID values cannot be added.

7. A major factor determining the response of cells and tissues to hyperthermia is their environment and, in particular, their pH.

8. Other factors such as individual tissue susceptibility, the presence of various modifying agents, and the previous history of the tissue will also affect the response of the TID to hyperthermia.

Table 3.4 Factors other than the TID affecting biological response.

Intrinsic cell or tissue susceptibility
Spatial distribution of temperature
Duration of heating
Rate of heating and cooling
Temporal fluctuations in temperature
Environmental conditions (such as pH and nutrient levels)
Cellular factors (cell cycle distribution)
Presence of other modifying factors
Combination with other modalities
Previous history

It must be pointed out, however, that many of the uncertainties are of a similar nature to those when dose is used in radiotherapy or chemotherapy. This can be seen by comparing Tables 3.1 and 3.2 with Table 3.4, which lists some of the problems associated with the use of TID.

3.6. *Summary and conclusions*

1. In the hypothetical situation of hyperthermia being defined at a specific and constant temperature, duration of heating is an acceptable and satisfactory method of describing thermal dose.
2. In the real situation where temperature is far from constant we are obliged to use a biologically equivalent isoeffect dose to account for such variations.
3. Various attempts have been made to solve this problem. At present, the most feasible approach appears to be the use of the isoeffect relationship (the Dewey formula), to define a TID. In practice, this is integrated throughout the duration of treatment.
4. The TID provides a practical and reasonable method of comparing hyperthermia treatments under conditions likely to be met in practice, i.e. moderate variations about a reasonably steady temperature. However, the formula would certainly be invalidated by a very long warm-up time, substantial alterations in temperature during treatment or very long treatment times.
5. There are a number of uncertainties and difficulties in using this formula, in particular, the uncertainty in the transition temperature, interactions with other modalities and how to deal with fractionated exposures.
6. In view of the difficulties and uncertainties in the use of TID it is not recommended that it is used in clinical decision making. Nevertheless, in view of the lack of a suitable alternative, it is very useful in the evaluation of laboratory data and, possibly, in the retrospective analysis of clinical data.

References

Atkinson, E. R. (1977) Hyperthermia dose definition, *Journal of Bioengineering*, 1, 487–492.

Dewey, W. C., Hopwood, L. E., Sapareto, S. A. and Gerweck, L. E. (1977) Cellular responses to combinations of hyperthermia and radiation, *Radiology*, 123, 463–474.

Field, S. B. (1987) Studies relevant to a means of quantifying the effects of hyperthermia, *International Journal of Hyperthermia*, 3, 291–296.

Field, S. B. and Morris, C. C. (1983) The relationship between heating time and temperature: its relevance to clinical hyperthermia, *Radiotherapy and Oncology*, 1, 179–183.

Field, S. B. and Morris, C. C. (1985) Application of the relationship between heating time and temperature for use as a measure of thermal dose, in Overgaard, J. (Ed.), *Hyperthermic Oncology* 1984, Vol. 1, London, Taylor and Francis, pp. 183–186.

Gerner, E. W. (1987) Thermal dose and time–temperature factors for biological responses to heat shock, *International Journal of Hyperthermia*, 3, 319–327.

Hahn, G. M. (1982) *Hyperthermia and Cancer*, New York, Plenum Press.

Hahn, G. M., Ning, S. C., Elizaga, M., Kapp, D. S. and Anderson, R. L. (1989) A comparison of thermal responses of human and rodent cells, *International Journal of Radiation Biology*, 56, 817–825.

Hume, S. P. and Marigold, J. C. L. (1985) Time–temperature relationships for hyperthermal radiosensitization in mouse intestine. Influence of thermotolerance, *Radiotherapy and Oncology*, 3, 165–171.

Sapareto, S. A. and Dewey, W. C. (1984) Thermal dose determination in cancer therapy, *International Journal of Radiation Oncology, Biology and Physics*, 10, 787–800.

4 Effects of hyperthermia on blood flow and metabolism

H. S. Reinhold and A. P. van den Berg

4.1. Introduction

There are many properties in which tumour tissue differs from its tissue of origin or that into which it has infiltrated. One group of such properties relates to the vascular system and the circulation physiology. The inefficient chaotic supply of blood and nutrients in tumours results in the development of hypoxic acidotic areas in which the tumour cells are resistant to ionizing radiation. On the other hand, such properties make the tumour potentially sensitive to hyperthermia. Therefore, this chapter will first discuss some of the peculiarities of the tumour microcirculation and then describe the changes that take place during, and after, hyperthermia.

4.2. Tumour vascularization and microcirculation

Most of our information on the vascularization and circulation physiology of tumours has been obtained from experimental tumours. Generally, experimental tumours in use originally arose as spontaneous cancers, or were induced in rodents and subsequently have been transplanted many times from one animal to another. In addition, depending upon the requirements for a given investigation, such tumours are transplanted into various different sites in the host animal. For hyperthermia, frequently used sites are the lower part of the leg, or footpad. The transplantation may be by means of an injection of a cell suspension or, more crudely, implantation of a small piece of tumour material. Such an inoculum will unavoidably contain some blood vessels or endothelial cells. There is usually an initial period of regression of the viable tumour tissue following transplantation, after which the tumour starts growing and thereby induces its own typical vascular supply.

77

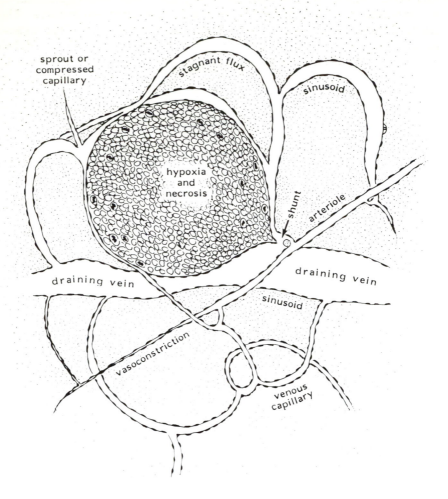

Figure 4.1 Schematic representation of some of the typical characteristics of the vascular supply of tumours. For clarity, only one 'tumour cord'-like structure is drawn, with its hypoxic and necrotic centre. Further widened sinusoids (which may have either a high or a stagnant flux), a shunt and a vessel with vasoconstriction are indicated.

Some tumours, such as adenocarcinomata (lung, colon, mammary gland) grow in a lobular fashion with a tendency to form tubules. In many instances the vascular supply is located at the perimeter of the tubular structures. Mammary tumours generally contain many widened sinusoidal vessels surrounding the lobules of tumour cells. On the other hand, sarcomata usually form intermingling bundles of cells. The capillaries supplying these tumours are both stretched and compressed by the growing mass of tumour cells, resulting in an ischaemic tumour centre and other regions which subsequently become necrotic. Squamous cell carcinomata, in contrast, tend to form a kind of 'inverted surface'. They normally produce keratin, which is excreted into the centre of the tumour, or in multiple small focii, forming (central) areas containing debris. Sometimes sinusoidal loops can be observed which extend into the necrotizing centre. These loops are frequently dilated, probably due to a reduced pressure in the necrotic centre.

One of the typical aspects of the microcirculation in tumours is the apparently random way in which the blood vessels and also blood flow run through the tumours (Reinhold, 1971). An impression of the chaotic tumour microcirculation is given in Figure 4.1. Various authors have used different expressions for this phenomenon, such as 'the flow has the characteristics of flow through a porous medium' (Endrich *et al.*, 1979), or that the vascular system of the tumour consists of a 'tortuous system' (Jain, 1987, 1988). In general, the rate of blood flow, or its velocity, decreases from the perimeter to the centre (Endrich *et al.*, 1979; Oleson *et al.*, 1989). With increasing tumour size interstitial pH decreased (Jain and Ward-Hartley, 1984; Jain *et al.*, 1984; Kallinowski and Vaupel, 1988; Thistlethwaite *et al.*, 1985), and probably the interstitial pressure increases (Young *et al.*, 1950, Wiig *et al.*, 1982). In addition, the distribution of oxygenation shifts to lower values (Vaupel, 1977) with a reduction in blood flow. For extensive recent reviews see Jain (1987, 1988). The findings can be grouped, and are given in Table 4.1.

Changes in the various parameters when a tumour increases in size, have been studied by Grunt *et al.* (1986), using scanning electron

Table 4.1 Some properties of tumour microcirculation.

1. Heterogeneous architecture
Chaotic vascular pattern
Elongated or compressed capillaries
Wide sinusoids
Areas with insufficient vascular supply

2. Heterogeneous flow and perfusion
Areas with high perfusion
Areas with low perfusion
Shunts
Sometimes discontinuous flow

3. Inferior composition of vessel wall
Oedema of endothelial cells
Sites without proper endothelial lining
Blood cell extravasation
Sometimes lack of basal lamina

4. Inferior milieu of the tumour tissue
Increased interstitial pressure
Areas with hypoxia/anoxia
Areas with low pH
Necrotic areas

After Warren (1979), Eddy (1980), Hammersen *et al.* (1985) and Peterson (1979).

**Table 4.2 Alterations of several microcirculatory
parameters during tumour growth.**

Parameter	Alteration
Total tumour blood flow	Decreases
Blood flow in sinusoids	Decreases
Local capillary perfusion pressure	Decreases
Vascular density	Decreases
Relative capillary volume per tumour cell	Decreases
Intercapillary distance	Increases
Mean vessel diameter	Increases
Mean vascular volume	Constant
Overall vascular surface area	Decreases
Overall vessel length	Decreases
Interstitial tissue pressure	Increases
Blood pressure in venules	Increases
Vascular resistance	Increases

After Grunt *et al.* (1986).

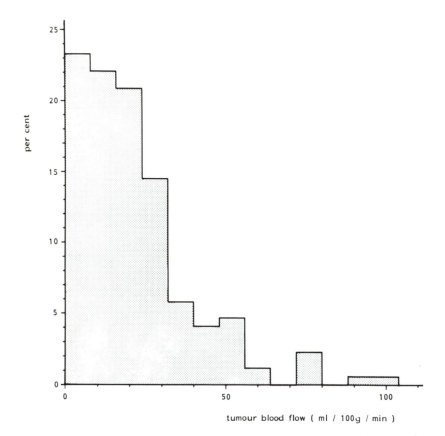

**Figure 4.2 Distribution of human tumour blood flow values, as collected from the
literature. Low blood flow value dominates this distribution (Reinhold, 1988).**

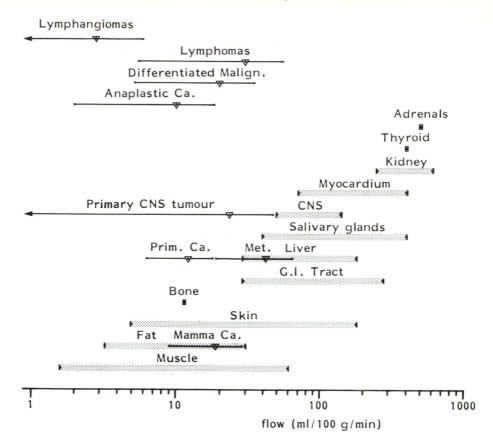

Figure 4.3 Composite diagram of blood flow in several human normal tissues and human tumours. Data collected from various sources (Reinhold, 1988; with permission). In general, but not always, the tumour blood flow is less than that of the tissue in which it is projected.

microscopy observations on the Lewis lung carcinoma. The changes observed are listed in Table 4.2. The majority of the investigations into tumour vasculature and tumour microcirculation, as discussed, have been performed on animals. It is likely that the same mechanisms hold for human tumours, be it that the sites of growth, as well as the interdependence on tumour histological type and size, may differ.

It is known that human tumours contain hypoxic areas, although this does not necessarily imply the existence of viable hypoxic tumour cells. The initial histological observation, leading to the concept of tumour cell hypoxia, was made by Thomlinson and Gray (1955). Since then a number of investigators have demonstrated indications of tumour hypoxia in human tumours by various physiological means (Kolstad, 1968; Mueller-Klieser *et al.*, 1981; Endrich *et al.*, 1982). These findings indicate that the

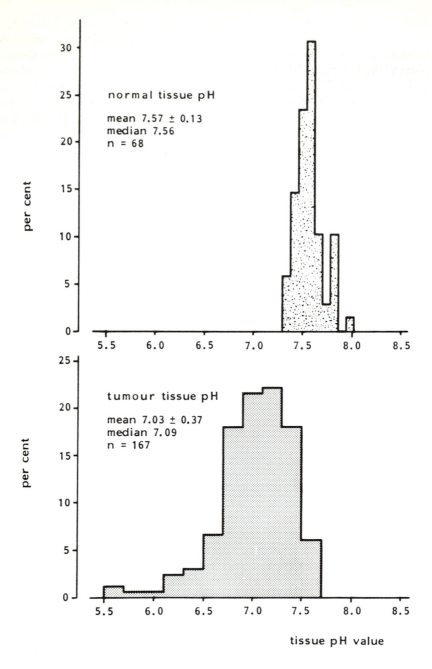

Figure 4.4 Distributions of published data on human tumour pH, as compared with human normal tissue pH (subcutis and muscle). The average tumour pH is significantly lower than the normal tissue pH. (After data by Wike–Hooley *et al.* (1984a) and Thistlethwaite and Leeper (1988).)

average blood flow in human tumours tends to be generally lower than that in normal tissues (Figures 4.2 and 4.3). The methods used have been thoroughly discussed by Jain and Ward-Hartley (1984) and more recently by Jain (1988). The inefficient microcirculation of tumours is also reflected by their low pH. This has been known for a long time, and the available data have been reviewed by Wike-Hooley *et al.* (1984a, b, 1985) and more recently by Thistlethwaite and Leeper (1988). The accumulated data are given in Figure 4.4. The question of whether the tumour pH changes during heating will be discussed later.

4.3. Effects of hyperthermia on the tumour vasculature and microcirculation

As early as 1961, observations performed for different purposes indicated that the microcirculation in experimental tumours is particularly vulnerable to elevated temperatures (Scheid, 1961). However, it was not until the mid-1970s that several investigators, apparently unaware of the earlier observations, became interested in determining what would happen to the tumour microcirculation when the tumour was subjected to hyperthermia: Would the circulation increase or decrease? The answers obtained were initially far from unequivocal: some authors found first an increase followed by a decrease (Sutton, 1976; Johnson, 1976; Reinhold *et al.*, 1978), while others found no, or inconsistent, changes in the tumour vascular function (Song, 1978; Gullino, 1980) and some found a continuous decline (von Ardenne and Reitnauer, 1982) (for reviews see Emami and Song, 1984; Endrich, 1988a, b; Song *et al.*, 1984; Reinhold and Endrich, 1986; Dewhirst, 1988). Gradually, it became clear that the effects depended on not only the tumour type (the Walker carcinoma probably having the most-resistant tumour microcirculation of the experimental tumours), but also on the exposure temperature and duration as well as the rate of heating (Dewhirst *et al.*, 1984a, b). During the course of a hyperthermia treatment a number of changes take place. These will be discussed below.

An example of the changes in tumour microcirculation that may occur during and after a hyperthermic treatment is illustrated in Figure 4.5. It is apparent from this series of photographs that, at 42.5°C for 1 h exposure, the microcirculation in this particular tumour is completely intact and functioning, but that with prolonged exposure a diminution of the microcirculation sets in, eventually resulting in ischaemia and necrosis. These observations were made in thin, transparent 'observation window' chambers and Figure 4.6 demonstrates the time course of a number of parameters that can be followed, or measured in a human tumour xenograft under similar conditions. Figure 4.7 shows the results of using a very simple optical system to measure the velocity of erythrocytes

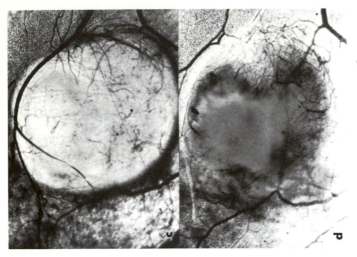

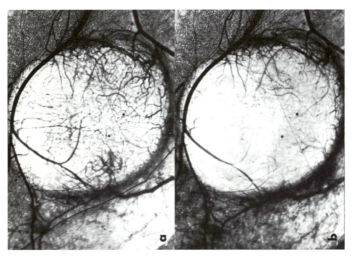

Figure 4.5 Sequence of events with a hyperthermic treatment (42.5°C) on the microcirculation of the experimental rhabdomyosarcoma BA 1112 in a 'sandwich' observation chamber (diameter, 3 mm; thickness, 200 μm). (a) After 1 h at 42.5°C there is good, visible circulation in all blood vessels. (b) After 2 h at 42.5°C the microcirculation has deteriorated but there is still good, visible circulation in some vessels. (c) After 3 h at 42.5°C there is vascular collapse in most of the centrally located vessels and development of some areas of decreased viability (cloudiness); end of treatment and return to 33.5°C. (d) Two days later note those (central) areas showing vascular stasis in (c) now show extensive necrosis. Note that the periphery is largely unaffected. (From Reinhold *et al.* (1978) with permission.)

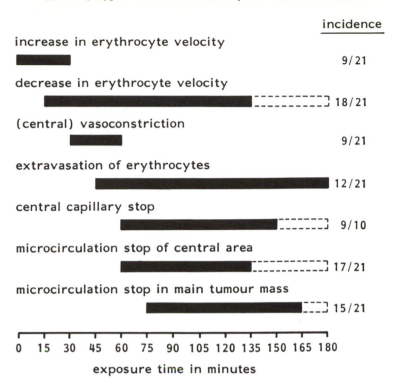

increase in erythrocyte velocity

incidence

9/21

decrease in erythrocyte velocity

18/21

(central) vasoconstriction

9/21

extravasation of erythrocytes

12/21

central capillary stop

9/10

microcirculation stop of central area

17/21

microcirculation stop in main tumour mass

15/21

0 15 30 45 60 75 90 105 120 135 150 165 180

exposure time in minutes

Figure 4.6 Schematic diagram of the observed time ranges for several events taking place in the microcirculation of a xenografted human colon carcinoma during a hyperthermic treatment at 42.5°C for 3 h. (van den Berg–Blok and Reinhold (1987) with permission.)

in selected capillaries in the tumour. It initially increases and then starts to decrease after about 1 h. In this type of experiment, about 3 h at 43°C is required to inhibit all flux in the tumours. It should be pointed out, however, that although simple (optical) observation of a tumour tissue microcirculation may be quite instructive about a number of changes that take place, it does not by any means represent the whole story. There are many different morphological, physiological and metabolic changes that occur in the tumour and its microcirculation during the time of observation, and most of these can only be measured by entirely different methods. For instance, there have been many determinations performed on tumour blood flow during hyperthermia. Most of these were made with methods using radioactive tracers or by 'thermal washout'. Such methods represent, by nature, determinations on a few single time points per individual tumour. Only continuous observation methods of the type mentioned, or the modern laser-doppler erythrocyte flux methods (Song *et al.*, 1987; Vaupel *et al.*, 1989; Oleson *et al.*, 1989) can provide continuous registration.

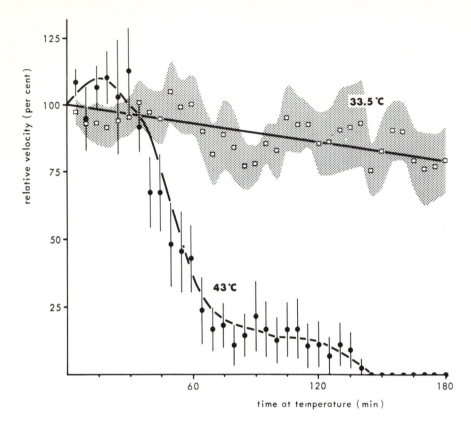

Figure 4.7 The effect of a hyperthermic treatment on the velocity of the erythrocytes in the capillaries of an experimental rhabdomyosarcoma during prolonged treatment at 43°C. (Reinhold and van den Berg–Blok (1983) with permission.)

As has been mentioned already, we are dealing with a complex therapeutic stimulus, the effect of which depends both upon the treatment temperature and on the duration of the exposure (see for example Figure 4.7). Moreover, the observed effects may be quite different between tumours and the surrounding normal tissues, the normal tissues being more resistant to changes (Dewhirst, 1988; Law, 1988). The complex nature of these parameters is schematically depicted in Figure 4.8. This figure attempts to explain that, depending upon temperature and exposure time, a simple monotonic response to the hyperthermic stimulus does not occur, but that there is first a 'reactive' phase which is then followed by a 'destructive' phase. These mechanisms, which depend very much on the tissue involved, the temperature and the exposure time, may be in part responsible for the large differences observed by various authors. However, within a given system, the response of the tumour

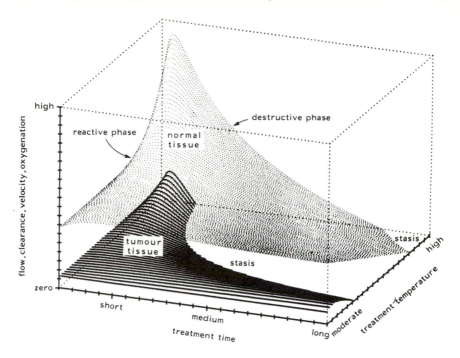

Figure 4.8 Conceptual model of the response of the circulation to hyperthermia.
This composite model is meant only to visualize the complex relationships between
temperature, exposure time and physiological response. Abscissae: temperature
and time on a relative scale. Ordinate: flow, expressed as 'flow' or 'clearance',
'velocity' of erythrocytes or 'oxygenation'. When the heat dose increases (either the
temperature, the exposure time or a combination of both), the flow in the tumour
will first increase and then decrease. The first is envisaged as a 'reactive phase', the
second as a 'destructive phase' ending, if the heat 'dose' is high enough, in vascular
stasis. For normal tissues the same tendency can be observed but only at higher
treatment levels. (Reinhold and Endrich (1986).)

vasculature is relatively reproducible and predictable. This is shown in
Figure 4.9. Here the time required for stoppage of the microcirculation in
half of the exposed tumours (the ST_{50}) is, for different temperatures,
plotted on a log scale against the exposure temperature. It appears that, as
with the majority of determinations on the relationship between
temperature and time to obtain an isoeffect relationship, it can be
expressed according to the Arrhenius equation (Field, 1987). However, if
all available published data are plotted in a single diagram, a picture
emerges that is less reassuring about the general consistency of response
(Figure 4.10). While some tumours respond to a mild hyperthermic
treatment with a rapid inhibition of blood flow, others seem to require
much higher temperatures and/or duration of heating. Recently, the
importance of a vascular shut down for the overall thermal sensitivity of
tumours was stressed by Hill *et al.* (1989). Their data indicate that only

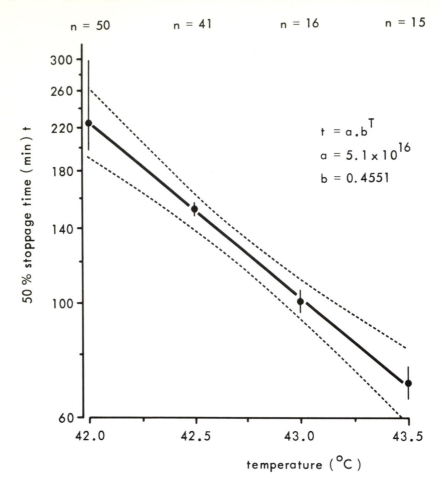

Figure 4.9 Time–temperature relationship for the development of vascular stasis in the rat rhabdomyosarcoma BA 1112 in 'sandwich' observation chambers. The slope of this curve is very similar to the slopes determined for many other end-points such as cell killing (cell survival) and normal tissue response. *t*, time; *a*, intercept; *b*, slope. (From Van den Berg–Blok and Reinhold (1984) with permission.)

those treatments which gave >60% reduction in blood flow yielded significant growth delays or thermal sensitization. Of the many tumours investigated, the Walker carcinoma and some of the human carcinomata must be called resistant. Waterman *et al.* (1987) investigated the changes in tumour blood flow, as determined by a thermal clearance method, in a number of superficially located human tumours and found no evidence of vascular stasis, as is observed in most rodent tumours after comparable treatment (60 min of heating up to 44 °C). On the other hand, Oleson *et al.* (1989) found evidence that blood perfusion did diminish in some soft-tissue

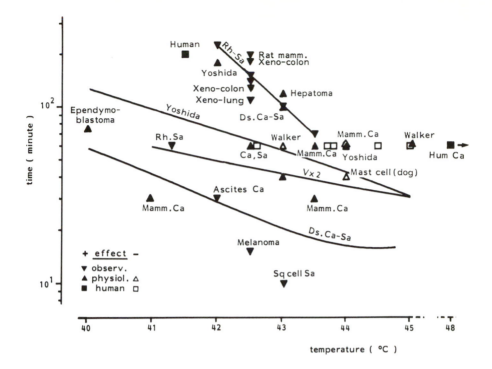

Figure 4.10 Time–temperature plot of the available data on the reduction of the microcirculation. (Updated from Reinhold and Endrich (1986) and Reinhold (1988).)

sarcomata. A similar observation has been made by van der Zee *et al.* (1983) and J. M. Cosset (personal communication). Also, F. M. Waterman (personal communication) has more recently observed that a sharp decrease in blood flow occurred in a tumour heated to 48–50°C during the first 30 min. This was also reflected in the necessity to decrease the microwave power at that time to keep the temperature from escalating. Therefore, the experimental data have sporadically been confirmed in the clinic. Probably, the causes of these differences in observed responses must be sought in differences in the underlying mechanisms between the various species, tumours and localizations.

4.4. *Underlying mechanisms*

It will be clear to the reader that, during a hyperthermic treatment, a large number of parameters will change. Some of these changes will interact with other alterations, possibly resulting in an amplification of the effect. Although a substantial amount of information has been gathered with experimental systems, no generally accepted picture has yet emerged. Rather, the interpretations of the findings, and in particular the speculations by von Ardenne and Krüger (1980) and von Ardenne and Reitnauer (1982), on how hyperthermia (in conjunction with glucose application) might exert an influence on tumour metabolism and microcirculation have, in recent years, been superseded by different interpretations. A summary of these, restricted to tumour metabolic and

Table 4.3 Proposed mechanisms involved in hyperthermia-induced vascular stasis in tumours.

Physiological changes	References
1. Changes in the vessel wall	
Changes in endothelial cells	Eddy (1980), Endrich and Hammersen (1986, 1988), Endrich (1988a, b), Fajardo *et al.* (1985), Badylak *et al.* (1985)
Leukocyte sticking	Endrich *et al.* (1979)
No changes in permeability	Song (1978), Jain and Ward–Hartley (1984)
2. Changes in blood rheology	
Decrease of flexibility of erythrocytes	Not accessible *in vivo*
Aggregation of erythrocytes	Endrich and Hammersen (1986, 1988), Endrich (1988a, b)
Thrombosis in capillaries	Dewhirst *et al.* (1984), Nishimura *et al.* (1988)
Fibrin formation	Lee *et al.* (1985)
3. Changes in microhaemodynamics	
Increased vascular resistance	Vaupel *et al.* (1980)
Decrease of arteriolar–venular pressure gradient	Endrich and Hammersen (1986, 1988), Endrich (1988a, b)
Arteriolar–venular shunting	Dewhirst *et al.* (1984a)
Arteriolar vasoconstriction	Eddy (1980), Endrich *et al.* (1984)
Oedema formation	Vaupel *et al.* (1983), Endrich *et al.* (1984)

After Reinhold and Endrich (1986).

Table 4.4 Possible sequence of events.

Haemodynamic changes	Morphological changes
Changes in metabolism	
	Vascular dilatation
Increase in venular pressure	
	Sticking of leukocytes
Increased shunt flow	
Decreased arteriolar–venular pressure gradient	
Decreased nutritional flow	
pO_2 and pH decrease	
	Aggregation of erythrocytes
Fibrinogenen gelling	
Increased vascular resistance	
	Thrombosis
Further decrease in flow	
	Endothelial changes
Vasoconstriction	
Stasis	
	Tumor cell destruction

circulation physiology, are listed in Tables 4.3 and 4.4. The various factors will be considered below.

4.4.1. Changes in metabolism

Hyperthermia influences cellular metabolism in several ways. In the first place the rate of chemical reactions increases with increasing temperature. Cellular reactions, however, are catalysed by enzymes whose function depend on their spacial configuration. The active sites of the enzymes to which the molecules participating in the reaction are to be attached usually have a very specific shape. Alterations in this shape may cause the reaction to become inhibited. Such structural transitions in proteins may occur at temperatures between 40 and 45°C. At still higher temperatures protein structure may be altered in a process called denaturation (which is what happends when an egg is boiled).

Cellular metabolism also depends on the cellular micro-environment. Changes in the composition of the extracellular fluid causes the cell to take action in order to maintain its well-balanced internal composition and the gradients of ions, such as H^+, K^+ or Ca^{2+}. Since such action consumes energy a severe or long-lasting environmental stress may lead to exhaustion of the cell if nutrients are insufficiently supplied, which may be the case in inefficiently vascularized tissues or regions which are likely to occur in tumours. Conversely, the cellular micro-environment may be influenced by changes in the cellular metabolism. The accelerated reactions which produce acid substances may increase the acidity of the extracellular fluid if these substances are not adequately removed by diffusion or by interstitial fluid transport. In particular, this may occur in tumours which show an elevated rate of lactic acid formation by anaerobic breakdown of glucose. This reaction constitutes the dominant route for generation of energy in the absence of oxygen. Since a lack of oxygen results from inadequate blood supply and thus poor blood perfusion, it will inevitably be accompanied by inefficient removal of waste products. Moreover, an increasing acidity of the environment forces the cell to invest extra energy in maintaining the proton balance (Haveman and Hahn, 1981). This requires further utilization of glucose. In this way, a cascade of events may occur. Hyperthermia represents one of the ways to trigger or enforce such a cascade.

Although the investigation of the effects of hyperthermia on cellular metabolism has involved many different aspects, a major effort has been directed towards the sensitizing effects of pH on hyperthermic treatment. Particular attention has been paid to the relevance of pH+oxygen distribution+glucose metabolism in tumours to the cytotoxic effects of hyperthermia (Vaupel and Kallinowski, 1987; Streffer, 1985; Streffer and van Beuningen, 1987). A dominant contribution to the energy metabolism of tumour cells by the glycolytic pathway has been thought to be the cause of the relative acidity of tumour tissue compared with normal tissues (Warburg *et al.*, 1924; Dickson and Calderwood, 1983; Wike-Hooley *et al.*, 1984a, b; Thistlethwaite and Leeper, 1988). This dominance of glycolysis may be intrinsic to the tumour cell or may be induced by hypoxia. At present, it is recognized that certain tumours do indeed show a characteristically abnormal energy metabolism (such as liver tumours, Weber, 1983), but also that an appreciable variability exists between different tumours, even of similar type and location. The main cause of the increased acidity of tumour tissue is believed to be insufficient vascularization which leaves certain areas of the tumour inadequately supplied with oxygen and causes an accumulation of lactic acid, the end product of anaerobic glycolysis. The heterogeneity of oxygen and pH distribution within animal tumours has been demonstrated by Vaupel *et al.* (1981), whereas the variability of extracellular pH between human as well as experimental tumours has been demonstrated and reviewed by Wike-Hooley *et al.* (1984b).

As explained above, the effects of hyperthermia on metabolism may be direct or indirect. A temperature increase leads to elevated reaction rates. Increased oxygen consumption has been demonstrated by Durand (1978) and increased glucose turnover has been demonstrated by Streffer (1982) and Schubert *et al.* (1982). There are several ways of exploiting this increased metabolism. On the one hand the accumulation of acidic metabolites may increase the sensitivity of the tumour cells to hyperthermia so that an additional supply of glucose may further stimulate acidification. This has formed the basis of the 'cancer multistep therapy' proposed by von Ardenne (1972). Recently, this idea has been preliminary tested in the clinic by Thistlethwaite *et al.* (1987) who showed that oral administration of 100 g glucose induced pH decreases of 0.05–0.5 units in non-diabetic patients. On the other hand, if the glucose reservoir becomes exhausted during hyperthermia an extensive degradation of lipids may add to the toxic effects of hyperthermia in hypoxic cells which are provided with just enough oxygen to support respiratory energy metabolism (Tamulevicius *et al.*, 1984). A different approach would be to inhibit glycolysis in order to deprive cells of the substrates needed to satisfy their increased energy demand. This could be particularly effective if the cells depend primarily on glycolysis for their energy supply. Such inhibition can be obtained with glucose analogues, such as 5-thio-D-glucose (Song *et al.*, 1977; Kim *et al.*, 1978).

Although enzymic reactions tend to become inhibited at temperatures above 40°C, this does not seem to be the case for glycolysis (Streffer *et al.*, 1981; Hengstebeck, 1983). Increasing the hyperthermic stress into the clinically relevant range (1 h at 43°C) increases the production of lactic acid, particularly under hypoxic conditions. Moreover, Streffer (1982) and Schubert *et al.* (1982) have shown that the citric acid cycle may become damaged under these conditions, which may lead to further acidification due to the formation of β-hydroxybutyrate and acetoacetate. The absolute amounts of these metabolites, however, are substantially lower than that of the lactic acid formed (Lee *et al.*, 1986).

It has become evident that the acidity of the tumour interstitial fluid depends on several intricately related factors, but that the overall effect of hyperthermia seems to be a further acidification of the tumour tissue. Although human tumours have been shown to have reduced tissue pH (van den Berg *et al.*, 1982; Wike–Hooley *et al.*, 1984b; Thistlethwaite *et al.*, 1985), this low initial pH value does not seem to be related to improved clinical tumour response after hyperthermic treatment (van den Berg *et al.*, 1989). Recent studies suggest that this may be due to adaptation and/or selection of tumour cells after chronic exposure to a low pH environment (Hahn and Shiu, 1986; Chu and Dewey, 1988; Cook and Fox, 1988). It has been well established that tumour cells *in vitro* are acutely sensitized to hyperthermia when put into an acidic medium (for a review see Calderwood and Dickson, 1983). Further efforts to optimize the degree of acidification *during* hyperthermic treatment, therefore, could play an important role in improvement of the clinical results of hyperthermia.

4.4.2. Vascular dilation and increase in venular pressure

The phenomenon of vascular dilation can be seen as one of the normally occurring reactions to elevated temperatures (Endrich and Hammersen, 1986). Apart from attempts by the body to increase the cooling rate, it may also be a reaction to the increased metabolic demand. All investigators working with 'observation windows' are familiar with the phenomenon (Dewhirst *et al.*, 1984a, b). This vasodilation goes hand in hand with changes in venular pressure (Endrich and Hammersen, 1986). This is probably due to an increased peripheral shunt perfusion.

4.4.3. Sticking of leukocytes

In most 'observation-window' types of systems one can occasionally observe leukocytes rolling along the walls of the venules. Endrich *et al.* (1979) made the initial observation that, during a hyperthermic treatment, these leukocytes stick to the vessel wall, and in this way may obstruct the flow of blood. This finding was confirmed by Dudar and Jain (1984).

4.4.4. Increased shunt flow and decreased arteriolar–venular pressure gradient

In the same fashion as heat-induced vascular dilation, it appears that blood vessels that connect the arterial and venous sides of the blood vessel

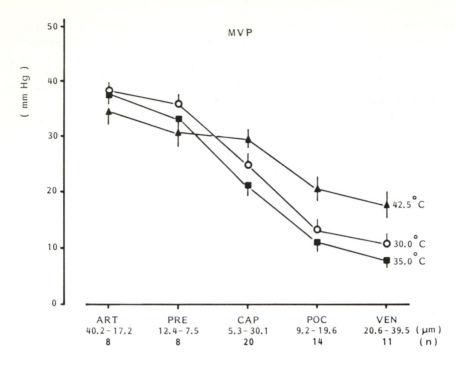

Figure 4.11 Changes of intravascular pressure (MVP) under local hyperthermia (mean values±SEM). The microcirculation was classified according to Zweifach. ART, arterioles; PRE, precapillaries; CAP, capillaries; POC, postcapillaries; VEN, collecting venules; *n*, number of single determinations. (Endrich and Hammersen (1986) with permission.)

network (the so-called 'shunts'), which are normally closed, may open up with heating. This important issue was investigated by Endrich (Endrich and Hammersen, 1986) and by Dewhirst *et al.* (1984a, b). A summary of the intravascular pressure measurements made by Endrich and Hammersen (1986) is depicted in Figure 4.11. The tumour under investigation was an amelanotic melanoma of the hamster. As can be seen from Figure 4.11, the ateriolar–venular pressure gradient was significantly reduced at 42.5°C, compared to 30°C and 35°C. The precapillary pressure was lower, whereas the draining vessels exhibited a much higher hydrostatic pressure. The net result was an increase of the rheologic resistance of the microvascular bed. Increased shunt flow was observed by Dewhirst *et al.* (1984a, b) using an experimental mammary adenocarcinoma.

4.4.5. Changes in pO_2 and pH

Changes in tissue oxygenation have been observed in tumours after hyperthermia (Bicher *et al.*, 1980; Bicher and Vaupel, 1980; Vaupel *et al.*,

1983). Generally, first an increase in pO_2 is observed, which is followed by a decrease during prolonged exposures. In addition, a number of investigators have reported a decrease in tumour pH in experimental tumours as a result of heating (Song *et al.*, 1980a, b; Vaupel, 1982; Vaupel *et al.*, 1983; Rhee *et al.*, 1984; Kallinowski *et al.*, 1986; Kallinowski and Vaupel, 1986).

In most investigations the changes in pH resulting from hyperthermia were deduced from the differences between pre- and post-treatment pH values. In most cases, a decrease in the order of about 0.1–0.25 pH units has been observed after treatments with temperatures above 42°C.

Vaupel (1982) and Rhee *et al.* (1984) have both demonstrated that the mean values of a large series of determinations of pH shifted to lower values after treatment at these temperatures. At lower temperatures no appreciable changes have been noted, except by Thistlethwaite *et al.* (1985) who found an increase in tumour pH in some clinical cases. Recent observations by Hetzel *et al.* (1988) indicate that irreversible changes in pH in a C3H mammary adenocarcinoma require a hyperthermic exposure of 2 h at 43.5°C, which is a high 'heat dose'. This result again stresses that not only the treatment temperature but also the exposure time is of importance, not only for cell killing, but also for inducing physiological changes (e.g. see Vaupel and Kallinowski, 1987).

The hyperthermia-induced changes in pH in human tumours during whole-body hyperthermia are, however, less striking (van der Zee *et al.*, 1983). Furthermore, the net result of a combined treatment consisting of multiple sessions of radiation therapy and hyperthermia given over several weeks was that the initially acidic tumour pH returned to more normal values (Wike–Hooley *et al.*, 1984b). The accurate measurement of tissue pH during hyperthermia is far from simple. All electrode-type probes are severely disturbed by electromagnetic heating. Recently, very thin fibre-optic systems have become available which are based on measuring the colour changes of a pH-sensitive dye, contained in a semi-permeable membrane of very small dimensions. Figures 4.12 and 4.13 are derived from such measurements. An immediate change in pH value as a response to an increase of temperature can be discerned. This reflects a temperature-induced shift in the Donnan ionic equilibrium and therefore represents a real, but small, decrease in pH value that should not be considered as a physiologically induced change in pH, as would be the result of, for example, the accumulation of lactic acid. In the rat tumour the pH value after treatment remains about 0.1 pH unit lower than the initial (equilibrium) level. This is likely to be caused by changes in the tumour physiology. The pH changes in the human tumour mirror, almost perfectly, the changes in temperature. Further work should demonstrate whether more extensive and irreversible changes will occur at higher temperatures, keeping in mind Vaupel's (1982) observation that, in animal tumours, the temperature should exceed 42°C in order to induce a

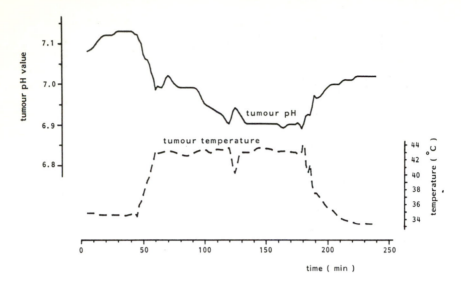

Figure 4.12 Changes in tumour pH in a rat rhabdomyosarcoma during microwave heating. The tumour pH mirrors the temperature rise, but the eventual pH after treatment remains below the (stabilized) pretreatment level, indicating a lasting physiological effect of hyperthermia. (Courtesy S. A. van de Merwe, unpublished data (1988).)

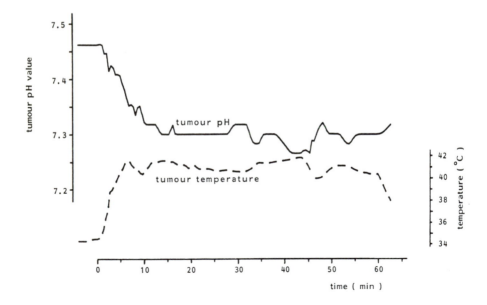

Figure 4.13 Changes in tumour pH in a human mammary carcinoma during a clinical hyperthermia treatment with 433 MHz microwaves. The tumour pH in this case mirrors the changes in temperature, indicating a real, but transient, decrease in tumour pH. (Courtesy S. A. van de Merwe, unpublished data (1988).)

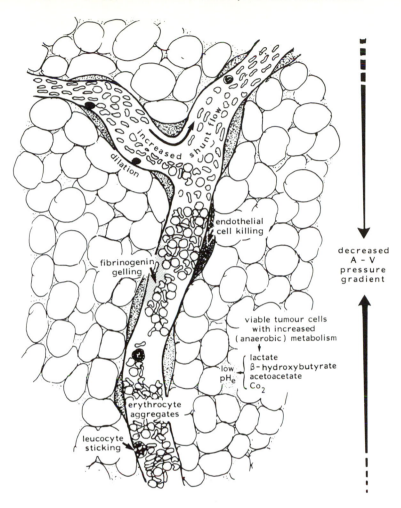

Figure 4.14 Diagrammatic representation of some of the changes that may occur in the circulation and metabolism of tumours during hyperthermia.

decrease in pH. However, a gradual decrease in tumour pH in the case of a human tumour has already been observed after only 30 min at 40.5 °C (S. A. van de Merwe, personal communication).

4.4.6. Aggregation of erythrocytes

The aggregation of red blood cells was described by Endrich and Hammersen (1986). In hamster melanoma at 42.5 °C it was observed that, in the postcapillaries, aggregation of erythrocytes occurred, which in these conditions were not 'washed away'. The erythrocytes form clusters, and tend to obstruct the lumen. This phenomenon should not be confused with the mechanism of erythrocyte stiffening, as was claimed by von Ardenne and Krüger (1980). Von Ardenne refers to experiments *in vitro*, where under prolonged incubation of erythrocytes at very low pH values, a stiffening of the erythrocytes gradually develops. It is not likely that these circumstances will ever occur, as pH values as low as 6.0 and exposure times of several hours are required to obtain this effect (Eddy, 1980; Crandall *et al.*, 1978).

4.4.7. Fibrinogen gel clotting

Increase in fibrinogen uptake in hyperthermia-treated tumours was reported by Lee *et al.* (1985) following studies on thrombosis by Copley (1980). The fibrinogen in the blood has its highest concentration near the vessel wall because of the rheological properties of the fibrinogen. It was found that the rigidity of the fibrinogen layers was markedly increased at pH below 6.5. It was therefore suggested by Copley that the conversion of fibrinogen into the so-called 'fibrinogenin', rather than increased erythrocyte rigidity, would be the dominating factor in causing vascular obstruction. Such a conversion occurs at the low pH of 6.5, and is a non-enzymic process. It cannot presently be excluded that the described mechanisms, proposed by Copley, play an important role in causing vascular obstruction in tumours under hyperthermic treatment.

4.4.8. Thrombosis

Thrombosis in tumour vessels as a result of hyperthermic treatment is frequently observed (Dewhirst *et al.*, 1984a, b; Nishimura *et al.*, 1988). Vascular stasis alone will not lead to thrombus formation (Dewhirst *et al.*, 1984a, b). Apart from stasis of blood corpuscles, coagulation of the blood proteins will occur. Dewhirst *et al.* (1984a, b) demonstrated the formation of thrombi in some tumour vessels during heating at temperatures between 42 and 43 °C. These observed thrombi did not result *per se* in total vascular occlusion within the observed vessels, but it was recorded that they only developed in tumour vessels, and not in those of normal tissues. Heat-induced vascular thrombosis may therefore, under limited conditions, be regarded as a tumour-specific factor.

4.4.9. Endothelial changes

There can be little doubt that most, if not all, phenomena related to a reduction of vascular function in heated tumour tissue could follow endothelial cell damage. Here one should distinguish between heat-induced clonogenic cell sterilization, as investigated by Rhee and Song

(1984) and by Fajardo *et al.* (1985), and direct cell breakdown. The results of the investigations on clonogenic cell sterilization are somewhat conflicting. Rhee and Song (1984) found the sensitivity of bovine aortic endothelial cells to be less than that of other mammalian cells and therefore the vascular system would — based on these criteria — not be the most vulnerable tissue component for hyperthermia. In contrast, however, a higher sensitivity for murine and human capillary endothelial cells was observed by Fajardo *et al.* (1985) compared to human fibroblasts. The issue is therefore not resolved (Reinhold and Van den Berg–Blok, 1989) but clonogenic cell survival of endothelial cells may be of lesser importance in hyperthermic response than the induced structural and morphological changes. Recently, Hill *et al.* (1989) found indications that the endothelial cell proliferation rate may be a determining factor for the difference in response between various tumours.

Morphological changes in endothelial cells of tumours during, or directly after, a hyperthermic treatment have been reported by Fajardo *et al.* (1980), Eddy (1980), Endrich and Hammersen (1984) and Badylak *et al.* (1985). Recently, the interesting observation was made that not only the endothelium, but also the presence or absence of connective tissue elements in the blood vessel wall, are of importance for the development of heat-induced vascular changes (Nishimura *et al.*, 1988). This would indicate that some tumour types in which the tumour vessels are supported by a connective tissue band may be less sensitive to heat than those without such support. Endothelial cell destruction is probably one of the final steps in the development of vascular alterations, and precedes stasis. Once stasis has developed, extensive studies have shown it to be an irreversible phenomenon, which is inevitably followed by necrosis (see e.g. Figure 4.5).

With the development of necrosis, a number of different processes may follow. In the first place, the pH of necrotic tumour tissue tends to return to 'normal', or even shift to alkaline values, in part due to the autolysis of any remaining metabolizing cells (Vaupel, 1982). Further, a process of removal of cellular debris and regrowth of any surviving tumour cells, with accompanying blood vessels into the necrotic zones is likely to occur (Nishimura *et al.*, 1988). Factors such as these may provide an explanation for the observation by several authors (Rhee *et al.*, 1984; Vaupel *et al.*, 1983) that, by about 1 day after a hyperthermia-induced acidification of tumour tissue the pH returns to pre-treatment levels.

4.5. Additional modifiers

The observation that tumour blood flow is in many instances lower than that of the surrounding tissues, and that the interstitial pH of tumours is lower than that of normal tissues has led to many investigators searching for methods to exploit further these differences. The lowering of tumour pH has been advocated for many years by von Ardenne (1972), who has proposed application of high doses of glucose. Despite the fact that much experimental work has been devoted to this approach (for a recent review see Ward and Jain, 1988) and that initial measurements on the effect of glucose on human tumour pH have been performed (Thistlethwaite *et al.*, 1987) no therapeutic application of this approach in the clinic has emerged as yet. Most of the experimental investigations on rodent tumours show an impressive effect on tumour pH, provided that the dose of glucose is high enough (for review, see Ward and Jain, 1988). However, the amount of glucose applied is frequently in the order of 6 g/kg body weight, a dose which approaches toxic levels (Vaupel *et al.*, 1989). This would relate to a human dose of about 0.5 kg of glucose, given as a single, intravenous bolus! In humans, very high doses of glucose in a rather short time (hours) cannot be given as 'normal' intravenous infusions of an isotonic solution, because it would require too much fluid. Therefore, application of high concentrations (e.g. 20%) must be given as infusions into large veins (e.g. the subclavian vein) in which the flow rate is high. The disturbances which follow in the osmotic balance of the blood and of the ionic equilibria have not yet been sufficiently investigated. Neither is it known to what extent such a treatment would decrease the pH of human tumours, and after what period of time. The anaesthetized rat is probably a poor model for predicting the responses of such agents in a conscious patient. The same is true for the study of the role of vaso-active agents as discussed below.

4.6. *The potential for vaso-active agents*

In view of the generally accepted view that tumour blood flow is lower than that of the surrounding tissues, attempts have been made to further manipulate this difference for therapeutic advantage. Many investigations were, in the past, aimed at improving tumour blood flow in the hope of increasing tumour oxygenation, and hence radiation sensitivity or accessibility to chemotherapeutic agents (Maeta *et al.*, 1989). More recently, the emphasis has been on the potential for decreasing tumour blood flow, while increasing that of the normal tissues (Roemer *et al.*, 1988) in order to obtain maximal heating of the tumour tissue (Voorhees and Babbs, 1982; Babbs *et al.*, 1982; Horsman *et al.*, 1989; Shrivastav *et al.*, 1985). Also, attempts have been made to selectively influence the tumour pH by the use of vaso-active agents, in combination with other measures such as glucose infusion and carbogen breathing (Tobari *et al.*, 1988). The most promising agents presently under investigation are hydralazine (Horsman *et al.*, 1989; Tobari *et al.*, 1988; Roemer *et al.*, 1988) and flunarazine (Jirtle, 1988).

Jain and Ward–Hartley (1984) reviewed the available data critically and exhaustively, encompassing the whole range of the effects of vaso-active drugs on tumour blood flow, and judged that it was 'hard to draw general conclusions'. This is not that surprising, because the majority of vaso-active drugs are very short-acting. If one then takes into account that the determinations of tumour blood flow (and, in most investigations, also of some normal tissues) were made at fixed intervals, then it is likely that important temporary effects may have either been missed, or misunderstood. Changes will also depend on the site of tumour inoculation, which is mostly restricted to the subcutis or muscle. Moreover, most of the vaso-active agents have a severe impact on the systemic blood pressure. Therefore, in animal experiments it is imperative that, at least, the animal's core temperature and the arterial blood pressure are monitored during the infusion, and that, in addition, the application of glucose and/or vaso-active drugs is performed by continuous infusion, with re-adjustments for feedback mechanisms that will take place. Finally, measurement of the effect (i.e. the evaluation of the influence of the additional modifiers on tumour heating, blood flow or tumour pH) will have to be performed on a continuous basis.

After establishing the optimum dosage and timing, choice of agents, a moderate but important therapeutic gain may be achieved by combining these modalities with radiation therapy allied with hyperthermia.

4.7. Summary

Tumours are characterized by a large variety of vascular abnormalities. This causes inefficient microcirculation, with areas which are devoid of a proper blood supply and nutrients and in which an accumulated level of waste products exist. This causes, among other things, tumour tissue hypoxia and acidosis.

When a tumour is heated a change in the forementioned parameters takes place. The intensity of these changes is not a simple reflection of the applied heat dose. The latter consist of a combination of heating temperature and exposure time. In the case of circulatory-related phenomena, first an improvement of some of the physiological values are observed, later this changes into a deterioration, with eventually a complete stasis of the tumour microcirculation. This is a generally observed phenomenon in experimental tumours in rodents. In human tumours this phenomenon has been observed several times, but not as a general rule. It is likely that, for inducing vascular stasis in the human situation, a higher treatment level is required.

References

Babbs, C.F., DeWitt, D.P., Voorhees, W.D., McCaw, J.S. and Chan, R.C. (1982) Theoretical feasibility of vasodilator-enhanced local tumour heating, *European Journal of Cancer and Clinical Oncology*, **18**, 1137–1146.

Badylak, S.F., Babbs, C.F., Skojac, T.M., Voorhees, W.D. and Richardson, R.C. (1985) Hyperthermia-induced vascular injury in normal and neoplastic tissue, *Cancer*, **56**, 991–1000.

Bicher, H.I. and Vaupel, P. (1980) Physiological responses of intratumour pO_2 and pH to hyperthermia and radiation treatment, *British Journal of Cancer*, 41(Suppl. IV), 99.

Bicher, H.I., Hetzel, F.W., Sandhu, T.S., Frinak, S., Vaupel, P., O'Hara, M.D. and O'Brian, T. (1980) Effects of hyperthermia on normal and tumour microenvironment, *Radiology*, **137**, 523–530.

Calderwood, S.K. and Dickson, J.A. (1983) pH and tumour response to hyperthermia, *Advances in Radiation Biology*, **10**, 135–186.

Chu, G.L. and Dewey, W.C. (1988) The role of low intracellular or extracellular pH in sensitization to hyperthermia, *Radiation Research*, **114**, 154–167.

Cook, J.A. and Fox, M.H. (1988) Effects of chronic pH 6.6 on growth, intracellular pH, and response to 42.0°C hyperthermia of Chinese hamster ovary cells, *Cancer Research*, **48**, 2417–2420.

Copley, A.L. (1980) Fibrinogen gel clotting, pH and cancer therapy, *Thrombosis Research*, **18**, 1–6.

Crandall, E.D., Critz, A.M., Osher, A.S., Keljo, D.J. and Forster, R.E. (1978) Influence of pH on elastic deformability of the human erythrocyte membrane, *American Journal of Physiology*, **235**, 269–278.

Dewhirst, M.W. (1988) Physiological effects of hyperthermia, in Paliwal, B.R., Hetzel, F.W. and Dewhirst, M.W. (Eds), *Biological, Physical and Clinical Aspects of Hyperthermia 1988*, New York, American Institute of Physics, pp. 16–56.

Dewhirst, M., Gross, J.F., Sim, D., Arnold, P. and Boyer, D. (1984a) The effect of rate of heating or cooling prior to heating on tumour and normal tissue microcirculatory blood flow, *Biorheology*, **21**, 539–558.

Dewhirst, M.W., Sim, D.A., Gross, J. and Kundrat, M.A. (1984b) Effect of heating rate on tumour and normal tissue microcirculatory function, in Overgaard, J. (Ed.), *Hyperthermic Oncology 1984*, Vol. 1, London, Taylor and Francis, pp. 177–180.

Dickson, J.A. and Calderwood, S.K. (1983) Thermosensitivity of neoplastic tissue *in vivo*, in Storm, F.K. (Ed.), *Hyperthermia in Cancer Therapy*, Boston, GK Hall, pp. 63–140.

Dudar, T.E. and Jain, R.K. (1984) Differential response of normal and tumor microcirculation to hyperthermia, *Cancer Research*, **44**, 605–612.

Durand, R.E. (1978) Potentiation or radiation lethality by hyperthermia in a tumor model: effects of sequence, degree, and duration of heating, *International Journal of Radiation Oncology, Biology and Physics*, **4**, 401–405.

Eddy, H.A. (1980) Alterations in tumor microvasculature during hyperthermia, *Radiology*, **137**, 515–521.

Emami, B. and Song, C.W. (1984) Physiological mechanisms in hyperthermia: a review, *International Journal Radiation Oncology, Biology and Physics*, **10**, 289–295.

Endrich, B. (1988a) Hyperthermia and microcirculatory effects of heat in animal tumors, *Recent Results in Cancer Research*, **109**, 96–108.

Endrich, B. (1988b) Hyperthermie und tumormikrozirkulation, *Contributions to Oncology*, Vol. 31, Basel, Karger, pp. 1–138.

Endrich, B. and Hammersen, F. (1984) Hyperthermia induced changes on capillary ultrastructure, *International Journal of Microcirculation, Clinical and Experimental*, **3**, 498.

Endrich, B. and Hammersen, F. (1986) Morphologic and hemodynamic alterations in capillaries during hyperthermia, in Anghileri, L.J. and Robert, J. (Eds), *Hyperthermia in Cancer Treatment 1986*, Vol. II, Boca Raton, CRC Press, pp. 17–47.

Endrich, B., Reinhold, H.S., Gross, J.F. and Intaglietta, M. (1979) Tissue perfusion inhomogeneity during early tumour growth in rats, *Journal of the National Cancer Institute*, **62**, 387–395.

Endrich, B., Voges, J. and Lehmann, A. (1984) The microcirculation of the amelanotic melanoma A-Mel-3 during hyperthermia, in Overgaard, J. (Ed.), *Hyperthermic Oncology 1984*, Vol. 1, London, Taylor and Francis, pp. 137–140.

Endrich, B., Götz, A and Messmer, K. (1982) Distribution of microflow and oxygen tension in hamster melanoma, *International Journal of Microcirculation, Clinical and Experimental*, 1, 81–99.

Endrich, B., Hammersen, F. and Messmer, K. (1988) Hyperthermia-induced changes in tumor microcirculation, *Recent Results in Cancer Research*, 107, 44–59.

Fajardo, L. F., Egbert, B., Marmor, J. and Hahn, G. M. (1980) Effects of hyperthermia in a malignant tumor, *Cancer*, 45, 613–623.

Fajardo, L. F., Schreiber, A. B., Kelly, N. I. and Hahn, G. M. (1985) Thermal sensitivity of endothelial cells, *Radiation Research*, 103, 276–285.

Field, S. B. (1987) Hyperthermia in the treatment of cancer, *Physics in Medicine and Biology*, 32, 789–811.

Grunt, T. W., Lametschwandtner, A., Karrer, K. and Staindl, O. (1986) The angioarchitecture of the lewis lung carcinoma in laboratory mice, *Scanning Electron Microscopy* 1986/II, 557–573.

Gullino, P. M. (1980) Influence of blood supply on thermal properties and metabolism of mammary carcinoma, *Annals of the New York Academy of Science*, 335, 1–17.

Hahn, G. M. and Shiu, E. (1986) Adaptation to low pH modifies thermal and thermo-chemical responses of mammalian cells, *International Journal of Hyperthermia*, 2, 379–387.

Hammersen, F., Endrich, B. and Messmer, K. (1985) The fine structure of tumor blood vessels. I. Participation of non-endothelial cells in tumor angiogenesis, *International Journal of Microcirculation, Clinical and Experimental*, 4, 31–43.

Haveman, J. and Hahn, G. M. (1981) The role of energy in hyperthermia-induced mammalian cell inactivation: A study of the effects of glucose starvation and an uncoupler of oxidative phosphorylation, *Journal of Cellular Physiology*, 107, 237–241.

Hengstebeck, S. (1983) Untersuchungen zum Intermediärstoffwechsel in der Leber und in einem Adenocarcinom der Maus nach Hyperthermie, **Dissertation**, Universität–Gesamthochschule Essen.

Hetzel, F. W., Avery, K. and Chopp, M. (1988) Hyperthermic "dose" dependent changes in intralesional pH, *International Journal of Radiation Oncology, Biology and Physics*, 16, 183–186.

Hill, S. A. Smith, K. A. and Denekamp, J. (1989) Reduced thermal sensitivity of the vasculature in a slowly growing tumour, *International Journal of Hyperthermia*, 5, 359–370.

Horsman, M. R., Christensen, K. L. and Overgaard, J. (1989) Hydralazine-induced enhancement of hyperthermic damage in a C3H mammary carcinoma *in vivo*, *International Journal of Hyperthermia*, 5, 123–136.

Jain, R. K. (1987) Tumour blood flow response to hyperthermia and pharmacological agents in Radiation Research, in Fielden, E. M., Fowler, J. F., Hendry, J. H. and Scott, D. (Eds), *Proceedings of the 8th International Congress of Radiation Research*, London, Taylor and Francis, pp. 813–818.

Jain, R. K. (1988) Determinants of tumor blood flow, *Cancer Research*, 48, 2641–2658.

Jain, R. K. and Ward–Hartley, K. (1984) Tumor blood flow — characterization, modifications and role in hyperthermia, *IEEE Transactions on Sonics and Ultrasonics*, SU-31, 504–526.

Jain, R. K., Shah, S. A. and Finney, P. L. (1984) Continuous noninvasive monitoring of pH and temperature in rat Walker 256 carcinoma during normoglycemia and hyperglycemia, *Journal of the National Cancer Institute*, 73, 429–436.

Jirtle, R. L. (1988) Chemical modification of tumour blood flow, *International Journal of Hyperthermia*, 4, 355–371.

Johnson, R. J. R. (1976) Effects of hyperthermia on tumor blood flow, in Robinson, J. E. and Wizenberg, M. J. (Eds), *Proceedings of the International Symposium on Cancer Therapy by Hyperthermia and Radiation*, Washington, DC, American College of Radiology, pp. 154–155.

Kallinowski, F. and Vaupel, P. (1986) Concurrent measurements of O_2 partial pressures and pH values in human mammary carcinoma xenotransplants, *Advances in Experimental Medicine and Biology*, 200, 609—621.

Kallinowski, F. and Vaupel, P. (1988) pH distributions in spontaneous and isotransplanted rat tumours, *British Journal of Cancer*, **58**, 314–321.

Kallinowski, F., Kluge, M. and Vaupel, P. (1986) Impact of localized ultrasound hyperthermia on pH distributions in xenotransplanted human mammary carcinoma, *International Journal of Hyperthermia*, **2**, 403.

Kim, S. H., Kim, J. H. and Hahn, E. W. (1978) Selective potentiation of hyperthermia killing of hypoxic cells by 5-thio-D-glucose, *Cancer Research*, **38**, 2935–2938.

Kolstad, P. (1968) Intercapillary distance, oxygen tension and local recurrence in cervix cancer, *Scandinavian Journal of Clinical Laboratory Investigation*, **22**(suppl. 106), 145–157.

Law, M. P. (1988) The response of normal tissues to hyperthermia, in Urano, M. and Douple, E. (Eds), *Thermal Effects on Cells and Tissues*, Utrecht, VSP Press, pp. 121–159.

Lee, S. Y., Song, C. W. and Levitt, S. H. (1985) Change in fibrinogen turnover in tumors by hyperthermia, *European Journal of Cancer and Clinical Oncology*, **21**, 1507–1513.

Lee, S. Y., Ryu, K. H., Kang, M. S. and Song, C. W. (1986) Effect of hyperthermia on the lactic acid and β-hydroxybutyric acid content in tumour, *International Journal of Hyperthermia*, **2**, 213–222.

Maeta, M., Karino, T., Inoue, Y., Hamazoe, R., Shimizu, N. and Koga, S. (1989) The effect of angiotensin II on blood flow in tumours during localized hyperthermia, *International Journal of Hyperthermia*, **5**, 191–197.

Mueller–Klieser, W., Vaupel, P., Manz, R. and Schmidseder, R. (1981) Intracapillary oxyhemoglobin saturation of malignant tumors in humans, *International Journal of Radiation Oncology, Biology and Physics*, **7**, 1397–1404.

Nishimura, Y., Hiraoka, M., Jo, S., Akuta, K., Yukawa, Y., Shibamoto, Y., Takahashi, M. and Abe, M. (1988) Microangiographic and histologic analysis of the effects of hyperthermia on murine tumor vasculature, *International Journal of Radiation Oncology, Biology and Physics,* **15**, 411–420.

Oleson, J. R., Dewhirst, D. V. M., Harrelson, J. M., Leopold, K. A., Samulski, T. V. and Tso, C. Y. (1989) Tumor temperature distributions predict hyperthermia effect, *International Journal of Radiation Oncology, Biology and Physics*, **16**, 559–570.

Peterson, H. I. (1979) Tumor blood flow compared with normal tissue blood flow, in Peterson, H. I. (Ed.), *Tumor Blood Circulation 1979*, Boca Raton, CRC Press, pp. 103–114.

Reinhold, H. S. (1971) Improved microcirculation in irradiated tumours, *European Journal of Cancer*, **7**, 273–280.

Reinhold, H. S. (1988) Physiological effects of hyperthermia, *Recent Results in Cancer Research*, **107**, 32–43.

Reinhold, H. S. and Endrich, B. (1986) Invited review: tumour microcirculation as a target for hyperthermia, *International Journal of Hyperthermia*, **2**, 11–137.

Reinhold, H. S. and van den Berg–Blok, A. E. (1983) Hyperthermia-induced alteration in erythrocyte velocity in tumors, *International Journal of Microcirculation, Clinical and Experimental*, **2**, 285–295.

Reinhold, H. S. and van den Berg–Blok, A. E. (1989) Differences in the response of the microcirculation to hyperthermia in five different tumours, *European Journal of Cancer Clinical Oncology*, **25**, 611–618.

Reinhold, H. S., Blachiewicz, B. and van den Berg–Blok, A. (1978) Decrease in tumor microcirculation during hyperthermia, in Streffer, C. (Ed.), *Cancer Therapy by Hyperthermia and Radiation*, Munich, Urban and Schwarzenberg, pp. 231–232.

Rhee, J. G. and Song, C. W. (1984) Thermosensitivity of bovine aortic cells in cultures, *in vitro* clonogenicity study, in Overgaard, J. (Ed.), *Hyperthermic Oncology 1984*, Vol. 1, London, Taylor and Francis, pp. 157–160.

Rhee, J. G., Kim, T. H., Levitt, S. H. and Song, C. W. (1984) Changes in acidity of mouse tumor by hyperthermia, *International Journal of Radiation Oncology, Biology and Physics*, **10**, 393–399.

Roemer, R. B., Forsyth, K., Oleson, J. R., Clegg, S. T. and Sim, D. A. (1988) The effect of hydralazine dose on blood perfusion changes during hyperthermia, *International Journal of Hyperthermia*, **4**, 401–415.

Scheid, P. (1961) Funktionelle Besonderheiter der Mikrozirkulation im Karzinom, *Bibliotheca Anatomica*, **1**, 327–335.

Schubert, B., Streffer, C. and Tamulevicius, P. (1982) Glucose metabolism in mice during and after whole-body hyperthermia, *National Cancer Institute Monographs*, **61**, 203–205.

Shrivastav, S., Joines, W. T. and Jirtle, R. L. (1985) Effect of 5-hydroxytryptamine on tissue blood flow and microwave heating of rat tumors, *Cancer Research*, **45**, 3203–3208.

Song, C. W. (1978) Effect of hyperthermia on vascular functions of normal tissues and experimental tumors: brief communication, *Journal of the National Cancer Institute*, **60**, 711–713.

Song, C. W., Clement, S. S. and Levitt, S. H. (1977) Cytotoxic and radiosensitizing effects of 5-thio-D-glucose on hypoxic cells, *Radiology*, **123**, 201–205.

Song, C. W., Kang, M. S., Rhee, J. G. and Levitt, S. H. (1980a) Effect of hyperthermia on vascular function in normal and neoplastic tissues, *Annals of the New York Academy of Science*, **335**, 35–47.

Song, C. W., Kang, M. S., Rhee, J. G. and Levitt, S. H. (1980b) The effect of hyperthermia on vascular function, pH, and cell survival, *Radiology*, **137**, 795–803.

Song, C. W., Lokshina, A., Rhee, J. G., Patten, M. and Levitt, S. H. (1984) Implication of blood flow in hyperthermic treatment of tumors, *IEEE Transactions on Biomedical Engineering*, **31**, 9–16.

Song, C. W., Rhee, J. G. and Haumschild, D. J. (1987) Continuous and non-invasive quantification of heat-induced changes in blood flow in the skin and RIF-1 tumour of mice by laser doppler flowmetry, *International Journal of Hyperthermia*, **3**, 71–77.

Streffer, C. (1982) Aspects of biochemical effects by hyperthermia, *National Cancer Institute Monographs*, **61**, 11–16.

Streffer, C. (1985) Metabolic changes during and after hyperthermia, *International Journal of Hyperthermia*, **1**, 305–319.

Streffer, C. and van Beuningen, D. (1987) The biological basis for tumour therapy by hyperthermia and radiation, *Recent Results in Cancer Research*, **104**, 24–70.

Streffer, C., Hengstebeck, S. and Tamulevicius, P. (1981) Glucose metabolism in mouse tumor and liver with and without hyperthermia, *Henry Ford Hospital Medical Journal*, **29**, 41–44.

Sutton, C. H. (1976) Necrosis and altered blood flow produced by microwave-induced tumor hyperthermia in a murine glioma, *Proceedings of the 12th Annual Meeting of the American Society for Clinical Oncology*, **17**, 63.

Tamulevicius, P., Würzinger, U., Luscher, G. and Streffer, C. (1984) Lipid metabolism in mouse liver and adenocarcinoma following hyperthermia, in Overgaard, J. (Ed.), *Hyperthermic Oncology 1984*, London, Taylor and Francis, pp. 23–26.

Thistlethwaite, A. J. and Leeper, D. B. (1988) in Paliwal, B. R., Hetzel, F. W. and Dewhirst, M. W. (Eds), *Biological, Physical and Clinical Aspects of Hyperthermia 1988*, Measurement and modification of human tumor pH, and the potential role of pH, New York, American Institute of Physics, pp. 279–298.

Thistlethwaite, A. J., Leeper, D. B., Moyland III, D. J. and Nerlinger, R. E. (1985) pH Distribution in human tumors, *International Journal of Radiation Oncology, Biology and Physics*, **11**, 1647–1652.

Thistlethwaite, A. J., Alexander, G. A., Moyland, D. J. and Leeper, D. B. (1987) Modification of human tumor pH by elevation of blood glucose, *International Journal of Radiation Oncology, Biology and Physics*, **13**, 603–610.

Thomlinson, R. H. and Gray, L. H. (1955) The histological structure of some human lung cancers and the possible implications for radiotherapy, *British Journal of Cancer*, **9**, 539–549.

Tobari, Ch., van Kersen, I. and Hahn, G. M. (1988) Modification of pH of normal and malignant mouse tissue by hydralizine and glucose, with and without breathing of 5% CO_2 and 95% air, *Cancer Research*, **48**, 1543–1547.

van den Berg–Blok, A. E. and Reinhold, H. S. (1984) Time–temperature relationship for hyperthermia induced stoppage of the microcirculation in tumors, *International Journal of Radiation Oncology, Biology and Physics*, **10**, 737–740.

van den Berg–Blok, A. E. and Reinhold, H. S. (1987) Experimental hyperthermic treatment of a human colon xenograft. The thermal sensitivity of the tumour microcirculation, *European Journal of Cancer and Clinical Oncology*, **23**, 1177–1180.

van den Berg, A. P., Wike–Hooley, J. L., van den Berg–Blok, A. E., van der Zee, J. and Reinhold, H. S. (1982) Tumour pH in human mammary carcinoma, *European Journal of Cancer and Clinical Oncology*, **18**, 457–462.

van den Berg, A. P., Wike–Hooley, J. L., Broekmeyer–Reurink, M. P., van der Zee, J. and Reinhold, H. S. (1989) The relationship between the unmodified initial tissue pH of human tumours and the response to combined radiotherapy and local hyperthermia treatment, *European Journal of Cancer and Clinical Oncology*, **25**, 73–78.

van der Zee, J., van Rhoon, G. C., Wike–Hooley, J. L., Faithfull, N. S. and Reinhold, H. S. (1983) Whole-body hyperthermia in cancer therapy: a report of a phase I–II study, *European Journal of Cancer and Clinical Oncology*, **19**, 1189–1200.

Vaupel, P. (1977) Hypoxia in neoplastic tissue, *Microvascular Research*, **13**, 399–408.

Vaupel, P. (1982) Einfluss einer lokalisierten Mikrowellen-Hyperthermie auf die pH-Verteilung in bösartigen Tumoren, *Strahlentherapie*, **158**, 168–173.

Vaupel, P. and Kallinowski, F. (1987) Physiological effects of hyperthermia, *Recent Results in Cancer Research*, **104**, 71–109.

Vaupel, P., Ostheimer, K. and Müller–Klieser, W. (1980) Circulatory and metabolic responses of malignant tumors during localized hyperthermia, *Journal of Cancer Research and Clinical Oncology*, **98**, 15–29.

Vaupel, P. W., Frinak, S. and Bicher, H. I. (1981) Heterogenous oxygen partial pressure and pH distribution in C3H mouse mammary adenocarcinoma, *Cancer Research*, **41**, 2008–2013.

Vaupel, P., Mueller–Klieser, W., Otte, J., Manz, R. and Kallinowski, F. (1983) Blood flow, tissue oxygenation, and pH distribution of malignant tumors upon localized hyperthermia. Basic pathophysiological aspects and the role of various thermal doses, *Strahlentherapie*, **159**, 73–81.

Vaupel, P., Okunieff, P. and Kluge, M. (1989) Response of tumour red blood cell flux to hyperthermia and/or hyperglycaemia, *International Journal of Hyperthermia*, **5**, 199–210.

von Ardenne, M. (1972) Selective multiphase cancer therapy: conceptual aspects and experimental basis. *Advances in Pharmacology and Chemotherapy*, **10**, 339–380.

von Ardenne, M. and Krüger, P. (1980) The use of hyperthermia within the frame of cancer multistep therapy, *Annals of the New York Academy of Science*, **335**, 356–361.

von Ardenne, M. and Reitnauer, P. G. (1978) Gewebe-Uebersäurung und Mikrozirkulation, *Archiv für Geschwulstforschung*, **8**, 729–749.

von Ardenne, M. and Reitnauer, P. G. (1982) Die manipulierte selektive Hemmung der Mikrozirkulation im Krebsgewebe, *Journal of Cancer Research and Clinical Oncology*, **103**, 269–279.

Voorhees III, W. D. and Babbs, C. F. (1982) Hydralazine-enhanced selective heating of transmissible venereal tumor implants in dogs, *European Journal of Cancer and Clinical Oncology*, **18**, 1027–1033.

Warburg, O., Posener, K. and Negelein, E. (1924) Ueber den Stoffwechsel der Carcinomzelle, *Biochemisch Zeitschrift*, **152**, 309–344.

Ward, K. A. and Jain, R. K. (1988) Response of tumours to hyperglycaemia: characterization, significance and role in hyperthermia, Invited review, *International Journal of Hyperthermia*, **4**, 223–250.

Warren, B. A. (1979) The vascular morphology of tumors, in Peterson, H. I. (Ed.), *Tumor Blood Circulation 1979*, Boca Raton, CRC Press, pp. 1–47.

Waterman, F. M., Nerlinger, R. E., Moyland, D. J. and Leeper, D. B. (1987) Response of human tumor blood flow to local hyperthermia, *International Journal of Radiation Oncology, Biology and Physics*, **13**, 75–82.

Weber, G. (1983) Biochemical strategy of cancer cells and the design of chemotherapy: GHA Glowes memorial lecture, *Cancer Research*, **43**, 3466–3492.

Wiig, H., Tveit, E., Hultborn, R., Reed, R. K. and Weiss, L. (1982) Interstitial fluid pressure in DMBA-induced rat mammary tumor, *Scandinavian Journal of Clinical Laboratory Investigation*, **42**, 159–164.

Wike–Hooley, J. L., van der Zee, J., van Rhoon, G. C. van den Berg, A. P. and Reinhold, H. S. (1984a) Human tumor pH changes following hyperthermia and radiation therapy, *European Journal of Cancer and Clinical Oncology*, **20**, 619–623.

Wike–Hooley, J. L., Haveman, J. and Reinhold, H. S. (1984b) The relevance of tumor pH to the treatment of malignant disease, *Radiotherapy and Oncology*, **2**, 343–366.

Wike–Hooley, J. L., van den Berg, A. P., van der Zee, J. and Reinhold, H. S. (1985) Human tumour pH and its variation, *European Journal of Cancer and Clinical Oncology*, **21**, 785–791.

Young, J. S., Lumsden, C. E. and Stalker, A. L. (1950) The significance of the "tissue pressure" of normal testicular and of neoplastic (Brown–Pearce carcinoma) tissue in the rabbit, *Journal of Pathologic Bacteriology*, **62**, 313–335.

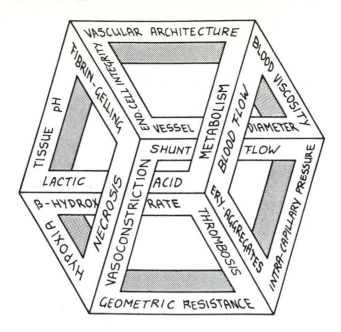

5 Hyperthermia and chemotherapeutic agents

O. Dahl and O. Mella

5.1. Introduction

A positive interaction between heat and radiation is generally accepted (see Chapters 1, 2 and 6). This chapter updates the basis for the combination of hyperthermia and chemotherapeutic drugs. Extensive literature reviews of experimental studies have been made previously (Hahn, 1982; Dahl, 1986, 1988; Engelhardt, 1987). Here, the general principles are presented as an introduction to this field of research. Possible mechanisms for the interaction of heat and drugs will be discussed together with some effects of hyperthermia on drug pharmacology. Finally, early clinical experience is reviewed. It is hoped that this may give ideas for further rational clinical application of hyperthermia in conjunction with cytotoxic agents.

5.2. Mechanisms of interaction

Anticancer drugs have various mechanisms of action. In this paper only mechanisms possibly involved in the interaction with hyperthermia will be mentioned. The main target for most drugs is the DNA macromolecule whose replication is hampered by strand breaks, cross linkage or structural changes due to incorporation of antimetabolites. For example, the vinca alkaloids impair replication through a destruction of the mitotic spindle apparatus necessary for cell division.

The exact mechanism of action of hyperthermia alone is still open for discussion. Cellular membrane damage is probably one of the main causes of cell killing due to hyperthermia alone (Wallach, 1977), as discussed in Chapter 1. Other targets seem to be protein and nucleic acid synthesis as well as damage to sensitive structural and functional proteins (enzymes).

5.2.1. Pharmacological changes

The increased effect of some drugs at elevated temperatures may be related to altered drug pharmacokinetics or pharmacodynamics. The term 'increased drug uptake' is often given as the underlying mechanism. Cellular drug uptake is only one of a cascade of processes from drug administration, disposition, uptake, metabolism (activation or inactivation), transport to binding at a final target.

5.2.1(a) Pharmacokinetic factors

Aqueous *solubility* of solid drugs increases with increasing temperatures (Ballard, 1974). A rise in temperature to 39–43°C resulted in more rapid hydrolysis of nitrosoureas (bis-chloro-ethylnitrosourea or carmustine (BCNU), cyclohexylnitrosourea or lanustine (CCNU) and methyl-cyclohexylnitrosourea or semustine (methyl-CCNU)), which was more pronounced at the higher temperatures, and was an exponential function of time (Hahn, 1978). The half-life of melphalan was 1.5 and 1.9 times shorter in phosphate buffer and plasma *in vitro* at 41°C compared to 37°C (Honess *et al.*, 1985). In canine serum *in vitro*, cisplatin was more rapidly changed to free, ultrafiltrable cisplatin at 43°C compared to 37°C (Riviere *et al.*, 1986). Chlorambucil is also rapidly decomposed in aqueous solution, the stability being influenced by pH, temperature and chloride ion concentration, but the stability is higher in plasma due to extensive protein binding (for references see Eksborg and Ehrsson, 1985). Preheating of 16 different drugs for 90 min at 43°C in saline did not reduce their cytotoxicity except for etoposide (VP 16-213) (Voth *et al.*, 1988).

Plasma *protein binding* may reduce the free, reactive fraction of drugs. Several drugs, e.g. nitrosoureas (Hahn, 1978), methotrexate (Warren and Bender, 1977), cisplatin (Hecquet *et al.*, 1985; van der Vijgh and Klein, 1986) and chlorambucil (Eksborg and Ehrsson, 1985), have been reported to bind to proteins. Generally, such drug-protein binding tends to be reduced at higher temperatures (Ballard, 1974). Hyperthermia may, however, change the binding in an unpredictable manner: elevation of the temperature to 43°C increased the aquation of cisplatin leading to increased protein binding *in vitro*. In dogs, however, whole-body hyperthermia (42°C) permitted an increased tissue binding of free ultrafiltrable cisplatin at the elevated temperature thereby actually reducing the binding to plasma proteins together with a reduction of free cisplatin in serum (Riviere *et al.*, 1986).

Enzymatic processes involved in drug activation and detoxification may be critically dependent on temperature. The activation of cyclophosphamide by rat liver microsomes led to reduced alkylating activity at 40.5°C (Clawson *et al.*, 1981). In liver slices, reduced alkylating activity was not observed at 41.8 °C, but did decrease at 43–44 °C (Clawson *et al.*, 1981). In contrast, the production of alkylating derivatives of mitomycin C was increased at 41–44 °C when heated in an anoxic milieu (Teicher *et al.*, 1981). For doxorubicin, altered drug metabolism with a higher aglycone fraction and an unknown metabolite have been observed in tumours heated at 43 °C for 60 min (Magin *et al.*, 1980). With perfusion of rat liver, bile flow and excretion of doxorubicin ceased at 42 °C (Skibba *et al.*, 1982). Hyperthermia had no significant effect on the uptake of doxorubicin by rat liver slices and only minor effect on metabolism *in vitro* (Dodion *et al.*, 1986). Fluorouracil must be bound to a sugar moiety in order to produce a nucleotide with an antimetabolic effect. Impairment of normal cellular enzymes at elevated temperatures may explain why most antimetabolites do not seem to be potentiated by hyperthermia.

Pharmacokinetic measurements of the dynamic level of active drug in serum reflect biotransformation, distribution and excretion. From perfusion of rat liver an increased plasma half-life of doxorubicin was observed, due to reduced biliary excretion at 41 °C (Skibba *et al.*, 1982). In mice, identical plasma clearance curves for [^{14}C]doxorubicin were observed in plasma and tumours as controls and also in tumours heated to 43 °C for 60 min (Magin *et al.*, 1980). Also, in rabbits the rate of clearance of doxorubicin from plasma was similar during whole-body hyperthermia (42.3 °C) and in control animals (Mimnaugh *et al.*, 1978). Higher plasma and tumour melphalan concentrations were measured during systemic heating at 41 °C for 45 min in C3H unheated mice, but in terms of the AUC (area under the concentration–time curve) no elevation in tumour compared to plasma melphalan exposure was observed (Honess *et al.*, 1985). In dogs, systemic hyperthermia at 42 °C significantly increased the clearance, volume of distribution and, somewhat surprisingly, the half-life of free, ultrafiltrable cisplatin and the parent drug (Riviere *et al.*, 1986). After intraperitoneal injection in dogs, total concentrations of cisplatin seen in serum or urine did not differ significantly between animals receiving regional heating at 41.5 °C for 60 min and those that were untreated (Zakris *et al.*, 1987). However, the AUC in serum was significantly lower for free, ultrafiltrable cisplatin in dogs receiving hyperthermia. The reason might be that reactive metabolites were retained intraperitoneally.

In two patients, peak unchanged cyclophosphamide and serum alkylating activity was unchanged, but urinary excretion of unchanged parent drug increased and alkylating activity decreased when measured after whole-body hyperthermia at 41.8 °C compared to previous treatment with the drug alone (Ostrow *et al.*, 1982). In man, no change in half-life of total cisplatin was seen by infusing the drug during whole-body hyperthermia (Gerad *et al.*, 1983), but an enhanced nephrotoxicity was observed.

5.2.1(b) Drug uptake

ALKYLATING AGENTS

In pilot studies using radiolabelled *thio-TEPA* (triethylenthiophosphoramide) and *nitrogen mustard*, an increased uptake was demonstrated in intact dog limbs (for references see Suzuki, 1967). Also, the uptake in Yoshida sarcoma in rats following intraperitoneal injection of radiolabelled *TSPA* was raised at higher temperatures (Suzuki, 1967). In culture, the intracelluar concentration of *melphalan* in V79 cells increased by a factor of 2–3 at 41°C, consistent with doubling of cell killing. However, in C3H mice, whole-body hyperthermia at 41°C for 45 min caused an increase of melphalan plasma and tumour concentration, but the elevation was not greater in tumour compared to plasma (Honess *et al.*, 1985). Both *thio-TEPA* and *cyclophosphamide* had greater antitumour effect in a murine bladder tumour after heating despite reduced drug concentration (Longo *et al.*, 1983).

CISPLATIN

After an initial similar uptake in P388 leukaemia cells in mice at 37 and 42.3°C whole-body hyperthermia, the heated cells accumulated more radiolabelled cisplatin after 1h of heating (Alberts *et al.*, 1980). Interestingly, P388 leukaemia cells accumulated more drug than normal bone marrow cells. Hyperthermia at 43°C also increased the intracellular cisplatin concentration in both cisplatin sensitive and resistant strains of CHO cells by factors of 1.5 and 2.2, respectively (Wallner *et al.*, 1986). The corresponding ratios for cell killing at 10% survival were between 5.5 and 6.5 at 43°C compared to 37°C. In a human squamous cell line, no difference in intracellular accumulation of cisplatin was observed for 1 h of heating at 37–45°C (Herman *et al.*, 1988a).

DOXORUBICIN

Doxorubicin (Adriamycin) is a fluorochrome which can easily be measured in single cells by flow cytophotometry. Increased drug uptake associated with increased cell killing has been demonstrated in Chinese hamster cells (Hahn and Strande, 1976; Bates and Mackillop, 1986; Rice and Hahn, 1987), mouse leukaemia cells (van der Linden, 1984; Toffoli *et al.*, 1989) and Ehrlich ascites cells (Yamane *et al.*, 1984), but also unchanged cell killing despite increased uptake in L1210 leukaemia cells (Block *et al.*, 1975). In two studies, doxorubicin given at temperatures ranging from 43 to 50°C did not alter drug uptake or cell killing, i.e. in MAM 16/C mouse mammary carcinoma (van der Linden, 1984) and CHO cells (drug-resistant subclone) (Bates and Mackillop, 1986). Doxorubicin-resistant and normal-sensitive MAM 16/C carcinoma cells accumulated similar drug concentrations. Increased cell killing has been reported in a mammary tumour in mice, without a change in drug uptake (Magin *et al.*,

1980). An excellent recent study showed that the intracellular doxorubicin concentration did increase with elevated temperature (45.5°C) (Rice and Hahn, 1987). However, when they sorted the cells according to intracellular drug concentration, there was excess cell killing in the cells exposed to hyperthermia. This could imply that drug concentration alone does not account for the cell killing, and that drug uptake is one of the several factors involved.

The complexity of drug uptake is illustrated for doxorubicin, which passes into cells by passive diffusion. The drug then either goes into lysosomes or is actively pumped out of the cell by an ATP-dependent mechanism. In P388 leukaemia cells, hyperthermia (42°C) does not modify intracellular drug chemosensitivity, but increases the passive influx without affecting the doxorubicin efflux kinetics (Toffoli *et al.*, 1989).

BLEOMYCIN

The uptake of [^{14}C]bleomycin in Chinese hamster cells (Braun and Hahn, 1975) and [^{57}Co]bleomycin in HeLa cells (Hassanzadeh and Chapman, 1982) was measured when the cells were exposed to hyperthermia (43°C) during and before drug exposure in culture. In both experiments hyperthermia reduced the drug concentration in the cells. In rats, the uptake of [^{57}Co]bleomycin was not changed after heating at 43°C for 60–90 min (Lin *et al.*, 1983).

ACTINOMYCIN D

Actinomycin D no longer contributed to cell killing beyond 30 min at 43°C. In fact, there was a heat-induced progressive protection against actinomycin D (Donaldson *et al.*, 1978). When these workers preheated cells for 90 min at 43°C, a transient increased uptake of actinomycin D was observed when the drug was added, followed by lower intracellular drug concentration after 60 min. However, preheating caused the cells to become resistant to the drug. Thus, drug resistance was not correlated with reduced permeability to actinomycin D.

ANTIMETABOLITES

Hyperthermia at 43 and 45°C increased the uptake of [^{14}C]*fluorouracil* in Ehrlich tumours in mice (Fujiwara *et al.*, 1984). The drug concentration was higher in tumours than normal muscle. The drug levels persisted longer in tumours at 43°C than at 45°C.

Methotrexate concentration was not increased in bladder cancer tumours (Tacker and Anderson, 1982) and Lewis lung tumours (Weinstein *et al.*, 1979) at 42°C. In murine leukaemia cells at 41.5°C (Klein *et al.*, 1979) and Chinese hamster cells at 43°C (Herman *et al.*, 1981) an initial increased uptake was found at the elevated temperature, followed by reduced drug content after 1–2 h.

MEMBRANE-ACTIVE DRUGS

Local anaesthetics and alcohols, which act on membranes, are more toxic when combined with hyperthermia (for references see Hahn, 1982; Dahl, 1986). The fungicide polyene antibiotic amphotericin B also becomes cytotoxic for mammalian cells at elevated temperatures, probably by interaction with sterols in the membranes. These experimental data imply a central role for the cell membranes, possibly by a common membrane target.

Increased uptake of chemotherapeutic agents has been repeatedly quoted as the principal mechanism underlying the enhanced effect of combining hyperthermia and drugs. However, there is not a positive interaction for all drugs where increased intracellular concentration has been demonstrated. In several experiments increased cell killing was observed despite similar or even reduced drug concentrations. There may be variations between different tumours and the topic is further complicated by the fact that chemotherapeutic agents differ in their uptake mechanism (active transport, passive diffusion) or exclusion (passive diffusion, active efflux). The drug charge may change due to temperature and pH shift, which also affect uptake.

5.2.2. DNA effects

5.2.2(a) DNA damage

When reactive parts of the chemotherapeutic drugs bind covalently to macromolecules, the process is called *alkylation*. Although most highly reactive intermediates bind to different sites, the nucleophilic N^{-7} position of guanine is the preferred reaction site of DNA leading to impaired DNA replication. Low activation energies which suggest an alkylating process, were found by Arrhenius analysis of survival curves for hyperthermia combined with thio-TEPA (Johnson and Pavelec, 1973) and nitrosoureas (Hahn, 1978). *Single-strand breakage* of DNA was demonstrated by alkaline sucrose sedimentation analyses after combination of hyperthermia and the monofunctional alkylating agent methyl methanesulphonate (Bronk *et al.*, 1973; Ben-Hur and Elkind, 1974), bleomycin (Meyn *et al.*, 1979; Kubota *et al.*, 1979; Smith *et al.*, 1986), cisplatin (Meyn *et al.*, 1980; Herman *et al.*, 1988a) and peptichemio (Djordjevic *et al.*, 1978). Changes in DNA structure were also detected by electron spin resonance after bleomycin treatment (Chapman *et al.*, 1983). Using the more-sensitive technique of neutral nucleoid sedimentation, early DNA damage was shown when bleomycin (Smith *et al.*, 1986) or doxorubicin (Daugherty *et al.*, 1988) were combined with hyperthermia.

5.2.2(b) DNA repair

Reduced repair of single-strand breaks at elevated temperatures has been repeatedly demonstrated (Bronk *et al.*, 1973; Ben-Hur and Elkind, 1974; Djordjevic *et al.*, 1978; Kubota *et al.*, 1979; Meyn *et al.*, 1979; Smith *et al.*, 1986) for which the *DNA repair enzyme* polymerase β may be the target enzyme responsible (Spiro *et al.*, 1982). In a recent study, polymerase activity was shown to be inhibited by hyperthermia (42°C for 1–3 h) but increased after doxorubicin exposure (1–10 μM for 30 min) (Daugherty *et al.*, 1988). In this study, polymerase activity did not correlate with cell survival. Exposure of HeLa cells and Chinese hamster cells to drugs (novobiocin, nalidixic acid) which inhibit the nuclear enzyme topoisomerase II, resulted in sensitization of both cell lines to hyperthermia at 41 and 45°C (Warters and Brizgys, 1988). Indirectly, this supports the role of DNA damage as a key target.

5.2.3. Oxygen radicals

Several investigations have implicated that *oxygen radical* production may be one of the basic mechanisms of action of several anticancer drugs, i.e. BCNU, doxorubicin, bleomycin, mitomycin (for references see Arrik and Nathan, 1984; Issels *et al.*, 1988), chlorambucil (Wang and Tew, 1985) and cisplatin (Hromas *et al.*, 1987). Radiation and hyperthermia can change the cellular content of glutathione (GSH) which can function as an oxygen radical detoxifier (Arrik and Nathan, 1984; Shrieve *et al.*, 1986). However, the effect of glutathione seems to depend on the concentration as well as the relative amount of reduced GSH and oxic forms (GSSG) of this molecule in a very complex manner (Issels *et al.*, 1988; Anderstam, 1988; Anderstam and Harms-Ringdahl, 1988). There is experimental evidence for an important role of GSH in modulating doxorubicin-induced hydroxyl radical formation via the scavenging of hydrogen peroxide by GSH peroxidase, thus impeding a potential interaction with DNA (Dusre *et al.*, 1989). Reduction of GSH by agents such as buthionine sulphoximine enhances the effect of several drugs when combined with hyperthermia, but its potential role for clinical use is not clear.

NADH dehydrogenase in mitochondria may contribute to production of semiquinone free-radical intermediates of anthracyclines (Gervasi *et al.*, 1986). The disruption of mitochondrial membranes is the characteristic pathologic picture of anthracycline cardiac toxicity (Gervasi *et al.*, 1986). It is of interest that mitochondrial changes were found recently to be particularly significant, appearing early and correlating well with the loss of viability and metabolic functioning after heat treatment (Wheatley *et al.*, 1989). The new drug lonidamine, which has an effect on mitochondria, particularly under acidic conditions, seems to interfere with the electron transport (Kim *et al.*, 1984).

5.2.4. Membrane damage

Increased calcium influx, decreased Ca^{2+}-ATPase activity and cyclic-AMP concentration has been observed in Ehrlich carcinoma cells submitted to hyperthermia (45°C) (Anghileri *et al.*, 1985). The significant augmentation of mitochondrial calcium simultaneous to an inhibition of the calcium pump, could indicate that a heat-induced failure of the system of calcium extrusion might be responsible for cell inactivation. A similar unidirectional influx of calcium has also been demonstrated in Chinese hamster cells (Stevenson *et al.*, 1987). As with data implying increased drug uptake, these studies are consistent with membrane changes either in the plasma membrane or in mitochondria being the principal mechanism of cell killing by hyperthermia. For futher details on changes in membrane functions, amino acid transport and electrolyte balance, see Wallach (1977).

5.2.5. Conclusion

Membrane changes leading to selective enhancement of drug delivery to tumours during hyperthermia may contribute to the observed interactions of hyperthermia with some drugs in some cell lines, but there is also evidence for DNA damage as a primary event. Some drugs might exhibit their killing activity by free-radical production, similar to hyperthermia itself and radiation. At present one can only conclude that probably several mechanisms are involved in the observed increased cytotoxicity when hyperthermia is combined with drugs.

5.3. *Experimental studies on cell killing*

5.3.1. Alkylating agents

When *nitrogen mustard* (5 mg/kg) was combined with local hyperthermia (40–46°C for 30 min) in rats bearing Yoshida sarcoma no enhancement was observed at 40°C, but at 42°C tumours gradually regressed and some tumours disappeared completely (Suzuki, 1967). The monofunctional alkylating agent *methyl methanesulphonate* also increased cell killing of Chinese hamster cells in culture with a decrease in the shoulder of the survival curve and above 41°C also an increase in the final slope (Ben-Hur and Elkind, 1974). The underlying mechanism appeared to be the enhanced production of DNA single-strand breaks and also their reduced repair (Bronk *et al.*, 1973; Ben-Hur and Elkind, 1974).

The cytotoxicity of *thio-TEPA* increased proportionally with temperature between 40 and 45°C, producing linear survival curves on semilogarithmic co-ordinates (Johnson and Pavelec, 1973). From

Arrhenius analysis of reaction rates, they found much lower activation energies for drug alone and drug combined with hyperthermia than for hyperthermia alone, consistent with an alkylating reaction as the cause of the increased cell killing by combined treatment.

In mice bearing Lewis lung carcinoma, the TCD_{50} was reduced from 170 mg/kg for *cyclophosphamide* alone to 100 mg/kg when cyclophosphamide was followed by heat (42.5°C for 30 min) (Hazan *et al.*, 1981). In murine RIF-1 tumours, a similar dose-modifying factor was observed: the growth delay caused by 150 mg/kg cyclyphosphamide in unheated animals was the same as that caused by 100 mg/kg cyclophosphamide in heated animals (whole-body hyperthermia, 45 min at 41°C) (Honess and Bleehen, 1982). However, compared with the bone marrow stem cell survival, the latter authors concluded that there was no therapeutic gain at the doses studied (75–200 mg/kg) as there was even more pronounced normal tissue damage. In our rat BT_4A tumours transplanted on the hind leg of rats, we observed seven complete reponses in eight rats after local water bath hyperthermia (44°C for 60 min), combined with 200 mg/kg cyclophosphamide, whereas the drug alone produced only nine partial responses in 10 rats. There were no partial responses after hyperthermia alone (Dahl and Mella, 1983). We observed no increase of normal tissue damage above that from hyperthermia alone. It is also of interest that four permanent tumour controls were observed in the combined group, which was not obtainable with drug alone at any dose. A more than additive effect was also recorded in mice bearing osteosarcomas in legs heated locally to 42.5°C for 30 min, combined with cyclophosphamide (200 mg/kg) (Hiramoto *et al.*, 1984). These authors used serum alkaline phosphatase as a measure of tumour burden and not common end-points such as growth delay or cell survival.

Melphalan had a more pronounced effect at elevated temperatures in several cell lines *in vitro* (Giovanella *et al.*, 1970; Goss and Parsons, 1977; Zupi *et al.*, 1984) These data were confirmed in B16 melanoma tumours when graded doses of melphalan were combined with 43°C for 1 h *in vivo*, the assay being cell survival (Joiner *et al.*, 1982). The Cambridge Group which has tested many drugs combined with hyperthermia, compared the degree of potentiation of KHT and/or RIF-1 tumours grown intramuscularly in the leg with bone marrow cells in mice, and only found a therapeutic gain for melphalan (Honess and Bleehen, 1985a,b).

In EMT6 oxygenated tumour cells *in vitro*, an additive effect was observed when the alkylating antibiotic *mitomycin C* was combined with elevated temperatures (41–43°C for 1–6h) (Teicher *et al.*, 1981). A synergistic effect was observed in hypoxic tumour cells. These authors also showed a 30–50% increase in rate of formation of reactive alkylating species under anaerobic conditions at 41–43°C compared to the rate at 37°C in control preparations. The data did not, however, show any difference between 42 and 43°C. In Chinese hamster ovary cells, mitomycin C cytotoxicity was enhanced by hyperthermia (42–43.5°C), both in sublines sensitive to and resistant to mitomycin C (Wallner *et al.*, 1987). These workers suggested that hyperthermia may be a method of circumventing drug resistance. An increased uptake of mitomycin C at 43.5°C was found in both cell lines, but the difference in uptake was less pronounced in the resistant cell line. In a human adenocarcinoma cell line, hyperthermia abolished the initial slope portion and steepened the subsequent exponential part of the survival curve with a dose-modifying factor, at the 10% survival level, of 1.5 at 41°C and 2.6 at 42°C (Barlogie *et al.*, 1980).

5.3.2. Nitrosoureas

Chinese hamster cells in culture showed a temperature-dependent reduction in survival at 41–43°C when the concentration of BCNU, CCNU and methyl-CCNU was corrected for increased hydrolytic decay at the elevated temperatures (Hahn, 1978). The activation energies (18–24 kcal/mol) calculated from Arrhenius analysis were consistent with the killing being caused by alkylation. Enhanced effectiveness of BCNU at 42.4–43.3°C was later reported in Chinese hamster ovary cells (Herman, 1983a). When B16 melanoma tumours were treated *in vivo* and assayed as surviving fraction in culture, there was increased cell killing at different drug doses at 43°C using a water bath (Joiner *et al.*, 1982).

In the EMT6 mouse tumour, a combination of BCNU (20 mg/kg) and local water bath heating at 43°C for 1 h yielded much greater effects than either modality alone, whether the end-point was tumour cure, growth delay or *in vitro* assay of the clonogenic fraction (Twentyman *et al.*, 1978). We have confirmed this in our neurogenic rat tumour BT$_4$A where the combination of BCNU (20 mg/kg) and local hyperthermia (44°C for 60 min) produced objective tumour regression in all rats in the combined group, with eight of 10 tumours permanently controlled (Dahl and Mella, 1982). When the B16 mouse melanoma and C24 human malignant melanoma lines were treated in mice by the water-soluble nitrosourea ACNU (10 mg/kg) and hyperthermia at (43°C for 30 min) at three repeated doses every other day, a marked synergistic effect was observed (Yamada *et al.*, 1984). However, experiments in which CCNU was combined with whole-body hyperthermia above 41°C for 30 min failed to demonstrate improved therapeutic responses in the Lewis lung carcinoma (Rose *et al.*, 1979). Similarly, no potentiation of methyl-CCNU was observed in nude mice bearing human colon cancer xenografts when given in conjunction with local hyperthermia (42.5°C for 3×60 min) or whole-body hyperthermia (41°C for 45 min) (Osieka *et al.*, 1978). No therapeutic gain was observed in mice with nitrosourea-resistant RIF-1 fibrosarcoma or the more-sensitive KHT sarcoma treated with BCNU (10–15 mg/kg) or CCNU

(5–15 mg/kg) combined with systemic hyperthermia (41 °C for 45 min), when the effect on tumours was compared with that on bone marrow stem cells (Honess and Bleehan, 1982, 1985a,b).

5.3.3. Cisplatin

In a mouse tumour model, normal bone marrow cells and P388 lymphocytic leukaemia cells were assayed using the spleen colony technique after local heating of the femurs in a water bath at 42.3 °C for 30–60 min (Alberts *et al.*, 1980). Hyperthermia added to cisplatin administration increased cisplatin inhibition of leukaemia colony formation by a factor of 100. At the same time, no synergistic effect against normal bone marrow colonies was observed. There was no evidence for any difference in tumour drug uptake during the first 30 min after exposure, but at 1 h post-exposure tumour cells from the heated femurs contained significantly more cisplatin, as did normal bone marrow cells but at lower levels. There was also a significant enhancement of cisplatin cytotoxicity in Chinese hamster cells in culture (Barlogie *et al.*, 1980; Fisher and Hahn, 1982; Herman, 1983b) and a human squamous cell carcinoma. By alkaline elution, it was shown that the number of DNA lesions (cross-linking) was significantly increased at 42–43 °C (Meyn *et al.*, 1980; Herman *et al.*, 1988a). In Chinese hamster cells and squamous carcinoma cells which were resistant to cisplatin, hyperthermia could be used to circumvent cisplatin resistance (Wallner *et al.*, 1986; Herman *et al.*, 1988a). Mild hyperthermia (41–42 °C) markedly potentiated the effect of cisplatin in human osteosarcoma cells and fibroblasts with normal human karyotype (Yamane *et al.*, 1986). In an *in vivo–in vitro* model, cisplatin combined with hyperthermia had an additive effect while *trans*-diamminedichloroplatinum(II), which is non-cytotoxic at physiological temperature, became cytotoxic *in vivo* and *in vitro* at 43 °C (Murthy *et al.*, 1984). In our BT_4A tumours where cisplatin and hyperthermia alone had only a transient growth-retarding effect, tumour controls were observed after combined treatment (Mella, 1985). The effect was greater at 44 °C than at 41 °C and also drug-dose dependent within the dose range tested (2–4 mg/kg). Histological examination showed that hyperthermia alone resulted in almost complete central tumour necrosis while the combination of cisplatin and hyperthermia produced additional cell death at the tumour periphery. Thus, a complementary spatial co-operation probably took place *in vivo*. We observed, however, an increased systemic toxicity at higher drug doses (4 mg/kg). In a separate study we showed that even a mild increase of core temperature to 41 °C produced a substantial increase in cisplatin-induced renal damage (Mella *et al.*, 1987). Similar results have been reported in rats (Wondergem *et al.*, 1988) as well as in three patients treated with cisplatin and whole-body hyperthermia (Gerad *et al.*, 1983).

Hyperthermia at 40.4 and 41.8°C markedly enhanced the efficacy of cisplatin in a human acute lymphoblastic leukaemia cell line *in vitro* (Cohen and Robins, 1987). When cisplatin and the analogues carboplatin and iproplatin were studied *in vitro* in a human ovarian cancer cell line, a 3°C incremental increase in temperature was associated with a 10-fold decrease in the drug dose necessary to obtain similar effects for each drug (Xu and Alberts, 1988). We have also shown an increased effect of carboplatin in BT_4A tumours in rats (B. C. Schem, unpublished findings). Cisplatin is also thought to be a candidate for combination with both hyperthermia and radiation as a trimodality therapy (see below).

5.3.4. Anthracyclines

Increased cytotoxicity has been reported when doxorubicin (adriamycin) was combined with hyperthermia in several cell lines: L1210 leukaemia at 42°C (Block *et al.*, 1975), Chinese hamster cells at 43°C (Hahn and Strande, 1976), mouse leukaemia and a mammary carcinoma cell line at 42 and 43°C, respectively (Mizuno *et al.*, 1980), the neurogenic BT_44C cell line at 41 and 42°C (Dahl, 1982), Chinese hamster V79 cells at 42.5°C (Roizin–Towle *et al.*, 1982), human malignant melanoma and Burkitt's lymphoma cell line at 42°C (Ohnoshi *et al.*, 1985), and in a drug-sensitive Chinese hamster ovary cell line and a pleiotropic drug-resistant mutant cell line (Bates and Mackillop, 1986). Hahn and Strande (1986) demonstrated that Chinese hamster cells after 30 min of combined treatment at 43°C became refractory to doxorubicin, and also showed that this resistance was caused by heating. Our BT_44C cell line showed a similar response at 41°C, but was extremely sensitive to the combined treatment at 42°C (Dahl, 1982). In P388 mouse leukaemia cells, there was no synergistic cytotoxicity despite a 2–3-fold increased uptake in a sensitive cell line and 1–2 times more drug uptake in a resistant cell line (van der Linden *et al.*, 1984).

The initial reports *in vivo* showed that doxorubicin was only potentiated in murine mammary carcinomas at drug doses higher than those compatible with survival of the host (Overgaard, 1976; Marmor *et al.*, 1979). It was shown later that the application of local tumour hyperthermia (43.3°C for 60 min) enhanced the effectiveness of doxorubicin (10 mg/kg iv) against the doxorubicin-sensitive 16C mouse mammary carcinoma (Magin *et al.*, 1980). These authors noted 25% cures in animals simultaneously treated with local hyperthermia and doxorubicin, while no cures were seen in any of the controls or other treatment groups. In BT_4A tumours on the hind legs of rats, an enhanced effect occurred when doxorubicin (7 mg/kg) and the analogue drug 4-epidoxorubicin (7 mg/kg) were given 30 min prior to local water bath hyperthermia (44°C for 1 h) (Dahl, 1983). Thus, we could confirm the potentiation seen *in vitro* in our animals, but we did not observe any permanent cures at a drug dose at about the LD_{10} level. There was no difference in doxorubicin sensitivity in mice implanted with P388 leukaemia when doxorubicin at different doses was combined with whole-body hyperthermia about 41 °C for 30 min (Rose *et al.*, 1979). It is difficult to draw any firm conclusion from these experimental data. It may be that there is varying sensitivity in different cell lines. Many authors have tried to relate the response at elevated temperature to increased uptake, but the sensitivity to doxorubicin is not necessarily correlated with intracellular drug levels (Bates and Mackillop, 1986; Rice and Hahn, 1987; for further references see Section 5.2.1(b)). Despite these conflicting data presented, a promising human pilot study has been reported (see below) (Arcangeli *et al.*, 1980).

In Chinese hamster ovary cells in culture, a synergistic effect on cell killing was observed after exposure for 1 h to various concentrations of *mitoxanthrone* at 42.4 °C. This was confirmed in a colon carcinoma cell line at 42 °C in culture (Wang *et al.*, 1987). These authors noted that the amount of drug incorporated in treated tumours was higher at 43 °C. However, they could not demonstrate that preheating induced any tolerance to mitoxanthrone and hyperthermia. Examining the combined cytotoxicity of hyperthermia (42 °C) and different anthracycline antibiotics in a human malignant melanoma and a Burkitt lymphoma line, hyperthermia was shown to potentiate the cytotoxic effect of doxorubicin, daunorubicin and quelamycin, but did not enhance that of aclacinomycin (Ohnoshi *et al.*, 1985; Herman, 1983b). Hyperthermia at 42 °C did not potentiate the action of daunomycin on cultured mouse leukaemia cells, and was also hardly sensitized at 43 °C in a mouse mammary carcinoma (Mizuno *et al.*, 1980). The related drug AMSA had substantially less cell killing ability at 42.4 °C than at 37 °C in Chinese hamster ovary cells (Herman, 1983a).

5.3.5. Bleomycin

Survival curves for bleomycin alone at 37 °C usually have an initial sensitive part, then a resistant tail. Generally, bleomycin causes only a slightly enhanced effect at 40–41 °C in culture, but at about 42–43 °C its effect is markedly potentiated in Chinese hamster cells (Braun and Hahn, 1975; Herman, 1983b; Roizin–Towle *et al.*, 1982), HeLa cells (Rabbani *et al.*, 1978), EMT6 cells (Har–Kedar, 1975; Marmor *et al.*, 1979; Morgan and Bleehen, 1980; Mircheva *et al.*, 1986). Recently, the sensitivity for bleomycin was shown to be increased in mouse leukaemia cells and Chinese hamster V79 cells in both the parent-sensitive cell lines and in bleomycin-resistant sublines at a temperature of 40 °C (Kano *et al.*, 1988).

In mice with KHT tumours, an enhanced effect of bleomycin (7 and 15 mg/kg) was observed at 42 and 43 °C for 30 min, with some cures at the higher temperature (Marmor *et al.*, 1979). When EMT6 tumours were heated *in vivo* and assayed *in vitro* a synergistic interaction was demonstrated. A positive interaction has also been demonstrated in other mouse carcinomas (von Szczepanski and Trott, 1981; Ma *et al.*, 1985), Lewis lung tumours (Magin *et al.*, 1979), murine squamous carcinoma (Hassanzadeh and Chapman, 1982), and the BT$_4$A neurogenic tumour (Dahl and Mella, 1982). The combined treatment of bleomycin and hyperthermia (44°C for 60 min) did not increase the skin damage in mice (Honess and Bleehen, 1980) or rats (Dahl and Mella, 1982a), but we observed a significantly increased weight reduction. Due to its threshold effect (Hahn, 1982) bleomycin should be combined with local hyperthermia where higher temperatures can be achieved. Using this approach Arcangeli *et al.* (1980) have obtained promising clinical data (see below). As this drug lacks bone marrow toxicity, bleomycin can be combined with hyperthermia in a schedule where hyperthermia can be given as repeated doses at optimal intervals.

5.3.6. Actinomycin D

In EMT6 cells, no enhancement was observed when actinomycin D was combined with hyperthermia at 42°C (Har–Kedar, 1975). However, leukaemic cells were more sensitive to the drug at 42–47°C (Giovanella *et al.*, 1970; Mizuno *et al.*, 1980). In plateau-phase Chinese hamster cells, the interaction seems complex, as actinomycin D had an increased effect on the survival curve at 43°C for up to 30 min, but beyond this time the drug no longer contributed to cell killing (Donaldson *et al.*, 1978). In fact, there was a progressive protection against actinomycin D. In mice bearing SV45 sarcoma, a synergistic interaction of local hyperthermia at 42.3 and 43.6°C for 30 min marginally increased the median survival time at the lower temperature, but substantially prolonged survival at the higher temperature (Yerushalmi, 1978). The author also stated that there was more effective control of local tumour growth than with either agent used singly, but no supporting data were given.

5.3.7. Antimetabolites and vinca alkaloids

In a Chinese hamster ovary cell line resistant to *methotrexate*, exposure to 43°C (but not 41 or 42°C) for 1 h sensitized the cells to methotrexate so that a 50% cell killing in excess to that caused by hyperthermia alone occurred (Herman *et al.*, 1981). These workers provided evidence for a mechanism by which hyperthermia induced cessation of protein synthesis, thereby causing the activity of the enzyme dihydrofolate reductase to fall as a function of its half-life. However, other Chinese hamster cells (HA-1 strain) were not sensitized to methotrexate at 43°C for 1 h (Hahn and Shiu, 1983), as was also reported for rats bearing VX2 carcinoma (Muckle and Dickson, 1973).

For *5-fluorouracil* some investigators have reported a positive interaction. There was significantly increased necrosis in a murine ependymoblastoma *in vivo* after 5-fluorouracil treatment at 50 mg/kg combined with hyperthermia at 42 °C (Sutton, 1971), and more complete regressions were observed in a human pancreatic carcinoma for combined treatment at 43.5 °C than with 5-fluorouracil alone (Shiu *et al.*, 1983). Others have presented negative results in leukaemia cells exposed to hyperthermia and 5-fluorouracil (Mizuno *et al.*, 1980) and the closely related drug 5-fluoro-2'-deoxyouridin (Giovanella *et al.*, 1970). In Ehrlich ascites cells *in vitro*, a weak enhancement of cytotoxicity was observed for the drug combined with hyperthermia, but cell killing was not greater than for 5-fluorouracil alone at 43.5 °C for 45 min *in vivo* (Tsumura *et al.*, 1988). In a human T-lymphoblast cell line exposed to 5-fluorouracil and hyperthermia (42 °C for 1 and 2 h), a simultaneous treatment resulted in subadditive or additive cell killing for different drug doses (Mini *et al.*, 1986). After heating, a prolonged exposure to fluorouracil (42 °C for 4 and 8 h) decreased the cell survival compared with the shorter periods. Their conclusion was that the exposure time and sequence of administration of hyperthermia and fluorouracil were critical determinants of cytotoxicity *in vitro*.

Other antimetabolites, e.g. *cytocin arabinocid* (Rose *et al.*, 1979; Mizuno *et al.*, 1980) and phosphoracetyl aspartate (*PALA*) (Roizin–Towle *et al.*, 1982) did not show potentiation when combined with hyperthermia.

Vinblastine was not enhanced at 42 °C in L1210 leukaemia cells (Giovanella *et al.*, 1970) or in a panel of seven transplantable human tumour cells (Neumann *et al.*, 1985). Negative results were observed for *vincristine* (Muckle and Dickson, 1973; Mizuno *et al.*, 1980) although four of seven cell lines indicated enhancement at 40.5 °C in one study (Neumann *et al.*, 1985).

Thus, the experimental data do not indicate that antimetabolites are suitable candidates for combination with hyperthermia, although, generally, the drugs have been tested at the lower temperature range (41–42 °C). The role of antimetabolites combined with local hyperthermia is therefore still open for further investigation.

5.3.8. Lonidamine

Increased cell killing by hyperthermia has been associated with experimental conditions where the cellular ATP content has been decreased by drugs. There is also experimental evidence for decreased cellular ATP content during hyperthermia *in vivo*, which correlates with survival of the cells (Lilly *et al.*, 1984). Lonidamine is a new anticancer agent. Its main effects seem to be inhibition of glycolysis and respiration, resulting in a significant reduction of ATP, but the drug has also several other effects, such as changes in membrane transport of lactate (Kim *et al.*, 1984; Caputo and Silvestrini, 1984). Lonidamine was substantially potentiated in a mouse fibrosarcoma at 41 and 42°C during acidic conditions (pH 6.5) and was also markedly potentiated at 41.6°C for 90 min and 42.6°C for 65 min (Kim *et al.*, 1984). Lonidamine has been administered to patients with local 'antiblastic perfusions', but no conclusions can at present be drawn (Cavaliere *et al.*, 1984).

5.3.9. Interferons and tumour necrosis factor

By using recombinant techniques several immune response modifiers have now become available for clinical research. *Interferon* and interferon inducers (polyinocinicpolyribocytidylic acid and Newcastle disease virus) were also more effective in murine Lewis lung carcinomas heated at 42.7°C for an hour 6 h after each treatment (Yerushalmi *et al.*, 1982). In a human bladder cancer cell line in culture, mild hyperthermia (39.5–41°C) markedly increased the cell killing of human α-interferon, but not β-interferon (Groveman *et al.*, 1984). The antiproliferative activity of recombinant murine β-interferon was enhanced at 40 and 43°C in a mouse malignant glioma cell line both *in vitro* and *in vivo* (Kuroki *et al.*, 1987). These workers noted higher cytotoxicity at 43°C than at 40°C *in vitro*, in contrast to an earlier report (Robins, 1984) where an increased effect was observed at 40—40.5°C, but no further increase at 41–41.5°C in mice bearing leukaemia cells.

Recently, a synergistic interaction between recombinant human *tumour necrosis factor* and hyperthermia (incubation at 38.5 and 40°C) was observed *in vitro* and *in vivo* in mouse tumourgenic fibroblasts (L-M cells (Niitsu *et al.*, 1988; Watanabe *et al.*, 1988). This was possibly related to an accelerated turnover rate of tumour necrosis factor receptor complex under elevated temperatures (Niitsu *et al.*, 1988). In other studies higher temperatures (43–44°C) caused an increased effect in a carcinosarcoma, probably mediated through damage of tumour vasculature (F. Kalinowski, personal communication).

5.3.10. Miscellaneous drugs

Surface-active agents, such as the polyene antibiotics and local anaesthetics as well as alcohols and polyamines, are more cytotoxic at elevated temperatures (for further references see Hahn, 1982; Dahl, 1986). As these drugs are not considered cytotoxic agents, a further discussion of them falls outside this text, but the search for heat sensitizers, that is drugs which are selectively cytotoxic at elevated temperatures, is a fascinating area of research.

Based on laboratory studies, the alkylating agents, including nitrosoureas, seem generally to have a more pronounced effect at elevated temperatures. There are more varying results, especially *in vivo*, for anthracyclines and bleomycin. The antimetabolites do not seem to be candidates for combination with hyperthermia.

5.4. *Methods for increasing drug uptake*

As discussed above [Section 5.2.1(b)], hyperthermia can increase the intracellular drug concentration of some drugs in some cell lines. Several investigators have therefore sought means to deliver more drug at the specific area where the cells are heated. Methods used are as follows:

5.4.1. Intra-arterial drug infusion

By local intra-arterial drug infusion, much higher local drug concentrations can be achieved. However, this can only be used in practice when there is a single or few supplying arteries. This approach has been used in extremity tumours and in liver (see Section 5.8.2).

5.4.2. Liposomes

Liposomes are small unilamellar vesicles which can be designed to have specific temperature-dependent phase transition points. This means that the vesicles disrupt and thereby release their contents at specific temperatures (Yatvin *et al.*, 1978; Weinstein *et al.*, 1979). By this method both a selective increase of drug in tumour areas, and the positive interaction between hyperthermia and certain drugs, can be used to increase tumour cell killing. Also, by reducing the systemic release of drug the dose-limiting systemic side-effects can be avoided. Whereas there was no increased uptake of methotrexate in a murine bladder cancer when combined with hyperthermia at 42°C, there was a 12-fold increase of uptake when administered in liposomes (Tacker and Anderson, 1982; Bassett *et al.*, 1986). However, neither free methotrexate nor liposome-delivered methotrexate caused significant antitumour effect (Bassett *et al.*, 1988). When doxorubicin was delivered systemically or entrapped in liposomes, there was an increased effect of the drug in liposomes on liver metastases, but no increased effect on subcutaneously grown tumours (Mayhew *et al.*, 1987). This could be related to the selective release of the liposomal content in liver, as has been noted in most investigations. In human hepatoma xenografts, liposomes were more effective and permitted a higher dose of doxorubicin (7.5 mg/kg against 4–5 mg/kg in free form) probably due to reduced distribution to the heart (Konno *et al.*,

1987). There was also a significant difference in suppression of tumour growth and prolonged survival in rats receiving temperature-sensitive liposomes containing bleomycin together with local hyperthermia (water bath at 44 °C for 20 min), compared to groups receiving hyperthermia alone, water solution of bleomycin alone, or a combination of both, in ascites tumours transplanted in rats (Maekawa *et al.*, 1987).

5.4.3. Degradable starch microspheres

By local administration of degradable starch microspheres a temporary blockade of the microcirculation can be achieved (Lindberg *et al.*, 1984). When administered before regional hyperthermia in pigs, a larger increase in intrahepatic temperature was achieved (Akuta *et al.*, 1987). This method could, after the administration of drugs, trap the drug within the heated area, possibly increasing the effect on tumours. Although no experimental data are available, this method has been used in a pilot study in human bladder cancer (Bichler *et al.*, 1987).

5.5. Trimodality therapy

Douple *et al.* (1982) reported that the combination of cisplatin, radiation and heat was better than that of only two modalities. Cisplatin, mitomycin C and, especially, bleomycin markedly enhanced the effect of hyperthermia (43 °C) when combined with radiation in two human bladder cancer lines (Nakajima and Hisazumi, 1987). Herman and Teicher and Herman *et al.* (1988a, b) have recently reviewed the concept of using a combination of drugs, hyperthermia and radiation in a trimodality approach. They suggest using the maximally tolerated radiation dose as the mainstay of therapy because it is the single most-effective treatment option for local disease which is not surgically amenable. The combination of hyperthermia and anticancer drugs is therefore considered an adjuvant to radiation. Before trimodality therapy is started, however, we need more data on the effects of single drugs and hyperthermia alone to select the optimal combination for such an adjuvant therapy. It will be very difficult to define the role of hyperthermia alone or of drugs alone in a multimodality combination in the clinic without these basic data.

5.6. The pH effect in thermochemotherapy

Low extracellular pH increases hyperthermic cell killing (Overgaard, 1975; Gerweck, 1977). The effect of some cytotoxic drugs (BCNU, bleomycin, cisplatin) on malignant cells *in vitro* is modified by changes in extracellular pH, especially at hyperthermic temperatures, with increased cytotoxicity at decreased pH (Hahn and Shiu, 1983; Urano *et al.*, 1988). Lonidamine is also more cytotoxic at low pH (Kim *et al.*, 1984). This is particularly true for acutely lowered pH, while chronic acidity seems to result in adaption and less increase in the drug-sensitizing effect (Hahn and Shiu, 1986). The mechanisms underlying the pH enhancement of thermochemotherapy seem complex. Transport of drugs across the cell membrane by active or passive routes may be changed (Mikkelsen and Wallach, 1982; Hahn, 1982). Alterations in pH may influence enzymes involved in drug degradation and formation of reactive intermediates (probably important for BCNU, where the degradation of the short-lived, reactive chloroethyldiazonium ion is base catalysed). Increased cell killing may also be simply the result of the inability of cells to survive the simultaneous stress of both hyperthermia, increased alkylation by increased reaction rates at elevated temperatures (as for BCNU, cyclophosphamide and cisplatin) and the rapid adaptation to increased proton concentrations (Hahn and Shiu, 1986). Amiloride analogues which influence the sodium ion/proton exchange system in plasma membrane, thereby regulating intracellular pH, potentiate the effect of hyperthermia (Miyakoshi *et al.*, 1987).

Tumours *in vivo* generally are thought to be at lower pH than the surrounding tissues, partly the result of a marginal vascular supply and poor microcirculation (for reviews see Wike–Hooley *et al.*, 1984; Reinhold and Endrich, 1986; Chapter 4). This opens the possibility of acutely and preferentially reducing tumour pH by using vasoactive drugs or hypertonic glucose.

Vasoactive drugs may influence thermal sensitivity and thermal tolerance (Molla *et al.*, 1987). Hydralazine is a peripheral vasodilator which is able to reduce tumour blood flow and increase the degree of tumour hypoxia. When combined with hyperthermia (41.5–43.5 °C), high-dose hydralazine (5 mg/kg) significantly enhanced the effect of hyperthermia (Horsman *et al.*, 1989). This drug is currently used as an antihypertensive agent and may have clinical potential together with conventional chemotherapeutics combined with hyperthermia, if sufficiently high doses can be administered to the patients.

Glucose has repeatedly been shown to decrease tumor pH *in vivo* and has been advocated as a means of increasing hyperthermia effects *in vivo*. Hyperglycaemia itself may in some experimental tumours reduce clonogenicity of cells. Hyperglycaemic enhancement of thermo-chemotherapy with cyclophosphamide was shown *in vivo* by Urano and Kim (1983) and with BCNU in parallel *in vitro* and *in vivo* experiments by Schem *et al.* (1989). The effect of the reduced pH is, however, difficult to estimate, as hyperglycaemia may exert a number of other effects *in vivo*, such as reduced tumour perfusion through microcirculatory effects, reduced cardiac output, redistribution of blood flow, and osmotic haemo-concentration if glucose is given intraperitoneally (Vaupel and Okunieff,

1988). As reduced pH following hypertonic glucose administration is to a great extent due to reduced tumour blood flow (Calderwood and Dickson, 1980), reduced rates of cytotoxic drug uptake may result which could counteract the potential benefit of low pH. To our knowledge, there are no clinical data available addressing these concepts.

5.7. *Preheating and thermochemotherapy*

In the clinic, hyperthermia combined with chemotherapy probably has to be given repeatedly to achieve tumour control. The previously given treatment could interfere with the interaction between drug and heat. Previously applied hyperthermia may alter the physiological environment, which may cause both altered temperature distributions and possibly drug penetration in different parts of the tumour at the subsequent heat and drug treatment. In addition, cellular properties, such as the thermal tolerance stage, cell cycle distribution and cellular membrane transport of cytotoxic agents may be changed (Hahn and Strande, 1976).

Increased normothermic BCNU cytotoxicity in CHO cells was demonstrated by preheating at 43°C, an interactive effect that was lost after 24 h (Hahn, 1982). The period of interaction between BCNU and heat after preheating was reduced when the temperature during BCNU exposure was elevated to 41 and 43°C. In EMT6 cells made thermotolerant by heating for 3 h at 40°C, normothermic cytotoxicity of BCNU, bleomycin or doxorubicin was not influenced (Morgan *et al.*, 1979). However, preheating reduced the effect of thermochemotherapy at 43°C for 1 h using bleomycin and BCNU, but not doxorubicin. Preheating at a higher temperature level (43°C) sensitized the cells to subsequent normothermic BCNU and bleomycin cytotoxicity, but protected against the toxicity of doxorubicin. Thermotolerance induced at a high temperature (45°C) resulted in reduced thermochemotherapy sensitivity for bleomycin at 43°C, while cisplatin thermochemosensitivity was retained (Neilan *et al.*, 1986). High-temperature heating, followed by lower-temperature hyperthermia (step-down heating) also increases thermochemosensitivity (Herman *et al.*, 1984). These *in vitro* data show that preheating may influence both normothermic cytotoxicity and thermochemotherapy sensitivity of drugs, depending both on the preheating temperature and duration and also on the cytotoxic drug utilized. In culture, most workers have found simultaneous exposure to hyperthermia and drugs to be the most effective, as clearly shown for BCNU (Hahn, 1979) and cisplatin (Wallner *et al.*, 1986).

The optimal timing of a single treatment with hyperthermia and drugs *in vivo* is also to apply both modalities simultaneously (Dahl and Mella,

1983; Mella and Dahl, 1985). An interactive effect between heat and drug (cyclophosphamide, BCNU and cisplatin) was seen especially with the two modalities given up to 2 h apart, while there was no obvious decrease in chemothermic drug sensitivity when thermotolerance can be demonstrated up to 24 hours after hyperthermia.

Although fractionated studies have been performed *in vivo*, they have usually not been designed to evaluate both thermotolerance kinetics and normothermic drug sensitivity or thermochemotherapy sensitivity. In the BT_4A rat glioma *in vivo*, we found no clear relationship between thermotolerance and normothermic sensitivity to BCNU during normothermia (unpublished data). Thermochemosensitivity to the drug was reduced parallel with the induction of thermotolerance. For the thermochemotherapy effect, the thermotolerant state of the tumour was obviously more important than normothermic drug sensitivity at a given time after preheating.

The relationship between preheating and subsequent thermochemotherapy sensitivity and the mechanisms are not well understood at the present time. However, there is a need for empirical studies in animal models where the relationship (although no causal relation can be concluded) between thermotolerance and thermochemotherapy with different drugs can be explored. Such data are needed to help design rational clinical studies of fractionated thermochemotherapy.

5.8. Clinical experience

5.8.1. Whole-body hyperthermia

Heating the whole body to about 42°C for several hours is discussed in Chapter 8. The rationale for this approach is that most malignancies are currently considered as systemic diseases. Hyperthermia may preferentially destroy malignant cells, although this is not the generally accepted view (see Chapter 1).

Bull (1984a, b) gave BCNU, in escalating doses from 75 to 250 mg/m², combined with whole-body hyperthermia to 12 patients with malignant melanomas and observed two complete responses and one partial response. The two complete responses lasted 6 and 16 months after stopping treatment respectively. Engelhardt *et al.* (1982) treated 15 patients with small cell lung carcinoma with a standard chemotherapy regimen of doxorubicin, cyclophosphamide and vincristine combined with 40–41°C for 1 h in three out of six chemotherapy cycles. They observed eight complete and five partial responses, an overall response rate of 87%. In a later phase III trial (Engelhardt, 1987), doxorubicin and vincristine seemed to be enhanced by hyperthermia: the response rate was 77% in the hyperthermia arm ($n=17$) compared to 50% in the normothermia arm ($n=15$). The 1-year survival rate was also increased from 0 to 29%. Again, bone marrow suppression seemed to be enhanced by hyperthermia. But one should bear in mind that long-term survival of 20–30% generally is obtained by chemotherapy alone in small cell lung cancers.

An objective response of 36% (two CRs and two PRs out of 11 patients) was observed when doxorubicin (45 mg/m²) was administered at the beginning of whole-body hyperthermia (41.8–43.0°C for 2 h) followed by cyclophosphamide (1000 mg/m²) after 6 h for soft-tissue sarcomas (Gerad *et al.*, 1984). However, similar results (28–41%) can be observed by giving doxorubicin alone in adequate doses. In a large Japanese series including different advanced malignancies (mostly lung tumours), cisplatin (28.6% of 63 patients), doxorubicin (21.9% of 32 patients) and mitomycin C (21.7% of 23 patients) showed antitumour effects when combined with systemic heating (Maeta *et al.*, 1987). In gastrointestinal cancers resistant to combination chemotherapy, whole-body hyperthermia (41.5°C for 3–5 h) was combined with the same drugs and three objective responses were observed among 17 patients, but no effect on survival was seen (Koga *et al.*, 1985). A difficulty with this study is that the authors also increased the dose of drugs or added cyclophosphamide to some patients, when combined with hyperthermia.

It is also of concern that hyperthermia seems to increase drug-specific side-effects: cardiac arrhythmias with doxorubicin (Kim *et al.*, 1979), kidney damage with cisplatin (Gerad *et al.*, 1983; Engelhardt, 1987), bone marrow suppression (Bull *et al.*, 1979; Engelhardt, 1987), recall skin damage with bleomycin (Kukla and McGuire, 1982) and possibly pulmonary damage with mitomycin C (Bull, 1984a, b). A therapeutic gain has not yet been established for the combination of whole-body hyperthermia and drugs.

5.8.2. Regional hyperthermia

A pilot study of regional perfusion of head and neck cancers by blood heated to 41–42°C combined with alkylating agents (nitrogen mustard, cyclophosphamide, thio-TEPA, A139) was reported in 1960 (Woodhall *et al.*, 1960). Two of 20 patients demonstrated a temporary complete regression.

Administration of cytotoxic drugs by hyperthermic regional perfusion has been widely used, but its role is still not unequivocally proved in clinical trials. The rationale for this combination is the following: (1) hyperthermia alone has an antitumour effect; (2) by regional perfusion, 5–10 times higher drug concentration can be achieved compared with systemic administration (Hafström and Jönsson, 1980; Walther *et al.*, 1984); (3) hyperthermia increases the effect of the cytotoxic drug *in vitro* and in animal experiments.

5.8.2(a) Malignant melanoma of limbs

Isolated extracorporal perfusion of heated blood (40–41 °C for 1–2 h) has been combined with melphalan (0.6–1.5 mg/kg for upper limbs, 0.45–1.4 mg/kg for lower limbs) in primary advanced or recurrent malignant melanomas (Stehlin *et al.*, 1975; Hafström and Jönsson, 1980; Krementz, 1983; Rege *et al.*, 1983; Vaglini *et al.*, 1985; Ghussen *et al.*, 1988). Although promising survival data have been obtained (Stehlin *et al.*, 1979), and the method is employed routinely by some centres (Di Filippo *et al.*, 1988), it should still be considered experimental as it is not yet clear whether hyperthermia added to perfusion of cytotoxic drugs improves survival (Rege *et al.*, 1983).

Data on objective tumour regressions are rare. Based on two small series of combined hyperthermia and melphalan treatments on 10 and 32 patients, a response was seen in eight out of every 10 patients (Hafström and Jönsson, 1980; Vaglini *et al.*, 1985). Lejeune *et al.* (1983) reported 65% complete and 26% partial responses in 23 patients with malignant melanoma after regional perfusion (skin temperature about 39 °C) with melphalan. Stehlin *et al.* (1975) also stated that hyperthermia added to perfusion with melphalan increased the objective response from 35 to 80%, but the actual numbers on which this statement was based were not given. In order to prove the role of thermochemotherapy in malignant melanoma, Ghussen *et al.* (1988) administered hyperthermic melphalan perfusion after radical surgery for melanomas of the extremities in a randomized trial. They were able to demonstrate a significant difference in both disease-free and overall survival. This does not, however, show whether the difference was caused by hyperthermia *per se*, high drug dose in the limbs or a positive interaction of the two modalities.

5.8.2(b) Limb sarcomas

There are optimistic reports of perfusion of limbs for different soft tissues and bone sarcomas. Better local tumour control, permitting limb-saving procedures, has been claimed (Stehlin *et al.*, 1977; Di Filippo *et al.*, 1988) but no data on objective regressions have been presented. For sarcomas, Stehlin *et al.* (1985) added adjuvant radiotherapy. The first patients of Di Filippo *et al.* (1988) probably also had adjuvant systemic chemotherapy (Santori *et al.*, 1980), and these workers also administered adjuvant perfusion with doxorubicin as well as radiotherapy. Thus, better survival does not necessarily prove the efficacy of thermochemotherapy in these studies. In a small series of 15 different limb sarcomas given hyperthermic perfusion (39–40 °C for 1 h with melphalan with or without actinomycin D), recurrence was seen in four of eight recurrent sarcomas, but none in six primary sarcomas after 30 months follow-up (Lejeune *et al.*, 1988).

Temporary or permanent nerve injuries have been observed in five (0.7%) patients following isolated hyperthermic perfusions with cytostatic agents (for references see Busse *et al.*. 1983). However, all 16 patients treated by hyperthermic lower-limb perfusions with cisplatin, experienced nerve damage, particularly of the lower leg and foot. There was no significant association with drug dose, but a significant correlation with temperature of the perfusate (range, 39–41 °C).

5.8.2(c) Abdominal tumours

During hyperthermia of the liver, large parts of the abdomen also will be heated, resulting in regional therapy. Twelve patients with metastases, mostly from colorectal cancers, were treated with hyperthermic liver perfusion (39–40 °C) for 1 h combined with fluorouracil. A median survival of 15 months was observed in contrast to 8 months for 12 patients given only fluorouracil (Walther *et al.*, 1984). A pilot study using regional liver heating (40.8–41.5 °C for 60 min by Magnetrode) for 5 days combined with dimethyl-friazene-imidazolcarboxamide or dacarbazine (DTIC), suggested that the combination of intra-arterial DTIC had a higher response rate than intravenous DTIC (Storm *et al.*, 1982).

Recently, a feasibility study using various chemotherapeutic agents in advanced abdominal and pelvic tumours has been published (Issels *et al.*, 1988). The heating pattern was reasonably good in pelvic tumours, but it was more difficult to heat tumours to more than 42.5 °C at other sites in the abdomen (see Chapter 6). Pain was the major limiting factor. Such therapy has to be improved technically before firm conclusions on the interaction of hyperthermia and drugs in this clinical setting can be reached.

5.8.3. Local hyperthermia

When local hyperthermia is combined with systemic drug therapy a selective targeting of the cytostatic effect may occur through selective enhancement of drug uptake and selective enhancement of the cytotoxic effect of the combined modalities. Arcangeli and colleagues (1979) adopted this approach to treat neck node metastases using local hyperthermia (42–43°C for 45 min) every other day for a total of 10 treatments. Hyperthermia was combined with doxorubicin, 5 mg injected intravenously twice a day at intervals of 6–8 h, every other day for a total dose of 100 mg, or with bleomycin, 7.5 mg injected intravenously twice a day at intervals of 6–8 h every other day, for a total of 115 mg. When given alone, doxorubicin induced one complete and two partial responses in six patients. Combined with hyperthermia the objective responses increased to three complete and five partial responses of eight patients having doxorubicin and three complete and four partial responses in seven patients having bleomycin. No acute epidermitis or mucositis were seen. Thus, a marked enhancement of objective responses was seen at this low-dose fractionated cytotoxic regimen when combined with heat.

Sixty-five patients with vaginal and vulva malignancies were randomly assigned to a chemotherapy regimen alone or combined with microwave hyperthermia, i.e. 43°C measured in the tumour tissue (Kohno *et al.*, 1984). The chemotherapeutic regimen was bleomycin 5 mg/day for seven consecutive days and mitomycin C 10 mg on the eighth day, given as a 1 h infusion. In primary cancer Kohno *et al.* (1984) obtained two partial responses in five patients in the chemotherapy group against four complete and nine partial responses among 16 patients in the thermochemotherapy group. Also, in the recurrent cancer group there was a more pronounced effect in the thermochemotherapy group above that of the chemotherapy group alone: one complete and eight partial responses among 21 patients against three partial responses among 21 patients, respectively. After the primary thermochemotherapy or chemotherapy, patients had surgery or radiotherapy. Also, the 2-year survival rate was higher in the thermochemotherapy group, 71.5% against 39.4%, respectively.

Recently, a phase I/II study in nine patients with malignant tumours which were refractory to conventional treatment was reported. Hyperthermia (42°C for 60 min) was combined with cisplatin (40–60 mg/m^2) (Green *et al.*, 1989). Four patients had short-lived partial responses of up to 3 months duration, two of these had progressive disease outside the treatment fields. We have published a similar study in different superficial tumours having cisplatin (50 mg/m^2) given immediately before hyperthermia (43°C for 60 min) (Dahl *et al.*, 1987). We observed no response in seven tumours without hyperthermia against three of nine tumours when combined with hyperthermia.

5.9. General conclusions

It is well documented that hyperthermia potentiates the action of the alkylating agents, cisplatin, nitrosoureas and the antibiotics doxorubicin and bleomycin in experimental studies. We still need further phase I/II studies to establish which drugs are most effective in different human malignancies. As radiotherapy and hyperthermia can both be applied as local modalities, patients with multiple subcutaneous tumours can be included in trials designed to have four treatment arms: controls, radiotherapy alone, hyperthermia alone, or combined radiotherapy and hyperthermia. As drugs are administered systemically, there will be no untreated tumours in these patients. A trial should therefore be designed in which patients with multiple tumours have lesions randomized to either drug alone or drug combined with local hyperthermia, to answer the biological question whether hyperthermia improves the effect of the chemotherapeutic agents. It seems at present unlikely that systemic hyperthermia will gain general acceptance, but further controlled studies using regional hyperthermic perfusions should be reassessed in randomized trials.

References

Akuta, K., Hiraoka, M., Jo, S., Ma, F., Nishimura, Y., Takahashi, M., Abe, M., Malmqvist, M., Lindbom, L.-O. and Lindblom, R. (1987) Regional hyperthermia combined with blockade of the hepatic arterial blood flow by degradable starch microspheres in pigs, *International Journal of Radiation Oncology, Biology and Physics*, **13**, 239–242.

Alberts, D. S., Peng, Y.-M., Chen, G. H.-S., Moon, T. E., Cetas, T. C. and Hoescheie, J. D. (1980) Therapeutic synergism of hyperthermia–*cis*-platinum in a mouse tumor model, *Journal of the National Cancer Institute*, **65**, 445–461.

Anderstam, B. (1988) Mechanisms for hyperthermic and chemotherapeutic sensitization. Role of glutathione and membrane lipids, Thesis, Stockholm, University of Stockholm.

Anderstam, B. and Harms–Ringdahl, M. (1988) Increased antineoplastic activity of combined hyperthermic and bleomycin treatments in an adenocarcinoma after glutathione depletion *in vivo*, *International Journal of Hyperthermia*, **4**, 279–306.

Anghileri, L. J., Escanye, M.-C., Marchal, C. and Robert, J. (1985) Calcium, calcium-ATPase and cyclic AMP changes in Ehrlich ascites cells submitted to hyperthermia, *Archiv Geschwulstforschung*, **55**, 171–175.

Arcangeli, G., Cividalli, A., Mauro, F., Nervi, C. and Pavin, G. (1979) Enhanced effectiveness of adriamycin and bleomycin combined with local hyperthermia in neck node metastases from head and neck cancers, *Tumori*, **65**, 441–486.

Arcangeli, G., Cividalli, A., Lovisolo, G., Mauro, F., Creton, G., Nervi, C. and Pavin, G. (1980) Effectiveness of local hyperthermia in association with radiotherapy or chemotherapy: Comparison of multimodality treatments on multiple neck node metastases, in Arcangeli, G. and Mauro, F. (Eds), *Proceedings of the 1st meeting of the European Group of Hyperthermia in Radiation Oncology*, Milan, Masson Italia Editori, pp. 257–265.

Arrick, B. A. and Nathan, C. F. (1984) Glutathione metabolism as a determinant of therapeutic efficacy: a review, *Cancer Research*, **44**, 4224–4232.

Ballard, B. E. (1974) Pharmacokinetics and temperature, *Journal of Pharmacological Science*, **63**, 1345–1358.

Barlogie, B., Corry, P. M. and Drewinco, B. (1980) *In vitro* thermochemotherapy of human colon cancer cells with *cis*-dichlorodiammineplatinum(II) and mitomycin C, *Cancer Research*, **40**, 1165–1168.

Bassett, J. B., Anderson, R. U. and Tacker, J. R. (1986) Use of temperature-sensitive liposomes in the selective delivery of methotrexate and *cis*-platinum analogues to murine bladder tumor, *Journal of Urology*, **135**, 612–615.

Bassett, J. B., Tacker, J. R., Anderson, R. U. and Bostwick, D. (1988) Treatment of experimental bladder cancer with hyperthermia and phase transition liposomes containing methotrexate, *Journal of Urology*, **139**, 634–636.

Bates, D. A. and Mackillop, W. J. (1986) Hyperthermia, adriamycin transport, and cytotoxicity in drug-sensitive and -resistant Chinese hamster ovary cells, *Cancer Research*, **46**, 5477–5481.

Ben–Hur, E. and Elkind, M. M. (1974) Thermal sensitization of Chinese hamster cells to methyl methanesulfonate: relation of DNA damage and repair to survival response, *Cancer, Biochemistry, Biophysics*, **1**, 23–32.

Bichler, K. H., Flüchter, S. H., Erdmann, W. D., Strohmaier, W. L. and Steinmann, J. (1987) Combined treatment of advanced bladder cancer, *Recent Results in Cancer Research*, **107**, 222–225.

Block, J. B., Harris, P. A. and Peale, A. (1975) Preliminary observations on temperature-enhanced drug uptake by leukemic leukocytes *in vitro*, *Cancer Chemotherapy Reports, Part 1*, **59**, 985–988.

Braun, J. and Hahn, G. M. (1975) Enhanced cell killing by bleomycin and 43°C hyperthermia and the inhibition of recovery from potentially lethal damage, *Cancer Research*, **35**, 2921–2927.

Bronk, B. V., Wilkins, R. J. and Regan, J. D. (1973) Thermal enhancement of DNA damage by an alkylating agent in human cells, *Biochemical and Biophysical Research Communications*, **52**, 1064–1071.

Bull, J. M. (1984a) An update on the anticancer effects of a combination of chemotherapy and hyperthermia, *Cancer Research*, **44** (Suppl.), 4853s–4856s.

Bull, J. M. C. (1984b) A review of systemic hyperthermia, *Frontiers of Radiation Therapy and Oncology 1984*, Basel, Karger, pp. 171–176.

Bull, J. M., Lees, D., Schuette, W., Whang–Peng, J., Smith, R., Atkinson, E. R., Gottdiener, J. S., Gralnick, H. R., Shawker, T. H. and Devita Jr, V. (1979) Whole body hyperthermia: a phase-I trial of a potential adjuvant to chemotherapy, *Annals of Internal Medicine*, **90**, 317–323.

Busse, O., Aigner, K. and Willimzig, H. (1983) Peripheral nerve damage following isolated extremity perfusion with *cis*-platinum, *Recent Results in Cancer Research*, **86**, 264–267.

Calderwood, S. K. and Dickson, J. A. (1980) Effect of hyperglycemia on blood flow, pH, and response to hyperthermia (42°C) of the Yoshida sarcoma in the rat, *Cancer Research*, **40**, 4728–33.

Caputo, A. and Silvestrini, B. (1984) Lonidamine, a new approach to cancer therapy, *Oncology*, **41** (Suppl. 1), 2–6.

Cavaliere, R., Di Filippo, F., Varanese, A., Carlini, S., Calabro, A., Aloe, L. and Piarulli, L. (1984) Lonidamine and hyperthermia: clinical experience in melanoma, preliminary results, *Oncology*, **41** (Suppl. 1), 116–120.

Chapman, I. V., Leyko, W., Gwozdzinski, K., Koter, M., Grzelinska, A. and Bartosz, G. (1983) Hyperthermic modification of bleomycin–DNA interaction detected by electron spin resonance, *Radiation Research*, **96**, 518–522.

Clawson, R. E., Egorin, M. J., Fox, B. M., Ross, L. A. and Bachur, N. R. (1981) Hyperthermic modification of cyclophosphamide metabolism in rat hepatic microsomes and liver slices, *Life Sciences*, **28**, 1133–1137.

Cohen, J. D. and Robins, H. I. (1987) Hyperthermic enhancement of *cis*-dammine-1, 1-cyclubutane dicarboxylate platinum(II) cytotoxicity in human leukemia cells *in vitro*, *Cancer Research*, **47**, 4335–4337.

Dahl, O. (1982) Interaction of hyperthermia and doxorubicin on a malignant, neurogenic rat cell line (BT₄C) in culture, *National Cancer Institute Monographs*, **61**, 251–253.

Dahl, O. (1983) Hyperthermic potentiation of doxorubicin and 4′-epi-doxorubicin in a transplantable neurogenic rat tumor (BT₄A) in BD IX rats, *International Journal of Radiation Oncology, Biology and Physics*, **9**, 203–207.

Dahl, O. (1986) Hyperthermia and drugs, in Wathmough, D. J. and Ross, W. M. (Eds), *Hyperthermia 1986*, Glasgow, Blackie, pp. 121–153.

Dahl, O. (1988) Interaction of hyperthermia and chemotherapy, *Recent Results in Cancer Research*, **107**, 157–169.

Dahl, O. and Mella, O. (1982) Enhanced effect of combined hyperthermia and chemotherapy (bleomycin, BCNU) in a neurogenic rat tumour (BT₄A) *in vivo*, *Anticancer Research*, **2**, 359–364.

Dahl, O. and Mella, O. (1983) Effect of timing and sequence of hyperthermia and cyclophosphamide on a neurogenic rat tumour (BT₄A) *in vivo*, *Cancer*, **52**, 983–987.

Dahl, O., Mella, O., Mehus, A. and Liavaag, P. G. (1987) Clinical hyperthermia combined with drugs and radiation. A phase I/II study, *Strahlentherapie und Onkologie*, **163**, 446–448.

Daugherty, J. P., Simpson Jr, T. A. and Mullins Jr, D. W. (1988) Effect of hyperthermia and doxorubicin on nucleoid sedimentation and poly(ADP-ribose) polymerase activity in L1210 cells, *Cancer, Chemotherapy and Pharmacology*, **21**, 229–232.

Di Filippo, F., Calabro, A. M., Cavallari, A., Carlini, S., Buttini, G. L., Moscarelli, F., Cavaliere, F., Piarulli, L. and Cavaliere, R. (1988) The role of hyperthermic perfusion as a first step in the treatment of soft tissue sarcoma of the extremities, *World Journal of Surgery*, **12**, 332–339.

Djordjevic, O., Kostic, L. and Brkic, G. (1978) The combined effects of hyperthermia and a chemotherapeutic agent on DNA in isolated mammalian cells, in Streffer, C. (Ed.), *Cancer Therapy by Hyperthermia and Radiation 1978*, Munich, Urban and Schwarzenberg, pp. 278–280.

Dodion, P., Riggs, C. E., Akman, S. R. and Bachur, N. R. (1986) Effect of hyperthermia on the *in vitro* metabolism of doxorubicin, *Cancer Treatment Reports*, 70, 625–629.

Donaldson, S. S., Gordon, L. F. and Hahn, G. M. (1978) Protective effect of hyperthermia against the cytotoxicity of actinomycin D on Chinese hamster cells, *Cancer Treatment Reports*, 62, 1489–1495.

Douple, E. B., Strohbehn, J. W., de Sieyes, D. C., Alborough, D. P. and Trembley, B. S. (1982) Therapeutic potentiation of *cis*-dichlorodiammineplatinum(II) and radiation by interstitial microwave hyperthermia in a mouse tumor, *National Cancer Institute Monographs*, 61, 259–262.

Dusre, L., Mimnaugh, E. G., Myers, C. E. and Sinha, B. K. (1989) Potentiation of doxorubicin cytotoxicity by buthionine sulfoximine in multidrug-resistant human breast tumor cells, *Cancer Research*, 49, 511–515.

Eksborg, S. and Ehrsson, H. (1985) Drug level monitoring: cytostatics, *Journal of Chromatography*, 340, 31–72.

Engelhardt, R. (1987) Hyperthermia and drugs, *Recent Results in Cancer Research*, 104, 136–203.

Engelhardt, R., Neumann, H., von der Tann, M. and Löhr, G. W. (1982) Preliminary results in the treatment of oat cell carcinoma of the lung by combined application of chemotherapy (CT) and whole body hyperthermia, in Gauthérie, M. and Albert, E. (Eds), *Biomedical Thermology 1982*, New York, Alan R. Liss, pp. 761–765.

Fisher, G. A. and Hahn, G. M. (1982) Enhancement of *cis*-platinum(II) diamminedichloride cytotoxicity by hyperthermia, *National Cancer Institute Monographs*, 61, 255–257.

Fujiwara, K., Kohno, I., Miyao, J. and Sekiba, K. (1984) The effect of heat on cell proliferation and the uptake of anti-cancer drugs into tumour, in Overgaard, J. (Ed.), *Hyperthermic Oncology 1984*, Vol. 1, Taylor and Francis, pp. 405–408.

Gerad, H., Egorin, M. J., Whitacre, M., van Echo, D. A. and Aisner, J. (1983) Renal failure and platinum pharmacokinetics in three patients treated with *cis*-diamminedichloroplatinum(II) and whole-body hyperthermia, *Cancer Chemotherapy and Pharmacology*, 11, 162–166.

Gerad, H., van Echo, D. A., Whitacre, M., Ashman, M., Helrich, M., Foy, J., Ostrow, S., Wiernik, P. H. and Aisner, J. (1984) Doxorubicin, cyclophosphamide, and whole body hyperthermia for treatment of advanced soft tissue sarcoma, *Cancer*, 53, 2585–2591.

Gervasi, P. G., Agrillo, M. R., Citti, L., Danesi, R. and del Tacca, M. (1986) Superoxide anion production by adriamycinol from cardiac sarcosomes and by mitochondrial NADH dehydrogenase, *Anticancer Research*, 6, 1231–1236.

Gerweck, L. E. (1977) Modification of cell lethality at elevated temperatures: the pH effect, *Radiation Research*, 70, 224–235.

Giovanella, B. C., Lohman, W. A. and Heidelberger, C. (1970) Effects of elevated temperatures and drugs on the viability of L1210 leukemia cells, *Cancer Research*, 30, 1623–1631.

Ghussen, F., Krüger, I., Groth, W. and Stutzer, H. (1988) The role of regional hyperthermic cytostatic perfusion in the treatment of extremity melanoma, *Cancer*, 61, 654–659.

Goss, P. and Parsons, P. G. (1977) The effect of hyperthermia and melphalan on survival of human fibroblast strains and melanoma cell lines, *Cancer Research*, 37, 152–156.

Green, D. M., Burton, G. V., Cox, E. B., Hanson, D., Moore, J. and Oleson, J. R. (1989) A phase I/II study of combined cisplatin and hyperthermia treatment for refractory malignancy, *International Journal of Hyperthermia*, 5, 13–21.

Groveman, D. S., Borden, E. C., Merrit, J. A., Robins, I. H., Sevens, R. and Bryan, G. T. (1984) Augmented antiproliferative effects of interferons at elevated temperatures against human bladder carcinoma cell lines, *Cancer Research*, 44, 5517–5521.

Hafström, L. and Jönsson, P.-E. (1980) Hyperthermic perfusion of recurrent malignant melanoma on the extremities, *Acta Chirurgica Scandinavica*, 146, 313–318.

Hahn, G. M. (1978) Interactions of drugs and hyperthermia *in vitro* and *in vivo*, in Streffer, C. (Ed.), *Cancer Therapy by Hyperthermia and Radiation 1978*, Munich, Urban and Schwarzenberg, pp. 72–79.

Hahn, G. M. (1979) Potential for therapy of drugs and hyperthermia, *Cancer Research*, **39**, 2264–2268.

Hahn, G. M. (1982) *Hyperthermia and Cancer 1982*, New York, Plenum Press, pp. 1–285.

Hahn, G. M. and Shiu, E. C. (1983) Effect of pH and elevated temperatures on the cytotoxicity of some chemotherapeutic agents on Chinese Hamster cells *in vitro*, *Cancer Research*, **43**, 5789–5791.

Hahn, G. M. and Shiu, E. C. (1986) Adaption to low pH modifies thermal and thermo-chemical responses of mammalian cells, *International Journal of Hyperthermia*, **2**, 379–387.

Hahn, G. M. and Strande, D. P. (1976) Cytotoxic effects of hyperthermia and adriamycin on Chinese hamster cells, *Journal of the National Cancer Institute*, **57**, 1063–1067.

Har–Kedar, I. (1975) Effect of hyperthermia and chemotherapy in EMT6 cells, in Wizenberg, M. J. and Robinson, J. E. (Eds), *Proceedings International Symposium on Cancer Therapy by Hyperthermia and Radiation 1975*, Washington, DC, pp. 91–93.

Hassanzadeh, M. and Chapman, I. V. (1982) Thermal enhancement of bleomycin-induced growth delay in a squamous carcinoma of CBA/Ht mouse, *European Journal of Clinical Oncology*, **18**, 795–797.

Hazan, G., Ben–Hur, E. and Yerushalmi, A. (1981) Synergism between hyperthermia and cyclophosphamide *in vivo*: the effect of dose fractionation, *European Journal of Cancer*, **6**, 681–684.

Hecquet, B., Meynadler, J., Bonneterre, J., Adenis, L. and Demaille, A. (1985) Time dependency in plasmatic protein binding of cisplatin, *Cancer Treatment Reports*, **69**, 79–83.

Herman, T. S. (1983a) Effect of temperature on the cytotoxicity of vindesine, amsacrine, and mitoxantrone, *Cancer Treatment Reports*, **67**, 1019–1022.

Herman, T. S. (1983b) Temperature dependence of adriamycin, cis-diamminedichloroplatinum, bleomycin, and 1,3-bis(2-chloroethyl)-1-nitrosourea cytotoxicity *in vitro*, *Cancer Research*, **43**, 365–369.

Herman, T. S. and Teicher, B. A. (1988) Sequencing of trimodality therapy (cis-diamminedichloroplatinum(II)/hyperthermia/radiation) as determined by tumor growth delay and tumor cell survival in the FSaIIc fibrosarcoma, *Cancer Research*, **48**, 2693–2697.

Herman, T. S., Cress, A. E., Sweets, C. and Gerner, E. W. (1981) Reversal of resistance to methotrexate by hyperthermia in Chinese hamster ovary cells, *Cancer Research*, **41**, 3840–3843.

Herman, T. S., Henle, K. J., Nagle, W. A., Moss, A. J. and Monson, T. P. (1984) Effect of step-down heating on the cytotoxicity of adriamycin, bleomycin, and cis-diamminedichloroplatinum, *Cancer Research*, **44**, 1823–1826.

Herman, T. S., Teicher, B. A., Cathcart, K. N. S., Kaufmann, M. E., Lee, J. B. and Lee, M.-H. (1988a) Effect of hyperthermia on cis-diamminedichloroplatinum(II) (Rhodamine 123)2(tetrachloroplatinum(II)) in a human squamous cell carcinoma line and a cis-diamminedichloroplatinum(II)-resistant subline, *Cancer Research*, **48**, 5101–5105.

Herman, T. S., Teicher, B. A., Jochelson, M., Clark, J., Svensson, G. and Coleman, C. N. (1988b) Rationale for use of local hyperthermia with radiation therapy and selected anticancer drugs in locally advanced human malignancies, *International Journal of Hyperthermia*, **4**, 143–158.

Hiramoto, R. N., Gantha, V. K. and Lilly, M. B. (1984) Reduction of tumor burden in a murine osteosarcoma following hyperthermia combined with cyclophosphamide, *Cancer Research*, **44**, 1405–1408.

Honess, D. J. and Bleehen, N. M. (1980) Effects of the combination of hyperthermia and cytotoxic drugs on the skin of the mouse foot, in Arcangeli, G. and Mauro, F. (Eds), *Proceedings of the First Meeting of the European Group of Hyperthermia in Radiation Oncology 1982*, Milan, Masson Italia Editori, pp. 151–155.

Honess, D. J. and Bleehen, N. M. (1982) Sensitivity of normal mouse marrow and RIF-1 tumour to hyperthermia combined with cyclophosphamide or BCNU: a lack of therapeutic gain, *British Journal of Cancer*, **46**, 236–248.

Honess, D. J. and Bleehen, N. M. (1985a) Thermochemotherapy with cisplatinum CCNU, BCNU, chlorambucil and melphalan on murine marrow and two tumours: therapeutic gain for melphalan only, *British Journal of Radiology*, **58**, 63–72.

Honess, D. J. and Bleehen, N. M. (1985b) Potentiation of melphalan by systemic hyperthermia in mice: therapeutic gain for mouse lung microtumours, *International Journal of Hyperthermia*, **1**, 57–68.

Honess, D. J., Donaldson, J., Workman, P. and Bleehen, N. M. (1985) The effect of systemic hyperthermia in melphalan pharmacokinetics in mice, *British Journal of Cancer*, **51**, 77–84.

Horsman, M. R., Christensen, K. L. and Overgaard, J. (1989) Hydralazine-induced enhancement of hyperthermic damage in a C3H mammary carcinoma *in vivo*, *International Journal of Hyperthermia*, **5**, 123–136.

Hromas, R. A., Andrews, P. A., Murphy, M. P. and Burns, P. C. (1987) Glutathione depletion reverses cisplatin resistance in murine L1210 leukemia cells, *Cancer Letters*, **34**, 9–13.

Issels, R. D., Wadepohl, M., Tiling, K., Müller, M., Sauer, H. and Wilmanns, W. (1988) Regional hyperthermia combined with systemic chemotherapy in advanced abdominal and pelvic tumors: First results of a pilot study employing an annular phased array applicator, *Recent Results in Cancer Research*, **107**, 236–243.

Johnson, H. A. and Pavelec, M. (1973) Thermal enhancement of thio-TEPA cytotoxicity, *Journal of the National Cancer Institute*, **50**, 903–908.

Joiner, M. C., Steel, G. C. and Stephens, I. C. (1982) Response of two mouse tumours to hyperthermia with CCNU or melphalan, *British Journal of Cancer*, **45**, 17–26.

Kano, E., Furukawa—Furuya, M., Kajimoto—Kinoshita, K. N. T., Picha, P., Sugimoto, K., Ohtsubo, T., Tsuji, K., Tsubouchi, S. and Kondo, T. (1988) Sensitivities of bleomycin-resistant variant cells enhanced by 40°C hyperthermia *in vitro*, *International Journal of Hyperthermia*, **4**, 547–553.

Kim, J. H., Kim, S. H., Alfieri, A., Young, C. W. and Silvestrini, B. (1984) Lonidamine: a hyperthermic sensitizer of HeLa cells in culture and of the Meth-A tumor *in vivo*, *Oncology*, **41**, 30–35.

Kim, Y. D., Lees, D. E., Lake, C. R., Whang—Peng, J., Schuette, W., Smith, R. and Bull, J. (1979) Hyperthermia potentiates doxorubicin-related cardiotoxic effects, *Journal of the American Medical Association*, **241**, 1816–1817.

Klein, M. E., Kowal, C. D., Kamen, B. A., Frayer, K. A. and Berting, J. K. (1979) Alteration in methotrexate (MTX) transport by hyperthermia, *Clinical Research*, **27**, 491A.

Koga, S., Maeta, M., Shimizu, N., Osaki, Y., Hamazoe, R., Oda, M., Karino, T. and Yamane, T. (1985) Clinical effects of total-body hyperthermia combined with anticancer chemotherapy for far-advanced gastrointestinal cancer, *Cancer*, **55**, 1641–1647.

Kohno, I., Kaneshige, E., Fujiwara, K. and Sekiba, K. (1984) Thermochemotherapy (TC) for gynecologic malignancies, in Overgaard, J. (Ed.), *Hyperthermic Oncology 1984*, Vol. 1, London, Taylor and Francis, pp. 753–756.

Konno, H., Suzuki, H., Tadakuma, T., Kumai, K., Yasuda, T., Kubota, T., Ohta, S., Nagaike, K., Hosokawa, S., Ishibiki, K., Osahiko, A. and Saito, K. (1987) Antitumor effect of adriamycin entrapped in liposomes conjugated with anti-human α-fetoprotein monoclonal antibody, *Cancer Research*, **47**, 4471–4477.

Krementz, E. T. (1983) Chemotherapy by isolated regional perfusion for melanoma of the limbs, *Recent Results in Cancer Research*, **86**, 193–203.

Kubota, Y., Nishimura, R., Takai, S. and Umeda, M. (1979) Effect of hyperthermia on DNA single-strand breaks induced by bleomycin in HeLa cells, *Gann*, **70**, 681–685.

Kukla, L. and McGuire, W. P. (1982) Heat-induced recall of bleomycin skin changes, *Cancer*, **50**, 2283–2284.

Kuroki, M., Tanaka, R. and Hondo, H. (1987) Antitumour effect of interferon combined with hyperthermia against experimental brain tumour, *International Journal of Hyperthermia*, **3**, 527–534.

Lejeune, F. J., Deloof, T., Ewalenko, P., Fruhling, J., Jabri, M., Mathieu, M., Nogaret, J.-M. and Verhest, A. (1983) Objective regression of unexcised melanoma in-transit metastases after hyperthermic isolation perfusion of the limbs with melphalan, *Recent Results in Cancer Research*, **86**, 268–276.

Lejeune, F. J., Lienard, D. and Ewalenko, P. (1988) Hyperthermic isolation perfusion of the limbs with cytostatics after surgical excision of sarcomas, *World Journal of Surgery*, **12**, 345–348.

Lilly, B. M., Ng, T. C., Evanochko, W. T., Katholi, C. R., Kumar, N. G., Elgavish, G. A., Durant, J. R., Hiramoto, R., Ghanta, V. and Glickson, J. D. (1984) Loss of high-energy phosphate following hyperthermia demonstrated by *in vivo* [31] P-nuclear magnetic resonance spectroscopy, *Cancer Research*, **44**, 633–638.

Lin, P.-S., Cariani, P. A., Jones, M. and Kahn, P. C. (1983) Work in progress: the effect of heat on bleomycin cytotoxicity *in vitro* and on the accumulation of [57] Co-bleomycin in heat treated rat tumors, *Radiology*, **146**, 213–217.

Lindberg, B., Lote, K. and Teder, H. (1984) Biodegradable starch microspheres — a new medical tool, In Davis, S. S., Illum, L., McVie, J. G. and Tomlinson, E. (Eds), *Microspheres and Drug Therapy, Pharmaceutical, Immunological and Medical Aspects 1984*, Amsterdam, Elsevier, pp. 153–188.

Longo, F. W., Tomashefsky, P., Rivin, B. D. and Tannenbaum, M. (1983) Interaction of ultrasonic hyperthermia with two alkylating agents in a murine bladder tumor, *Cancer Research*, **43**, 3231–3235.

Ma, F., Hiraoka, M., Jo, S., Akuta, K., Nishimura, Y., Takahashi, M. and Abe, M. (1985) Response of mammary tumors of C3H/He mice to hyperthermia and bleomycin *in vivo*, *Radiation Medicine*, **3**, 230–233.

Maekawa, S., Sugimachi, K. and Kitamura, M. (1987) Selective treatment of metastatic lymph nodes with combination of local hyperthermia and temperature-sensitive liposomes containing bleomycin, *Cancer Treatment Reports*, **71**, 1053–1059.

Maeta, M., Koga, S., Wada, J., Yokoyama, M., Kato, N., Kawahara, H., Sakai, T., Hino, M., Ono, T. and Yuasa, K. (1987) Clinical evaluation of total-body hyperthermia combined with anticancer chemotherapy for far-advanced miscellaneous cancer in Japan, *Cancer*, **59**, 1101–1106.

Magin, R. L., Sisik, B. I. and Cysyk, R. L. (1979) Enhancement of bleomycin activity against Lewis lung tumors in mice by local hyperthermia, *Cancer Research*, **39**, 3792–3795.

Magin, R. L., Cysyk, R. L. and Litterst, C. L. (1980) Distribution of adriamycin in mice under conditions of local hyperthermia which improve systemic drug therapy, *Cancer Treatment Reports*, **64**, 203–210.

Marmor, J. B., Kozak, D. and Hahn, G. M. (1979) Effects of systemically administered bleomycin or adriamycin with local hyperthermia on primary tumor and lung metastases, *Cancer Treatment Reports*, **63**, 1279–1290.

Mayhew, E. G., Goldrosen, M. H., Vaage, J. and Rustum, Y. M. (1987) Effects of liposome-entrapped doxorubicin on liver metastases of mouse colon carcinomas 26 and 38, *JNCI*, **78**, 707–713.

Mella, O. (1985) Combined hyperthermia and *cis*-diamminedichloroplatinum in BD IX rats with transplanted BT$_4$A tumours, *International Journal of Hyperthermia*, **1**, 171–183.

Mella, O. and Dahl, O. (1985) Timing of combined hyperthermia and 1,3-bis-(2-chloroethyl)-1-nitrosourea or *cis*-diamminedichloroplatinum in BD IX rats with BT$_4$A tumours, *Anticancer Research*, **5**, 259–264.

Mella, O., Eriksen, R., Dahl, O. and Laerum, O. D. (1987) Acute systemic toxicity of combined *cis*-diamminedichloroplatinum and hyperthermia in the rat, *European Journal of Cancer and Clinical Oncology*, **23**, 365–373.

Meyn, R. E., Corry, P. M., Fletcher, S. E. and Demetriades, M. (1979) Thermal enhancement of DNA strand breakage in mammalian cells treated with bleomycin, *Radiation Oncology, Biology, Physics*, **5**, 1487–1489.

Meyn, R. E., Corry, P. M., Fletcher, S. E. and Demetriades, M. (1980) Thermal enhancement of DNA damage in mammalian cells treated with *cis*-diamminedichloroplatinum(II), *Cancer Research*, **40**, 1136–1139.

Mikkelsen, R. B. and Wallach, D. (1982) Transmembrane ion gradients and thermochemotherapy, *Progress in Clinical and Biological Research*, **107**, 103–107.

Mimnaugh, E. G., Waring, R. W., Sikic, B. I., Magin, L., Drew, R., Litterst, C. L., Gram, T. E. and Guarino, A. M. (1978) Effect of whole-body hyperthermia on the disposition and metabolism of adriamycin in rabbits, *Cancer Research*, **38**, 1420–1425.

Mini, E., Dombrowski, J., Moroson, B. A. and Bertino, J. R. (1986) Cytotoxic effects of hyperthermia, 5-fluorouracil and their combination on a human leukemia T-lymphoblast cell line, CCRF-CEM, *European Journal of Clinical Oncology*, **22**, 927–934.

Mircheva, J., Smith, P. J. and Bleehen, N. M. (1986) Interaction of bleomycin, hyperthermia and a calmodulin inhibitor (trifluoperazine) in mouse tumour cells: I. *In vitro* cytotoxicity, *British Journal of Cancer*, **53**, 99–103.

Miyakoshi, J., Hirata, M., Oda, W., Cragoe Jr, E. J. and Inagaki, C. (1987) Sensitization to heat by amiloride analogues in Chinese hamster cells, *Japanese Journal of Pharmacology*, **45**, 281–283.

Mizuno, S., Amagai, M. and Ishida, A. (1980) Synergistic cell killing by antitumor agents and hyperthermia in cultured cells, *Gann*, **71**, 471–478.

Molla, M. R., Yoshiga, K. and Takada, K. (1987) Influence of methoxamine, 5-hydroxytryptamine and physically induced low pH on thermal sensitivity and thermotolerance, *Neoplasma*, **34**, 397–406.

Morgan, J. E. and Bleehen, N. M. (1980) A comparison of the interaction between hyperthermia and bleomycin on the EMT6 tumour as a monolayer or spheroids *in vitro* and *in vivo*, in Arcangeli, G. and Mauro, F. (Eds), *Hyperthermia in Radiation Oncology*, Milan, Masson Italia Editori, pp. 165–170.

Morgan, J. E., Honess, D. J. and Bleehen, N. M. (1979) The interaction of thermal tolerance with drug cytotoxicity *in vitro*, *British Journal of Cancer*, **39**, 422–428.

Muckle, D. S. and Dickson, J. A. (1973) Hyperthermia (42°C) as an adjuvant to radiotherapy and chemotherapy in the treatment of the allogeneic VX2 carcinoma in the rabbit, *British Journal of Cancer*, **27**, 307–315.

Murthy, M. S., Khandekar, J. D., Travis, J. D. and Scanlon, E. F. (1984) Combined effect of hyperthermia (HT) and platinum compounds *in vivo* and *in vitro* on murine and human tumor cells, in Overgaard, J. (Ed.), *Hyperthermic Oncology 1984*, Vol. 1, London, Taylor and Francis, pp. 421–424.

Nakajima, K. and Hisazumi, H. (1987) Enhanced radioinduced cytotoxicity of cultured human bladder cells using 43°C hyperthermia or anticancer drugs, *Urological Research*, **15**, 255–260.

Neilan, B. A., Henle, K. J., Nagle, W. A. and Moss, A. J. (1986) Cytotoxicity of hyperthermia combined with bleomycin or *cis*-platinum in cultured RIF-cells: modification of thermo*tolerance* by polyhydroxy compounds, *Cancer Research*, **46**, 2245–2247.

Neumann, H. A., Fiebig, H. H., Löhr, G. W. and Engelhardt, R. (1985) Effects of cytostatic drugs and 40.5°C hyperthermia on human bone marrow progenitors (CFU-C) and human clonogenic tumor cells implanted into mice, *Journal of the National Cancer Institute*, **75**, 1059–1066.

Niitsu, Y., Watanabe, N., Umeno, H., Sone, H., Neda, H., Yamauchi, N., Maeda, M. and Urushizaki, I. (1988) Synergistic effects of recombinant human tumor necrosis factor and hyperthermia on *in vitro* cytotoxicity and artificial metastasis, *Cancer Research*, **48**, 654–657.

Ohnoshi, T., Ohnuma, T., Beranek, J. T. and Holland, J. F. (1985) Combined cytotoxicity effect of hyperthermia and anthracycline antibiotics on human tumor cells, *JNCI*, **74**, 275–281.

Osieka, R., Magin, R. L. and Atkinson, E. R. (1978) The effect of hyperthermia on human colon cancer xenografts in nude mice, in Streffer, C. (Ed.), *Cancer Therapy by Hyperthermia and Radiation 1978*, Munich, Urban and Schwarzenberg, pp. 287–290.

Ostrow, S., van Echo, D., Egorin, M., Whitacre, M., Grochow, L., Aisner, J., Colvin, M., Bachur, N. and Wiernik, P. H. (1982) Cyclophosphamide pharmacokinetics in patients receiving whole-body hyperthermia, *National Cancer Institute Monographs*, **61**, 401–403.

Overgaard, J. (1975) Effect of pH on response to hyperthermia, in Wizenberg, M. J. and Robinson, J. E. (Eds), *Proceedings of the International Symposium on Cancer Therapy by Hyperthermia and Radiation 1975*, Washington, DC, p. 43.

Overgaard, J. (1976) Combined adriamycin and hyperthermia treatment of a murine mammary carcinoma *in vivo*, *Cancer Research*, **36**, 3077–3081.

Rabbani, B., Sondhaus, C. A. and Swingle, K. F. (1978) Cellular response to hyperthermia and bleomycin: effect of time sequencing and possible mechanisms, *Proceedings of the International Symposium on Cancer Therapy by Hyperthermia and Radiation, American College of Radiology, Washington, DC, 1975*.

Rege, V. B., Leone, L. A., Soderberg Jr, C. H., Coleman, G. V., Robidoux, H. J., Fijman, R. and Brown, J. (1983) Hyperthermic adjuvant perfusion chemotherapy for stage I malignant melanoma of the extremity with literature review, *Cancer*, **52**, 2033–2039.

Reinhold, H. S. and Endrich, B. (1986) Tumour microcirculation as a target for hyperthermia, *International Journal of Hyperthermia*, **2**, 111–137.

Rice, G. C. and Hahn, G. M. (1987) Modulation of adriamycin transport by hyperthermia as measured by fluorescence-activated cell sorting, *Cancer Chemotherapy and Pharmacology*, **20**, 183–187.

Riviere, J. E., Page, R. L., Dewhirst, M. W., Tyczkowska, K. and Thrall, D. E. (1986) Effect of hyperthermia on cisplatin pharmacokinetics in normal dogs, *International Journal of Hyperthermia*, **2**, 351–358.

Robins, H. I. (1984) Role of whole-body hyperthermia in the treatment of neoplastic disease: its current status and future prospects, *Cancer Research*, **44**, 4878s–4883s.

Roizin–Towle, L., Hall, E. J. and Capuano, L. (1982) Interaction of hyperthermia and cytotoxic agents, *National Cancer Institute Monographs*, **61**, 149–151.

Rose, W. C., Veras, G. H., Laster Jr, W. R. and Schabel Jr, F. M. (1979) Evaluation of whole-body hyperthermia as an adjunct to chemotherapy in murine tumors, *Cancer Treatment Reports*, **63**, 1311–1325.

Santori, F. S., Cavaliere, R., De Chiara, N. and Di Filippo, F. (1980) Hyperthermic antiblastic perfusion in the treatment of malignant bone tumors, in Arcangeli, G. and Mauro, F. (Eds), *Hyperthermia in Radiation Oncology 1980*, Milano, Masson Italia Editori, pp. 171–177.

Schem, B. C., Mella, O. and Dahl, O. (1989) Potentiation of combined BCNU and hyperthermia by pH reduction and hypertonic glucose in the BT_4 rat glioma *in vitro* and *in vivo*, *International Journal of Hyperthermia*, **5**, 707–715.

Shiu, M. H., Cahan, A., Fogh, J. and Fortner, J. G. (1983) Sensitivity of xenografts of human pancreatic adenocarcinoma in nude mice to heat and heat combined with chemotherapy, *Cancer Research*, **43**, 4014–4018.

Shrieve, D. C., Li, G. C., Astromoff, A. and Harris, J. W. (1986) Cellular glutathione, thermal sensitivity, and thermotolerance in Chinese hamster fibroblasts and their heat-resistant variants, *Cancer Research*, **46**, 1684–1687.

Skibba, J. L., Jones, F. E. and Condon, R. E. (1982) Altered hepatic disposition of doxorubicin in the perfused rat liver at hyperthermic temperatures, *Cancer Treatment Reports*, **66**, 1357–1363.

Smith, P. J., Mircheva, J. and Bleehen, N. M. (1986) Interaction of bleomycin, hyperthermia and a calmodulin inhibitor (trifluoroperazine) in mouse tumour cells: II. DNA damage, repair and chromatin changes, *British Journal of Cancer*, **53**, 105–114.

Spiro, I. J., Denman, D. L. and Dewey, W. C. (1982) Effect of hyperthermia on CHO DNA polymerase α and β, *Radiation Research*, **89**, 134–149.

Stehlin Jr, J. S., Giovanella, B. C., de Ipolyi, P. D., Muenz, L. R. and Anderson, R. F. (1975) Results of hyperthermic perfusion for melanoma of the extremities, *Surgery, Gynecology and Obstetrics*, **140**, 339–348.

Stehlin Jr, J. S., Giovanella, B. C., de Ipolyi, P. D., Muenz, L. R., Anderson, R. F. and Gutierrez, A. A. (1977) Hyperthermic perfusion of extremities for melanoma and soft tissue sarcomas, in Rossi–Fanelli, A., Cavaliere, R., Mondovi, B. and Moricca, G. (Eds), *Selective Heat Sensitivity of Cancer Cells 1977*, Berlin, Springer–Verlag, pp. 171–185.

Stehlin Jr, J. S., Giovanella, B. C., de Ipolyi, P. D. and Anderson, R. F. (1979) Results of eleven years' experience with heated perfusion for melanoma of the extremities, *Cancer Research*, **39**, 2255–2257.

Stevenson, M. A., Calderwood, S. K. and Hahn, G. M. (1987) Effect of hyperthermia (45 °C) on calcium flux in Chinese hamster ovary HA-1 fibroblasts and its potential role in cytotoxicity and heat resistance, *Cancer Research*, **47**, 3712–3717.

Storm, F. K., Kaiser, L. R., Goodnight, J. E., Harrison, W. H., Elliott, R. S., Gomes, A. S. and Morton, D. L. (1982) Thermochemotherapy for melanoma metastases in liver, *Cancer*, **49**, 1243–1248.

Sutton, C. H. (1971) Tumor hyperthermia in the treatment of malignant gliomas of the brain, *Transactions of the American Neurological Association*, **96**, 195–199.

Suzuki, K. (1967) Application of heat to cancer chemotherapy — experimental studies, *Nagoya Journal of Medical Science*, **30**, 1–21.

Tacker, J. R. and Anderson, R. U. (1982) Delivery of antitumor drug to bladder cancer by use of phase transition liposomes and hyperthermia, *The Journal of Urology*, **127**, 1211–1214.

Teicher, B. A., Kowal, C. D., Kennedy, K. A. and Sartorelli, A. C. (1981) Enhancement by hyperthermia of the *in vitro* cytotoxicity of mitomycin C toward hypoxic tumor cells, *Cancer Research*, **41**, 1096–1099.

Toffoli, G., Bevilaqua, C., Franceschin, A. and Boiocchi, M. (1989) Effect of hyperthermia on intracellular drug accumulation and chemosensitivity in drug-sensitive and drug-resistant P388 leukemia cell lines, *International Journal of Hyperthermia*, **5**, 163–172.

Tsumura, M., Yoshiga, K. and Takada, K. (1988) Enhancement of antitumour effects of

1-hexylcarbomyl-5-fluorouracil combined with hyperthermia on Ehrlich ascites tumor *in vivo* and Nakahara–Fukuoka sarcoma cell *in vitro*, *Cancer Research*, **48**, 3977–3980.

Twentyman, P. R., Morgan, J. E. and Donaldson, J. (1978) Enhancement by hyperthermia of the effect of BCNU against the EMT6 mouse tumor, *Cancer Treatment Reports*, **62**, 439–443.

Urano, M. and Kim, M. S. (1983) Effect of hyperglycemia on thermochemotherapy of a spontaneous murine fibrosarcoma, *Cancer Research*, **43**, 3041–3044.

Urano, M., Kahn, J. and Kenton, L. A. (1988) Effect of bleomycin on murine tumor cells at elevated temperatures and two different pH values, *Cancer Research*, **48**, 615–619.

Vaglini, M., Andreola, S., Attili, A., Belli, Marolda, R., Nava, M., Prada, A., Santinami, M. and Cascinelli, N. (1985) Hyperthermic antiblastic perfusion in the treatment of cancer of the extremities, *Tumori*, **71**, 355–359.

van der Linden, Sapareto, S. A., Corbett, T. H. and Valeriote, F. A. (1984) Adriamycin and heat treatments *in vitro* and *in vivo*, in Overgaard, J. (Ed.), *Hyperthermic Oncology 1984*, Vol. 1, London, Taylor and Francis, pp. 449–452.

van der Vijgh, J. F. and Klein, I. (1986) Protein binding of five platinum compounds. Comparison of two ultrafiltration systems, *Cancer, Chemotherapy and Pharmacology*, **18**, 129–132.

Vaupel, P. W. and Okunieff, P. G. (1988) Role of hypovolemic hemoconcentration in dose-dependent flow decline observed in murine tumors after intraperitoneal administration of aglucose or mannitol, *Cancer Research*, **48**, 7102–7106.

von Szczepanski, L. and Trott, K. R. (1981) The combined effect of bleomycin and hyperthermia on the adenocarcinoma 284 of the C3H mouse, *European Journal of Cancer and Clinical Oncology*, **17**, 997–1000.

Voth, B., Sauer, H. and Wilmanns, W. (1988) Thermostability of cytostatic drugs *in vitro* and thermosensitivity of cultured human lymphoblasts against cytostatic drugs, *Recent Results in Cancer Research*, **107**, 170–176.

Wallach, D. F. H. (1977) Basic mechanisms in tumor thermotherapy, *Journal of Molecular Medicine*, **17**, 381–403.

Wallner, K. E., DeGregorio, M. W. and Li, G. C. (1986) Hyperthermic potentiation of *cis*-diamminedichloroplatinum(II) cytotoxicity in Chinese hamster ovary cells resistant to the drug, *Cancer Research*, **46**, 6242–6245.

Wallner, K. E., Banda, M. and Li, G. C. (1987) Hyperthermic enhancement of cell killing by mitomycin C in mitomycin C resistant Chinese hamster ovary cells, *Cancer Research*, **47**, 1308–1312.

Walther, H., Aigner, K. R., Helling, H. J. and Schwemmle, K. (1984) Hyperthermic isolated perfusion with cytotoxics of the liver — dog experiments and first clinical results, in Overgaard, J. (Ed.), *Hyperthermic Oncology 1984*, Vol. 1, London, Taylor and Francis, pp. 457–460.

Wang, A. L. and Tew, K. D. (1985) Increased glutathione-*S*-transferase activity in a cell line with acquired resistance to nitrogen mustards, *Cancer Treatment Reports*, **69**, 677–682.

Wang, B. S., Lumanglas, A. L., Silva, J., Ruszala–Mallon, V. M. and Durr, F. E. (1987) Effect of hyperthermia on the sensitivity of human colon carcinoma cells to mitoxantrone, *Cancer Treatment Reports*, **71**, 831–836.

Warren, R. D. and Bender, R. A. (1977) Drug interactions with antineoplastic agents, *Cancer Treatment Reports*, **61**, 1231–1241.

Warters, R. L. and Brizgys, L. M. (1988) Effect of topoisomerase II on hyperthermic cytotoxicity, *Cancer Research*, **48**, 3932–3938.

Watanabe, N., Niitsu, Y., Umeno, H., Sone, H., Neda, H., Yamauchi, N., Maeda, M. and Urushizaki, I. (1988) Synergistic cytotoxic and antitumor effects of recombinant human tumor necrosis factor and hyperthermia, *Cancer Research*, **48**, 650–653.

Weinstein, J. N., Magin, R. L., Yatvin, M. B. and Zaharko, D. S. (1979) Liposomes and local hyperthermia: selective delivery of methotrexate to heated tumors, *Science*, **204**, 188–191.

Wheatley, D. N., Kerr, C. and Gregory, D. W. (1989) Heat-induced damage to HeLa-S3 cells: correlation of viability, permeability, osmosensitivity, phase-contrast light-, scanning electron- and transmission electron-microscopical findings, *International Journal of Hyperthermia*, **5**, 145–162.

Wike–Hooley, J. L., Haveman, J. and Reinhold, R. S. (1984) The relevance of tumour pH to the treatment of malignant disease, *Radiotherapy and Oncology*, **2**, 343–366.

Wondergem, J., Bulger, R. E., Strebel, F. R., Newman, R. A., Travis, E. L., Stephens, L. C. and
Bull, J. M. C. (1988) Effect of cis-diamminedichloroplatinum(II) combined with whole
body hyperthermia on renal injury, *Cancer Research*, **48**, 440–446.
Woodhall, B., Pickrell, K. L., Georgiade, N. G., Mahaley Jr, M. S. and Dukes, H. T. (1960)
Effect of hyperthermia upon cancer chemotherapy; application to external cancers of
head and face structures, *Annals of Surgery*, **151**, 750–759.
Xu, M. J. and Alberts, D. S. (1988) Potentiation of platinum analogue cytotoxicity by
hyperthermia, *Cancer Chemotherapy and Pharmacology*, **21**, 191–196.
Yamada, K., Someya, T., Shimada, S., Ohara, K. and Kukita, A. (1984) Thermochemotherapy
for malignant melanoma: Combination therapy of ACNU and hyperthermia in mice,
Journal of Investigative Dermatology, **82**, 180–184.
Yamane, T., Koga, S., Maeta, M., Hamazoe, R., Karino, T. and Oda, M. (1984) Effects of *in
vitro* hyperthermia on concentration of adriamycin in Ehrlich ascites cells, in
Overgaard, J. (Ed.), *Hyperthermic Oncology 1984*, Vol. 1, London, Taylor and Francis,
pp. 409–476.
Yamane, T., Nakabayashi, H., Ito, S., Sumii, H. and Kawai, A. (1986) *In vitro* thermochemo-
therapy of human osteosarcoma cells with cis-dichlorodiammineplatinum(II), *Hiroshima
Journal of Medical Sciences*, **35**, 191–195.
Yatvin, M. B., Weinstein, J. N., Dennis, W. H. and Blumenthal, R. (1978) Design of liposomes
for enhanced local release of drugs by hyperthermia, *Science*, **202**, 1290–1293.
Yerushalmi, A. (1978) Combined treatment of a solid tumour by local hyperthermia and
actinomycin D, *British Journal of Cancer*, **37**, 827–832.
Yerushalmi, A., Tovey, G. M. and Gresser, I. (1982) Antitumor effect of combined interferon
and hyperthermia in mice, *Proceedings of the Society for Experimental Biology and
Medicine*, **169**, 413–415.
Zakris, E. L., Dewhirst, M. W., Riviere, J. E., Hoopes, P. J., Page, R. L. and Oleson, J. R. (1987)
Pharmacokinetics and toxicity of intraperitoneal cisplatin combined with regional
hyperthermia, *Journal of Clinical Oncology*, **5**, 1613–1620.
Zupi, G., Badaracco, G., Cavaliere, R., Natali, P. G. and Greco, C. (1984) Influence of sequence
on hyperthermia and drug combination, in Overgaard, J. (Ed.), *Hyperthermic Oncology
1984*, Vol. 1, London, Taylor and Francis, pp. 429–432.

6 Clinical hyperthermic practice: non-invasive heating

D. S. Kapp and J. L. Meyer

6.1. Introduction

The medical use of hyperthermia as an adjuvant in cancer therapy requires attention to numerous clinical and technical factors that may potentially influence the treatment outcome. Specific concerns include the treatment site and tumour accessibility, the tumour histology and size, and the composition and anatomy of the surrounding normal tissues — as well as the overall status of the patient and any pertinent health limitations or limitations in positioning during treatment that the patient may have. The details of the radiation dose and fractionation, the radiation–hyperthermia sequencing, and the hyperthermia method (including the resulting distribution of temperatures, and the duration of treatment and fractionation) are significant concerns. This chapter will review from a clinician's perspective the practical aspects of non-invasive electromagnetic (EM) and ultrasonic (US) heating of superficial and deep-seated tumours. Hyperthermia factors of prognostic importance in treatment outcome will be delineated to aid in equipment selection for patient treatment.

6.2. Superficially located tumours

Lesions less than 3–5 cm from the surface have provided the greatest experience with human tumour response to hyperthermia. Methods to induce localized hyperthermia from external EM and US wave sources have been improving rapidly, but considerable further development is needed if most tumours greater than a few centimeters in size or depth are to be treated adequately. The clinical work instituted in the mid 1970s using EM and US applicators was limited initially to the treatment of quite small (<3–4 cm diameter) neoplasms. Microwave (MW) applicators with much larger effective field diameters are currently available, and arrays of MW and of US sources are being developed that provide increasingly broader field sizes, segmental control of applicator power output for greater temperature homogeneity, improved power delivery to areas of limited access, ability to avoid sensitive adjacent normal tissues, greater conformity to curved treatment surfaces and improved patient comfort during treatment. These aspects of applicator design are discussed in Chapters 11 and 14. In the selection of the apparatus for treatment of a superficially located tumour, a variety of prognostic parameters need to be taken into account, as follows.

6.2.1. Factors influencing hyperthermic treatment approach and outcome

6.2.1(a) Tumour volume

Larger tumours are more difficult to control by irradiation than smaller ones and this is also true when irradiation is combined with hyperthermia (Table 6.1). Valdagni *et al.* (1986) treated N_3 metastatic neck nodes with radiotherapy and hyperthermia and observed complete response rates of 75% with nodes <6 cm, but only 36% with larger nodes. Van der Zee *et al.* (1986) also demonstrated an inverse correlation of tumour volume with response and showed that the calculated minimum thermal doses tended to decrease with increasing volume (while the maximum and mean temperatures increased with volume). Oleson *et al.* (1984) showed that tumour volume had a significant inverse correlation with response rate. Larger volumes were correlated with lower minimum tumour temperatures (but higher maximum tumour temperatures). Kapp *et al.* (1987) found that the average minimum tumour temperatures obtained in superficial tumours >5 cm^3 were lower than for tumours <5 cm^3, and that there was a correspondingly dramatic difference in complete response rates (25% versus 81%). Thus, the decrease in the effectiveness of hyperthermia in larger tumours may result from the greater intratumoural thermal gradients and lower minimum temperatures, as well as the greater number of tumour cells.

However, the negative effect of tumour size may be even more pronounced for irradiation alone. Dewhirst *et al.* (1984) treated 130 spontaneous large animal tumours with irradiation alone or combined with hyperthermia in a prospectively randomized trial. Large tumour volumes (>100 cm^3) responded more poorly than smaller volumes to combined therapy (33% versus 73% complete response), but far more poorly to irradiation alone (9% versus 69% complete response). A possible explanation for this comes from experimental work: larger tumours may have a relatively greater percentage of poorly perfused, hypoxic cells at low pH that could have higher sensitivity to hyperthermia or could focally heat to higher temperatures.

Table 6.1 Prognostic significance of tumour size in trials of radiation therapy and hyperthermia.

Reference	No. evaluable patients (sites)	Response criteria	Tumour size	Results	
				XRT+HT(%)	XRT alone (%)
Kim et al. (1982)[1]	38 (108)[1]	CR (at completion of treatment)	≤10 cm³	73	58
			>10 cm³	65	19
Dewhirst and Sim (1984, 1986)[2]	227	CR (of ≥1 month duration)	< 1.8 cm³	76	60
			1.8–8.3 cm³	54	60
			8.3–49 cm³	48	15
			>49 cm³	48	10
Hiraoka et al. (1984a)	36 (40)	CR (at time of maximum response)	< 4 cm	100	ND
			4–7 cm	65	ND
			7–10 cm	44	ND
			>10 cm	0	ND
Luk et al. (1984)	133	CR (initial)	≤ 3 cm	61	ND
			> 3 cm	44	ND
Arcangeli et al. (1985)	38 (81)	CR (at completion of treatment)	1–4 cm³	—	75
			7–11 cm³	—	60
			12–24 cm³	—	39
			25–61 cm³	—	14
			1–4 cm³	100	—
			5–10 cm³	100	—
			13–24 cm³	60	—
			24–53 cm³	69	—
Dewhirst et al. (1985)[2]	43[1]	CR	< 2 cm³	86	50
			2–10 cm³	86	0
			>10 cm³	57	13
Perez et al. (1986)	48 (XRT plus HT) 116 (XRT alone)	Local control (in field)	1–3 cm	79	48
			> 3 cm	65	28
Kapp et al. (1987)	18 (38)	Local control (in field)	< 5 cm³	81	ND
			> 5 cm³	25	ND

CR, complete response; XRT, radiation therapy; HT, hyperthermia treatment; ND, not done.
[1] Malignant melanomas only.
[2] Large animal spontaneous tumours.

In patients with melanomas, Kim *et al.* (1982) also found that tumour size was less important for combined modality treatment than for irradiation alone. In 38 melanoma cases treated with irradiation with or without inductive hyperthermia, tumour volumes >10 cm³ achieved nearly the same tumour control rates as those <10 cm³ with combined therapy (65% versus 73%). But with irradiation alone, the larger tumours had far poorer control rates than smaller ones (19% versus 58%). Arcangeli *et al.* (1985) have reported on the response rates of metastatic neck nodes from head and neck cancers. They demonstrated a decrease in the frequency of complete responses with increasing tumour volume for both irradiation alone and for irradiation plus hyperthermia, but less decrease for the combined therapy.

6.2.1(b) Tumour depth

The biophysical tissue absorption characteristics of MW and US radiations at therapeutically useful wavelengths dictate that heating to depths greater than about 2–3 cm will require multiple and/or focused sources, either fixed or scanning, or special lower-frequency EM techniques. The limitation of power penetration from single-wave sources is routinely observed in the clinic. Perez *et al.* (1983) and Perez and Meyer (1985) reported on 101 superficial tumours treated with irradiation and hyperthermia, usually from single 915 MHz MW applicators. Average temperatures at the deepest central point of the tumours were <42.5 °C in 67% of tumours ⩽2 cm deep and 65% of tumours 2.1 to 4 cm deep, but in only 36% of tumours >4 cm deep. This poorer ability to heat tumours >4 cm deep translated into poorer response; 51% of tumours ⩽4 cm deep had complete response compared to only 27% of tumours >4 cm deep. The effective penetration achieved by single unfocused US transducers is similar (Marmor *et al.*, 1979; Meyer, 1987).

6.2.1(c) Temperature

Because of the non-homogeneity of temperatures that may occur across the tumour and during the treatment time, three-dimensional temperature monitoring during the full treatment period needs to be performed routinely. Temperatures measured at a single point during treatment are not representative of the temperature distributions through tumours, and correlate poorly with response. For example, Corry *et al.* (1982a, b) monitored temperature at single points (tumour centre) and found little difference in the response rates between 43 and 44 °C (53% responded) and 45 and 47 °C (43% responded). Very similar findings are reported by Marmor *et al.* (1979). Using primarily 915 MHz MW equipment, Perez *et al.* (1983) reported single-point intratumoural temperatures (at tumour depth, central axis) against response rates. Rates did not differ for temperatures between 41 and 42°C (55% complete response) and ≥42.5°C (56%). However, single-point tumour temperature measurements are now felt to be very inadequate, and optimal treatment methods currently employ multiple temperature samplings from tumours during treatment. These are providing important correlations with response (Table 6.2).

Van der Zee *et al.* (1986) reviewed the results of 112 courses of irradiation and hyperthermia using RF or MW external sources. From multiple intratumoural temperature readings, they showed that the minimum tumour temperature correlated with response, no matter which of four different methods of calculating thermal dose was used. No response correlation could be found with the maximum or mean tumour temperatures. In their multivariate analysis, Oleson *et al.* (1984) found the minimum temperatures in tumours to be the overall most important variable in predicting complete response. There was no significant correlation between response and the maximum temperatures in the interstitially treated patients, where thermal dosimetry was the most extensive. This confirmed earlier work of Dewhirst *et al.* (1984) on spontaneous tumours in large animals.

Kapp *et al.* (1988) at Stanford University, California, analysed the results of hyperthermia and irradiation for breast cancer recurrent on the chest wall according to the thermal profiles from repeated temperature mappings during treatment. The duration of complete response correlated with the average measured minimum tumour temperature for all treatments in a course, and inversely correlated with the percentage of measured intratumoural temperatures <41°C. A previous study by this group (Kapp *et al.*, 1986b) also showed a correlation of complete response with the average mean intratumoural temperatures. Thus, the measured minimum tumour temperature may not be the only significant thermal predictor for response.

Corry *et al.* (1988) analysed the 'success' of treatment in terms of the percentage of intratumoural temperatures achieving a value greater than or equal to index temperatures from 39 to 50°C and an index percentage of the volume of the tumour heated. As more thermometry points were monitored, the number of 'successful' treatments declined. These results again emphasize the need for multiple-point or mapped thermometry to obtain adequate evaluation of heating.

6.2.1(d) Hyperthermia method

The physical method of inducing hyperthermia will determine the power distribution in tissue and thereby affect the tissue temperatures. The physical method may also influence patient tolerance to treatment, and

Table 6.2(a) Prognostically significant thermal parameters in trials of radiation therapy and hyperthermia: parameters correlating with response.

Reference	No. evaluable patients (sites)	Response criteria	Thermal parameters correlating with response
Dewhirst and Sim (1984)[1]	117	CR (for 1 month)	Minimum tumour temperature, 1st treatment (in equivalent min at 43°C) Minimum tumour temperature 2nd treatment (in equivalent min at 43°C) Non-site specific average minimum equivalent min at 43°C
Oleson et al. (1984)	144	CR (at time of best response)	Minimum tumour temperature measured averaged (over all treatments)
Luk et al. (1984)	133	CR (initial)	Lowest daily averaged tumour temperature
Hiraoka et al. (1984a)	36 (40)	CR (at time of maximum tumour regression)	Averaged tumour core temperature
van der Zee et al. (1984)	33 (44)	CR+PR (at 1–2 months)	Mean measured tumour temperature
Arcangeli et al. (1985)	38 (81)	CR (at completion of treatment)	Average overall equivalent min at 42.5°C (temperature measured in only one tumour point)
Kapp et al. (1985)	15 (31)	CR (at 3 weeks after completion of treatment)	Minimum tumour temperature, averaged over all treatments Average tumour temperature, all treatments Per cent measured tumour temperature <40°C, all treatments (negative correlation)
Sapozink et al. (1986)	43	CR+PR (time not specified)	Number of satisfactory heat treatments (achieving infra- or juxta-tumour temperatures of 42°C)
van der Zee et al. (1986)	112	CR+PR (for duration of at least 1 month)	Mean minimum mean temperature over all treatments Mean minimum maximum temperature over all treatments Total minimum equivalent minutes at 43°C (based on Sapareto-type formulation) Total minimum equivalent time at 43°C

CR, complete remission; PR, partial remission.
[1] Large animal spontaneous tumours.

Table 6.2(b) Prognostically significant thermal parameters in trials of radiation therapy and hyperthermia correlating with duration of complete response.

Reference	No. evaluable patients (sites)	Thermal parameters correlating with duration of complete response
Dewhirst and Sim (1984)[1]	117	Non-site-specific average minimum equivalent min at 43°C
Arcangeli et al. (1985)	38 (81)	Average overall equivalent min at 42.5°C (temperature measured at only one tumour point)
Kapp et al. (1986a)	29 (85)	Average minimum temperatures over all treatments Average (%) of temperature less than 41°C for all treatments (negatively correlated) Average of average temperatures over all treatments

[1] Large animal spontaneous tumours.

some methods may cause greater patient discomfort than others, forcing power reduction during the therapy session. Meyer *et al.* (1986a, 1988) compared 37 US treatments to MW treatments given to the same site in the same patient. Probably because of the greater effect of US on periosteal nerve fibres, US treatments were more poorly tolerated: pain prevented achieving a maximum tumour temperature of 42.5 °C in 38.4% of US treatments but only 7.9% of MW treatments. Across-tumour, time-averaged temperature profiles were also inferior with US in this study. As a result, the Stanford hyperthermia clinic now infrequently uses unfocused US for superficial treatment volumes, and especially avoids US for tumours overlying bone.

6.2.1(e) Number and frequency of hyperthermia treatments

Extensive laboratory work demonstrates that thermotolerance which develops during or after a thermal exposure can be a powerful mediator of response to subsequent treatment (Field, 1985; Hahn, 1982) (see also Chapters 1 and 2). The results of a limited amount of clinical work evaluating hyperthermia fractionation can also be explained on this basis (Table 6.3). Valdagni *et al.* (1986) have found no correlation between the response of metastatic neck nodes and the number of hyperthermia treatments when grouped as <6, 6 or >6 sessions. Dunlop and co-workers (1986) at Hammersmith Hospital, London analysed response by the number of treatments that gave a 20 min equivalent at 43 °C (Sapareto and Dewey, 1984) and found two treatments superior to none or one, but three or four treatments similar to two in terms of complete response.

Kapp *et al.* (1986a) have been investigating the response of superficial tumours to irradiation and hyperthermia with lesions stratified by tumour size, histology and extent of prior irradition, and prospectively randomized to two or six hyperthermia treatments. The hyperthermia treatments are separated by between 3 and 7 days. Results to date demonstrate nearly identical response rates at 3 weeks following treatment: complete responses occurred in 12/20 cases (60%) for two treatments compared to 14/23 (61%) for six; partial responses occurred in 7/20 (35%) for two treatments compared to 8/23 (35%) for six. Tumour size (<3 cm versus >3 cm) was prognostically important but response rates did not vary between the two hyperthermia schedules within these tumour-size groups. An update of this study now with 157 fields analysed reveals no difference in the response rates at 3 weeks or in the duration of local control between fields receiving hyperthermia for two treatments (78 fields) or for six treatments (79 fields) (D.S. Kapp *et al.*, unpublished data). These results are consistent with the laboratory findings of Meyer *et al.* (1986b).

Because of the kinetics of thermotolerance development and decay, the

Table 6.3 Influence of number of hyperthermia treatments on outcome in trials of radiation therapy (XRT) and adjuvant hyperthermia (HT).

Reference	No. evaluable patients (sites)	Hyperthermia treatment parameters						Results	
		Tumour temperature (Goal, °C)	Duration (min)	Treatments (per week)	Total no. of planned treatments (No.)	Outcome parameters	Actual No. HT treatments	CR (%)	Statistical significant difference in CR(%)
Arangeli *et al.* (1984)	25	43.5	45	1 or 2	5 or 10	CR at end of treatment; duration of CR	5, 10	64, 78	NA
Dewhirst and Sim (1984)[1]	116	44.0	30	1	4	CR at 1 month; duration of CR	1, 2, 3, 4	67, 79, 45, 58	NS
Hiraoka *et al.* (1984a)	36 (40)	41–44	30–60	2	2–12	CR at time of maximum regression (1–8 months)	2–7, 8–12	50, 53	NS
Valdagni *et al.* (1986)	27	42.5–43.5	30	2 or 3	2–12	CR at 3 months	< 6, 6, > 6	67, 37, 70	NS
Alexander *et al.* (1987)	44 (48)	42.5–44	60	1 vs 2	4 vs 8	CR	4, 8	42, 21	NA
Kapp *et al.* (1987)	43 (126)	43–45	45	1 vs 2	2 vs 6	CR at 3 weeks and duration of local control	2, 6	65[2], 73[2]	NS

NA, not available; NS, statistically significant differences in either CR or duration of CR not demonstrated. CR, complete remission.
[1] Large animal spontaneous tumours.
[2] Includes ongoing responses.

time period between treatments is very likely to be pertinent to tumour response. Alexander *et al.* (1987) have reported preliminary findings on a randomized trial of irradiation plus hyperthermia (42.5–44.0°C for 60 min) given one or two times each week for 4 weeks. Complete responses were more frequent with treatment once (42%) than twice (21%) each week, and objective responses were high in both groups (92% and 84% respectively). Somewhat different results were obtained by Arcangeli *et al.* (1984), who treated patients with multiple lesions with 5000 cGy over 5 weeks alone, or combined with once or twice weekly hyperthermia. Superior complete response rates were achieved at the end of therapy for the 10 compared to the five heat treatments (7/9, 78%; 9/14, 64%, respectively), and fewer tumours recurred after the 10 treatments. Further work with greater numbers of patients is important to clarify this difference.

6.2.1(f) Radiation–Hyperthermia sequencing

Since hyperthermia can sometimes induce changes in tissue blood flow during treatment (Waterman *et al.*, 1987), it has been conjectured that hyperthermia given immediately prior to irradiation might alter tumour hypoxia and sensitivity to irradiation. However, this has not been shown to be a clinically significant effect (Table 6.4). Kim *et al.* (1984) treated melanomas with hyperthermia before irradiation in 28 lesions, and with hyperthermia after irradiation in 22 lesions. The complete response rates were very similar for either sequence (64% versus 68%, respectively).

It has also been considered that a component of radiation damage that can be potentiated by hyperthermia might last longer for tumours than for normal tissues. Preliminary clinical evidence lends support to this. Arcangeli *et al.* (1984) delayed hyperthermia until 4 h after irradiation, and found that both normal-tissue reaction (measured by incidence of moist desquamation) and tumour response (measured by incidence of complete regression) had decreased compared to immediate irradiation–hyperthermia sequencing. However, the decrease had been greater for the normal-tissue than for the tumour response. This corroborates experimental work by Overgaard (1980) and suggests that specific radiation–hyperthermia time intervals can increase therapeutic gain. More recent clinical work by Overgaard and Overgaard (1984, 1987) further confirms this: melanoma lesions were treated with irradiation (500–1000 cGy fractions for three treatments) then hyperthermia (immediately after or delayed by 3–4 h). Tumour control rates were nearly identical in both groups, but normal-tissue reaction was less with a 3–4 h delay between treatments. Both Arcangeli and Overgaard used large radiation fractions, which are associated with increased normal-tissue reaction when combined with hyperthermia and may necessitate this delay between treatments. On the other hand, if tumours can be *selectively*

Table 6.4 Sequencing of hyperthermia (HT) and irradiation (XRT) in clinical trials.

Reference	No. evaluable patients (sites)	Tumour temperature (goal, °C)	Duration (min)	Treatments per week (No. (total))	Dose per fraction (cGy)	No. RXs per week (No. (total))	Timing of HT and XRT	Tumour response	Skin reaction	TER[1] Tumour	TER[1] Skin	TGF[1]
								CR(%)	*Moist desquamation (%)*			
Arcangeli et al. (1984)	25	42.5	45	2(8)	500	2(8)	XRT, immediate HT	77	64	2.05	1.80	1.14
							XRT, 4 h delay then HT	67	46	1.78	1.29	1.38
	15	45.0	30	2(5)	600	2(5)	XRT, immediate HT (skin cooled)	87	33	2.60	1.25	2.08
	26	42.5	45	3(7)	200–150–150 (4 h intervals)	3(36)	HT immediately after 2nd daily XRT fraction	73	42	1.72	1.10	1.57
								TCD_{50} (cGy)	*ED_{50} (cGy)*			
Overgaard and Overgaard (1984)	28 (65)[2]	43.0	30	2(3)	500–1000	2(3)	XRT immediate HT	600	650	1.5	1.4	1.0
							XRT, 3–4 h delay then HT	690	850	1.3	1.0	1.3
							XRT alone	880	850	1.0	1.0	1.0
								CR(%)				
Kim et al. (1984)	48[2]	42–43	30	2(10)	400	2(10)	HT–XRT (within 3–6 min)	57	No increase	NA	NA	NA
							XRT–HT (within 3–6 min)	63	No increase	NA	NA	NA
							XRT alone	30	—	NA	NA	NA
	49[2]	42–43	30	1(6)	660	1(6)	HT–XRT (within 3–6 min)	75	Greatest increase	NA	NA	NA
							XRT–HT (within 3–6 min)	72	Increased	—	—	—
							XRT alone	59	—	—	—	—

[1] TER, thermal enhancement ratio; TGF, therapeutic gain factor (see individual references for definitions); CR, complete response; TCD_{50}, dose per fraction for 50% CR; ED_{50}, dose per fraction for 50% rate of severe erythema; NA, not available.
[2] Malignant melanomas.

heated compared to the normal surrounding tissues, the efficacy of the hyperthermia may be higher for simultaneous treatments. The experimental results on this are discussed in Chapters 1 and 2.

6.2.2. Selection of applicators

The choice of the appropriate hyperthermia applicator in a given clinical situation must take into account the surface area, depth and accessibility of the tumour, as well as the type and location of the adjacent normal tissues. The clinical performances of a wide range of waveguide, horn and spiral microstrip MW, and planar US applicators have recently been reported (Kapp *et al.*, 1988) (see also Chapter 17). The thermal distributions obtained were tabulated as a function of the applicator used (independent of site), of all applicators used at a given tumour site and of individual applicators used at a given tumour site. Based on our knowledge of the limitations of the applicators which were available at that time, an extensive equipment development effort was instituted in the Department of Radiation Oncology at Stanford University, in California, to produce microstrip applicator arrays with improved heating patterns (Fessenden *et al.*, 1986, 1987; Lee *et al.*, 1986).

The first problem area identified for the commercially available applicators tested was the limited field volumes. Several investigators have demonstrated that the effective heating areas (arbitrarily chosen to be represented by the areas enclosed by the 50% isoSAR curves at depth) are significantly smaller than the aperture areas of the commercial microwave applicators (Nussbaum, 1984; Kapp *et al.*, 1988) and that the ratio of the aperture area to the area of the 50% isoSAR increases with increasing depth from the surface. If one chooses a perhaps more realistic value of the 70–80% isoSAR area to encompass the treatment volume, the situation is even worse. Several approaches have been utilized to increase the effective field size of applicators, including the use of mechanical scanning of microstrip antenna arrays (Kapp *et al.*, 1988) or the use of stationary arrays of multiple microstrip antennas (Lee *et al.*, 1989) or arrays of US transducers. An alternative approach is to use multiple overlapping fields to cover a large tumour volume. This approach has the potential risk of inducing thermotolerance in the tissues within the area of overlap, therefore limiting the effectiveness of subsequent treatments.

A second limitation of the currently available surface MW or US equipment is in the depth of heating obtained. In the MW frequency range of 300–915 MHz, effective heating depths of only 1–3 cm can be obtained with stationary waveguide or microstrip applicators (Kapp *et al.*, 1988; Nussbaum, 1984). Shifting to lower frequencies (e.g. 95 MHz) may help somewhat in improving the depth of heating, but for operation at lower frequencies large aperture sizes are generally required that may pose physical limitations of access to the tumour and other problems because of the applicator and coupling water bag size and weight. Alternative approaches employing scanned focused US beams have been developed to improve depth of heating (Hynynen *et al.*, 1987; Shimm *et al.*, 1988) and these will be reviewed in the section on heating of deep-seated tumours in Chapter 14.

One needs also to take into account tumour accessibility when choosing applicators for the treatment of surface tumours in areas of limited physical access. Such areas include the axilla, neck (particularly after radical surgery) and base of nose. Microstrip linear arrays or linearly scanning dipole antennas can be utilized for treatment in such sites (Fessenden *et al.*, 1988).

Special considerations also have to be made for the treatment of areas in close proximity to heat-sensitive normal structures, for example tumours adjacent to the lens of the eye. Custom-made multi-element microstrip arrays for the treatment of facial tumours have been developed that have employed separate saline-filled, circulating, temperature-controlled water bags to assure protection of the lens (Kapp *et al.*, 1989).

Curved surface areas with tumour involvement, as in cases where metastatic or primary tumours involve the circumference of an extremity, present additional difficulties for treatment with flat commercial applicators. Wrap-around conformal microwave blanket arrays have been developed for treatment of such lesions (Wilsey *et al.*, 1988). The additional problem of adequate applicator coupling needs to be considered in the treatment of surface tumours with uneven contours and overlying curved surfaces. Individually shaped or contoured coupling boluses have been utilized to facilitate treatment in these situations.

A final problem encountered in the treatment of superficially located tumours is the lack of uniform heating throughout the tumour volume. Temperature differences may be due to non-uniform power deposition from the applicators, variations in the dielectric constant throughout the tumour and differences in blood flow within the tumour. Central areas of the tumour may be necrotic, have minimal blood flow and heat very well. However, blood flow often can be greatest at the tumour–normal-tissue junction, and such regions can heat poorly (Samulski *et al.*, 1987a). Furthermore, blood flow rates may increase in the surrounding normal tissue and increase or decrease in the tumour during treatment (Waterman *et al.*, 1987).

It becomes apparent that some mechanism for power modulation needs to be employed to attempt to provide selected power deposition to various areas of the tumour, and to be able to respond to altered blood flow rates during the treatment. Two approaches are currently in use at the Stanford clinic. The first employs a power-modulated scanning double-spiral microstrip applicator (Kapp *et al.*, 1988). This device has a 360° circular scanned path with a 4 s scan period. Power adjustment during treatment is possible for each of the eight (45°) sectors in the scanned paths of the inner and outer spiral microstrip antennas. The second approach utilizes multi-element arrays of 915 MHz spiral microstrip antennas with independent power control to each of the antenna elements. Field sizes up to 17×17 cm can be encompassed with a 25-element microstrip array, and power can be independently modulated to each element of the array (Lee *et al.*, 1989).

6.2.3. Thermometry

The optimum utilization of applicators for hyperthermia requires a detailed knowledge of intratumoural and adjacent normal-tissue temperatures. Although there is no universal agreement currently on the optimum hyperthermia treatment exposure, most clinical studies attempt to obtain intratumoural temperatures of 43–45°C for superficially located tumours. Temperatures of normal tissue are usually kept below 43.0 or 43.5°C, and maximum intratumoural temperatures are often limited to 50°C. Treatment sessions of 45 to 60 min at steady state temperatures are customarily employed. Guidelines for minimum thermometry have been proposed (Shrivastava *et al.*, 1989) (see also Chapter 17).

The details for the required thermometry need to be fully delineated in any hyperthermia treatment protocol. In general, for superficial tumours, a temperature probe should be placed intratumourally at the location expected to be at the lowest temperature (e.g. at the tumour base along the periphery of the tumour) and other probes should be placed at the tumour–normal-tissue junctions. Additional sensors should be placed to monitor any areas of normal tissue or tumour at risk for overheating (e.g. skin folds, scars, necrotic regions of tumour). Manual or automatic mechanical mapping of temperatures along catheters placed within the treatment field is recommended. Alternatively, multisensor probes can be utilized to simultaneously record temperatures from multiple points within the tumour and adjacent normal tissues. Optimally, probe positions should be verified with the aid of orthogonal radiography and/or computed-tomography scans.

In summary, the number of temperature probes required will be a function of the tumour size and complexity, the character and sensitivity of the adjacent normal tissues included in the treatment volume and the complexity of the power control permitted by the hyperthermia applicator employed.

6.2.4. Toxicities associated with hyperthermia

6.2.4(a) Potentiation of radiation effects

Experimentally it is known that radiation damage to normal tissues is enhanced by hyperthermia (see Chapters 2 and 3). However, most clinical studies report that normal-tissue damage incurred in areas treated by combined hyperthermia and irradiation has not been significantly greater than that in areas treated by radiotherapy alone (Kim *et al.*, 1982; Perez *et al.*, 1983; Scott *et al.*, 1984). While some of these series used relatively low radiation doses, Kim and coworkers used four fractionation schedules with large radiation doses to treat metastatic melanoma. The addition of hyperthermia to second lesions in each patient resulted in no apparent enhancement of irradiation damage. Scott *et al.* found that sites that received combined therapy were indistinguishable from sites receiving radiotherapy alone, both at the end of therapy and at 6, 12, 18, and 24 months later despite the high doses of irradiation used.

Valdagni and colleagues (1986) treated 15 patients with metastatic neck nodes to total doses of 6000–7000 cGy plus hyperthermia, given 2–3 times each week during the first 4000 cGy. Normal tissue temperatures were kept at 43.5 °C or less; tumour temperatures were allowed to reach high levels, sometimes 55–57 °C. Nonetheless, the average skin reaction scored at the end of treatment and later was similar to that of contralateral nodes treated with radiotherapy alone. Residual fibrosis was noted in 3/34 lesions treated with combined therapy.

Many investigators have found that hyperthermia can induce higher temperatures in tumours than in normal tissues. This has been attributed to the energy deposition characteristics of the applicators, active skin cooling, or lower blood flow rates in the tumour than in the normal tissues (Meyer, 1984b). Thus, it is possible that the improved tumour responses with minimal normal tissue damage in part reflects that hyperthermia (and sometimes radiotherapy) could be physically localized preferentially in tumours.

Increased skin reaction after combined therapy has been reported when high daily radiation doses in single or multiple fractions were used (Arcangeli *et al.*, 1983). In one of their protocols, paired lesions were treated at 42.5 °C for 45 min immediately after each of eight fractions of 500 cGy, given twice per week (4000 cGy total dose). Skin cooling had not been provided. Substantial increase in the incidence of skin desquamation occurred with the combined therapy compared to the same irradiation alone (64% verses 36%). However, this was less than the improvement in the complete tumour control rate (77% versus 37%). Similarly, in the other protocols reported by these investigators, hyperthermia effects were consistently greater in tumours than in skin. Not surprisingly, the greatest therapeutic advantage was seen when skin was actively cooled whilst tumours received high-temperature treatment (45 °C for 30 min).

Gonzalez Gonzalez *et al.* (1986) and Overgaard and Overgaard (1984, 1987) have also reported enhanced skin reactions when hyperthermia was combined with large daily radiation fractions. Such studies can provide valuable information about the relative benefits and toxicities of different treatment regimens. However, the relative skin and tumour temperatures and the tissue tolerance to treatment will differ with the applicators used, the tumour lesions and specific sites treated, and the techniques of application used. Further, it is unlikely that the amount of hyperthermic enhancement of irradiation effect will be exactly the same for all normal tissues, and skin and subcutaneous tissues may not be representative of the potential effects in other major organs.

6.2.4(b) Direct effects of hyperthermia

Apart from the enhancement of irradiation effects, direct acute and subacute toxicities have been attributed to hyperthermia (Table 6.5) because they occurred as the treatment power was applied, or appeared typical of heat damage (Meyer, 1984b). Superficial burns (blisters) usually have been asymptomatic and have healed in 2–5 days. Deeper burns have been very uncommon, but have required much longer periods to heal. Ulcerations developing in treated tumours and/or normal tissues have occasionally persisted for weeks, but have usually been asymptomatic (Meyer *et al.*, 1986a; Kapp *et al.*, 1988) (see also Chapter 3). There are few reports of late toxicities despite follow-up of 2 or more years (Scott *et al.*, 1984).

A recently reported analysis of complications following combined hyperthermia–radiation therapy has attempted to delineate pretreatment and treatment factors predictive of complications (Kapp *et al.*, 1990). A total of 249 hyperthermia treatments with detailed thermometry were analysed (198 fields with superficially located tumours and 51 fields with deep-seated tumours). A total of 31 complications developed in 27 treatment fields containing superficial tumours (27/198, 13.6%). Complications included 17 tumour ulcerations, five normal-tissue ulcerations, four infections, two fields with induration or fibrosis, and one

Table 6.5 Reported toxicities of localized hyperthermia.

Reference	Treatment courses	Toxicity	Incidence (%)
U *et al.* (1982)	53	Superficial burn	13
		Deep burn	8
Corry *et al.* (1982a)	30	Pain	20
		Radicular pain	10
		Blistering	13
Scott *et al.* (1983)	100	Neurosensory reactions	5
Marmor (1982)	52	Pain	23
		Superficial burn	19
		Ulceration	10
		Systemic (vomiting, syncope)	2
Perez *et al.* (1983)	101	Second- and third-degree burns	6

field each with oedema, persistent pain and a late erythematous skin reaction. The complication rate was 18.6% in fields that received three or more heat treatments (24/129) compared with 7.5% (9/120) in fields that had received only one or two heat treatments ($P=0.017$). The average of the maximum-measured intratumoural temperatures was also higher in those fields which developed complications (45.9°C) compared to those that did not (44.6°C) ($P=0.005$). The multivariate model best predictive of the development of complications included the average of the maximum-measured intratumoural temperatures and the number of treatments ($P=0.00043$). These results, if confirmed by other studies, would suggest that attempts should be made to employ the minimum number of hyperthermia treatments and to limit extremely high temperatures in tumours during treatment.

6.3. Deep-seated tumours

For tumours situated more than 3 cm from the body surface, different treatment approaches are required that may greatly influence the tumour and normal-tissue temperatures that are obtainable, and the tumour responses and toxicities that result. Effective deposition of power at depth in tissues requires innovative equipment designs that result in heating patterns that are usually regional in distribution rather than localized. A greater volume of normal tissue is usually heated and this often limits the temperatures which can be achieved because of cardiovascular or other physiological changes (e.g. increased blood pressure and systemic temperature elevation) (see Chapter 8). There is higher probability of deep-seated normal-tissue damage, and patients symptomatically tolerate the treatments less well. The thresholds for damage of the various visceral normal tissues to hyperthermia or to combined hyperthermia and irradiation are still poorly understood. Substantial irradiation doses are often delivered to normal tissues around tumours in deep sites despite the best-possible treatment planning. Without the physical localization advantages both for hyperthermia and for irradiation that exist for superficial lesions, it is not yet possible to achieve the same efficacious and safe results in deep-tumour volumes as in superficial sites. Yet much progress has been made in this direction, and several clinical treatment approaches are currently available.

6.3.1. Radiofrequency capacitive methods

Perhaps the simplest method of delivering EM energy deep into the body is to have the target region sandwiched between two large-area electrodes that have an oscillating voltage between them. This technique of capacitive heating uses the patient's tissues as the dielectric between the capacitor plates. As the two plates alternate between positive and negative charges, an oscillating electric field exists virtually everywhere within that volume of the patient between the plates. The presence of layers of body fat in the patient is especially important for this modality; whenever there is an electric-field component perpendicular to a fat–muscle interface, the relative heating in fat with respect to muscle is increased by about a factor of 10 (Christensen and Durney, 1986). Since the electric field lines for capacitive heating must generally be transverse to the patient, fat heating is a very real limiting factor even in patients of modest weight and can be overcome only partially by conductive cooling at the surface (see Chapters 10 and 11).

The RF-8 Thermotron (Yamamoto Vinyter Co. Ltd., Osaka, Japan) is a versatile capacitive hyperthermia system using paired plates for deep heating that operates at a frequency of 8 MHz. For several years it has been the basis for pilot clinical studies in Japan (Abe *et al.*, 1986), and experience with its use in the USA was published recently (Song *et al.*, 1986). Hiraoka *et al.* (1984a) treated 40 superficial and deep tumours by this method. Electrodes varied from 4 to 30 cm in diameter. For deep tumours, pairs of large plates were used; for superficial ones, a plate at least 4 cm larger than the tumour was paired with a large plate. Single intratumoural temperatures, time-averaged over the 30–60 min of treatment, were 42 °C or higher in 37 of the 40 tumours reported. Normal-tissue temperatures adjacent to the tumour were monitored in eight cases, and were 1–4 °C cooler than the tumour temperature in seven of these. When combined with radiation therapy doses of 3200–6000 cGy (in 400 cGy fractions given twice each week), 53% of tumours responded completely and a further 40% responded partially. Tumour size was a significant factor — none of seven tumours over 10 cm diameter responded completely.

Hiraoka *et al.* (1984b) have reviewed the results of 24 deep-seated tumours. Tumour sites included the liver in nine cases, colorectum in four, pancreas in three, thorax in three, stomach in two and ischium in one. Across-tumour thermal profiles were obtained in 19 lesions, and maximum or centre tumour temperatures were maintained above 42 °C in 75% of treatments. This was true regardless of the treatment site. Temperatures typically varied between 1 and 5 °C within tumours, and heating was significantly limited in patients who had layers of subcutaneous fat of 2 cm or more thickness.

These investigators and coworkers have reported on further use of the Thermotron RF-8 in a seven-institution co-operative trial (Abe *et al.*, 1986). Sixty-three of 70 tumours could be heated to 42.5±0.5 °C intratumoural temperatures. Of these, 22 were <3 cm deep ('superficial'), 28 were 3–6 cm deep ('subsurface') and 20 were >6 cm deep ('deep'). In 13 deep tumours, maximum measured intratumoural temperatures were ≥42 °C. All seven patients with deep tumours that did not achieve 42 °C, had an adipose layer of 2 cm or greater. Discomfort during treatment was common but could be reduced by surface cooling. When large plates were used, an increase in systemic temperature of 1–2 °C, decrease in blood pressure and increase in pulse were noted in some patients. When hyperthermia for 40–60 min twice each week was given with radiation therapy of 4000 cGy (400 cGy twice each week) or 5000 cGy (200 cGy five times per week), 54% of subsurface tumours and 23% of deep tumours (of the 13/20 achieving the prescribed temperature) responded completely. Song *et al.* (1986) have reported obtaining similar temperatures using the same device in their small clinical series.

None of these investigations has reported normal-tissue damage resulting from the hyperthermia. Taken together, these results are encouraging in that some subsurface and deep tumours can be treated by this method when only a limited amount of body fat is present within the field. However, the results to date are preliminary and more specific investigations monitoring three-dimensional temperature distributions and clinical results in individual tumour sites are required.

6.3.2. Radiofrequency inductive methods

Inductive methods to induce hyperthermia have been extensively used for a range of superficial to deep sites using coils in various configurations. Magnetic induction hyperthermia at RF frequencies is accomplished by having one or more coils encircling or otherwise in close proximity to the target tumour volume. The resulting oscillating magnetic field induces electric fields that, in turn, set up eddy currents in the patient. This technique appears to have clinical advantages because the heating does not require direct contact of equipment with the patient to achieve heating. The technique favours preferential power deposition at the surface, however, so devices which contact the patient's surface to cool it are often required (see Chapters 10 and 11).

A commercial device called the Magnetrode (Henry Medical Electronics, Los Angeles, California), which consists of a single coil that surrounds the patient in the region of the tumour, and operates at a frequency of 13.56 MHz, has been used clinically for a very large number of patients (Storm *et al.*, 1985). Other magnetic induction configurations that attempt to position the regions of induced electric fields more advantageously may use one or a pair of coils placed near, but not actually encircling, the targeted body part (Oleson, 1984; Corry *et al.*, 1988).

Experience with the Magnetrode has been reported by Storm *et al.* (1985) for a wide range of deep pelvic, abdominal, extremity, body wall and head and neck cancers. Across-tumour thermal distributions generally were not measured in this study, and often tumour temperatures were not obtained at all. However, time-averaged tumour temperatures for at least one point, during at least one of the 30–60 min sessions of each course, were reported for 51% of the 960 completed courses. These were 42°C or higher in 198 (41% of those reported). This was a non-randomized study, but an effect of hyperthermia was shown since at least 25% tumour regression was evident in 23% of 142 tumours treated with heat alone. This included 21 tumours treated at quoted temperatures <45°C which responded in 67% of cases. Regression occurred in 55% of tumours treated with heat and radiotherapy, with or without chemotherapy. Second- or third-degree burns occurred in 32 cases, or about 3% of the treatment courses.

Baker *et al.* (1982) used the same concentric coil to treat 118 patients with tumours in many sites, including the abdomen in 31 (mostly liver and pancreas) and the pelvis in 28. Temperature monitoring was generally performed only during the first treatment, and often at only one tumour point. However, temperatures were reported above 42°C in many of the monitored tumours of the liver, abdomen, pelvis and chest wall. Temperatures in the subcutaneous fat were elevated, but generally were less than 43°C. Burns were uncommon. Responses were frequent but all patients received chemotherapy or radiotherapy in a non-randomized fashion.

While these studies do generally demonstrate elevation of intra-tumoural temperatures by this device, the work lacks evaluation of the volume of tumour heated (and of a number of other factors potentially affecting temperature distributions). It is difficult to judge the benefit gained by the hyperthermia treatments. We have learned from these studies that tissue toxicity has been uncommon despite some degree of deep-tumour heating.

Somewhat more promising results were reported by Corry *et al.* (1988), employing different types of magnetic induction equipment. Larger tumours (>5 cm) were found to have heated better. They noted a general tendency towards higher temperatures in the centre of tumours. Reduction in normal tissue toxicities and improved intratumoural temperature distributions were obtained with minor alterations in the incident axis of the magnetic field.

6.3.3. External radiating electromagnetic methods

Radiative devices operating at frequencies higher than those used for capacitive or inductive methods include annular arrays of multiple applicators that ring the patient's torso or limb and radiate EM waves towards the centre in a converging geometry. The array device most widely used is the Annular Phased Array System (APAS) (Turner, 1984; BSD Co., Salt Lake City, Utah). It consists of eight pairs of radiating horn-type resonant cavities in an octagonal geometry with an aperture of 50 cm diameter that is 46 cm deep. With the patient prone or supine, the APAS encircles the patient in a similar manner to that for CT scanning. An efficient coupling to the patient is accomplished by water filled 'bolus' bags between the octagonal ring and the patient surface. The APAS operates in the frequency range 55–100 MHz. The rapid attenuation of the energy radiated from each applicator is somewhat offset by the constructive interference among the converging waves, since all sources are operated coherently in phase. In practice, the region of maximum energy deposition is difficult to control and not well localized because of the large tissue wavelengths involved (approximately 40 cm). This device has been used in several clinical trials (Gibbs, 1984; Oleson *et al.*, 1986; Samulski *et al.*, 1987b; Kapp *et al.*, 1988). Particularly in the pelvic region, therapeutic temperatures are sometimes achieved over large regions with maximum temperatures in the tumour.

The often less than satisfactory energy deposition patterns obtained with the APAS can be improved at times by controlling the phase and amplitude of the individual radiating applicators in order to 'steer' the region of maximum heating toward the tumour volume (Sathiaseelan *et al.*, 1986; Samulski *et al.*, 1987b). A redesign of the APAS (Sigma 60), available also in a scaled-down version appropriate for the treatment of limbs (Mini-annular Phased Array; Turner, 1986), has been developed recently and is capable of being operated in a phase-and amplitude-controlled mode.

The need for thorough quality assurance programmes in facilities utilizing the APAS has been emphasized by the Hyperthermia Physics Center (Shrivastava *et al.*, 1988). Tissue-equivalent phantom and patient measurements of treatments utilizing the APAS at five facilities have demonstrated the following: significant sensitivity of the power distribution in the vertical axis to the frequency of operation; changes in the SAR as a function of load positioning in the APAS; variations in the SAR between the different APAS utilized in the five facilities; and errors in the power meter readings compared with the actual delivered power to the applicators. Furthermore, high levels of stray EM fields were noted that were highly sensitive to small changes in coupling between the applicator and the load as well as being affected by the general layout of cables, furniture, and the presence of persons in the treatment room (Shrivastava *et al.*, 1988). Frequent checks of the stray EM fields are recommended during treatments and attempts should be made to minimize the EM exposure to the operator and to the vital organs of the patient when these organs are outside the primary treatment volume. Guidelines for quality assurance in the utilization of the EM and RF systems for deep heating are available (Shrivastava *et al.*, 1988, 1989) (see also Chapter 17).

In the largest single-institution series reported using this device, 43 patients with deep pelvic malignancies were treated at the University of Utah (Sapozink *et al.*, 1986). Most patients had local–regional recurrences of primary pelvic malignancies, generally adeno-, squamous or transitional cell carcinomas. Temperatures across tumours and in multiple normal tissue sites were monitored throughout treatment. During 36 treatments, an average of 31% of the intratumoural points monitored exceeded a thermal dose of 15 equivalent minutes at 43°C (see Chapter 3).

Most toxicities during treatment were related to patient discomfort — pain developed in tissues within the applicator aperture in 74% of patients and outside the applicator in 33%; 9% of patients had pain radiating along nerve distributions. The systemic symptoms of stress, fatigue, nausea, anxiety and pruritus were often observed. Physiological changes during treatment often included an increase in body core temperature (average maximum 38.6°C) and an increase in pulse (average maximum 128 beats/min). Pain during treatment was the primary factor limiting the use of higher power levels in 41% of cases, and normal-tissue heating in 29%. Five patients developed musculoskeletal pain that persisted over 24 h.

Although the treatments were stressful during their administration, subsequent complications were uncommon despite the often large, invasive tumours being treated. Three patients developed superficial second-degree burns that required only conservative therapy; one developed an obstruction of the small bowel and one developed a rectal fistula (both were managed without surgery); and three patients developed problems probably resulting from tumour necrosis (bleeding in two, vaginal vault necrosis in one). Seventy-nine per cent of patients received concurrent radiotherapy, which may have contributed to these complications. Adverse effects that were typical of the radiotherapy being given also occurred but did not appear to be made worse by the addition of hyperthermia. In a subsequent publication, significant muscle necrosis was reported in one patient (Sapozink *et al.*, 1988).

Tumour response correlated with radiation dose and with the number of heat treatments in which an intratumoural temperature of 42°C or higher was obtained at some time at any point during the treatment. Other institutions have reported comparable degrees of heating and clinical results in small numbers of patients (Howard *et al.*, 1986; Yanagawa *et al.*, 1984; Emami *et al.*, 1984; Kapp *et al.*, 1988). These early experiences suggest that pelvic tumours can be treated with an acceptable margin of safety using this device. While temperature distributions have been far from optimal, there is a suggestion that the hyperthermia treatments have contributed to the effectiveness of the irradiation in achieving tumour response.

In 11 patients the temperatures induced by the APAS have been compared directly to those of the Magnetrode: all patients underwent a trial heating with one, then the other, device (Oleson *et al.*, 1986). Tumours treated had an average volume of 544 cm³ and were located in the pelvis in nine patients, abdomen in one and thorax in one. Temperatures were monitored at 4–16 intratumoural points, and the minimum and maximum steady state temperatures were higher for the APAS than for the Magnetrode. An average of 66% of tumour temperatures exceeded 41 °C with the array compared to 12% with the coil. These results corroborate a very similar clinical comparison by Sapozink *et al.* (1985), and together confirm theoretical estimates of central power deposition by Paulsen *et al.* (1985). These studies illustrate the benefit of comparative trials for evaluating deep-heating equipment, and such trials can guide our efforts in evaluating newer heating devices.

6.3.4. Ultrasound methods

In order to heat at depth, a means must be found to improve the ratio of energy deposition at the treatment depth compared to the patient's surface. One approach to this problem is to use a number of US transducers simultaneously with their beams directed in a converging geometry (see Chapter 14). This approach was developed at Stanford University in collaboration with the Hewlett–Packard Corporation (Palo Alto, California), and combined six transducers operating at a frequency of 365 kHz and fixed in an isospherical configuration (the isospherical ultrasound device, or IUD) (Hahn *et al.*, 1981; Fessenden *et al.*, 1984, 1985).

In a pilot study of 20 patients using the IUD (Fessenden *et al.*, 1984), the significance of reflection at soft-tissue–bone and soft-tissue–air interfaces within the human anatomy and the potential for overheating of bone were noted. These factors demand a high level of sophistication in the use of US, probably incorporating control of frequency and scanning of multiple focused beams as well as on-line high-quality imaging of the target and the temperature probes. Lele (1983) has pioneered studies in this direction, together with the research group from the University of Arizona (Swindell *et al.*, 1982; Hynynen *et al.*, 1987). A scanning multitransducer US system has been developed by Varian Associates (Seppi *et al.*, 1985).

Analysis of 15 treatment courses (57 treatments) using the Stanford IUD has been reported by Meyer *et al.* (1983, 1984a). Toxicities included pain in the tumour site during 46% of treatments, and pain radiating away from the treatment area in 19%. The presence of discomfort was highly dependent on the patient position relative to the US beams, and likely resulted from the beams encountering bones. Small adjustments in the angle of beam incidence, therefore, alleviated pain in about half of the uncomfortable treatments, but pain remained a significant problem. Such symptoms limited power so that maximum temperatures were below 42 °C in 23% of treatments and forced termination of treatment in 5% of sessions. Two patients developed small superficial blisters and one patient developed a small deep burn. No other complications occurred. Tumour responses were observed, but the series was too small to assess the contribution of hyperthermia compared with the other therapies concurrently given. This hyperthermia approach can provide a more localized treatment than occurs with EM techniques: in the Stanford series, when tumour temperatures above 42°C were achieved, the monitored cutaneous, subcutaneous, systemic and other normal tissues more than 5 cm away from the tumour were below 40°C in virtually all cases. A disadvantage is that heating can be focal even within tumours. The approach may be useful for the selective heating of a limited number of anatomically well-located tumours, especially with systems incorporating scanning and focusing of the US beams.

Thermal distributions obtained in 42 treatments using a scanned focused US system were recently reported (Shimm *et al.*, 1988). Tumours were located in the pelvis (nine patients), extremities (four patients), brain (two patients) and extracranial head and neck (two patients). Average maximum temperatures were 44.2, 44.7, 44.8 and 42.0°C within the scanned tumour volumes for the pelvic, extremity, brain and extracranial head and neck tumours, respectively. A mean of 10 points was monitored during each treatment (range 2–25) and 55, 45, 71 and 0% of the monitored tumour points exceeded 42.5°C for tumours in the pelvis, extremity, brain and extracranial head and neck sites, respectively. Pain limited the applied power in 15 of the 42 treatments. Patient responses to treatment were not presented.

6.4. *Conclusions*

Significant progress has been made in the clinical utilization of non-invasive EM and US heating systems. However, current commercially available systems can probably provide adequate heating to only selected small, superficially located tumours. Improvements in field sizes, depths of heating and power control are forthcoming. Carefully controlled site–specific trials with an emphasis on excellent quality assurance and thermometry are to be encouraged and will aid in the clinical implementation of hyperthermia. Improvements in localization of energy deposition employing phase and amplitude control for EM devices and scanned focus techniques for US devices are anticipated, and will require further phase I and II testing prior to phase III clinical studies on deep-seated tumours.

References

Abe, M., Hiraoka, M., Takahashi, M., Egawa, S., Matsuda, C., Onoyama, Y., Morita, K., Kakehi, M. and Sugahara, T. (1986) Multi-institutional studies on hyperthermia using an 8-MHz radiofrequency capacitive heating device (Thermotron RF-8) in combination with radiation for cancer therapy, *Cancer*, **58**, 1589–1595.

Alexander, G. A., Moylan, D. J., Waterman, F. M., Nerlinger, R. E. and Leeper, D. B. (1987) Randomized trial of 1 vs. 2 adjuvant hyperthermia treatments in patients with superficial metastases, *Abstracts of Papers for the 35th Annual Meeting of the Radiation Research Society*, p. 18.

Arcangeli, G., Cividalli, A., Nervi, C. and Creton, G. (1983) Tumor control and therapeutic gain with different schedules of combined radiotherapy and local external hyperthermia in human cancer, *International Journal of Radiation Oncology, Biology and Physics*, **9**, 1125–1134.

Arcangeli, G., Nervi, D., Cividalli, A. and Lovisolo, G. A. (1984) Problem of sequence and fractionation in the clinical application of combined heat and radiation, *Cancer Research*, **44** (Suppl.), 4857s–4863s.

Arcangeli, G., Arcangeli, G., Guerra, A., Lovisolo, G., Cividalli, A., Marino, C. and Mauro, F. (1985) Tumour response to heat and radiation: prognostic variables in the treatment of neck node metastases from head and neck cancer, *International Journal of Hyperthermia*, **1**, 207–217.

Baker, H. W., Snedecor, P. A., Goss, J. C., Galen, W. P., Gallucci, J. J., Horowitz, I. J. and Dugan, K. (1982) Regional hyperthermia for cancer, *American Journal of Surgery*, **143**, 586–590.

Christensen, D. A. and Durney, C. H. (1986) Hyperthermia production for cancer therapy: a review of fundamentals and methods, *Journal of Microwave Power*, **16**, 89–105.

Corry, P. M., Barlogie, B., Tilchen, E. J. and Armour, E. P. (1982a) Ultrasound-induced hyperthermia for the treatment of human superficial tumors, *International Journal of Radiation Oncology, Biology and Physics*, **8**, 1225–1229.

Corry, P. M., Spanos, W. J., Tilchen, E. J., Barlogie, B., Barkley, H. T. and Armour, E. P. (1982b) Combined ultrasound and radiation therapy treatment of human superficial tumors, *Radiology*, **145**, 165–169.

Corry, P. M., Jabboury, K., Kong, J. S., Armour, E. P., McCraw, F. J. and Leduc, T. (1988) Evaluation of equipment for hyperthermia treatment of cancer, *International Journal of Hyperthermia*, **4**, 53–74.

Dewhirst, M. W. and Sim, D. A. (1984) The utility of thermal dose as a predictor of tumour and normal tissue responses to combined radiation and hyperthermia, *Cancer Research*, **44** (Suppl.), 4772s–4780s.

Dewhirst, M. W. and Sim, D. A. (1986) Estimation of therapeutic gain in clinical trials involving hyperthermia and radiotherapy, *International Journal of Hyperthermia*, **2**, 165–178.

Dewhirst, M. W., Sim, D. A., Sapareto, S. and Connor, W. G. (1984) Importance of minimum tumor temperature in determining early and long-term responses of spontaneous canine and feline tumors to heat and radiation, *Cancer Research*, **44**, 43–50.

Dewhirst, M. W., Sim, D. A., Forsyth, K., Grochowski, K. J., Wilson, S. and Bicknell, E. (1985) Local tumor control and distant metastases in primary canine malignant melanomas treated with hyperthermia and/or radiotherapy, *International Journal of Hyperthermia*, **1**, 219–234.

Dunlop, P. R. C., Hand, J. W., Dickinson, R. J. and Field, S. B. (1986) An assessment of local hyperthermia in clinical practice, *International Journal of Hyperthermia*, **2**, 39–50.

Emami, B., Perez, C., Nussbaum, G. and Leybovich. (1984) Regional hyperthermia in treatment of recurrent deep-seated tumors: preliminary report, in Overgaard, J. (Ed.), *Hyperthermic Oncology 1984*, Vol. 1, London, Taylor and Francis, pp. 605–608.

Fessenden, P., Lee, E. R., Anderson, T. L., Strohbehn, J. W., Meyer, J., Samulski, T. V. and Marmor, J. B. (1984) Experience with a multitransducer ultrasound system for localized hyperthermia of deep tissues, *IEEE Transactions on Biomedical Engineering*, **31**, 126–135.

Fessenden, P., Meyer, J. L., Valdagni, R., Lee, E. R., Samulski, T. V., Kapp, D. S. and Bagshaw, M. A. (1985) Analysis of deep hyperthermia treatments using six ultrasound transducers in a fixed frequency/fixed geometry configuration, *Abstracts of Papers for the 33rd Annual Meeting of the Radiation Research Society*, p. 15.

Fessenden, P., Samulski, T. V. and Lee, E. R. (1986) New approaches and improvements in heating equipment for surface hyperthermia, *Abstracts of Papers for the 33rd Annual Meeting of the Radiation Research Society*, p. 13.

Fessenden, P., Kapp, D. S., Lee, E. R. and Samulski, T. V. (1988) Clinical microwave applicator design, in Paliwal, B. R., Hetzel, F. W. and Dewhirst, M. W. (Eds), *Biological, Physical and Clinical Aspects of Hyperthermia*, New York, American Institute of Physics, pp. 123–131.

Fessenden, P., Lee, E. R., Kapp, D. S., Tarczy–Hornoch, P., Prionas, S. D. and Sullivan, D. M. (1988) Improved microwave (MW) applicators for surface hyperthermia, *Abstracts of the 5th International Symposium on Hyperthermic Oncology, Kyoto, Japan*, p. 262.

Field, S. B. (1985) Clinical implications of thermotolerance, in Overgaard, J. (Ed.), *Hyperthermic Oncology 1984*, Vol. 2, London, Taylor and Francis, pp. 235–244.

Gibbs, F. A. (1984) Regional hyperthermia: a clinical appraisal of non-invasive deep-heating methods, *Cancer Research*, **44** (Suppl.), 4765s–4770s.

Gonzalez Gonzalez, D., van Dijk, J. D. P., Blank, L. E. C. M. and Rumke, Ph. (1986) Combined treatment with radiation and hyperthermia in metastatic melanoma, *Radiotherapy and Oncology*, **6**, 105–113.

Hahn, G. M. (1982) *Hyperthermia and Cancer*, New York, Plenum Press.

Hahn, G. M., Marmor, J. B. and Pounds, D. (1981) Induction of hyperthermia by ultrasound, *Bulletin Cancer, Paris*, **68**, 249–254.

Hiraoka, M., Jo, S., Dodo, Y., Ono, K., Takahashi, M., Nishida, H. and Abe, M. (1984a) Clinical results of radiofrequency hyperthermia combined with radiation in the treatment of radioresistant cancers, *Cancer*, **54**, 2898–2904.

Hiraoka, M., Jo, S., Takahashi, M. and Abe, M. (1984b) Thermometry results of RF capacitive heating for human deep-seated tumors, in Overgaard, J. (Ed.), *Hyperthermic Oncology 1984*, Vol. 1, London, Taylor and Francis, pp. 609–612.

Howard, G. C. W., Sathiaseelan, V., King, G. A., Dixon, A. K., Anderson, A. and Bleehen, N. M. (1986) Regional hyperthermia for extensive pelvic tumours using an annular phased array applicator: a feasibility study, *British Journal of Radiology*, **59**, 1195–1201.

Hynynen, K., Roemer, R., Anhalt, D., Johnson, C., Xu, Z. X., Swindell, W. and Cetas, T. C. (1987) A scanned, focused, multiple transducer ultrasonic system for localized hyperthermia treatments, *International Journal of Hyperthermia*, **3**, 21–35.

Kapp, D. S., Samulski, T. V., Meyer, J. L., Fessenden, P., Lee, E. R., Lohrbach, A. W. and Bagshaw, M. A. (1985) Metastatic breast cancer with chest wall recurrences in previously irradiated areas: management with low–moderate dose irradiation therapy and hyperthermia, *Abstracts of Papers for the 33rd Annual Meeting of the Radiation Research Society*, p. 29.

Kapp, D. S., Bagshaw, M. A., Meyer, J. L., Hahn, G. M., Samulski, T. V., Fessenden, P., Lee, E. R. and Lohrbach, A. W. (1986a) Hyperthermia as an adjuvant to radiation in the treatment of superficial metastases: a randomized trial of 2 vs. 6 treatments, *Abstracts of Papers for the 34th Annual Meeting of the Radiation Research Society*, p. 24.

Kapp, D. S., Samulski, T. V., Bagshaw, M. A., Fessenden, P., Meyer, J. L., Lee, E. R. and Lohrbach, A. W. (1986b) Hyperthermia techniques for the management of local-regional recurrences from adenocarcinoma of the breast, *International Journal of Radiation Oncology, Biology and Physics*, **12** (Suppl. 1), 156.

Kapp, D. S., Samulski, T. V., Fessenden, P., Bagshaw, M. A., Lee, E. R., Lohrbach, A. W. and Cox, R. S. (1987) Prognostic significance of tumor volume on response following local-regional hyperthermia (HT) and radiation therapy (XRT), *Abstracts of Papers for the 35th Annual Meeting of the Radiation Research Society*, p. 17.

Kapp, D. S., Fessenden, P., Samulski, T. V., Bagshaw, M. A., Cox, R. S., Lee, E. R., Lohrbach, A. W., Meyer, J. L. and Prionas, S. D. (1988) Stanford University institutional report: phase I evaluation of equipment for hyperthermia treatment of cancer, *International Journal of Hyperthermia*, **4**, 75–115.

Kapp, D. S., Cox, R. S., Fessenden, P., Meyer, J. L., Prionas, S. D., Lee, E. R. and Bagshaw, M. A. (1990) Parameters predictive for complications of treatment with combined hyperthermia and radiation therapy, *International Journal of Radiation Oncology, Biology and Physics*, in press.

Kapp, D.S., Lee, E.R., Tarczy-Hornoch, P. and Fessenden, P. (1989) Specially designed applicators for hyperthermia (HT) treatment of facial tumors, *Abstracts of Papers for the 37th Annual Meeting of the Radiation Research Society*, p. 8.

Kim, J.H., Hahn, E.W. and Ahmed, S. (1982) Combination hyperthermia and radiation therapy for malignant melanoma, *Cancer*, **50**, 478–482.

Kim, J.H., Hahn, S.A., Ahmed, S.A. and Kim, Y.S. (1984) Clinical study of the sequence of combined hyperthermia and radiation therapy of malignant melanoma, in Overgaard, J. (Ed.), *Hyperthermic Oncology 1984*, Vol. 1, London, Taylor and Francis, pp. 387–390.

Lee, E.R., Samulski, T.V., Fessenden, P. and Kapp, D.S. (1986) Controlled scan surface heating, *Abstracts of Papers for the 34th Annual Meeting of the Radiation Research Society*, p. 31.

Lee, E.R., Tarczy-Hornoch, P., James, D., Fessenden, P., Wilsey, T., Kapp, D.S. and McEuen, A.H. (1989) Body conformal microstrip array hyperthermia applicators, *Abstracts of Papers for the 37th Annual Meeting of the Radiation Research Society*, p. 103.

Lele, P.P. (1983) Physical aspects and clinical studies with ultrasonic hyperthermia, in Storm, F.K. (Ed.), *Hyperthermia in Cancer Therapy*, Boston, Hall, pp. 333–367.

Luk, K.H., Pajak, T.F., Perez, C.A., Johnson, R.J., Corner, N. and Dobbins, T. (1984) Prognostic factors for tumor response after hyperthermia and radiation, in Overgard, J. (Ed.), *Hyperthermic Oncology 1984*, Vol. 1, London, Taylor and Francis, pp. 353–356.

Marmor, J.B. (1982) Clinical use of localized hyperthermia, in Williams, C.J. and Whitehouse, J.M.A. (Eds), *Recent Advances in Clinical Oncology*, Edinburgh, Churchill Livingstone, pp. 35–44.

Marmor, J.B., Pounds, D., Postic, T.B. and Hahn, G.M. (1979) Treatment of superficial human neoplasms by local hyperthermia induced by ultrasound, *Cancer*, **43**, 188–197.

Meyer, J.L. (1984a) Ultrasound hyperthermia — the Stanford experience, *Frontiers of Radiation Therapy and Oncology*, **18**, 126–135.

Meyer, J.L. (1984b) The clinical efficacy of localized hyperthermia, *Cancer Research*, **44** (Suppl.), 4745s–4751s.

Meyer, J.L. (1987) Clinical and technical considerations in planning a course of hyperthermia and irradiation, *Frontiers of Radiation Therapy and Oncology*, **21**, 107–121.

Meyer, J., Fessenden, P., Bagshaw, M., Hahn, G., Samulski, T., Lee, E. and Lohrbach, A. (1983) Ultrasound hyperthermia for the treatment of deep tumor volumes, *International Journal of Radiation Oncology, Biology and Physics*, **9** (Suppl. 1), 82.

Meyer, J., Kapp, D., Bagshaw, M., Samulski, T., Fessenden, P. and Lee, E. (1986a) Complications from microwave and ultrasound hyperthermia: effect of treatment volume, *Abstracts of Papers for the 34th Annual Meeting of the Radiation Research Society*, p. 24.

Meyer, J.L., Van Kersen, I. and Hahn, G.M. (1986b) Tumor responses following multiple hyperthermia and X-ray treatments: role of thermotolerance at the cellular level, *Cancer Research*, **46**, 5691–5695.

Meyer, J., Kapp, D., Fessenden, P., Lee, E., Lohrbach, A. and Bagshaw, M.A. (1988) Direct comparison of ultrasound (US) and microwave (MW) treatment methods in the same treatment site, *Abstracts of the 5th International Symposium on Hyperthermic Oncology, Kyoto, Japan*, p. 218.

Nussbaum, G.H. (1984) Quality assessment and assurance in clinical hyperthermia: requirements and procedures, *Cancer*, **44** (Suppl.), 4811s–4817s.

Oleson, J.R. (1984) A review of magnetic induction methods for hyperthermia treatment of cancer, *IEEE Transactions on Biomedical Engineering*, **31**, 91–97.

Oleson, J.R., Sim, D.A. and Manning, M.R. (1984) Analysis of prognostic variables in hyperthermia treatment of 161 patients, *International Journal of Radiation Oncology, Biology and Physics*, **10**, 2231–2239.

Oleson, J.R., Sim, D.A., Conrad, J., Fletcher, A.M. and Gross, E.J. (1986) Results of a phase I regional hyperthermia device evaluation: microwave annular array versus radiofrequency induction coil, *International Journal of Hyperthermia*, **2**, 327–336.

Overgaard, J. (1980) Simultaneous and sequential hyperthermia and radiation treatment of an experimental tumor and its surrounding normal tissue *in vivo*, *International Journal of Radiation Oncology, Biology and Physics*, **6**, 1507–1517.

Overgaard, J. and Overgaard, M. (1984) A clinical trial evaluating the effect of simultaneous or sequential radiation and hyperthermia in the treatment of malignant melanoma, in Overgaard, J. (Ed.), *Hyperthermic Oncology 1984*, Vol. 2, London, Taylor and Francis, pp. 383–386.

Overgaard, J. and Overgaard, M. (1987) Hyperthermia as an adjunct to radiotherapy in the treatment of malignant melanoma, *International Journal of Hyperthermia*, 3, 483–501.

Paulsen, K. D., Strohbehn, J. W. and Lynch, D. R. (1985) Comparative theoretical performance for two types of regional hyperthermia systems, *International Journal of Radiation Oncology, Biology and Physics*, 10, 775–786.

Perez, C. A., Nussbaum, G., Emami, B. and Von Gerichten, D. (1983) Clinical results of irradiation combined with local hyperthermia, *Cancer*, 52, 1597–1603.

Perez, C. A. and Meyer, J. L. (1985) Clinical experience with localized hyperthermia and irradiation, in Overgaard, J. (Ed.), *Hyperthermic Oncology 1984*, Vol. 2, London, Taylor and Francis, pp. 181–198.

Perez, C. A., Kuske, R. R., Emami, B. and Fineberg, B. (1986) Irradiation alone or combined with hyperthermia in the treatment of recurrent carcinoma of the breast in the chest wall: a nonrandomized comparison, *International Journal of Hyperthermia*, 2, 179–187.

Samulski, T. V., Fessenden, P., Valdagni, R. and Kapp, D. S. (1987a) Correlations of thermal washout rate, steady state temperatures and tissue type in deep-seated recurrent or metastatic tumors, *International Journal of Radiation Oncology, Biology and Physics*, 13, 907–916.

Samulski, T. V., Kapp, D. S., Fessenden, P. and Lohrbach, A. (1987b) Heating deep seated eccentrically located tumors with an annular phased array system: a comparative clinical study using two annular array operating configurations, *International Journal of Radiation Oncology, Biology and Physics*, 13, 83–94.

Sapareto, S. A. and Dewey, W. (1984) Thermal dose determination in cancer therapy, *International Journal of Radiation Oncology, Biology and Physics*, 10, 787–800.

Sapozink, M. D., Gibbs, F. A., Thomson, J. W. and Stewart, J. R. (1985) A comparison of deep regional hyperthermia from an annular phased array and a concentric coil in the same patients, *International Journal of Radiation Oncology, Biology and Physics*, 11, 179–190.

Sapozink, M. D., Gibbs, F. A., Egger, M. J. and Stewart, J. R. (1986) Regional hyperthermia for clinically advanced deep-seated pelvic malignancy, *American Journal of Clinical Oncology*, 9, 162–169.

Sapozink, M. D., Gibbs, P., Gibbs, F. A. and Jolles, C. (1988) Myonecrosis following deep pelvic hyperthermia, *International Journal of Hyperthermia*, 4, 251–258.

Sathiaseelan, V., Iskander, M. F., Howard, G. C. W. and Bleehen, N. M. (1986) Theoretical analysis and clinical demonstration of the effect of power pattern control using the annular phased-array hyperthermia system, *IEEE Transactions on Microwave Theory and Techniques*, 34, 514–519.

Scott, R. S., Johnson, R. J. R., Kowai, H., Krishnamsetty, R. M., Story, K. and Clay, L. (1983) Hyperthermia in combination with radiotherapy: a review of five years experience in the treatment of superficial tumors, *International Journal of Radiation Oncology, Biology and Physics*, 9, 1327–1333.

Scott, R. S., Johnson, R. J. R., Story, K. V. and Clay, L. (1984) Local hyperthermia in combination with definitive radiotherapy: increased tumor clearance, a reduced recurrence rate in extended followup, *International Journal of Radiation Oncology, Biology and Physics*, 10, 2119–2123.

Seppi, E., Shapiro, E., Zitelli, L., Henderson, S., Wehlau, A., Wu, G. and Dittmer, C. (1985) A large aperture ultrasonic array system for hyperthermia treatment of deep-seated tumors, *Proceedings IEEE Ultrasonic Symposium*, pp. 942–948.

Shimm, D. S., Hynynen, K. H., Anhalt, D. P., Roemer, R. B. and Cassady, J. R. (1988) Scanned focused ultrasound hyperthermia: initial clinical results, *International Journal of Radiation Oncology, Biology and Physics*, 15, 1703–1708.

Shrivastava, P., Saylor, T. K., Matloubieth, A. Y. and Paliwal, B. R. (1988) Hyperthermia quality assurance results, *International Journal of Hyperthermia*, 4, 25–37.

Shrivastava, P., Luk, K., Oleson, J., Dewhirst, M., Paja, K. T., Paliwal, B., Perez, C., Sapareto, S., Saylor, T. and Steeves, R. (1989) Hyperthermia quality assurance guidelines, *International Journal of Radiation Oncology, Biology and Physics*, 10, 571–587.

Song, C. W., Rhee, J. G., Lee, C. K. K. and Levitt, S. H. (1986) Capacitive heating of phantom and human tumors with an 8 MHz radiofrequency applicator (Thermotron RF-8), *International Journal of Radiation Oncology, Biology and Physics*, **12**, 365–372.

Storm, F. K., Baker, H. W., Scanlon, E. F., Plenck, H. P., Meadows, P. M., Cohen, S. C., Olson, C. E., Thomson, J. W., Khandekar, J. D., Roe, D., Nizze, A. and Morton, D. L. (1985) Magnetic-induction hyperthermia. Results of a 5-year multi-institutional national cooperative trial in advanced cancer patients, *Cancer*, **55**, 2677–2687.

Swindell, W., Roemer, R. B. and Clegg, S. T. (1982) Temperature distributions caused by dynamic scanning of focused ultrasound transducers, *Proceedings of the Ultrasonics Symposium IEEE Group on Sonics and Ultrasonics, San Diego, 1982*, Vol. 2, pp. 750–753.

Turner, P. F. (1984) Regional hyperthermia with an annular phased array, *IEEE Transactions on Biomedical Engineering*, **31**, 106–115.

Turner, P. F. (1986) Mini-annular phased array for limb hyperthermia, *IEEE Transactions on Microwave Theory and Techniques*, **34**, 508–513.

U, R., Worde, B. T., Fishburn, R. I., Noell, K. T., Woodward, K. T., Miller, L. S. and Herskovic, A. M. (1982) Hyperthermia in cancer treatment: current and future prospects, *Gan-to-Kagakuryoho*, **9**, 343–356.

Valdagni, R., Kapp, D. S. and Valdagni, C. (1986) N_3 (TNM-UICC) metastatic neck nodes managed by combined radiation therapy and hyperthermia: clinical results and analysis of treatment parameters, *International Journal of Hyperthermia*, **2**, 189–200.

Van der Zee, J., Van Rhoon, G. C., Wike–Hooley, J. L., van den Berg, A. P. and Reinhold, H. S. (1984) Thermal enhancement of radiotherapy in breast carcinoma, in Overgaard, J. (Ed.), *Hyperthermic Oncology 1984*, Vol. 1, London, Taylor and Francis, pp. 345–348.

Van der Zee, J., Van Putten, W. L. J., Van den Berg, A. P., Van Rhoon, G. C., Wike Hooley, J. L., Broekmeyer–Reurink, M. P. and Reinhold, H. S. (1986) Retrospective analysis of the response of tumours in patients treated with a combination of radiotherapy and hyperthermia, *International Journal of Hyperthermia*, **2**, 337–349.

Waterman, F. M., Nerlinger, R. E., Moylan III, D. J. and Leeper, D. B. (1987) Response of human tumor blood flow to local hyperthermia, *International Journal of Radiation Oncology, Biology and Physics*, **13**, 75–82.

Wilsey, T. R., McEuen, A. H., Fessenden, P., Lee, E. R., Tanabe, E., Nelson, L. V., Schlitter, R. C. and Kapp, D. S. (1988) Arm cuff microwave microstrip array applicator, *Abstracts of Papers for the 36th Annual Meeting of the Radiation Research Society*, p. 15.

Yanagawa, S., Tsukiyama, I., Watai, K., Akine, Y. and Kakehi, M. (1984) Regional hyperthermia combined with radiation for locally advanced deep-seated malignancy, in Overgaard, J. (Ed.), *Hyperthermic Oncology 1984*, Vol. 1, London, Taylor and Francis, pp. 613–616.

7 Clinical hyperthermic practice: interstitial heating

C. T. Coughlin

7.1. Introduction

A great deal of interest over the past several years has developed in interstitial hyperthermia. This treatment involves placing catheters, needles or seeds directly into tumour tissue with the purpose of raising that tissue to a higher uniform temperature. Three major techniques have been developed: interstitial radio-frequency (RF) hyperthermia, interstitial microwave antenna-array hyperthermia (IMAAH) and interstitial ferromagnetic seed hyperthermia. The following discussion includes the practical aspects of these heating techniques from a clinical standpoint.

7.2. Radio-frequency heating

The first interstitial technique that was used for the delivery of hyperthermia consisted of an implanted array of needle electrodes (Doss and McCabe, 1976; Gerner *et al.*, 1975). In its simplest form, two parallel planes of electrodes were implanted on opposite sides of a tumour. RF voltage was then applied between two planes. Frequencies used in clinical practice ranged from 500 kHz to 13 MHz. The frequency must be high enough to avoid the biological effects of depolarization of muscle or nerve fibres. Once this threshold has been passed, the frequencies used to deliver hyperthermia are not critical.

The implanted needle-electrode system is similar to placing two capacitive electrode plates on opposite sides of a tumour and applying RF voltage between them. The needle electrodes create an RF current in the tumour volume that causes a rise in temperature throughout that volume. The interstitial needle system merely replaces the plates with a plane of needles. However, since the needles must be small in diameter (approximately 1 mm) the currents must diverge from these small sources. When this happens, the current density in the neighbourhood of the electrodes is high, leading to a high absorbed power density-specific absorption rate (SAR) near the electrodes. Therefore, thermal conduction is necessary for a smooth temperature distribution. In regions of high blood flow, there are large temperature gradients near the needles. This may be overcome by implanting the electrodes no more than 1 cm apart.

The needles must also be precisely parallel to one another in order to ensure a uniform temperature distribution. The current density tends to be much higher when the needles converge, resulting in much higher SAR patterns and a tendency for overheating. Conversely, the current density falls off quickly when the needles diverge, resulting in underheating of the tumour.

The systems designed for clinical use are much more complicated than the theoretical system described above. Thermometry data must be used to control the voltages to the electrodes in order to control the temperature over time. Rigid-needle electrodes have several drawbacks in clinical use; however, modification of these needles has made them more acceptable. The majority of implant volumes are insufficiently regular to be optimally treated with a simple two-plane implant. Even if the power deposited to the tumour volume was uniform, there would be large temperature gradients primarily due to variations in blood profusion in the tissue. A system can be designed to overcome this problem so that the power deposited in various regions of the tumour can be modified. With a separate power generator for each pair of electrodes (Cosset *et al.*, 1984, 1985), the current between each pair or groups of pairs can be controlled. It is also possible to time-multiplex the control voltage across pairs of electrodes (Astrahan and George, 1980; Astrahan and Norman, 1982). Controlling the amount of time the power is on across any electrode pair controls the temperature in the neighbourhood of any electrode pair. This system can then be connected to a switching system that can connect any two pairs of implanted electrodes. This comprehensive control system can bring all electrodes to the same temperature, affording better temperature control throughout the volume.

A problem faced by clinicians has been the fact that the first RF electrodes were stainless-steel needles. These bare stainless-steel needles deposited power along the total length of the needle. If a tumour was several centimeters below the skin, much normal tissue was unnecessarily heated. It would be much more desirable not to deliver power between the skin and the tumour. One approach to alleviating this difficulty would be to provide an insulating cover for the electrode and to remove the insulation in the region of the tumour.

Stainless-steel needles are also rigid and can be most uncomfortable in the non-sedated patient. This problem can be overcome by using flexible-needle electrodes. This reduces patient discomfort and allows the patient to interact with the physician during a hyperthermia treatment. This degree of interaction is an extra safeguard against overheating a tumour volume and reduces the complication rate of hyperthermia.

Finally, if brachytherapy is the radiation treatment of choice, the electrodes must be small enough to fit into flexible plastic catheters. In this way, the electrode can be removed after a hyperthermia treatment, and radioactive seeds can be afterloaded throughout the tumour volume.

The clinical studies in which RF interstitial heating was used have involved superficial tumour sites with a typical distribution of 10–12 needles in at least two planes. The primary sites tested have been head and neck, chest wall and extremity (Cosset *et al.*, 1984, 1985; Emami *et al.*, 1987; Joseph *et al.*, 1981; Manning *et al.*, 1982; Oleson and Cetas, 1982; Oleson *et al.*, 1984; Vora *et al.*, 1982, 1988; Yabumoto and Suyama, 1984). These areas are easiest to implant, with the constraint that the needles must be kept absolutely parallel. It is important to rely upon multipoint thermometry so that varying temperatures due to heterogeneous blood flow throughout the volume can be more adequately assessed. It is best to eliminate thermal gradients to the greatest degree possible so that uniform temperatures throughout the volume can be attained. A typical target dose of 42–43°C for 60 min appears to be optimal. Thermal gradients with higher temperatures in certain points of the tumour volume may be necessary due to heterogeneity of blood flow. These higher temperatures are acceptable only within the tumour volume. The heating of normal tissue to higher than therapeutic temperatures increases the risk of significant complications. The geometry of an irregular tumour volume may pose some difficulty with RF heating. This is especially true with the requirement that the electrodes should be parallel for uniform heating. While this system may be simpler to engineer, it has limited flexibility for clinical use. The substantial number of wires and thermometry probes may make the setup so awkward that adequate control of temperature throughout the hyperthermia sessions is difficult. Also, many of the electrodes currently available are not compatible with plastic catheters used in standard brachytherapy techniques. Future goals for RF heating would include: (1) the development of flexible electrodes to increase compatibility with brachytherapy techniques, (2) to create dielectrically coated electrodes appropriately sized for adjustable heating lengths, (3) to change the hardware and software for control of the thermometry to each electrode to a simpler system and (4) to integrate RF heating into three-dimensional planning for brachytherapy.

7.3. Microwave heating

Interest in interstitial microwave antennas as a means for localized hyperthermia for both superficial and deep-seated tumours arose in the

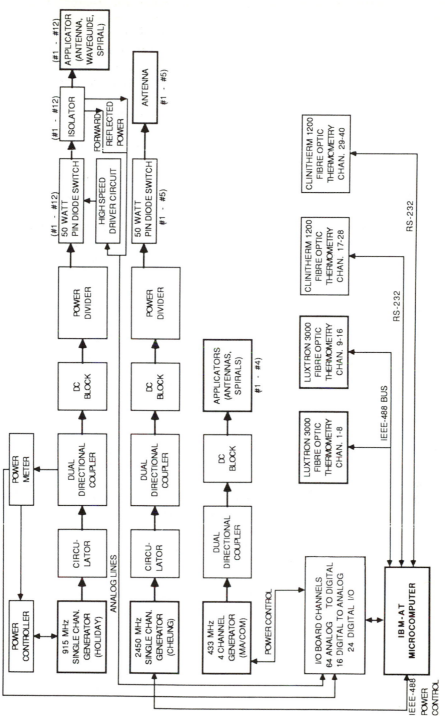

Figure 7.1 The IMAAH system used at Dartmouth College, Hanover, New Hampshire.

late 1970s (Mendecki *et al.*, 1977, 1978; Taylor, 1978, 1980). Much recent work has described the engineering aspects in this area (Strohbehn and Douple, 1984; Strohbehn and Mechling, 1986; Strohbehn, 1987). Figure 7.1 is a schematic representation of the IMAAH system currently used at Dartmouth College, Hanover, New Hampshire. The microwave power source can operate at a variety of frequencies between 433 and 2450 MHz. The power source is followed by a power divider and a set of power controllers that deliver microwave energy to each antenna. Antennas are normally inserted into catheters implanted in the tumour volume by the physician, using standard interstitial brachytherapy techniques. This system makes the hyperthermia entirely compatible with the brachytherapy treatment. The temperature throughout a tumour volume is measured by a set of sensors that feed the temperature information into the thermometry system. The temperature information is then used to set the levels of power to the antennas. This automated system provides instant feedback control to each of the antennas so that temperature can be maintained uniformly throughout that volume over time.

The most commonly used frequency is 915 MHz, although 433 and 2450 MHz systems are also available. The power deposition pattern along the longitudinal direction of the antenna is approximately one-half wavelength in tissue. Therefore, different frequencies can give longer or shorter heating patterns. Typical half-wavelengths are 2.5 cm at 2450 MHz, 7 cm at 915 MHz and 12 cm at 433 MHz. Most tumours treated in clinical work are of 4–8 cm diameter, making the 915 MHz system most appropriate.

The power from the microwave generator can go through a power divider that splits the power to each of the antennas. The number of antennas can vary from one to 12, though some systems can accommodate a much higher number of antennas. In the 915 MHz power configuration, the power from the divider then passes into a power controller which enables the power to each antenna to be set independently. Once again, this enables the operator to set the power control manually as a percentage of the power going to each antenna or automatically by a feedback system that regulates the temperature at a specific location in the tumour.

Another alternative is to use a number of independent microwave power sources, typically, either four or eight. If the number of antennas is equal to or less than the number of power sources, each antenna can be driven independently to set the total power level. If the number of antennas is greater than the number of power sources, each source can drive two or more antennas, with groups of antennas under independent control. In most systems, the independent power sources are driven from a single oscillator maintaining the coherence of the system.

Whichever system is used, the clinician can control the power independently to different regions of the tumour, shaping the power deposition pattern to control the temperature distribution within a given tumour volume. This degree of control is not available in RF systems.

The key element in the IMAAH system is the antenna, a coaxial cable with an extension of an inner conductor on the end. The antenna is inserted into an electrically insulating catheter. The theory for these antennas and their engineering constraints can be found in other references (Denman *et al.*, 1988a, b; Jones, 1987; Jones *et al.*, 1988; King *et al.*, 1983; Strohbehn and Mechling, 1986; Strohbehn, 1987; Trembly, 1982, 1985; Wong *et al.*, 1986). The radiation characteristics of such antennas are very sensitive to small changes in parameters, such as the wall thickness of the catheter into which the antennas are inserted and insertion depth. If the effect of these parameters is not well understood, the power deposition and temperature patterns produced may vary considerably from what the user expects, resulting in a poor treatment or significant complications.

The theory for a variety of microwave antennas is based on studies by King *et al.* (1983) and Trembly (1982, 1985). First, because of the coherent addition of the longitudinal component of the electrical fields, a maximum SAR value, both in theory and experimentally, is in the centre of the array. The 50% contour is well inside a 2×2 cm array in theory, as well as experimentally. It is also important to note that the SAR pattern is quite sensitive to insertion depth. This makes it very important for an engineering–physics team to have in-depth experience in controlling hot spots in a clinical setting.

While there have been improvements in microwave antennas, there remain three areas of concern. The first is to improve the power deposition near the tip of the antenna, the second is to make the power deposition patterns less sensitive to insertion depth and the third is to increase the length of the heating volume. Improvements in these areas are being actively investigated in a number of laboratories (Lee *et al.*, 1986; Lin and Wang, 1987; Satoh and Stauffer, 1988).

Microwaves were first used for interstitial heating in a clinical setting in 1983 (Coughlin *et al.*, 1983; Salcman and Samaras, 1983). Since that time, several centres have described their experience utilizing microwave antennas for the delivery of heat (Coughlin *et al.*, 1985; Emami and Perez, 1985; Emami *et al.*, 1987; Lam *et al.*, 1988; Puthawala *et al.*, 1985; Roberts *et al.*, 1986, 1987; Salcman and Samaras, 1983). The initial series involved small numbers of patients and primarily discussed the technical aspects of microwave-induced hyperthermia. More recently, the system's flexibility has become more widely accepted, and more recent series have included large numbers of patients.

The majority of clinical experiences with microwave antennas have used a 915 MHz power source. Other frequencies are available but this one is most popular because it yields a length of active heating of the order of 5–7 cm. A typical implant might involve from 2 to 20 interstitial antennas. The implant geometry and separation of catheters is governed by standard brachytherapy technology. As it turns out, iridium-192 dosimetry matches the thermal dosimetry of 915 MHz quite closely. Power sources of 2450 MHz have been used to treat smaller tumours, especially in the brain. Masses with a cross-sectional diameter of approximately 5 cm can be adequately covered with five antennas, one in each corner of a box array with a central antenna. The spacing of the antennas is 0.5–1 cm within the margin of the tumour volume. This allows at most a 2 cm separation between antennas. Thermometry can be reported in at least three points within each catheter and many points within a central catheter, which does not need an antenna due to the summation of microwave power in the centre of the array (Trembly, 1985). It is also standard practice at some centres to place a catheter perpendicular to the array for a thermal sweep to provide more complete three-dimensional thermal mapping.

Most programmes have had as a target dose 42–43°C for 60 min. Unfortunately, this is usually at an unspecified point within the tumour volume. Thermometry has been used in an effort to represent a thermal mapping with multipoint temperature distributions over time. Studying primarily superficial tumours with implant locations in the head and neck and the chest wall regions has made the assessment of responses relatively simple. However, several studies have extended interstitial techniques into deep-seated regions. The depth makes the response rate much more difficult to assess, requiring the use of indirect techniques, such as computed-tomography scanning, ultrasound or magnetic resonance imaging (MRI). A common shortcoming of the current clinical series has been the inability to record and assess the percentage of the tumour volume that was actually carried to a therapeutic temperature. In an attempt to overcome this problem, the Dartmouth group has tried to quantitate temperature by using a hyperthermia equipment performance (HEP) rating. This is defined as the percentage of the tumour volume that achieves some specific thermal end-point, e.g. 43°C. Another alternative is to attempt to correlate tumour response rate with tumour volume. Clearly, masses less than 4 cm in diameter have a much higher complete response rate than those greater than 4 cm in diameter (Emami *et al.*, 1987). The majority of patients with the smaller masses are treated with two hyperthermia sessions, one before and one after implantation of radioactive material, usually iridium-192. Iodine-125 has been used primarily in treating brain tumours.

There are several advantages to the use of implantable microwave antennas. They are compatible with the plastic catheters used in standard brachytherapy techniques. Power can be deposited in the tissue at some distance from the antennas and, therefore, relatively few sources must be inserted to adequately heat a given region. The actual thermal distribution measured depends critically upon blood flow, which can be quite heterogeneous. Blood flow turns out to be the most critical factor in the uniformity of temperatures achieved throughout a tumour volume.

The actual dosimetry is governed by the separation of the catheters necessary to deliver a uniform iridium-192 dose, establishing a feedback control system such that power can be varied to each antenna dynamically during a hyperthermia treatment, providing better control of the thermal distribution. This equipment is relatively easily obtained from several commercial vendors.

Clinically, there are relatively few disadvantages to the use of implantable microwave antennas. One is in the use of deep-seated implanted antennas, in which there may be heating at fixed points along the antenna shafts. It is important to measure thermal dose at several points outside of the tumour volume along the axis of the antenna. Care must be taken not to overheat the area where the antennas exit from the skin. Most implants must be left in place from 4 to 7 days to adequately deliver the prescribed radiation dose, allowing sufficient time to elapse such that thermotolerance is not a significant issue.

Interstitial techniques include relatively uniform heating within the implant volume, multipoint thermometry to more accurately define thermal mapping, and sparing of normal tissue, because the area implanted is the same as that to which the hyperthermia is delivered. Microwave antennas are compatible with standard brachytherapy techniques and can be used to deal with a wide variety of clinical circumstances.

Improvement in these techniques will depend on the development of antennas with insertion-depth-independent heating. It will also be important to increase the uniformity of heating along an antenna axis. Depending on the complexity of a given tumour volume, it may be desirable to create a family of different heating-length antennas in order to cover the entire tumour volume adequately. It will also be important to integrate three-dimensional treatment planning into the brachytherapy planning and the hyperthermia planning.

7.4. Curie point heating

A third alternative for interstitial hyperthermia is the implantation of ferromagnetic seeds throughout a tumour volume. The basic concept behind the use of ferromagnetic seeds is that if a patient is placed in a large magnetic coil, currents will be induced in the seeds, thus producing a temperature rise. Heat will then flow into the surrounding tissues due to thermal conduction. A major advantage to this technique is that once the seeds are implanted, heating can be conducted non-invasively. A magnetic coil could be placed around the volume of interest, and a hyperthermia session could be generated quite easily. However, since the temperature rise in the seeds is proportional to a variety of factors, for example the strength of the magnetic field, standard ferromagnetic materials would produce temperatures that would need to be monitored invasively. To solve this problem, several groups have been developing seeds whose Curie point is near the therapeutic temperature range. As a result, the temperature in the seeds is self-regulating and, potentially, thermometry would be unnecessary.

Curie point seeds have been developed over the past several years at the University of Arizona in Tucson and the University of Alabama in Birmingham. A potential advantage of ferromagnetic seeds is that this metal could be combined with an alloy used for radioactive materials containing iodine-125. These seeds could then be implanted within a tumour and left in place permanently, allowing the delivery of therapeutic doses of radiation through the natural isotopic decay of iodine-125. The seeds could be used in tumours for which seed implants are already standard practice within radiation oncology, especially carcinomas of the head and neck and the prostate.

Curie point heating could also have several other advantages. The seeds themselves could adjust their own heat at all depths, through thermal conduction away from each seed. A uniform seed distribution should be able to deliver uniform heating regardless of variable cooling, provided that seed distribution is sufficiently dense within the tumour mass; otherwise there would be cold spots. No internal connections to the heating source would be necessary. The seeds would be permanent surgical implants requiring only one manipulation and no other invasive procedures. Many heatings could then be delivered over a long period of time. There would be no restriction as to the actual number of heatings, so patient tolerance could be maximized. The dose along the seed implant would be uniform regardless of length. Heat delivery would be cylindrically as well as axially symmetrical, independent of the local tissue electrical properties. Depending upon the size of the mass, it would also be possible to alternate heat and radiation sources so there would be no restrictions as to the actual seed length. This alternation would, in theory, give the simplest thermometry and would require no computerization for thermal dose. Thermal regulation would be available on 100% of the sources, and many heatings could be delivered over an extended period of time.

Ferromagnetic seeds also have significant disadvantages. The first is that no power is deposited between the seeds. This leaves relative hot and cold spots within an implant volume. In order to change power deposition, the spacing between the seeds would need to be altered. In general, this would require spacing of less than 1 cm between seeds in each direction. Also, pretreatment planning would be absolutely critical as it would be impossible to adjust the sources or temperature distributions once the seeds are in place. If thermal distributions are inadequate, it would be impossible to redistribute the seeds. There would also be no real-time power control of individual seeds except through the Curie point regulation. If the patient's pain is not tolerable, the only alternative would be to terminate the treatment. Electromagnetic shielding would be required in the room, due to the strong magnetic fields necessary to reach therapeutic temperatures. The equipment is not portable and would require a significant initial investment. A high number of seeds would be required to cover a given tumour volume, for example a $5 \times 5 \times 5$ cm mass may require as many as 150 seeds with uniform spacing for adequate coverage. Also, once the seeds are in place it would be difficult to correct for seed migration after the implant. Migration is fairly common in a variety of tumour systems as the tumour either continues to grow or shrink in response to treatment.

There are several future goals for research with ferromagnetic seeds. The first would be to optimize the use of multitemperature ferromagnetic seeds within an array, determining whether each seed must be involved in the hyperthermia treatment or whether other spacings could be developed to adequately cover the tumour. Optimization of thermal distribution will require techniques for predetermining relative tumour blood flow and the optimal placement of variable temperature seeds. The treatment plan will have to remain static as changes in blood flow over time and with treatment cannot be accommodated. It may be possible to place these seeds within catheters or strings that can be moved once a treatment is complete. This would enable the clinician to vary thermal distribution over time to some degree. The trade off with seeds would obviously be that the ease of a given thermal treatment would be offset by the loss of the ability to contour thermal distributions to a tumour mass over time.

7.5. Conclusions

Interstitial techniques for the delivery of hyperthermia have gained a broader interest within the last several years. While each of these techniques has its major advantages, each suffers from its own set of technical problems. Microwave technology has improved more dramatically than either RF or seed technology, and has now gained the interest of multi-institutional co-operative group study, primarily through the Radiation Therapy Oncology Group (RTOG). Microwave technology has the advantage of being applicable to a broader spectrum of clinical situations in the field of oncology, yet further development of better antennas is necessary. The ability to heat larger portions of a tumour to therapeutic temperatures as well as to maintain real-time feedback control is also quite important. As thermal dosimetry becomes more easily manipulated, clinical trials can be developed for which hyperthermia can be integrated into primary therapy rather than used merely with palliative intent. Then the field of hyperthermia will gain the credibility necessary for its establishment as a curative modality. Only then will hyperthermia truly be integrated into multimodal approaches to the treatment of cancer.

References

Astrahan, M. A. and George, F. W. III (1980) A temperature regulating circuit for experimental localized field hyperthermia systems, *Medical Physics*, **7**, 362–364.

Astrahan, M. A. and Norman, A. (1982) A localized current field hyperthermia system for use with 192-iridium interstitial implants, *Medical Physics*, **9**, 419–424.

Cosset, J. M., Dutreix, J., Dufour, J., Janoray, P., Damia, E., Haie, C. and Clarke, D. (1984) Combined interstitial hyperthermia and brachytherapy: Institut Gustave-Roussy technique and preliminary results, *International Journal of Radiation Oncology, Biology and Physics*, **10**, 307–312.

Cosset, J. M., Dutreix, J., Haie, C., Gerbaulet, A., Janoray, P. and Dewar, J. A. (1985) Interstitial thermoradiotherapy: a technical and clinical study of 29 implantations performed at the Institut Gustave-Roussy, *International Journal of Hyperthermia*, **1**, 3–13.

Coughlin, C. T., Douple, E. B., Strohbehn, J. W., Eaton Jr, W. L., Trembly, B. S. and Wong, T. Z. (1983) Interstitial microwave-induced hyperthermia in combination with brachytherapy, *Radiology*, **148**, 285–288.

Coughlin, C. T., Wong, T. Z., Strohbehn, J. W., Colacchio, T. A., Belch, R. Z. and Douple, E. B. (1985) Intra-operative interstitial microwave-induced hyperthermia and brachytherapy, *International Journal of Radiation Oncology, Biology and Physics*, **11**, 1673–1678.

Denman, D. L., Elson, H. R., Lewis Jr, G. C., Breneman, J. C., Clausen, C. L., Dine, J. and Aron, B. S. (1988a) The distribution of power and heat produced by interstitial microwave antenna arrays: I. Comparative phantom and canine studies, *International Journal of Radiation Oncology, Biology and Physics*, **14**, 127–137.

Denman, D. L., Foster, A. E., Lewis Jr, G. C., Redmond, K. P., Elson, H. R., Breneman, J. C., Kereiakes, J. G. and Aron, B. S. (1988b) The distribution of power and heat produced by interstitial microwave antenna arrays: II. The role of antenna spacing and insertion depth, *International Journal of Radiation Oncology, Biology and Physics*, **14**, 537–545.

Doss, J. D. and McCabe, C. W. (1976) A technique for localized heating in tissue: an adjunct to tumor therapy, *Medical Instrumentation*, **10**, 16–21.

Emami, B. and Perez, C. A. (1985) Interstitial thermoradiotherapy: an overview, *Hyperthermia Oncology*, **1**, 35–40.

Emami, B., Perez, C. A., Leybovich, L., Straube, W. and Vongerichten, D. (1987) Interstitial thermoradiotherapy in treatment of malignant tumours, *International Journal of Hyperthermia*, **3**(2), 107–118.

Gerner, E. W., Connor, W. G., Boone, M. L. M., Doss, J. D., Mayer, E. G. and Miller, R. C. (1975) The potential of localized heating as an adjunct to radiation therapy, *Radiology*, **116**, 433–439.

Jones, K. M. (1987) *Evaluation and improvement of an interstitial microwave antenna array hyperthermia system*, ME Thesis, Thayer School of Engineering, Dartmouth College, Hanover, New Hampshire.

Jones, K. M., Mechling, J. A., Trembly, B. S. and Strohbehn, J. W. (1988) Theoretical and experimental SAR distributions for interstitial dipole antenna arrays for hyperthermia, *IEEE Transactions on Microwave Theory and Techniques*, **35**(10), 851–897.

Joseph, C. D., Astrahan, M., Lipsett, J., Archambeau, J., Forell, B. and George III, F. W. (1981) Interstitial hyperthermia and interstitial iridium 192 implantations: a technique and preliminary results, *International Journal of Radiation Oncology, Biology and Physics*, **7**, 827–833.

King, R. W. P., Trembly, B. S. and Strohbehn, J. W. (1983) The electromagnetic field of an insulated antenna in a conducting or dielectric medium, *IEEE Transactions on Microwave Theory and Techniques*, **31**, 574–583.

Lam, K., Astrahan, M., Langholz, B., Jepson, J., Cohen, D., Luxton, G. and Petrovich, Z. (1988) Interstitial thermoradiotherapy for recurrent or persistent tumours, *International Journal of Hyperthermia*, 4(3), 259–266.

Lee, D. J., O'Neill, M. J., Lam, K. S., Rostock, R. and Lam, W. C. (1986) A new design of microwave interstitial applicators for hyperthermia with improved treatment volume, *International Journal of Radiation Oncology, Biology and Physics*, 12, 2003–2008.

Lin, J. C. and Wang, Y. J. (1987) Interstitial microwave antennas for thermal therapy, *International Journal of Hyperthermia*, 3, 37–47.

Manning, M. R., Cetas, T. C., Miller, R. C., Oleson, J. R., Connor, W. G. and Gerner, E. W. (1982) Clinical hyperthermia: results of a phase I trial employing hyperthermia alone or in combination with external beam or interstitial radiotherapy, *Cancer*, 49, 205–216.

Mendecki, J., Friedenthal, E., Botstein, C., Sterzer, F., Paglione, R., Nowogrodzki, M. and Beck, E. (1977) Microwave applicators for localized hyperthermia in the treatment of malignant tumors, *Journal of Bioengineering*, 1, 511–518.

Mendecki, J., Friedenthal, E., Botstein, C., Sterzer, F., Paglione, R., Nowogrodzki, M. and Beck, E. (1978) Microwave induced hyperthermia in cancer treatment: apparatus and preliminary results, *International Journal of Radiation Oncology, Biology and Physics*, 4, 1095–1103.

Oleson, J. R. and Cetas, T. C. (1982) Clinical hyperthermia with RF currents, in Nussbaum, G. H. (Ed.), *Physical Aspects of Hyperthermia*, New York, American Institute of Physics, pp. 280–306.

Oleson, J. R., Sim, D. A. and Manning, M. R. (1984) Analysis of prognostic variables in hyperthermia treatment of 161 patients, *International Journal of Radiation Oncology, Biology and Physics*, 10, 2231–2239.

Puthawala, A. A., Syed, N., Sheikh, A. M., Khalid, M. A., Rafie, S. and McNamara, C. S. (1985) Interstitial hyperthermia for recurrent malignancies, *Endocurietherapy/Hyperthermia Oncology*, 1, 125–131.

Roberts, D. W., Coughlin, C. T., Wong, T. Z., Fratkin, J. D., Double, E. B. and Strohbehn, J. W. (1986) Interstitial hyperthermia and iridium brachytherapy in treatment of malignant glioma: a phase I clinical trial, *Journal of Neurosurgery*, 64, 581–587.

Roberts, D. W., Strohbehn, J. W., Coughlin, C. T., Ryan, T. P., Lyons, B. E. and Double, E. B. (1987) Iridium-192 brachytherapy in combination with interstitial microwave-induced hyperthermia for malignant glioma, *Applied Neurophysiology*, 50, 287–291.

Salcman, M. and Samaras, G. M. (1983) Interstitial microwave hyperthermia for brain tumors, *Journal of Neuro-Oncology*, 1, 225–236.

Satoh, T. and Stauffer, P. R. (1988) Implantable helical coil microwave antenna for interstitial hyperthermia, *International Journal of Hyperthermia*, 4, 497–512.

Strohbehn, J. W. (1987) Interstitial techniques for hyperthermia, in Field, S. B. and Franconi, C. (Eds), *Physics and Technology of Hyperthermia*, Dordrecht, Martinus Nijhoff, pp. 211–239.

Strohbehn, J. W. and Double, E. B. (1984) Hyperthermia and cancer therapy: a review of biomedical engineering contributions and challenges, *IEEE Transactions on Biomedical Engineering*, 31, 779–787.

Strohbehn, J. W. and Mechling, J. A. (1986) Interstitial techniques for clinical hyperthermia, in Hand, J. W. and James, J. R. (Eds), *Physical Techniques for Clinical Hyperthermia*, Letchworth, Research Studies Press, pp. 210–287.

Taylor, L. S. (1978) Devices for microwave hyperthermia, in Streffer, C. *et al.* (Eds), *Cancer Therapy by Hyperthermia and Radiation*, Munich, Urban and Schwarzenberg, pp. 115–117.

Taylor, L. S. (1980) Implantable radiators for cancer therapy by microwave hyperthermia, *Proceedings of the IEEE*, 68, 142–149.

Trembly, B. S. (1982) The electric field of an insulated linear antenna embedded in an electrically-dense medium, PhD Thesis, Dartmouth College, Hanover, New Hampshire.

Trembly, B. S. (1985) The effects of driving frequency and antenna length on power deposition within a microwave antenna array used for hyperthermia, *IEEE Transactions on Biomedical Engineering*, 32(2), 152–157.

Vora, N., Forell, B., Joseph, C., Lipsett, J. and Archambeau, J. D. (1982) Interstitial implant with interstitial hyperthermia, *Cancer*, 50, 2518–2523.

Vora, N., Luk, K., Forell, B., Findley, D. O., Lipsett, J. A., Penzer, R. D., Desae, K. R., Wong,

J. Y. C. and Hill, B. (1988) Interstitial local current field hyperthermia for advanced cancers of the cervix, *Endocurietherapy/Hyperthermia*, 4, 97–106.

Wong, T. Z., Strohbehn, J. W., Jones, K. M., Mechling, J. A. and Trembly, B. S. (1986) SAR patterns from an interstitial microwave antenna array hyperthermia system, *IEEE Transactions on Microwave Theory and Techniques*, **34**(5), 560–567.

Yabumoto, E. and Suyama, S. (1984) Interstitial radiofrequency hyperthermia in combination with external beam radiotherapy, in Overgaard, J. (Ed.), *Hyperthermic Oncology 1984*, Vol. 1, London, Taylor and Francis, pp. 579–582.

8 Clinical hyperthermic practice: whole-body hyperthermia

J. van der Zee, G. C. van Rhoon, N. S. Faithfull and A. P. van den Berg

8.1. Introduction

Whole-body hyperthermia is, at present, the only method with which a homogeneous temperature distribution can be reached within a deep-seated tumour. Disadvantages are that the tumour cannot be heated preferentially and that the maximum temperature tolerated lies around 41.8–42°C (Pettigrew *et al.*, 1974a). However, it has been shown that at this temperature level tumour cell mortality can be achieved.

The concept that tumour tissue can be destroyed by a temperature increase of only a few degrees above the normal physiological level has a long history, as mentioned in the introduction to this book. For example, Kluger (1980) described how, in the fourth century BC, Rufus of Ephesus had observed the beneficial role of fever, and had advocated the use of fever induction for, among other diseases, malignant tumours. In the nineteenth century, two publications reported tumour regression and even cure of patients following infection accompanied with high fever (Busch, 1866; Bruns, 1888). At the end of the nineteenth century, Coley was administering bacterial pyrogens to induce pyrexia in cancer patients. In a review, his daughter (Nauts, 1982b) states that complete regression and five year survival had occurred in 46% of 523 inoperable cases and in 51% of 374 operable cases. The beneficial effects of such fever therapy are attributed, according to some immunotherapists, to stimulation of the immunological system (Mastrangelo *et al.*, 1984). The observation that results improved when higher temperatures were attained (Nauts, 1982b), however, can be interpreted by the finding that heat itself is a cytotoxic agent.

8.2. *Methods employed for whole-body hyperthermia (WBHT)*

The problem of how to induce hyperthermia and how to control the quality and the safety of the treatment cannot be discussed without first discussing some basic thermophysiology (Ganong, 1971; Ingram and Mount, 1975; Emsly–Smith *et al.*, 1983). The body temperature normally is kept within narrow limits. In order to maintain thermal equilibrium, the heat gained by the body must be balanced by its losses. Each violation of this equilibrium will be followed by physiological reactions.

Heat loss to the environment from the skin surface produces a temperature gradient between the core and skin surface (shell). The magnitude of the temperature difference between the various parts of the body varies with the environmental temperature. The core temperature, represented by the rectal temperature, varies least with changes in environmental temperature.

8.2.1. Mechanisms for heat gain

Man gains heat mostly from metabolic sources, at rest (basal metabolism, food-induced thermogenesis), during exercise and while shivering. The most active sites of heat production are the skeletal muscles, the liver and other intra-abdominal organs and the brain. In cold environments, shivering can increase the metabolic rate by a factor of two to five. The main non-metabolic sources for heat gain are infra-red radiation, short-wave radiation from the sun, long-wave radiation from the surroundings, ingestion of hot food and drink, ventilation in a hot environment and conduction of heat, e.g. during immersion in hot water.

8.2.2. Mechanisms for heat loss

The loss of heat from the skin is greatly influenced by environmental conditions such as the ambient temperature, relative humidity and air movements. It also depends on the heat conductance of the skin, which is influenced by skin blood flow. Heat is lost by convection, radiation, conduction and evaporation. If the subject is in air, the loss takes place mainly by convection and radiation, the amount depending upon ambient temperature and air currents. Heat loss by conduction is important during immersion in cold water.

If these passive mechanisms for heat loss are insufficient for maintaining thermal equilibrium, then cooling through evaporation of water becomes an important contributor. Normally there is an 'insensible' water loss which takes place partly through the skin: water vapour diffuses continuously through the skin without wetting it. Water is also lost from the respiratory tract: water from the mucous membranes of the mouth and the respiratory passages vaporizes in the respiratory air and is lost with each expiration. The amount of the insensible water loss, hence heat loss, depends on the humidity of the environment and the inspired gases. Thermoregulatory sweating involves the evaporation of large amounts of water by the eccrine sweat glands. Eccrine sweat glands are capable of producing over 4 litres dilute salt solution per hour. The water in sweat evaporates on the skin, so reducing the skin temperature, depending on the ambient relative humidity and air currents. During complete thermal insulation of the body from its environment, the metabolic heat production can cause a temperature increase of about 1 °C per hour (van der Zee *et al.*, 1987).

8.2.3. Thermoregulation

Temperature regulation mechanisms appear to be controlled by a central 'set-point', anatomically located in and around the hypothalamus and connected with thermoreceptors in the skin, organs and tissues involved in the gain or loss of heat. Peripheral thermoregulation is centred on the thermal conductivity of the skin. Thermal conductivity of the skin is defined as the rate of change of heat per 1°C difference in temperature between that of the body and its surroundings. The thermal conductivity can be varied rapidly by redistribution of blood flow. Vasoconstruction reduces the transfer of heat from the core to the surface so that the surface of the skin becomes ischaemic and less heat is lost by convection, radiation and conduction. Cutaneous vasodilatation has the opposite effect. The superficial blood flow may increase to as much as 100 times the minimum flow.

8.2.4. Techniques for the induction of WBHT

In WBHT, the temperature of the patient's entire body is raised; to do this the thermoregulatory system has to be bypassed. The modern approach is to both introduce thermal energy into the body and reduce heat losses from the patient. Many methods have been and are still being used to apply WBHT. The oldest one is fever induction by administration of bacteria or bacterial toxins, which resets the patient's thermostat to a higher set-point (Coley, 1893; Nauts, 1982a). Methods that deposit energy into the body can be divided into two categories — invasive and non-invasive. The techniques involved are summarized in Figure 8.1. In non-invasive methods the energy is applied to the body surface. The energy is then absorbed by the blood and transported through the whole body. The non-invasive techniques that have been employed include the use of hot air, hot water or wax, either in direct contact with the skin or within bags, mattresses or suits; radiation such as infra-red or electromagnetic

	TRANSCUTANEOUS	*Heating rate (°C h)*
Molten wax (ANAESTHETIC, WAX)	Pettrigrew *et al.* (1974) Blair and Levin (1978) Greenlaw *et al.* (1980a	3–6 1.7–3.1
Water blanket or suit (ANAESTHETIC, WATER)	Barlogie *et al.* (1979) Bull *et al.* (1979) Larkin *et al.* (1977) Moricca *et al.* (1979) Herman *et al.* (1982) Gerad *et al.* (1984)	1.9 2.5–3.3 2.5–3.3 1.6–2.5 1.8 1–2.5
Infra-red (RADIANT + AIR)	Heckel and Heckel (1979) Robins *et al.* (1985)	1.1–3.3 4.8
Pomp–Siemens cabin (ANAESTHETIC, AIR, RF OR WATER)	Priesching (1976) Pomp (1978) Engelhardt *et al.* (1987) van der Zee *et al.* (1987) Kirsch and Schmidt (1966) Wüst *et al.* (1975)	3.5–5.3 2–3.6
Waterbath (ANAESTHETIC, WATER, HOT COLD)	von Ardenne (1980) Versteegh (1980)	6–10
	INVASIVE	
Femoral arterial–venous shunt (ANAESTHETIC, BLOOD, WATER)	Parks *et al.* (1979) Herman *et al.*, (1982) Lange *et al.* (1983)	13.4 3.8 2.5–4.9
Peritoneal irrigation (ANAESTHETIC, HEATED FLUID)	Priesching (1976)	

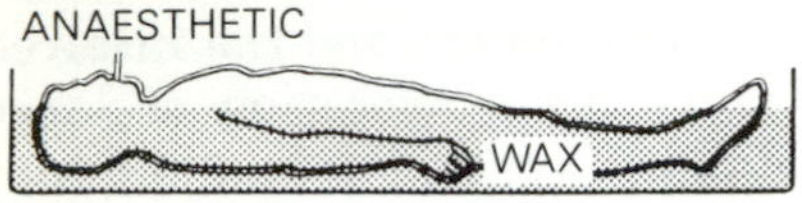

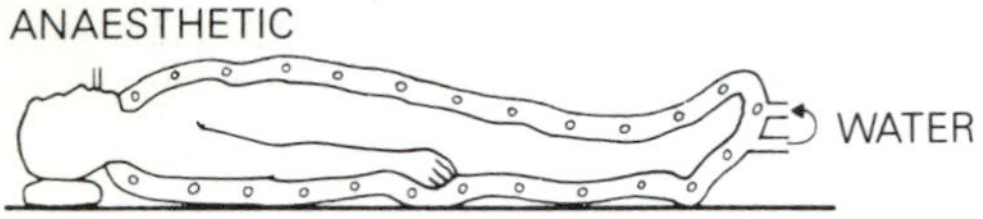

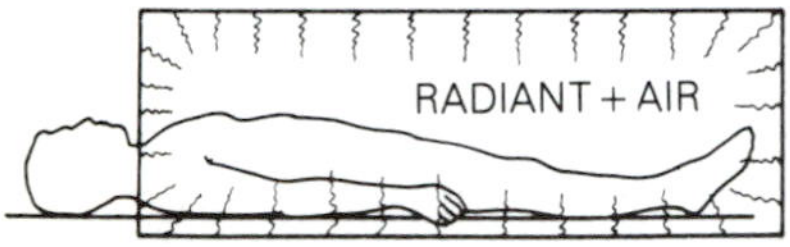

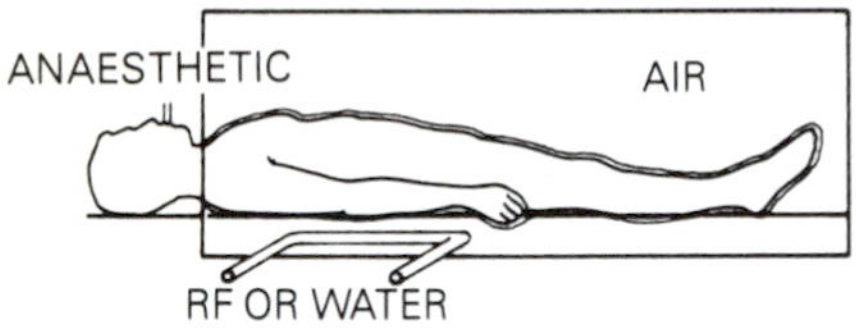

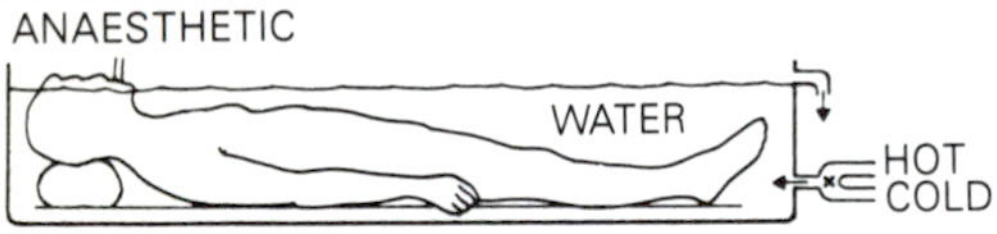

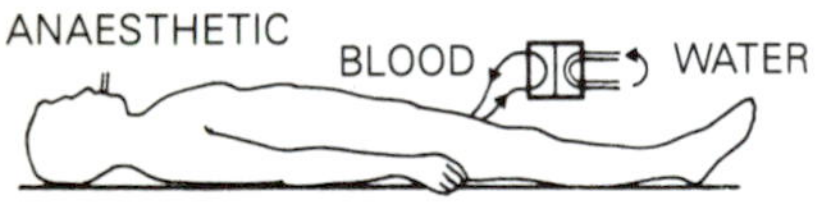

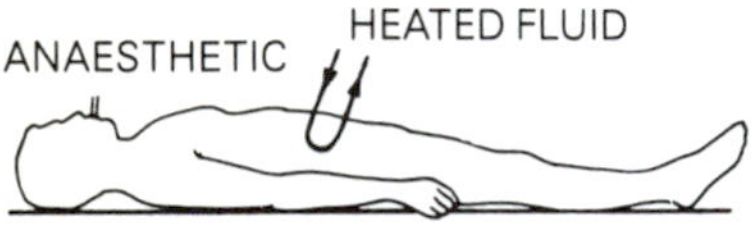

Figure 8.1 Methods used and in use for WBHT induction and heating efficiency. The various techniques for induction of WBHT are shown, including references. The heating rate achieved with each technique, as far as these are published, is given in °C per minute, either as a range or as a mean value.

radiation; or a combination of two or more of these methods. The immersion of the patient in hot water or wax includes simultaneous blockage of the patient's cooling mechanisms. With the alternative methods, additional isolation is necessary, e.g. with a plastic or aluminium sheet or a blanket. The importance of this additional isolation for the heating rate and the temperature distribution achieved is illustrated in Figure 8.2 (van der Zee, 1987). Invasive techniques for energy deposition are, for example, extracorporeal circulation or peritoneal irrigation with heated fluids.

Table 8.1 Heat dose administered during WBHT.

Reference	*Temperature (°C)*	*Time (h)*	*Combination*	*Number of patients*
High dose (temperatures above 41°C)				
Kirsch and Schmidt (1966)	42–44	0.5–2	None	48
Pettigrew (1975)	41.8	1.5–4	None, RT, CT	67
Mackenzie *et al.* (1975)	42.5 (max)	3–4	None	34
Larkin *et al.* (1977)	42	2	None	24
Bull *et al.* (1979)	41.8	4 (max)	None	14
Moricca *et al.* (1979)	41.8	3	CT	15
Robins *et al.* (1982, 1988, 1989a,b)	41.8	2.3 (max)	None, CT, RT, IT	54
Lange *et al.* (1983)	41.8	5–6	CT	14
Gerad *et al.* (1984)	41.8–43	2	CT	11
Koga *et al.* (1985)	41.5–42	3–4	CT	17
van der Zee (1987)	41.8–42	2–3	RT, CT, none	27
Maeta *et al.* (1987)	41.5–42	2–7	CT	168
Moderate dose (temperatures upto 41°C)				
Heckel and Heckel (1979)	39–41.4	1–11.7	None	46[1]
Wallach *et al.* (1982)	40–41	1 (min)	RT	9
Neumann *et al.* (1982)	40–41	1	CT	24

[1] Including 23 patients with non-malignant diseases.
RT, radiotherapy; CT, chemotherapy; IT, immunotherapy.

The temperature increase that can be achieved by these methods is restricted by the function and thermosensitivity of some critical systems, especially of heart and lungs, liver and brain. The maximum tolerable temperature is generally assumed to be 42°C, although treatments at higher temperatures have been reported (Herman *et al.*, 1982; Kirsch and Schmidt, 1966; Parks *et al.*, 1979; Gerad *et al.*, 1984). A review of heat doses administered during WBHT is given in Table 8.1.

8.2.5. Thermometry

Adequate thermometry is an essential requisite, even more so than in other fields of hyperthermia. Since temperature increases as small as 0.1°C become critical when the core temperature reaches 42°C,

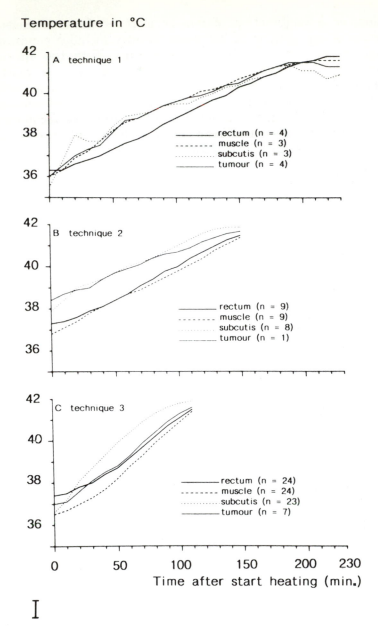

Temperature in °C

Figure 8.2 Temperature distribution during WBHT. This figure summarizes the temperature distribution achieved with the various techniques as have been used in Rotterdam (van der Zee, 1987). In panel I, the temperature distribution during the heating phase is shown, each graph represents the mean courses of temperature, with the number of treatments given in parentheses. In panel II, the temperature distribution during plateau phase is given, with the means and SDs together with

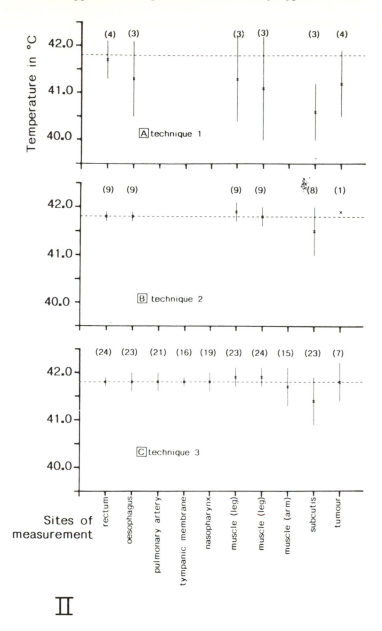

the number of treatments in parentheses. Technique 1: warm air only; technique 2: warm air and insulation with plastic foil; technique 3: warm air, plastic foil and warm-water-perfused water mattress. Additional insulation with plastic foil improves homogeneity of temperature distribution; additional heating through the skin of the back by the water mattress increases the heating rate.

inaccurate thermometry could be disastrous for the patient. Indeed, a lethal event due to thermometer failure has been reported by Pettigrew *et al.* (1974a). Therefore, all investigators using WBHT should use at least two independent thermometry systems. This can be readily done because, in addition to the specific hyperthermia thermometry system, many electrocardiogram monitoring systems have built-in thermometry facilities.

During a treatment with whole body hyperthermia, the accurate measurement of temperatures which are representative for the temperature in organs which are especially heat sensitive is of vital importance. Two relevant questions are therefore: 'where to measure?' and 'how to measure?'. The question on how to measure is extensively discussed in Chapter 15.

8.2.6. Sites of measurement

The choice of measurement sites depends on the temperature distribution in the body. Temperatures of the various tissues vary depending on the rate of metabolism, the efficiency of heat discharge and the distance to the body surface. During normothermia, or during a steady-state hyper- or hypothermia, the rectal temperature may be regarded as representative for core temperature. The response to changes in temperature, however, is relatively slow, which makes the use of rectal temperature less appropriate for control during the heating phase. The temperature in the region of the oesophagus which lies next to the aorta reflects the central blood temperature and is, therefore, a better representative of changes in core temperature than rectal temperature. The temperature within the pulmonary artery is of course the fastest responding and most reliable representative of core temperature.

The superficial temperatures may be very different from the temperatures in the deeper tissues. Skin temperature may be very high during cutaneous induction of WBHT, and may limit the rate of heating. During steady-state hyperthermia, skin temperature may be lower than core temperature (Figure 8.2). The tympanic-membrane temperature is expected to reflect hypothalamic temperature (Benzinger and Taylor, 1963; Cabanac and Caputa, 1979), and thus the temperature of the central nervous system. Cabanac and Caputa subjected healthy volunteers to heat stress, with or without facial cooling. When facial cooling was applied, the tympanic temperature decreased while the oesophageal temperature increased. An explanation for this finding is that there is a connection between the branches of the facial veins and the cavernous sinus (via the ophthalmic veins, the deep facial vein and the pterygoid plexus), through which cooler blood enters the skull and subsequently cools the brain tissue. The different techniques used for the induction of WBHT may result in characteristic temperature distributions. Temperature gradients during WBHT measured in experimental animals have only limited value for the clinical situation, as thermoregulation mechanisms generally are different. It is therefore important to measure temperatures at as many sites as possible during WBHT in patients, because only then it is possible to evaluate the overall temperature distribution, which is of importance with regard to patient toxicity and tumour response.

8.3. Combination of WBHT with other treatment modalities

It is generally accepted that hyperthermia up to 42 °C by itself is not able to destroy all tumour cells. In the Rotterdam Institute, the original design was to induce WBHT up to a level which was tolerated by an unanaesthetized patient and was safe with regard to normal tissue tolerance, and then to administer additional heating specifically to the tumour using microwave radiation at 433 MHz. This approach completely changed after the third treatment of the first patient. The first treatment of this patient, with extensive tumour involvement of the thoracic wall (recurrent breast cancer), had been given under sedation, to a maximum rectal temperature of 40.5 °C, combined with additional electromagnetic heating of the tumour mass. This treatment, which took over 8 h, was intolerable for the patient, and it was decided to give subsequent treatments under general anaesthesia. The second treatment involved a maximum rectal temperature of 41.4 °C and a maximum tumour temperature of 43.8 °C and yielded no problems. During the third treatment session, however, a very rapid increase in tumour temperature suddenly occurred. This resulted in the development of a severe third-degree burn of the thoracic wall, including necrosis of an area of rib cartilage. It is known now that the temperature distribution within an area heated by electromagnetic radiation may be very inhomogeneous and the possibility of development of a local undetected hot spot cannot be completely ruled out due to the limited number of temperature measurements. Alert patients with normal sensibility can warn when a local temperature increase is too high. It was decided to discontinue the use of local hyperthermia by electromagnetic heating in an anaesthetized patient, unless better temperature monitoring systems could be developed. From experimental studies, it is known that the effect of hyperthermia is complementary to that of radiotherapy and that both modalities can be given in combination without necessarily increasing toxicity (Overgaard, 1980) (see Chapter 2). At present, the possibility of achievement of therapeutic gain in a combination with chemotherapy is not yet fully understood. It is known that many drugs show an increased cytotoxicity under elevated temperatures (Bleehen, 1984; Landberg, 1985) (see Chapter 5), but it has not yet been shown that this should happen preferentially within tumour tissues. In fact, enhanced toxicity has been observed clinically, e.g. enhanced cardiotoxicity of doxorubicin (Kim *et al.*, 1979) and renal toxicity from *cis*-diamminedichloroplatinum(II) (Gerad *et al.*, 1983). It would be safest to give both treatment modalities sequentially, the action of hyperthermia is then purely additive and tumour debulking. However, several workers report having given chemotherapy and WBHT simultaneously without severe toxicity. In these cases, the drug dose is often reduced. Also, addition of a moderate level of WBHT (1 h at 40.5°C) to standard multichemotherapy (ACO: Adriamycin (doxorubicin), Cyclophosphamide and Vincristine) did not result in enhanced toxicity, as reported by R. Engelhardt and colleagues (presented at the fifth ISHO meeting in 1988).

8.4. Treatment

8.4.1. Pretreatment evaluation

Age is not considered to be a limiting factor, presuming the general condition of the patient is reasonable (Karnofsky performance status >50%, Maeta *et al.*, 1987). The cardiovascular and respiratory systems should be able to meet the demands imposed by WBHT. This can be investigated by previous history (no symptoms of myocardial hypoxia), radiography of the chest and ergometry. Carcinomatous lymphangitis of the lungs may result in post-hyperthermia problems, probably due to the development of oedema in the (diffusely spread) tumour metastases, resulting in increased obstruction of the airways (van der Zee, 1987). As the liver may be especially sensitive to heat (Pettigrew *et al.*, 1974b; Blair and Levin, 1978; Wike–Hooley *et al.*, 1983), patients with severe liver abnormalities, revealed by either laboratory tests, isotope scanning or diagnostic ultrasound, should be excluded. The presence of brain metastasis is a contra-indication, as this carries the risk of a fatal complication due to bleeding or development of oedema in the intracranial tumour (Priesching, 1976; Barlogie *et al.*, 1979; Herman *et al.*, 1982; Gerad *et al.*, 1984). Coagulation factors should be normal (Pettigrew *et al.*, 1974b; Barlogie *et al.*, 1979).

8.4.2. Preparation for WBHT

The rectum should be empty, e.g. by enema, to ensure reliable rectal temperature monitoring. During the preparatory phase, which inevitably takes some time, cooling of the patient should be prevented: hypothermia, accompanied by vasoconstriction in the skin, is an unfavourable starting situation for heating. Cooling can be prevented by using a warm-water-perfused mattress and/or blankets. When artificial respiration is used, the ventilation gases can be warmed. Infusion fluids can be passed through a heat exchanger. Development of pressure necrosis should be prevented by the avoidance of pressure sites during the treatment procedure. These can occur at the heels, the sacrum, the occiput, the elbows, and also at places where, for example, catheters or thermometry probes are attached to the skin. Following sedation or induction of anaesthesia, infusion and monitoring lines are attached and introduced for the following: intravenous fluid supply, electrocardiogram monitoring, cardiovascular dynamics, blood laboratory values, urine production, temperatures, etc. Finally, the installation of the heating equipment can be completed.

8.4.3. Anaesthesia

Most investigators applying WBHT to patients perform the treatment under general anaesthesia, although some treat the patients under sedation only (Bull *et al.*, 1979; Robins *et al.*, 1985). The reason for anaesthesia is that it alleviates the anxiety and discomfort. When general anaesthesia is applied, it is advisable to place all the required equipment after the patient is anaesthetized.

Anaesthesia applied during administration of WBHT varies widely. It has been noted that requirements for relaxants and volatile anaesthetic agents are increased during the hyperthermic phase and the early cooling phase (Faithfull *et al.*, 1984; Pettigrew and Ludgate, 1977).

8.5. Results

8.5.1. Efficiency of heating technique

During a WBHT treatment, three phases can be recognized. After the patient has been prepared (introduction of all monitoring lines and installation of heating equipment), heating can start. The efficiency of energy transfer is characterized by the heating rate (°C per hour) during this heating phase. After the desired temperature has been reached, the patient's temperature is controlled by adjustment of the equipment. The last phase is cooling, during which the patient's temperature is decreased, actively and/or passively, down to normal values.

8.5.2. Heating rate

Heating rates reported vary greatly (Figure 8.1). With transcutaneous methods, values ranging from 1 to 10°C per hour have been reported.

Pettigrew *et al.* (1974a) (heating with wax) report the achievement of a heating rate of 3–6°C per hour (depending on body weight) by using an epidural block which induces vasodilation. Vasodilation in the skin increases the heat conductance and in this way increases the energy transport through the skin into the body. Blair and Levin (1978) who also used wax for energy transfer found a considerably lower heating rate of 1.7–3.1°C per hour, the highest value being found when an epidural block was given. The major difference between the two techniques is that Pettigrew covered the patient with molten wax in immediate contact with the skin, whereas Blair and Levin used wax in bags which made the method cleaner but also may have resulted in an insulating layer between the wax and the skin of the patient.

For the methods using water-perfused blankets or suits, heating rates of 1–3.3°C per hour have been published, the mean of all values being 2.3°C per hour. Versteegh *et al.* (1981) found a mean heating rate of 7.3°C per hour using water bath heating in four patients during 10 treatments. The combination of hot air with radiofrequency as used by Engelhardt *et al.* (1982) results in a heating rate of 3.5–5.3°C per hour, but there are difficulties owing to disturbance of electronic monitoring systems and the risk of developing burns where sweat accumulates. With a combination of hot air, a warm-water mattress and insulation of the skin, a heating rate of 2–3.6°C has been reported by van der Zee (1987).

A considerably higher heating rate than those generally found with non-invasive techniques has been reported when heating was induced by femoral arteriovenous shunts. Parks *et al.* (1979) reported a mean heating rate of 13.4°C per hour (a temperature of 41.8°C was reached in only 22 min). This finding is much higher than reported by Lange *et al.* (1983) and Herman *et al.* (1982), being 2.5–4.4 and 3.8°C per hour, respectively. The major disadvantage of the invasive technique is that surgery is required before heating sessions can start.

8.5.3. Heat dose

The 'heat doses' (see Chapter 3) administered during WBHT can be broadly divided into two levels: (1) moderate, temperatures of up to 41°C, and (2) high, temperatures above 41°C. Most workers limit core temperatures during WBHT to a maximum of about 42°C, although there are also reports on the use of temperatures even above 43°C (Table 8.1).

8.5.4. Temperature distribution

Temperature distribution may vary largely depending on the method used. Differences in temperature between the various parts of the body can be critical for both tumour response and toxicity achieved.

Only a few workers report more than the measurement of core temperatures (rectal, oesophageal, bladder) during WBHT. With the use of water-perfused blankets, skin temperature was close to rectal and oesophageal temperatures during the heating phase, but became considerably lower than these temperatures after reducing the water temperature for the maintenance of core temperature at 41.9–42°C during plateau phase (Barlogie *et al.*, 1979). This pattern is somewhat different from the radiant-heat method (Robins *et al.*, 1985). Here skin temperature is up to 2°C higher during the heating phase, but 0.4–0.8°C lower than core temperature during plateau phase. Van der Zee (1987) reported important changes in temperature distribution with adjustment of the heating technique. Using warm air at 60–70°C only, the variation between temperatures at the various measuring sites was very large. At superficial sites (muscle, subcutis and tumour), the mean temperature during the heating phase was 0.5°C higher than the rectal temperature, but 0.4–1.1°C lower during the plateau phase. When insulation of the skin was added to the heating method, the maximum temperature difference between rectum and subcutaneous tissue became 0.3°C. This temperature distribution was not further improved by the addition of a warm-water mattress at the back of the patient, although the rate of heating increased (Figure 8.2).

8.6. Physiology

8.6.1. Cardiovascular changes

Within the Rotterdam institute, extensive measurement of cardiovascular function during WBHT was performed (Faithfull, 1983). This experience is presented here, followed by a short discussion of relevant reports from the literature.

During the heating phase the cardiac index rose significantly (P<0.001) due to increases in both heart rate and stroke volume, and marked decreases in both systemic and pulmonary vascular resistance were observed (P<0.001). Because pulmonary resistance decreased proportionately less than systemic resistance, the increases in right-ventricular work were much more than left-ventricular work. During the first half of the plateau period, further decreases in systemic vascular resistance were accompanied by very significant falls in mean systemic arterial pressures (P<0.001). Hence, left-ventricular work decreased (P<0.001) but no decrease in right-ventricular work was seen. From this point on, the left-ventricular work was not significantly more than before the induction of anaesthesia. This is in contrast to right-ventricular work which was very significantly raised above the pre-induction values throughout the plateau phase. During the second half of plateau there were no significant changes in cardiovascular haemodynamics, but once cooling commenced, the most obvious and constant change observed is a significant fall of mean arterial pressure at 15 min of cooling (P<0.01), followed by a significant rise at 30 min (P<0.05). Over the same period pulmonary artery pressure was steadily increased while cardiac index did not decrease to any marked extent. Hence left-ventricular work significantly decreased and right-ventricular work remained constant.

The majority of the WBHT series have been conducted under anaesthesia due mainly to the unacceptable psychological stress of heating. Recently, Robins *et al.* (1985) have developed a system in which only very slight sedation is employed. Barlogie *et al.* (1979), Blair and Levin (1977), Larkin *et al.* (1977), Parks *et al.* (1979), Pettigrew and Ludgate (1977) and Cronau *et al.* (1984) have employed general anaesthesia while Bull *et al.* (1979), Bynum *et al.* (1977) and Ostrow *et al.* (1981) have used a generalized sedation technique. Most workers report changes taking place in heart rate and, though absolute figures vary, fairly good correlation is obtained when the changes are expressed as change in the number of beats per minute per 1°C rise of body temperature (beats.$^{-1}$min^{-1}/°C). The mean value from the literature is 11.7 beats/min/°C (SEM=0.73).

Systemic arterial pressure changes during WBHT have been variable. Some authors report decreases in pressure (Dubois *et al.*, 1980; Larkin *et al.*, 1977; Barlogie *et al.*, 1979; Lees *et al.*, 1980; Kim *et al.*, 1979; Herman *et al.*, 1982) while others report increases (Pettigrew *et al.*, 1974b; Euler–Rolle *et al.*, 1978). Still others noted no changes (Moricca *et al.*, 1979; Ostrow *et al.*, 1981; Bynum *et al.*, 1977). Kim *et al.* (1979) presented results indicating cardiac indices increasing by an average of 72% and Cronau *et al.* (1984) report a doubling of cardiac output; our own figures indicate a 140% rise up to 41.8°C. From the figures of Lees *et al.* (1980) it may be calculated that, in their patients, left-ventricular work increased and systemic vascular resistance decreased by 25 and 51%, respectively.

There are few publications on the haemodynamic changes taking place in the pulmonary circulation under WBHT treatment. The filling pressure of the right ventricle (the central venous pressure) has been fairly widely reported but published results are very variable; Pettigrew and Ludgate (1977) comment that central venous pressure is not a good index of the need for fluid replacement. This agrees with our own experience that administration of colloidal solutions in an attempt to maintain central venous pressure at a certain level only leads to interstitial oedema without significant alteration of cardiovascular parameters. Parks *et al.* (1979) noted that pulmonary arterial pressures were little affected and it can be calculated that their patients experienced an average rise in right-ventricular work of 110% and a fall in pulmonary vascular resistance of 7%. In the Rotterdam investigation, right-ventricular work rose by 215% and pulmonary resistance fell by 50%.

The concept of vascular impedance was introduced by Randall and Stacy (1956) and was applied by Caro and McDonald (1961) to the pulmonary circulation. Milnor *et al.* (1966) in a study in dogs have demonstrated the effect of heart rate and pulmonary blood flow on the oscillatory component of hydraulic input power. At a fixed pulse rate, oscillatory power varies with the square of the flow. They demonstrated, however, that flow could be increased with less increase in input power if the pulse rate increased while the stroke volume remained constant. Hence, high pulse rates may prevent excessive rises of right-ventricular work in patients undergoing WBHT and one should be wary of trying to decrease heart rate by means of β-blockers (Euler-Rolle *et al.*, 1978; Moricca *et al.*, 1979). From the above brief discussion it is clear that the pulsatile nature of pulmonary vascular flow will tend to cause the increase in right-ventricular work to be proportionately more than is the case in the systemic circulation. In the latter the pulsatile component of work is much less and hence steady work output predominates.

In conclusion, attention should be drawn once again to the relative difference in amount of work performed by the left and right ventricles during WBHT and to reiterate that, though left-ventricular work was not significantly raised above pretreatment levels during the mid-plateau phase the right-ventricular work was very significantly raised at this stage. These important results may have considerable clinical relevance when assessing the fitness of patients to undergo WBHT treatment.

The Rotterdam experience is that following the hyperthermic treatment, in the intensive care unit, heart rate remains significantly elevated for 3 h after body temperature was normal. Pulmonary artery pressure does not return to normal values until 10 h after the beginning of cooling and though left-ventricular work was not significantly raised at any stage during the patients stay in the intensive care unit, the right-ventricular work did not return to normal until 11 h of cooling had elapsed. Thirteen hours after the cabin was opened the systemic vascular resistance returned to normal values. The only remaining 'abnormal' parameter was then the mean systemic arterial pressure. This remained significantly lower than before treatment commenced and even 18 h after the end of plateau (the end of this analysis) it was still significantly reduced — that is not to say that the pressure at this stage was at a level that caused any anxiety. At 18 h of cooling the mean pressure was 71.6 mmHg as opposed to 87.7 mmHg at the beginning of treatment.

Few workers have commented in detail on the cardiovascular state of patients following WBHT. Herman *et al.* (1982) reported 30 WBHT treatments — 16 using warming blankets and 14 using extracorporeal circulation — and in seven instances noted mean blood pressure decreases to 60–70 mmHg for up to 2.5 h after treatment. They also noted that decreases of pressure occurring at elevated temperatures were accentuated at the end of therapy when patients were actively cooled.

Pettigrew *et al.* (1974b) observed that a persistent tachycardia with low blood pressure could develop in patients with 'sensitive' tumours. Barlogie *et al.* (1979), commenting on the diastolic pressure (which had decreased during treatment), noted a rapid return to normal during the ensuing 12 h, while Bull *et al.* (1979) noted that five out of 14 patients had systolic blood pressures between 70 and 90 mmHg for 30 min to 3 h post-treatment. Post hyperthermic hypotension was also noted by Smith *et al.* (1980).

It would seem, from the above, that hypotension is a fairly frequently occurring phenomenon in the early recovery phase following WBHT. Apart from the comment by Pettigrew *et al.* (1974b) there have been no reports on pulse rate changes nor have there been reports of changes we noted in other cardiovascular parameters. The most likely explanation is that the appropriate investigations were never performed.

8.6.2. Oxygen transport and consumption

The amount of oxygen transferred from the lung to the cells of the body per minute is the product of the cardiac output and the oxygen content of the arterial blood, the latter being dependent on haemoglobin concentration and oxyhaemoglobin oxygen saturation. Figures for the oxygen combining power of haemoglobin vary from 1.39 ml/g (Braunitzer, 1963) through 1.34 (Prys-Roberts *et al.*, 1971) to 1.306 ml/g (Gregory, 1974). This latter figure was used for the results presented below. The small amounts of oxygen carried in solution, which will decrease as body temperature is raised, have been neglected in the calculation, but oxyhaemoglobin saturation was corrected for pH, temperature and pCO_2 using the computer routine described by Kelman (1966). Oxygen flux ($\dot{Q}O_2$) was calculated as the product of cardiac output and the arterial oxygen content, and oxygen consumption ($\dot{V}O_2$) as the product of cardiac output and the arterio-mixed venous oxygen content difference.

Both $\dot{Q}O_2$ and $\dot{V}O_2$ were unchanged by the induction of anaesthesia and both showed highly significant increases during heating and at plateau temperature. As cooling was instituted, both values fell and after half an hour of cooling oxygen consumption was still significantly greater than the pre-anaesthesia value. It is of great importance to know, not so much the absolute value for oxygen flux or oxygen consumption, but the ratio between the two. It should be noted that, during warming, there is a highly significant decrease in extraction indicating the proportionately greater increase in oxygen flux than that of oxygen consumption. Hence, $\dot{V}O_2$ requirements are very easily met during WBHT treatment.

8.7. Therapeutic effectiveness

Clinical results can be divided in two groups: (1) objective response, which implies measureable tumour reaction, and (2) subjective response, the change in quality of life. For the assessment of objective response the recommendations of the World Health Organisation (WHO) are generally used (Miller *et al.* 1981). These can be summarized as follows:

- Complete response (CR): disappearance of tumour for at least 4 weeks.
- Partial response (PR): a 50% decrease in tumour volume for at least 4 weeks.
- No response: a less than 50% decrease and less than 25% increase in tumour volume (no change=NC) or a more than 25% increase in tumour volume (progression).

An important parameter for subjective response is pain relief. Results from 17 reports on WBHT are summarized in Table 8.2.

8.7.1. Objective tumour response

After WBHT alone, complete response was never achieved in the studies listed in Table 8.2. Partial response was observed in an overall average of 29% of the patients. This agrees with the findings of experimental studies: a hyperthermia dose of 2–4 h at 41–42.2 °C may be sufficient to kill tumour cells in a thermosensitizing environment, e.g. low pH, but not the more heat-resistant tumour cells in well-perfused tumour regions. However, two cases in which complete response was obtained following WBHT alone were reported by Warren (1935) and Wüst (1975). Warren describes how all visible tumour nodules in a patient with metastatized hypernephroma disappeared over a period of 5 months following three WBHT treatments of 5 h at 41.5 °C. At that time, regrowth of brain metastases had occurred. Wüst reported a complete remission of 5 months duration obtained in a previously untreated (!) patient with Hodgkin's disease, following 12 WBHT treatments of 1 h at 40 °C. Results from 11 other patients were not so impressive, and he suggested that higher treatment levels would be necessary.

Following WBHT combined with chemotherapy, complete response was achieved in 7% of patients. The highest CR rate (53%) was reported by Engelhardt *et al.* (1982), who combined 1 h at 40.5 °C with multidrug therapy (ACO), in treating 15 patients with oat cell carcinoma of the lung. The 50% survival time in these patients was 12.3 months, which is a prolongation of 4.7 months in comparison with a historical control group treated with ACO under normothermic conditions. At the fifth ISHO meeting, Engelhardt (1988) presented the results of a randomized comparative trial in similar patients. The combined treatment resulted in

Table 8.2 Tumour response[1] following WBHT.

	WBHT alone			WBHT+chemo-immunotherapy			WBHT+radiotherapy		
	CR	Response	Pain relief	CR	Response	Pain relief	CR	Response	Pain relief
Pettigrew (1975)		22/49	24/49		12/15	10/15		3/3	3/3
Priesching (1976)					6/16				
Levin and Blair (1978)							1/14	8/14	++
Parks et al. (1979)				3/25	13/25	25/25			
Moricca et al. (1979)				0/15	1/15	11/15			
Bull et al. (1979)	0/13	1/13							
Barlogie et al. (1979)	0/5	0/5		0/6	0/6				
Hinkelbein et al. (1981)				2/11	4/11		2/9	4/9	
Herman et al. (1982)				0/10	6/10				
Engelhardt et al. (1982)				8/15	13/15				
Lange et al. (1983)				0/13	1/13				
van der Zee et al. (1983)	0/6	0/6	2/3	0/5	1/5	1/3	4/17	8/17	11/12
Gerad et al. (1984)				2/11	4/11				
Maeta et al. (1987)	0/6	0/6		4/126	34/126				
Robins et al. (1988, 1989b)				1/17	3/17		3/8	7/8	
Robins et al. (1989a)				0/17	2/17				
Overall	0/30	23/79	26/52	20/271	100/302	47/58	10/48	30/51	14/15
Percentage	0	29	50	7	33	81	21	59	93

[1] In this table, CR means complete response, and response means complete and partial response, according to the guidelines recommended by the WHO.

68% response, significantly higher than that (36%) following ACO at normothermia (Fisher's exact test, $P=0.034$). Following combined treatment, the duration of response was longer. In a phase I study, WBHT was combined with interferon (Robins *et al.*, 1989a). With this combination, partial response was achieved in 2/17 patients.

The combination of WBHT and radiotherapy results in the highest response rates: the overall complete response rate is 21%, and response is obtained in 59% of the patients. Response and even complete response has been obtained using radiation doses as low as 24 Gy or less (van der Zee *et al.*, 1983). The effect of WBHT in this combination, however, is only additive to that of the local radiotherapy. For the patient's benefit (see Section 8.8 on toxicity) the combination of radiotherapy with local hyperthermia is therefore preferable. There are, however, many problems in inducing local hyperthermia in deep-seated tumours, which have not yet been solved.

Remarkable results following WBHT and total-body irradiation (total dose 1.5–2 Gy) have been reported by Robins *et al.* (1989b). In patients with advanced B-cell neoplasms, complete response was achieved three times and partial response four times.

8.7.2. Palliation

Pain relief has been observed by many investigators following WBHT treatment, whether given alone or in combination with radiotherapy or chemotherapy. The reported percentage of patients experiencing pain relief is highest in the WBHT–radiotherapy group (92–100%). Some workers mention that pain disappeared immediately following treatment, even in patients in whom the tumour failed to respond (Levin and Blair, 1978; Parks *et al.*, 1979; Pettigrew *et al.*, 1974b; van der Zee, 1987). It has been suggested that the increase in β-endorphin blood levels, which was observed to be induced by WBHT (Robins *et al.*, 1987), might be responsible for this palliative effect. β-Endorphin is released by the pituitary gland under conditions of stress and has a strong analgesic action. Whatever the mechanism, for the patients this effect of pain relief means a worthwhile improvement of the quality of life.

8.8. Toxicity

Toxicity following WBHT does not only depend on the degree of hyperthermia applied during treatment. Many other factors influence the clinical outcome, including:

• The general condition of the patient and the condition of the organ systems (heart, lungs, liver, kidney, haemopoietic system).

- Physiological conditions during treatment, which may be influenced by anaesthetic procedures, artificial respiration, infusion schedule and heating-up time.
- Combination with (prior) chemotherapy or radiotherapy.
- Post-hyperthermic care.

Even (massive) tumour regression may cause clinical problems.

8.8.1. Toxicity at temperatures up to 41°C

Below a treatment temperature of 41°C, toxicity is minimal. The group in Freiburg experienced cardiac problems in two patients. A 68-year-old patient developed left-ventricular decompensation during her thirteenth WBHT, which was controlled without further problems (Hinkelbein *et al.*, 1981). In another patient, cardiac arrhythmia was observed during WBHT which necessitated termination of treatment. This complication was probably due to enhancement of the cardiotoxicity induced by doxorubicin, as the patient had received a cumulative dose of 620 mg (Neumann *et al.*, 1982). Normal rhythm returned when the patient was cooled to normothermia.

Other side-effects include decubitus and second-degree burns (Neumann *et al.*, 1979). Laboratory data for haematology and electrolyte blood levels

Table 8.3 Reported toxicity of WBHT.

Laboratory changes
Electrolyte disturbances:
  decreases in K^+, Ca^{2+}, Mg^{2+}, PO^{3-}
Haematology disturbances:
  decreases in haematocrit, platelets, fibrinogen; leukocytosis, disseminated intravascular coagulation
Total protein decrease
Enzyme increases:
  ALAT, ASAT, alkaline phosphatase, LDH, CPK
Kidney function disturbances:
  increases in urea and creatine, increases in urinary output of Na^+, K^+, decrease in Mg^{2+}

Clinical toxicity
Nervous system:
  hyperexcitability and agitation, hallucinations, myelopathy, epileptic insults, bleeding or oedema in metastatic tumour[1], brain oedema[1], peripheral neuropathy
Cardiovascular system:
  hypotension, cardiac arythmias and/or arrest[1]
Pulmonary system:
  capillary leak syndrome with pulmonary oedema[1], respiratory failure[1], adult respiratory distress syndrome[1]
Skin:
  decubitus, second- and third-degree burns, circumoral herpes simplex
Other organs:
  liver failure[1], nausea, vomiting and diarrhoea, disseminated intravascular coagulation[1]

[1] Reported lethality.

did not show significant changes during and following WBHT at a level up
to 41°C, and there were no indications for liver or kidney damage.

8.8.2. Toxicity at temperatures between 41 and 42°C

With treatment temperatures above 41°C, toxicity was much greater. A
summary of reported toxicities is given in Table 8.3. During WBHT,
cardiac dysarrythmia or circulation failure necessitated termination of
treatment in several cases. Hypotension during and following WBHT is
commonly observed. In cases where the cardiac output was measured
there was not usually any indication of decreased whole-body oxygen flux.

Decreases in the concentration of haemoglobin were seen by Pettigrew *et
al.* (1974b) and van der Zee *et al.* (1983). These changes were almost
certainly due to haemodilution consequent upon vasodilatation and net
movement of fluid into the vascular space. This view is also held by Larkin
et al. (1977). Many workers report a leukocytosis following WBHT.
Pettigrew *et al.* (1974a) reported a mean rise of 67% in white cell count
after 4 h WBHT, returning to normal within 24 h. There would appear to
be an absolute rise in the polymorphonuclear cells accompanied by
decrease of lymphocytes. Herman *et al.* (1982) found neutrophilic
leukocytosis when using the blanket technique but saw neutropenia in
some patients when employing the extracorporeal warming technique.
Parks *et al.* (1979), who also used extracorporeal heating, noted, in
contrast to most workers, that the leukocyte counts were little affected.
This might indicate differences caused by heating techniques *per se.* Van
der Zee (1987) found a significant increase in total white cell count for up to
4 h after cooling, and lymphocyte counts were depressed for 3 days
following WBHT. Leukocytosis induced by WBHT may have considerable
significance. Grogan *et al.* (1980) have demonstrated that polymorphonuclear cells from patients who have undergone WBHT, showed an
increase in bacteriocidal activity. Reduction in platelet counts frequently
occur following WBHT, though some workers (Bull *et al.*, 1977) have not
reported this effect. Pettigrew *et al.* (1974a) reported fatalities from
disseminated intravascular coagulation often associated with recent
tumour necrosis. Barlogie *et al.* (1979) also reported prolongation of the
prothrombin and partial thromboplastin times with mild to moderate
decrease of fibrinogen levels and increases in fibrin degradation products.
These trends were also reported by Herman *et al.* (1982), again
particularly in association with extracorporeal heating, and by van der
Zee (1987).

Electrolyte changes are common in association with WBHT: decreased
levels of magnesium, phosphate, calcium and potassium ions are frequently observed (Larkin *et al.* 1977; Barlogie *et al.*, 1979; Herman *et al.*,
1982; van der Zee *et al.*, 1983). The phosphate ion level tends to return to

normal after 24 h and that of magnesium ions after 48 h. The calcium ion decrease appears to be associated with the decreased total amount of protein, as the free calcium ion level remains normal (van der Zee, 1987). The urinary output of sodium, potassium and magnesium ions was found to be increased during 24 h following the start of WBHT (van der Zee, 1987).

Changes in kidney function do not generally pose a problem. During treatment oliguria may occur, but this recovers. Blood urea and creatinine levels have been found to be increased in the period 24–48 h following WBHT. Increases in liver enzyme levels (ALAT, ASAT, LDH), indicating liver damage, are frequently observed and in some cases liver damage causes a clinical problem. Complications concerning the central nervous system have also been reported: brain oedema, epileptic insults, and oedema or haemorrhage from brain metastases causing clinical symptoms. It is unclear, however, whether this results from hyperthermia-induced damage to the central nervous system or from failing oxygenation of the brain tissue (Sminia *et al.*, 1988). Acute myelopathy, probably due to synergism of prior radiotherapy and hyperthermia and further complicated by simultaneous administration of Carmustine and Metronidazole, has been described by Douglas *et al.* (1981). Peripheral neuropathy, in some cases possibly due to 'recall' vinca alkaloid toxicity also has been described (Barlogie *et al.*, 1979; Bull *et al.*, 1979).

Pressure sores and second- and third-degree burns occur more frequently with treatment temperatures above 41°C than at moderate-level WBHT. Development of pressure sores and burns can be partially prevented by taking measures to avoid pressure. Following WBHT many patients experience nausea, vomiting and diarrhea for 1–2 days. Herpes simplex infection occurs in many patients, according to some investigators only after the first WBHT (Gerad *et al.*, 1984; Pettigrew *et al.*, 1974a; Barlogie *et al.*, 1979; Moricca *et al.*, 1979), but others report that after subsequent treatments herpes simplex recurs (van der Zee, 1987).

WBHT has proved to be lethal for a number of patients. Death has been ascribed to cardiac failure, respiratory failure, massive liver necrosis, brain oedema, oedema or bleeding in intracranial tumours, disseminated intravascular coagulation, massive tumour necrosis causing an overload of toxic products to the circulation, and from abundant haemorrhage of (regressing) tumours (Priesching, 1976; Pettigrew *et al.*, 1974a; Larkin *et al.*, 1977; Moricca *et al.*, 1979; Bull *et al.*, 1979; van der Zee *et al.*, 1983). The general condition of patients appears to be of vital importance: Maeta *et al.* (1987) report that 19/24 patients who died as a result of treatment complications had a pretreatment performance (Karnofsky scale) of 50% or less. With proper pretreatment examination, some patients who are at risk of serious complications following WBHT can be excluded from treatment, for instance those patients who have decreased cardiac and/or respiratory reserve, or patients with tumour involvement in the liver or brain. Unfortunately, experience has shown that even extensive pre-WBHT screening may not be always sufficient to prevent serious complications or even lethality (van der Zee, 1987).

Almost all investigators reporting clinical results of WBHT have described some form of toxicity. Recently, there have been more optimistic reports from a group in Wisconsin (Robins *et al.*, 1985) treating conscious, sedated patients with WBHT using radiant heat to incresase body temperature up to 42°C, almost on an 'out-patient' basis. In up to 150 treatments, no significant toxicity was reported. Preliminary work from the same group has shown that WBHT can be added to (total-body) irradiation, chemotherapy and immunotherapy with negligible toxicity.

8.9. Future indications for WBHT?

It has been demonstrated that WBHT at 41.8–42°C is effectively lethal for tumour cells. When it is demonstrated that WBHT can be applied simply and safely, for instance by the 'radiant-heat method' (Robins *et al.*, 1985), WBHT, indeed, may play an important role in future cancer therapy. There are at least two categories of patients who may benefit from this approach.

1. *Patients with localized radio-resistant tumours.* These patients may be treated with WBHT at 42°C, or at 40–41°C in support of deep local tumour heating, in addition to radiotherapy. Since one of the problems with deep local heating is the cooling effect of the blood which enters the tumour at, normally low, core temperature. Experimental and clinical studies have demonstrated a correlation between the hyperthermia dose at the coldest spot in the tumour and the therapeutic result. The chance of complete response and local control increases when this 'minimum hyperthermia dose' increases (Dewhirst *et al.*, 1984; Oleson *et al.*, 1984; van der Zee *et al.*, 1986; Kapp *et al.* , 1989). Heating of the blood by means of WBHT guarantees a given minimum hyperthermic dose. In these patients, addition of hyperthermia to radiotherapy may improve both local control and cure.

2. *Patients with metastatized tumours.* These patients may be subjected to WBHT at 40–42°C. Administration of chemotherapy during WBHT treatment might render an increased therapeutic ratio. With regard to those (combinations of) drugs for which it is not yet clear whether simultaneous combination renders therapeutic gain, it is safer to give the two treatment modalities sequentially. In the sequential combination the hyperthermic treatment, at a high level, is used as a debulking agent, increasing the degree of tumour cell killing without influencing toxicity.

8.10. Summary

WBHT is one of the older methods for treatment of malignant tumours with heat. It is, at present, the only method with which a homogeneous temperature distribution can be achieved in deep-seated tumours. The techniques applied to induce WBHT include non-invasive (through the skin) and invasive (through the blood or peritoneum) approaches. The hyperthermia dose which may be given is limited by the function and thermosensitivity of some critical systems, especially of heart and lungs, liver and brain. The maximum tolerable temperature is generally assumed to be 42°C. Adequate thermometry during the treatment is an essential requisite and it is advisable to use at least two independent systems for monitoring core temperature. The treatment is often applied under general anaesthesia, although several workers have reported treatment sedation only without problems.

It is generally accepted that hyperthermia up to 42°C by itself is not able to destroy all tumour cells and therefore the treatment should be combined with another treatment modality. The highest objective response rates have been reported following combination with radiotherapy. At present some studies are in progress including the combination of WBHT with chemotherapy or immunotherapy. A remarkable, often observed effect of WBHT is the disappearance of pain immediately after the treatment. The toxicity of treatment with WBHT appears to be manageable, according to several reports. Therefore, WBHT may play a role in future cancer therapy, either in support of deep local tumour heating or as the only method for hyperthermia induction.

References

Barlogie, B., Corry, P. M., Yip, E., Lippman, L., Johnston, D. A., Khalil, K., Tenczynski, T. F., Reilly, E., Lawson, R., Dosik, G., Rigor, B., Hankenson, R. and Freireich, E. J. (1979) Total-body hyperthermia with and without chemotherapy for advanced human neoplasms, *Cancer Res.*, **39**, 1485–1493.

Benzinger, T. H. and Taylor, G. W. (1963) Cranial measurements of internal temperature in man, in Hertzfeld, C. M. and Hardy, J. D. (Eds), *Temperature; its Measurement and Control in Science and Industry*, Vol. III, New York, Reinhold, pp. 111–120.

Blair, R. and Levin, W. (1978) Clinical experience of the induction and maintenance of whole-body hyperthermia, in Streffer, C. (Ed.), *Cancer Therapy by Hyperthermia and Radiation*, Baltimore, Munich, Urban and Schwarzenberg, pp. 318–321.

Bleehen, N. M. (1984) Hyperthermia with drugs: current status, *Strahlentherapie*, **160**, 721–724.

Braunitzer, G. (1963) Molekulare struktur der Hämoglobine, *Nova Acta Acad. Caesar. Leop. Carol.*, **26**, 471.

Bruns, P. (1888) Die Heilwirkung des Erysipels auf Geschwülste, in Bruns, P. (Ed.), *Beiträge zur klinischen Chirurgie. Mitteilungen aus der chirurgischen Klinik zu Tübingen*, Vol. 3, Tübingen, Verlag der H., Laupp'schen Buchhandlung, pp. 443–466.

Bull, J. M., Lees, D., Schuette, W., Whang-Peng, J., Smith, R., Bynum, G., Atkinson, E. R., Gottdeiner, J. S., Grainick, H. R., Shawker, T. H. and DeVita, V. T. (1979) A phase I trial of a potential adjuvant to chemotherapy, *Annals of Internal Medicine*, **90**, 317–323.

Busch, W. (1866) Uber den Einfluss welchen heftigere Erysipeln zuweilen auf organisierte Neubildungen ausüben, *Verhandl. Naturh. Preuss Rheinl. Westphal.*, **23**, 28–33.

Bynum, G., Patton, J., Bowers, W., Leav, I., Wolfe, D., Hamlet, M. and Marsili, M. (1977) An anaesthetised dog heat stroke model, *J. Appl. Physiol.: Respirat. Environ. Exercise Physiol.*, **43**(2), 292–296.

Cabanac, M. and Caputa, M. (1979) Natural selective cooling of the human brain: evidence of its occurrence and magnitude, *J. Physiol.*, **286**, 255–264.

Caro, G.G. and McDonald, J. (1961) The relation of pulsatile pressure and flow in the pulmonary vascular bed. *J. Physiol.* **157**, 426–453.

Coley, W.B. (1893) The treatment of malignant tumors by repeated inoculations of erysipelas: with a report of ten original cases, *Am. J. Med. Sci.*, **105**, 486–511.

Cronau, Jr, L.H., Bourke, D.L. and Bull, J.M. (1984) General anesthesia for whole-body hyperthermia, *Cancer Res.*, **44**, (Suppl.) 4873s–4877s.

Dewhirst, M.W., Sim, D.A., Sapareto, S. and Connor, W.G. (1984) Importance of minimum tumor temperature in determining early and long-term responses of spontaneous canine and feline tumors to heat and radiation, *Cancer Research*, **44**, 43–50.

Douglas, M.A., Parks, L.C. and Bebib, J. (1981) Sudden myelopathy secondary to therapeutic total-body hyperthermia after spinal-cord irradiation, *New Engl. J. Med.*, **304**, 583–585.

Dubois, M., Sato, S., Lees, D.E., Bull, J.M., Smith, R., White, B.G., Moore, H. and Macnamara, T.E. (1980) Electro-encephalographic changes during whole body hyperthermia in humans, *Electroenceph. and Clin. Neurophys.*, **50**, 486–495.

Emslie-Smith, D., Lightbody, I. and Maclean, D. (1983) Regulation of body temperature in man, in Tinker, J. and Rapin, M. (Eds), *Care of the critically ill patient*, Berlin, Heidelberg, New York, Springer-Verlag, pp. 111–124.

Engelhardt, R., Neumann, H., von der Tann, M. and Löhr, G.W. (1982) Preliminary results in the treatment of oat cell carcinoma of the lung by combined application of chemotherapy (CT) and whole-body hyperthermia, in *Biomedical Thermology*, New York, Alan R. Liss, pp. 761–765.

Engelhardt, R., Neumann, H., Adam, G., Hinkelbein, W. and von der Tann, M. (1982) Möglichkeiten der Ganzkörperhyperthermie, *Strahlentherapie*, **159**, 99–103.

Euler-Rolle, J., Priesching, A., Vormittag, E., Tschakaloff, C. and Polterauer, P. (1978) Prevention of cardiac complications during whole body hyperthermia by beta receptor blockage, in Streffer, C. (Ed.), *Cancer Therapy by Hyperthermia and Radiation*, Baltimore, Munich, Urban and Schwarzenberg, pp. 302–305.

Faithfull, N.S. (1983) Hyperthermia. Physiological changes associated with whole body hyperthermia for cancer treatment, Thesis, Rotterdam, The Netherlands.

Faithfull, N.S., Reinhold, H.S., van den Berg, A.P., van Rhoon, G.C., van der Zee, J. and Wike-Hooley, J.L. (1984) Cardiovascular changes during whole body hyperthermia treatment of advanced malignancy, *Eur. J. Appl. Physiol.*, **53**, 274–281.

Ganong, W.F. (1971) *Review of Medical Physiology*, Los Altos, Calif., Lange Medical.

Gerad, H., Egonin, M.J., Whitacre, M., van Echo, D.A. and Aisner, J. (1983) Renal failure and platinum pharmocokinetics in three patients treated with *cis*-diaminedichloro-platinum(II) and whole body hyperthermia, *Cancer Chemother. Pharmacol.*, **11**, 162–266.

Gerad, H., van Echo, D.A., Whitacre, M., Ashman, M., Henrich, M., Foy, J., Ostrow, S.S., Wiernik, P.M. and Aisner, J. (1984) Doxorubicin, cyclophosphamide and whole body hyperthermia for treatment of advanced soft tissue sarcoma, *Cancer*, **53**, 2585–2591.

Greenlaw, R.H., Doyle, L.A., Loshek, D.D. and Swamy P.G. (1980) Cost analysis of whole body hyperthermia by the Pettigrew method, *The Third International Symposium on Cancer Therapy by Hyperthermia, Drugs and Radiation*, Fort Collins, Colorado.

Gregory, I.C. (1974) The oxygen and carbon monoxide capacities of foetal and adult blood, *J. Physio.*, **236**, 625–634.

Grogan, J.B., Parks, L.C. and Minaberry, D. (1980) Polymorphonuclear leucocute function in cancer patients with whole body hyperthermia, *Cancer*, **45**, 2611–2615.

Heckel, M. and Heckel, I. (1979) Beobachtungen an 479 Infrarothyperthermiebehande-lungen, *Die Medizinische Welt*, **30**, 971–975.

Herman, T.S., Zukoski, C.F., Andaerson, R.M., Hutter, J.J., Blitt, C.D., Malone, J.M., Larson, D.F., Dean, D.C. and Roth, H.B. (1982) Whole-body hyperthermia and chemotherapy for treatment of patients with advanced refractory malignancies, *Cancer Treat. Rep.*, **66**, 259–265.

Hinkelbein, W., Neumann, H., Engelhardt, R. and Wannenmacher, M. (1981) Kombinierte Behandlung nichtkleinzelliger Bronchialkarzinome mit Strahlentherapie und moderater Ganzkörper-Hyperthermie, *Strahlentherapie*, **157**, 301–304.
Ingram, D. L. and Mount, L. E. (1975) *Man and Animals in Hot Environments.*, Berlin, Heidelberg, New York, Springer-Verlag.
Kapp, D. S., Fessenden, P., Cox, R., Baghaw, M. A., Lee, E. R., Prionas, S. D. and Lohrbach, A. (1989) Combined hyperthermia and radiation therapy in the treatment of local regional recurrences from adenocarcinoma of the breast, in Sugahara, T. and Saito, M. (Eds), *Hyperthermic Oncology 1988*, Vol. 2, London, Taylor and Francis, pp. 374–377.
Kelman, G. R. (1966) Digital computer subroutine for the conversion of oxygen tension into saturation, *J. Appl. Physiol.*, **21**, 1375–1376.
Kim, Y. D., Lake, C. R., Lees, D. E., Schuette, W. H., Bull, J. M., Weise, V. and Kopin, I. J. (1979) Hemodynamic and plasma catecholamine responses to hyperthermic cancer therapy in humans, *Am. J. Physiol.*, **237**, H570–H574.
Kirsch, R. and Schmidt, D. (1966) Erste experimentelle und klinische Erfahrungen mit der Ganzkörper-Extrem-Hyperthermie, in Doerr, W., Linder, F. and Wagner, G. (Eds), *Aktuelle Probleme aus dem Gebiet der Cancerologie*, Berlin, Heidelberg, New York, Springer-Verlag, pp. 53–70.
Kluger, M. J. (1980) Historical aspects of fever and its role in disease, Cox, B., Lomax, P., Milton, A. S. and Schönbaum, E. (Eds), *Thermoregulatory Mechanisms and their Therapeutic Implications, Fourth International Symposium on the Pharmacology of Thermoregulation, Oxford, 1979*. Basel, S. Karger, pp. 65–70.
Koga, S., Maeta, M., Shimizu, N., Osaki, Y., Hamazoe, R., Oda, M., Karino, T. and Yamane, T. (1985) Clinical effects of total-body hyperthermia combined with anticancer chemotherapy for far-advanced gastrointestinal cancer, *Cancer*, **55**, 1641–1647.
Landeberg, T. (1985) Hyperthermia and cancer chemotherapy clinical results: A literature review, Proceedings of the Fourth International Symposium on Hyperthermic Oncology, Aarhus, Denmark, in Overgaard, J. (Ed.), *Hyperthermic Oncology 1984*, Vol. 2, London, Philadelphia, Taylor and Francis, pp. 169–179.
Lange, J., Zänker, K. S., Siewert, J.R., Eisler, K., Landauer, B., Kolb, E., Blümel, G. and Remy, W. (1983) Extrakorporeal induzierte Ganzkörperhyperthermie bei Konventionell inkurablen Malignompatienten, *Deutsche Mediz. Wochenscrift*, **108**, 504–509.
Larkin, J. M., Edwards, W. S., Smith, D. E. and Clark, P. J. (1977) Systemic thermotherapy: description of a method and physiologic tolerance in clinical subjects, *Cancer*, **40**, 3155–3159.
Lees, D. E., Kim, Y. D., Bull, J. M., Wang-Peng, J., Schuette, W., Smith, R. and Macnamara, T. (1980) Anesthetic management of whole body hyperthermia for the treatment of cancer, *Anesthesiology*, **52**, 418–428.
Levin, W. and Blair, R. M. (1978) Clinical experience with combined whole-body hyperthermia and radiation, in Streffer, C. (Ed.), *Cancer Therapy by Hyperthermia and Radiation*, Baltimore, Munich, Urban and Schwarzenberg, pp. 322–325.
Mackenzie, A., McLeod, K., Cassels-Smith, A. J. and Dickson, J. A. (1975) Total body hyperthermia: techniques and patient management *Proceedings of the International Symposium on Cancer Therapy by Hyperthermia and Radiation, Washington, DC, April 1975*, pp. 272–281.
Maeta, M., Koga, S., Wada, J., Yokoyama, M., Kato, N., Kawahara, H., Sakai, T., Hino, M., Ono, T. and Yuasa, K. (1987) Clinical evaluation of total-body hyperthermia combined with anticancer chemotherapy for far-advanced miscellaneous cancer in Japan, *Cancer*, **59**, 1101–1106.
Mastrangelo, M. J., Berd, D. and Maguire, H. C. (1984) Current condition and prognosis of tumor immunotherapy: a second opinion, *Cancer Treatment Rep.*, **68**, 207–219.
Miller, A. B., Hoogstraten, B., Staquet, M. and Winkler, A. (1981) Reporting results of cancer treatment, *Cancer*, **47**, 207–214.
Milnor, W. R., Bergel, D. H. and Bargainer, J. D. (1966) Hydraulic power associated with pulmonary blood flow and its relation to heart rate, *Circulation Res.*, **19**, 467–480.
Moricca, G., Cavaliere, R., Lopez, M. and Caputo, A. (1979) Combined whole-body hyperthermia and chemotherapy in the treatment of advanced cancer with diffuse pain, *Advances in Pain Research and Therapy*, **2**, 195–210.
Nauts, H. C. (1982a) Bacterial products in the treatment of cancer: past, present and future,

in Jeljaszewics, J., Pulverer, G. and Roskowski, W. (Eds), *Bacteria and Cancer*, London, New York, Academic Press, pp. 1–25.

Nauts, H. C. (1982b) Bacterial pyrogens: beneficial effects on cancer patients, in Gautherie, M. and Albert, E. (Eds), *Biomedical Thermology, Progress in Clinical Biological Research*, New York, Alan R. Liss, pp. 687–696.

Neumann, H., Engelhardt, R., Fabricius, H. S., Stahn, R. and Löhr, G. W. (1979) Klinisch-chemische Untersuchungen an tumorpatienten unter Zytostatica- und Ganzkörper-hyperthermiebehandlung, *Klin. Wochenschr.* **57**, 1311–1315.

Neumann, H., Fabricius, H. A., Burmeister, D., Stahn, R., von der Tann, M., Engelhardt, R. and Löhr, G. W. (1982) Experience with diathermia-induced whole-body hyperthermia, in *Biomedical Thermology*, New York, Alan R. Liss, pp. 697–704.

Oleson, J. R., Sim, D. A. and Manning, M. R. (1984) Analysis of prognostic variables in hyperthermia treatment of 161 patients, *Int. J. Radiation Oncology Biol. Phys.*, **10**, 2231–2239.

Ostrow, S., van Echo, D., Whitacre, M., Aisner, J., Simon, R. and Wiernik, P. H. (1981) Physiologic response and toxicity in patients undergoing whole body hyperthermia for the treatment of cancer, *Cancer Treatment Rep.*, **65**, 323–325.

Overgaard, J. (1980) Simultaneous and sequential hyperthermia and radiation treatment of an experimental tumor and its surrounding normal tissue *in vivo*, *Int. J. Radiation Oncology Biol. Phys.*, **6**, 1507–1517.

Parks, L. C., Minaberry D., Smith, D. P. and Neely, W. A. (1979) Treatment of far-advanced bronchogenic carcinoma by extracorporeally induced systemic hyperthermia, *J. Thorac. Cardiovasc. Surg.*, **78**, 883–892.

Pettigrew, R. T. (1975) Cancer therapy by whole body heating, in *Proceedings of the International Symposium on Cancer Therapy by Hyperthermia and Radiation, American College of Radiology, Washington, DC*, pp. 282–288.

Pettigrew, R. T. and Ludgate, C. M. (1977) Whole body hyperthermia. A systemic treatment for disseminated cancer, in Rossi-Fanelli, A., Cavaliere, R., Mondovi, B. and Moricca, E. (Eds), *Recent Results in Cancer Research*, Vol. 59, Berlin, Heidelberg, New York, Springer-Verlag, pp. 153–170.

Pettigrew, R. T., Galt, J. M., Ludgate, C. M., Horn, D. N. and Smith, A. N. (1974a) Circulatory and biochemical effects of whole body hyperthermia, *Br. J. Surg.*, **61**, 727–730.

Pettigrew, R. T., Galt, J. M., Ludgate, C. M. and Smith, A. N. (1974b) Clinical effects of whole-body hyperthermia in advance malignancy, *Br. Med. J.*, **4**, 679–682.

Pomp, H. (1978) Clinical application of hyperthermia in gynecological malignant tumors, in Streffer, C. (Ed.), *Cancer Therapy by Hyperthermia and Radiation*, Baltimore, Munich, Urban and Schwarzenberg, pp. 326–327.

Priesching, A. (1976) Hyperthermie in der Krebsbehandlung? In *Prophylaxe und Therapie von Behandlungsfolgen bei Karzinomen der Frau; 2. Oberaudorger Gespräch, Oktober 1975*, Hrsg. von D. Schmähl, Stuttgart, Thieme Verlag, pp. 56–64.

Prys-Roberts, C., Foex, P. and Hahn, C. E. W. (1971) Calculation of blood O_2. Anesthesiology **34**, 581–582.

Randall, J. E. and Stacy, R. W. (1956) Mechanical impedance of the dog's hind leg to pulsatile blood flow, *Amer. J. Physiol.*, **187**, 94–98.

Robins, H. I., Dennis, W. H., Neville, A. J., Shecterie, L. M., Marion, P. A., Grossman, J., Davis, T. E., Neville, S. R., Gillis, W. K. and Rusy, B. F. (1985) A nontoxic system for 41.8°C whole-body hyperthermia: results of a phase I study using a radiant heat device, *Cancer Res.*, **45**, 3937–3944.

Robins, H. I., Kalin, N. H., Shelton, S. E., Shecterle, L. M., Barksdale, C. M., Martin, P. A., and Marshall, J. (1987) Neuroendocrine changes in patients undergoing whole body hyperthermia, *Int. J. Hyperthermia*, **3**, 99–105.

Robins, H. I., Longo, W. L., Lagoni, R. K., Neville, A. J., Hugander, A., Schmitt, C. L. and Riggs, C. (1988) Phase I trial of lonidamine with whole body hyperthermia in advanced cancer, *Cancer Research*, **48**, 6587–6592.

Robins, H. I., Sielaff, K. M., Storer, B., Hakins, M. J. and Borden, E. C. (1989a) A phase I trial of interferon-Ly with whole body hyperthermia in advanced cancer, *Cancer Research*, **49**, 1609–1615.

Robins, H. I., Longo, W. L., Schmitt, C. L., Lagoni, R. L., Neville, A. J., Giese, W. and Steeves, R. A. (1989b) Whole body hyperthermia and total body irradiation in the treatment of favorable B-cell neoplasms, in Sugahara, T. and Saito, M. (Eds), *Hyperthermic Oncology 1988*, Vol. 2, London, Taylor and Francis, pp. 507–508.

Sminia, P., van der Zee, J., Wondergem, J. and Haveman, J. (1988) A review on the effect of hyperthermia on the central nervous system, submitted for publication.

Smith, R., Bull, J. M., Lees, D. E. and Schuette, W. H. (1980) Whole body hyperthermia: nursing management and intervention, *Cancer Nursing*, June 1980, 185–189.

van der Zee, J., van Rhoon, G. C. Wike-Hooley, J. L., Faithfull, N. S. and Reinhold, H. S. (1983) Whole-body hyperthermia in cancer therapy: a report of a phase I–II study, *Eur. J. Cancer Clin. Oncol.*, **19**, 1189–1200.

van der Zee, J., van Putten, W. L. J., van den Berg, A. P. van Rhoon, G. C., Wike-Hooley, J. L., Broekmeyer-Reurink, M. P. and Reinhold, H. S. (1986) Retrospective analysis of the response of tumours in patients treated with a combination of radiotherapy and hyperthermia, *Int. J. Hyperthermia*, **2**, 337–349.

van der Zee, J. (1987) Whole body hyperthermia. The development of and experience with a clinical method, Thesis, Rotterdam, The Netherlands.

van der Zee, J., Faithfull, N. S., van Rhoon, G. C. and Reinhold, H. S. (1987) Whole body hyperthermia as a treatment modality, in Field, S. B. and Franconi, C. (Eds), *Physics and Technology of Hyperthermia*, Dordrecht, Boston, Lancaster, Martinus Nijhoff, pp. 420–440.

Versteegh, P. M. H., van Hoogen, R. H. W. M. and Zwaveling, A. (1981) Systemic hyperthermia by the immersion bath method, *Neth. J. Surg.*, **33**, 195–199.

von Ardenne, M. and Krüger, W. (1980) The use of hyperthermia within the frame of cancer multistep therapy. Thermal characteristics of tumor: applications in detection and treatment, *Annals of the New York Academy of Science*, **335**, 356–361.

Wallach, D. F. H., Madoc-Jones, H., Sternick, E. S., Santaro, J. J. and Curran, B. (1982) Moderate-temperature whole-body hyperthermia in the treatment of malignant disease, in *Biomedical Thermology*, New York, Alan R. Liss, pp. 715–720.

Warren, S. L. (1935) Preliminary study of the effect of artificial fever upon hopeless tumor cases, *Amer. J. Roentgenol.*, **33**, 75–87.

Wike-Hooley, J. L., Faithfull, N. S., van der Zee, J., van den Berg, A. P. (1983) Liver damage and extraction of isocyanine green under whole body hyperthermia, *Eur. J. Appl. Physiol.*, **51**, 269–270.

Wüst, G., Dreiling, H. and Meister, R. (1975) Experimentelle und klinische Befunde zur Tumortherapie mit Hyperthermie, *Verh. Dtsch. Ges. Inn. Med.*, **81**, 1618–1621.

9 The rationale for clinical trials in hyperthermia

J. Overgaard

9.1. Introduction

At present hyperthermia is widely practised in the treatment of a variety of tumours and in different combined approaches. It has been considered by some to be an established therapy despite lack of scientifically based controlled clinical trials, which increases the demand for such studies.

The present chapter will deal only with the use of localized hyperthermia, primarily as an adjuvant to radiotherapy. In addition to a discussion of the indications for hyperthermia the practical aspects of the design and conduct of clinical trials will be reviewed. Similar problems have been discussed elsewhere (Kapp, 1986; Overgaard, 1985, 1989b).

9.2. Types of clinical trials

Clinical evaluation of a new treatment advances through a number of steps in order to evaluate its potential role and place in cancer therapy. Similar to other modalities, studies involving hyperthermia will include the following:

Phase I. These investigations determine the relationship between toxicity and 'dose' of the treatment. In studies with hyperthermia this also includes analysis of technical feasibility related to heating of the tumour in question. The side-effects are evaluated both as heat damage and, when given with another modality, also as a possible enhancement of the toxicity of this modality. Phase I studies are often performed in patients with advanced disease outside the range of conventional therapy.

Phase II. In these studies the treatment is normally tumour-type specific and the aim is to identify the tumour types suitable for treatment. Early clinical studies with hyperthermia are often a combination of phase I and II studies and are strongly influenced by the technical problems related to application of the heat treatment.

Phase III. This is the controlled clinical trial where the new treatment will be compared with conventional therapy. Patients are allocated to each treatment on a randomized basis. In phase II studies a new treatment is evaluated as to whether it is or is not more effective than conventional therapy for equal morbidity, or whether it is equally effective but gives more or less morbidity.

Generally, toxicity and dose problems are evaluated in phase I and II studies and therefore clarified prior to the controlled clinical trial. In practice, however, the unpredictability of the heating patterns demands that some variability in the distribution of heat will have to be dealt with in phase III trials.

The purpose of phase III studies also includes an analysis of the effects of a new treatment relative to the natural history of the disease, e.g. does improved local control influence the survival in patients with the tumour type in question?

The phase III clinical trial is the ultimate evaluation of years of biological, technical and preliminary clinical research. Within the technical limitations the biological rationale should be applied to a clinical situation where, to the best of our knowledge, hyperthermia is expected to yield an improved therapeutic effect. The design and conduct of such trials must therefore be exercised with extreme care. Thus, the validity of the clinical question raised, the suitability of the heating equipment to fulfil the desired treatment and the strength of the biological basis must all be carefully considered and balanced against each other in order to secure an appropriate study. Although it seems obvious, the trial should be designed so that the question(s) asked can, in practice, be answered. Too often trials have been conducted in such a way that the results are inconclusive due to the wrong choice of end-point or poor design. In particular, multicentre studies bear an additional risk of error because many persons are involved in the planning, and local interests may influence against an overall balanced design.

9.3. Protocol design

All clinical trials should be based on a protocol which describes the rationale, background and objectives of the study, and the following should all be considered: the selection of patients; the design of the study and description of the various treatments; how to deal with toxicity and deviation from the planned treatment; the description of end-points and other criteria for evaluating the treatment effect; quality control; handling of data; the ethical and statistical considerations. It is obvious that successful design and conduct of clinical trials requires careful planning. Some of the specific problems related to hyperthermia are discussed below. The reader is referred to the literature on clinical trials in general (Buyse *et al.*, 1984; Leventhal and Wittes, 1988; Meinert and Tonascia, 1986; Overgaard, 1985; Peto *at al.*, 1976a, b).

9.3.1. Feasibility of heating tumours

The question of technical feasibility in hyperthermia is often controversial. Heating of superficial tumours has been possible, often not homogeneously, but in a way that allows a significant heating of tumours, frequently to a temperature above that of the surrounding tissue. Nevertheless, it is evident that in most treatments there will be 'cold spots', and any biological evaluation of a hyperthermic treatment may be influenced by technical insufficiencies. Thus, a failure may not necessarily be due to the lack of biological efficacy but rather to insufficient heating of critical tumour areas. However, given proper quality assurance (see Chapter 17), it should be possible, e.g. by detailed thermometry, to roughly outline the temperature distribution, and thereby also to secure that a given degree of heating has been achieved in the target area. Hyperthermia may therefore be applied in clinical trials with some caution, although the technical difficulties must not be forgotten.

9.3.2. Stratification and prognostic parameters

Current knowledge allows identification of a number of prognostic parameters that may influence the outcome of combined hyperthermia and radiation (Arcangeli *et al.*, 1985a, 1987; Luk *et al.*, 1983, 1984; Overgaard, 1985, 1987a; Valdagni *et al.*, 1988b). These parameters will therefore need to be stratified in clinical trials. They include tumour histology and tumour localization and tumour volume, as large tumours may be relatively more heat sensitive than small tumours (Figure 9.1) (although also more difficult to heat). There may also be differences in response between primary and recurrent tumours and the lesion may or may not have received previous treatment with some other modality.

9.3.3. End-points and evaluation of results

The evaluation of response should always reflect the damage to the *most resistant tumour cells* and the damage to the *most-sensitive normal tissue*.
    Complete and persistent disappearance of the tumour (i.e. control) is the major end-point in curative studies of locally advanced tumours (Kapp, 1986; Overgaard, 1985, 1987a). The use of 'partial response' should be avoided as it has no biological implication and can be very misleading in palliative cases. Growth delay can be a useful parameter when comparing

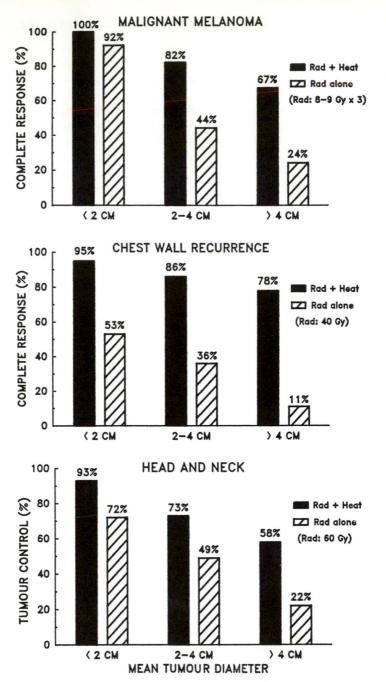

Figure 9.1 Relationship between tumour volume and response to radiotherapy given alone or combined with hyperthermia. The volume dependency is most pronounced after radiotherapy alone in melanoma, recurrent carcinoma of the breast, and head and neck carcinoma. (Modified from Overgaard (1988, 1989c,d).)

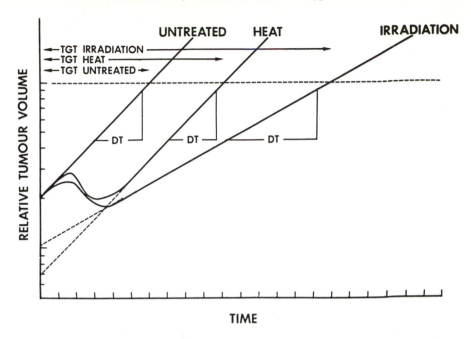

TIME

Figure 9.2 Schematic drawing showing growth patterns of an untreated tumour, after treatment with hyperthermia and after treatment with radiation, respectively. Note that hyperthermia does not influence the tumour volume doubling time (DT) whereas irradiation induces slower growth due to the so-called 'tumour bed effect'. This may give a false impression of a better response although the heat damage may have been more severe. (From Overgaard (1987a).)

different heat schedules alone (Tait and Carnochan, 1987) but, when hyperthermia is combined with radiation, differences in reponse of tumour vasculature to heat or radiation, 'the tumour bed effect' must be taken into consideration (see Figure 9.2). However, it is important to realize that a significant effect may be obtained which is not reflected in tumour growth delay (Overgaard *et al.*, 1987a). This is related to the preferential heat sensitivity of non-proliferating chronic hypoxic and acidic tumour cells. Hyperthermic treatment can also result in a phenomenon called 'persistent regression'. This is a slow, but continuous, regression, often leaving a small residue of tumour (Hofman *et al.*, 1984; Overgaard and Overgaard, 1987). On surgical excision this remaining part appears to be without viable cells. It may persist for more than a year, and may cause confusion if not realized. It should be noted that after combined treatment the likelihood of recurrence may be less than after radiation alone. The observation time is therefore important (Overgaard, 1987a; Perez *et al.*, 1989; Valdagni *et al.*, 1988b)

The evaluation of normal tissue damage should include both early and, if possible, late damage. The reaction will mainly be expressed as modification of radiation- or chemotherapy-induced damage, but analysis should also focus on specific heat-induced complications, such as burns or necrosis. It should be noted that heat damage and radiation damage are different, and they cannot be described in terms of the same parameters, nor do they necessarily occur at the same time (Field and Bleehen, 1979; Overgaard, 1983a). The usual graded scoring system used for radiation may therefore not be applicable to heat-induced normal-tissue damage.

9.3.4. Thermal enhancement ratio

The efficiency of adjuvant hyperthermia is frequently reported as the 'thermal enhancement ratio' (TER). Experimentally, TER values are almost always based on an *isoeffect* (Overgaard, 1985, 1987a) i.e.

$$\text{TER} = \frac{\text{radiation dose alone to achieve a given end-point}}{\text{radiation dose with heat to achieve the same end-point}}$$

In clinical studies there has been a tendency to describe the TER based on *isodose*, i.e. the response, such as complete response, after radiation alone, relative to the response after the same dose of radiation given together with heat (Arcangeli *et al.*, 1985b; Overgaard, 1985, 1989a). Such values of TER are frequently greater than those obtained based on isoeffect (Figure 9.3). The isodose definition is less useful because it depends on a number of

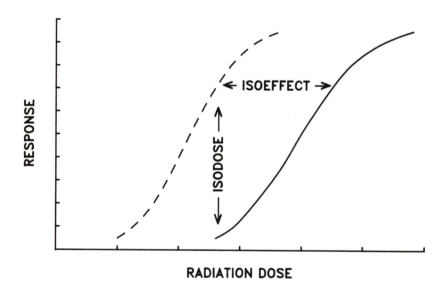

Figure 9.3 Dose–response curves illustrating the principles of estimating thermal enhancement ratios based on either isoeffect or isodose values defined as:

$$\text{TER(isoeffect)} = \frac{\text{radiation dose alone to achieve a given end-point}}{\text{radiation dose with heat to achieve the same end-point}}$$

$$\text{TER(isodose)} = \frac{\text{frequency of response after radiation with heat}}{\text{frequency of response after same radiation dose alone.}}$$

factors which are, in principle, independent of the effect of adjuvant hyperthermia. These include the gradient of the dose–response curve, the level of response to radiation alone, which can be influenced by the radiation dose, etc. If possible, it is preferable to generate isoeffect data in clinical studies if the desire is to analyse the mechanisms of the combined treatment. Of course, the isodose TER should not be neglected because it represents the clinical gain, provided no enhancement of normal tissue damage occurs in the combined schedule. In the clinical situation one would expect the maximal radiation dose to be given and therefore the potential gain of the treatment will relate to the isodose TER.

The design of protocols should preferably be made in such a way that dose–response data can be achieved, i.e. by randomizing to two or more different dose levels in the different arms of the study. This allows estimation of isoeffect in both tumour and normal tissue (Overgaard, 1989a; Overgaard and Overgaard, 1987).

9.3.5. Time–Temperature relationship ('thermal dose')

A description of the hyperthermic treatment (temperature and time) must be included in the evaluation. Thermal dose is a means of expressing a biological heat trauma in equivalent minutes at a given temperature (e.g. $43\,^{\circ}$C), based on measured time–temperature isoeffect relationships (Field and Morris, 1983; Sapareto and Dewey, 1984). Biologically, the use of a hyperthermic isoeffect to describe a heat dose remains controversial (see Chapter 3). The meaning of this expression is particularly vague with fractionated schedules (Overgaard, 1987b) and the difficulties with heterogeneous tumour heating and thermotolerance make the use of isoeffective thermal dose questionable in practical clinical therapy. Nevertheless, the use of the minimum tumour temperature and the maximum normal-tissue temperature may be helpful. Also, several clinical studies have shown a relationship between response and 'heat dose' (see Overgaard 1989a; Valdagni *et al.*, 1988b), although this probably indicates that good heating (i.e. a relatively large 'heat dose') is better than poor heating. This is consistent with the concept of needing at least one good heating.

Since the problem related to physical and biological definitions of thermal dosimetry is far from solved, one should not convert time-temperature relationships into simple equivalent thermal dose values without retaining the original data.

9.3.6. Reporting and handling of data

In order to secure uniform and optimal handling of data it is recommended that this is performed by an established clinical trials office. These are organized within large international or national research groups and also in some major university clinics. The trials office should handle the randomization procedure in order to allow uniform criteria for inclusion, to minimize error and avoid criticism of bias (Meinert and Tonascia, 1986).

When dealing with a new modality it is important to collect the necessary data in a uniform way. For this purpose it is strongly recommended that a standardized data base be used (e.g. Sapareto and Corry, 1989). These are now available and are supported by major research institutions. Such data bases should be in a form which allows direct transfer into major commercial statistical programmes. Suppliers of commercial equipment should be urged to use standard data bases and all calculations should be 'open' in such a way that modifications of formulae can be performed according to common guidelines.

Although it is well established that a radiotherapeutic treatment can only be properly described as a function of total dose, number of fractions and overall treatment time, as well as giving tumour site, histology and volume, there is a strong tendency in the literature not to describe properly the radiotherapeutic part of the combined heat and radiation treatment. Using an adjuvant treatment certainly does not rule out the importance of radiation fractionation, and it should be a fundamental requirement that these parameters are specified.

9.3.7. Quality assurance

The topic of quality assurance is dealt with in Chapter 17. Briefly, the treatment quality is of utmost importance and careful recording of relevant parameters is fundamental for proper evaluation of a clinical study. Quality control should include all important parameters (hyperthermia, radiotherapy, chemotherapy, patient data, end-point evaluation, etc.) and should not only be concerned with hyperthermia.

The temperature heterogeneity obtained when heating human tumours makes the question of the technical feasibility of applying hyperthermia rather controversial. The biological evaluation of all hyperthermic treatments might therefore be strongly influenced by technical difficulties. Furthermore, we are faced with the problem of using different treatment techniques and equipment in multicentre studies. An intense effort should be made to compare the different types of equipment, and the treatment description by thermometry should be as detailed and standardized as possible. Careful quality assurance, including detailed description of treatment and guidance on how to measure temperature, are a 'must' for all protocols (Hand *et al.*, 1989; Shrivastava *et al.*, 1988, 1989). Efforts have been made to create international guidelines which, together with multi-institutional co-operation, include visits by a quality assurance committee. Only if such guidelines are adhered to will it be possible to compare the results from various protocols, and thereby secure optimal evaluation of clinical studies.

9.3.8. Statistical considerations

Statistical evaluation of the results requires a sufficient number of patients to be included in the various studies in order to answer the questions raised. Depending on the treatment design, the tumour type, etc., a phase III clinical trial is expected to require at least 100 evaluable patients/tumours to provide satisfactory answers (Leventhal and Wittes, 1988; Meinert and Tonascia, 1986). In most cases at least twice that number is likely to be needed. This illustrates the need for multicentre studies, since it is rather unlikely that an adequate number of patients will be available from a single institution. As an example, the three first ESHO (European Society for Hyperthermia Oncology) protocols aim for an intake of 120 evaluable patients based on an expected improvement in local control of 30%. Thus, if the true frequency of an event was changed by 30%, e.g. from 50 to 80%, the likelihood of a significant difference ($P<0.05$) being observed is more than 90% (Overgaard, 1987c). In contrast, if the expected gain is estimated to be *15%*, as in the MRC (Medical Research Council) protocols, a substantially larger number of patients, in the region of 400, needs to be included. The probability of a given event should be tested by Kaplan–Meier or log–rank analysis (Lee, 1980; Peto *et al.*, 1976b), especially in patient groups where the observation time may be relatively short.

The use of more-sophisticated design (e.g. group sequential analysis) may reduce the need for high patient numbers, but involves various theoretical and practical problems (Lee, 1980). It is crucially important under all circumstances to consider carefully the statistical problems and likely patient accrual rate prior to the start of any trial.

9.3.9. Ethical aspects

The purpose of clinical studies is to improve the treatment of fellow human beings who suffer from a serious disease. To secure ethical conduct of clinical investigations, all studies must be performed in strict accordance with the guidelines and rules laid down in the Helsinki Declaration. In addition, all protocols should be approved by local ethical committees and adhere to existing legislation. The principal investigator bears the full responsibility for the protocol conduct and should carefully consider the benefits and risk. No scientific goal may be pursued in a clinical trial other than the aim of improving treatment, hopefully with control of the cancer and even cure of the patient. The scientist who participates in clinical studies must always be guided by an uncompromised ethical attitude towards the work.

It should be stressed that the indication for hyperthermia exists only when simpler and/or better means of treatment have been ruled out. Hyperthermia is a costly procedure both in time and manpower and is slightly unpleasant for the patient. It should therefore be preceded by a proper evaluation of the various therapeutic alternatives prior to its use. Also, an acceptably large benefit from the heat treatment should be expected in order to institute such therapy.

9.4. Hyperthermia alone

It is generally agreed that local hyperthermia alone has no role in the curative treatment of malignant tumours. Regardless of treatment schedules, the response rate to heat alone seems to be about 50%, with a complete response rate of 10–15% (Table 9.1), but typically of a short response duration. Localized tumours can only be sterilized at high temperatures and not without causing significant damage to the surrounding normal tissue. This is consistent with biological studies suggesting that the sensitivity of normal tissues and of tumour cells in a well-vascularized area under normal physiological conditions is approximately the same (see Overgaard, 1981, 1983b; Overgaard et al., 1987b). Therefore, tumour cells situated in a normal physiological environment will only be destroyed by a treatment which also causes necrosis of the normal tissues. However, in palliative situations local hyperthermia may prove useful as it has been shown to be effective with regard to pain relief, control of bleeding, etc.

Clinical studies with heat alone therefore play a limited role, and they should only be performed on patients with a short life expectancy or where conventional treatment is contraindicated. The investigations should focus on specific questions such as time–temperature relationships, alterations in tissue vascularization and analysis of the effect of thermotolerance. Such studies will preferably be carried out on patients with multiple tumours which can serve as each other's control. Also, there are reports suggesting that hyperthermia may have a potential in certain benign lesions such as benign prostatic hypertrophy (Sapozink et al., 1989; Yerushalmi et al., 1985). The evaluation of such treatment should also be performed according to the guidelines given above.

Table 9.1 **Effect of local hyperthermia alone.**

Reference	Number of tumours	Response			Response rate (CR+PR)(%)
		CR	PR	NR	
Luk *et al.* (1981)	11	1	2	8	27
U *et al.* (1980)	6	0	3	3	50
Overgaard (1981, 1983b)	13	0	5	8	38
Fazekas and Nerlinger (1981)	4	1	0	3	25
Kim and Hahn (1979)	19	4	6	9	53
Perez *et al.* (1981)	5	2	0	3	40
Marmor *et al.* (1982)	44	5	14	25	43
Corry *et al.* (1982)	28	5	11	12	57
Israel and Besenval (1980)	36	1	13	22	39
Marchal *et al.* (1982)	12	0	1	11	8
Abe *et al.* (1982)	6	0	1	5	17
Okada *et al.* (1977)	69	15	28	26	62
Hall *et al.* (1974)	35	4	19	12	66
Hiraoka *et al.* (1984)	9	0	2	7	22
Lele (1983)	36	6	20	10	72
Dunlop *et al.* (1986)	9	1	2	6	33
Manning *et al.* (1982)	11	2	3	6	45
Dubois *et al.* (1983)	27	1	8	18	33
Lauche *et al.* (1987)	9	0	1	6	11
Jaulerry *et al.* (1989)	18	0	0	18	0
Sannazzari *et al.* (1989)	30	3	6	21	10
All studies	437	51 (12%)	145 (33%)	239 (55%)	(45%)

CR, complete response; PR, partial response; NR, no response.

9.5. Hyperthermia and chemotherapy

This topic has been dealt with in Chapter 5 and will therefore be mentioned only briefly here.

Hyperthermia presently assumes its major role as an adjuvant to chemotherapy in whole-body hyperthermia and limb perfusion. The clinical use of local hyperthermia in combination with chemotherapy is limited (Dahl, 1986; Hahn, 1982; Herman *et al.*, 1988), but the most evident indication is in patients with local failure after a previous full course of radiotherapy. In this situation, clinical phase I and II trials are useful in an attempt to determine both tumour response and, especially, how local hyperthermia influences tumour pharmacokinetic parameters. Such studies can only be properly dealt with if detailed knowledge on drug metabolism and distribution within the heated area is available. Unfortunately, this has only drawn minimal attention so far and should be clarified in experimental studies prior to clinical application. However, several small phase I and II trials indicate potential benefit of combining heat with a number of different drugs, such as cisplatin and blomycin (Arcangeli *et al.*, 1979; Calabresi *et al.*, 1989; Pilepich *et al.*, 1989).

9.6. Hyperthermia and radiotherapy

The use of hyperthermia as an adjuvant to radiation in the treatment of local and regional disease is the area where hyperthermia currently seems to offer the greatest advantages, and it is consequently a situation where there is a great need for well-conducted clinical trials defining the potentials and limitations of such combined treatments. A more detailed analysis of the problems related to clinical studies with combined heat and radiation will therefore be given below.

9.6.1. Rationale for combining hyperthermia and radiation

The requirement for clinical studies of combined hyperthermia and local radiation should be seen in the light of the current status of local tumour treatment. First, local tumour control is a very important parameter and is required to cure a patient from cancer. About one-third of all cancer deaths may be due to lack of local control (Suit, 1982). An improvement of local and regional treatment is expected therefore, in many situations, to improve disease-free survival (Kapp, 1986; Overgaard, 1985, 1987a; Suit, 1982; Suit and Westgate, 1986). Secondly, radiation is effective against most solid tumours, but not always effective enough. The lack of local control in some tumours is frequently due to the fact that tumour sterilization cannot be achieved without increasing the radiation dose to a level which will cause unacceptable damage to the surrounding normal tissue. Thirdly, we now have abundant evidence from clinical phase I and II studies of combined heat and radiation showing that heat may enhance the radiation effect to a significant degree (see Overgaard, 1989a) (Table 9.2). Current clinical experience is sufficiently promising for a detailed analysis of the value of adjuvant hyperthermia given together with radiation in the primary treatment of advanced local and regional disease. Furthermore, there is a strong need for clinical analyses of some of the well-known biological phenomena related to fractionation (e.g. thermotolerance), as there may be quantitative differences between responses in human and animal tissues and tumours.

The indications for hyperthermia in clinical radiotherapy are two-fold (Arcangeli *et al.*, 1988; Overgaard, 1983b, 1989a). First, hyperthermia may be included in the primary treatment of locally advanced tumours where improved local control is expected to result in an improved cure rate and survival. Secondly, hyperthermia has a role in palliative treatment, especially of recurrent tumours in previously irradiated sites. This situation is currently the only 'established' place for hyperthermia in general cancer therapy. In both the curative and palliative situations, clinical experience is limited by the inadequacy of heat delivery other than to very superficial lesions. One of the most crucial factors limiting clinical experience is the ability to homogeneously heat a well-described tumour volume. This is especially true for deeply seated tumours (Field and Franconi, 1987; Marchal *et al.*, 1985; Petrovich *et al.*, 1989; Sapozink *et al.*, 1986; Shimm *et al.*, 1988).

Table 9.2 Tumour response after treatment with the same radiation dose given alone or combined with heat.

Site and reference	Number of tumours	Radiation alone (%)	Radiation and heat (%)	Isodose TER
Chest wall recurrence				
Perez *et al.* (1986)	35	43	86	2.00
van der Zee *et al.* (1985)	40	0	24	>>1
Scott *et al.* (1984)	34	47	94	2.00
Gonzalez Gonzalez *et al.* (1988)	18	33	78	2.36
Lindholm *et al.* (1989)	66	35	70	2.00
Li *et al.* (1985)	42	36	73	2.03
Dunlop *et al.* (1986)	32	67	82	1.22
Overgaard (1988)	14	40	78	1.94
Malignant melanoma				
Overgaard and Overgaard (1987)	63	57	90	1.58
Gonzalez Gonzalez *et al.* (1986)	24	17	83	4.88
Kim *et al.* (1977, 1978, 1982a,b, 1984)	149	43	67	1.56
Arcangeli *et al.* (1984)	38	53	76	1.43
R. Valdagni *et al.*[3]	35	20	80	4.00
Emami *et al.* (1988b)	116	24	59	2.45
Shidnia *et al.* (1987)	185	33	64	1.94
Primary advanced breast				
Morita *et al.* (1988)	22	50	100	2.00
Head and neck				
Arcangeli *et al.* (1985a)	81	42	79	1.88
Scott *et al.* (1984)	18	22	88	4.00
Valdagni *et al.* (1988a)	36	37	82	2.22
Emami *et al.* (1988a)	160	13	38	2.92
Goldobenko *et al.* (1987)	65	86	100	1.16
Datta *et al.* (1990)	65	32	55	1.72
Uterine cervix				
Hornback *et al.* (1986)	66	35	72	2.06
Datta *et al.* (1987)	53	58	74	1.18
Morita *et al.* (1988)	24	71	90	1.26
Gastro-intestinal				
Muratkhodzhaev *et al.* (1987)	313	25[1]	63[1]	2.52
Hiraoka *et al.* (1984)	24	10[2]	43[2]	4.30
Goldobenko *et al.* (1987)	48	0	11[1]	>>1
Mentesjasjvili *et al.* (in Tsyb and Berdov, 1987)	117	33[1,2]	69[1,2]	2.09

[1] Complete response.
[2] Partial response.
[3] Unpublished data.

9.7. *Tumours suitable for controlled clinical trial*

With currently available heating techniques, studies must be limited to superficially situated tumours which can be treated reasonably either by interstitial or external hyperthermia. Clinical trials on deeper-seated lesions are certainly of interest, but should not be performed until more detailed knowledge and experience have been gained with regard to proper heating of such tumours (Arcangeli *et al.*, 1988).

These limitations leave us with two different groups of tumours in which the questions to be analysed and the design of the trials will differ. This situation involves either the use of hyperthermia as an adjuvant to radiotherapy in the primary treatment of advanced tumours, or combined heat and radiation treatment in recurrent or metastatic lesions. In the latter case, tumours may also be used to explore biological principles related to the interaction between the two modalities, whereas in the treatment of primary tumours the combined treatment is applied to determine whether or not hyperthermia (when applied properly) significantly improves local control and curability.

9.7.1. Primary (curative) treatment

Based on the above considerations, there now seems to be sufficient justification for applying adjuvant hyperthermia in the primary treatment of cancer, the prime purpose being local control of bulky primary lesions. Among the tumours which seem to fulfil the requirement for phase III clinical trials are locally advanced breast carcinoma, advanced head and neck tumours, and, with improved techniques, also pelvic tumours (e.g. uterine cervix tumours, recurrent or inoperable colorectal carcinoma).

9.7.1(a) Advanced breast carcinoma

Locally advanced breast carcinoma is a tumour type which clearly fulfils the requirements for a site-specific phase III clinical trial. Such tumours are frequently treated primarily with radiation, though with limited success due to the bulk of the lesion (Arriagada *et al.*, 1985). In this tumour type, there are clinical data pointing towards a significant benefit of hyperthermia. A composite dose–response curve showing an isoeffect TER of about 1.5 is given in Figure 9.4 indicating a high probability of control of such advanced tumours. However, these advanced lesions are currently rather difficult to heat, so that adequate hyperthermic treatment is still difficult to achieve (Hofman *et al.*, 1989; Scott *et al.*, 1988).

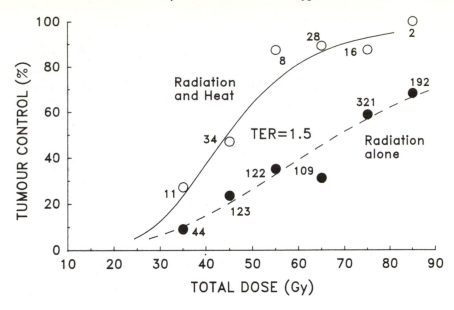

Figure 9.4 Dose–response relationship for advanced breast carcinoma treated with radiation or combined radiation and hyperthermia. The total doses were normalized according to Withers *et al.* (1983) assuming daily fractions of 2 Gy and an α:β ratio of 25 Gy. (From Overgaard (1989a).)

9.7.1(b) Head and neck carcinoma

There are several studies of the use of hyperthermia in the treatment of advanced neck nodes which have all shown a TER greater than 1 (Table 9.2). Dose–response curves for the two treatments can be created demonstrating an isoeffect TER of 1.6 (Figure 9.5). In addition, neck nodes show a well-defined relationship between complete response and persistent disappearance (which tends to diverge as the observation time increases). A few studies of neck nodes have focused on the details of the response and demonstrated that both the amount of heat applied (see Chapter 3) and the tumour volume are prognostic parameters (Arcangeli *et al.*, 1985a; Overgaard, 1989d; Scott *et al.*, 1988; Valdagni *et al.*, 1986, 1988b).

The role of hyperthermia in treatment of the primary site (T-position) has also been explored. A small randomized trial by Datta *et al.* (1990) on advanced head and neck cancer has shown significantly improved local control which, consequently, is expressed as improved survival.

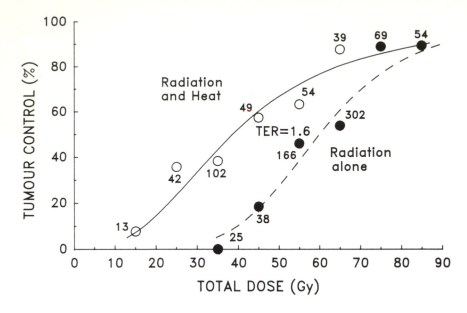

Figure 9.5 Dose–response relationship for advanced neck nodes treated with radiation alone or combined radiation and hyperthermia. The doses were estimated as described in Figure 9.4. (From Overgaard (1989a).)

9.7.1(c) Gastrointestinal tumours

Information on the effect of combined heat and radiation of deeply sited tumours is naturally sparse due to the problems of deep heating (Table 9.2). However, there are data from such tumour types, especially those associated with interstitial heating of oesophageal carcinoma. A large study from the USSR of more than 300 patients has shown both improved local control and a, consequently, increased survival rate (Muratkhodzhaev *et al.*, 1987). With improved heating techniques, oesophageal cancer seems to be a good candidate for future studies (Kai *et al.*, 1988; Z. Petrovich, personal communication). Colorectal tumours are also well suited for hyperthermia, provided proper heating can be achieved (see Overgaard and Overgaard, 1988; Overgaard, 1989a). It has been well documented that control of recurrent colorectal tumours in the pelvis requires large doses of radiotherapy, normally to a level which cannot be reached without causing unacceptable normal tissue damage. Since there is no reason not to expect a TER of the same magnitude as that obtained with other tumours, combined therapy might achieve substantial improvement in control in these cases (Overgaard and Overgaard, 1988). Early reports on advanced colorectal tumours treated with combined therapy point toward marked improvement of the radiation response (Table 9.2). The high incidence of pelvic recurrences associated with the frequency of this disease suggests that a substantial effort should be made in this regard.

A special application of hyperthermia may be in conjunction with intraoperative radiotherapy. Although this is far from being common practice, the operative procedure allows direct heating of inoperable bulky tumours. Early experience is encouraging, although the techniques are difficult and increase the risk of side-effects (Goldson *et al.*, 1987).

9.7.1(d) Uterine cervix carcinoma

Adjuvant hyperthermia has given indications of both improved local control and survival in advanced uterine cervix tumours (Table 9.2), notably in a small randomized study by Datta *et al.* (1987). More recently, these data have been supported by the results from several Chinese studies, some of which were controlled clinical trials (Hu *et al.*, 1989).

9.7.1(e) Other tumour sites

The potential for improving the effect of radiation by heat has also been investigated in other tumour sites, although to a lesser degree. These include soft-tissue sarcomas, urogenital tumours, brain tumours and carcinoma of the lung (for references see Overgaard, 1989a). Of special interest are the studies of hyperthermia as part of the radiotherapeutic management of retinoblastoma and ocular melanoma (Astrahan *et al.*, 1987; Lagendijk *et al.*, 1989; Riedel *et al.*, 1985). In this anatomical region a relatively homogeneous heating pattern may be possible, partly due to the lack of blood vessels in the vitreous body.

As a whole, clinical studies have shown that hyperthermia improves local control of tumours, especially those that are advanced. So far, almost all studies have shown an improved TER, and in no situation has the reverse effect been reported.

9.7.1(f) Pre-operative hyperthermia

Several clinical trials in the USSR (Table 9.3) have reported on the use of hyperthermia in connection with pre-operative radiotherapy. In both advanced malignant melanoma and colorectal carcinoma the 5-year survival rates are improved. A similar large randomized trial comparing pre-operative radiotherapy alone with combined radiotherapy and hyperthermia of stage II and III breast cancer showed a significantly improved local control rate, but as expected the survival rate was only marginally improved. Although detailed information is lacking, these trials are encouraging as they have demonstrated a significant

Table 9.3 Effect of pre-operative irradiation with or without hyperthermia on the 5-year survival rate (Savchenko *et al.*, 1987).

Tumour type	Number of tumours	Radiation alone (%)	Radiation and heat (%)	Isodose TER
Malignant melanoma	247	48[2]	73	1.52
Advanced breast	507	77[1]	87	1.13
Colorectal carcinoma	101	55[1]	71	1.29

[1] 20 Gy/5 fx.
[2] 50–52 Gy, 4–5 Gy/fx.

improvement of pre-operative combined treatment compared with pre-operative radiotherapy alone.

9.7.1(g) Strategy for applying hyperthermia with curative radiotherapy

When dealing with primary treatment of malignant lesions in patients with a relatively long life expectancy, optimal treatment must be the aim. We must consequently add hyperthermia to irradiation treatment schedules known to be as effective as possible. This implies that the adjuvant hyperthermia must not sensitize the radiation effect in the involved normal tissue as the radiation dose will be given to normal tissue 'tolerance'. The treatment principle should therefore be based on conventional radiotherapy using a standard fractionation schedule with daily radiation of approximately 2 Gy per fraction, and investigating whether additional heat treatment given once or twice a week improves the tumour-destructive effect. The argument for this technique has been discussed in detail, especially with regard to the development of thermotolerance and the interval between the hyperthermic fractions (Overgaard, 1987b, 1989a; Overgaard and Nielsen, 1983; Overgaard *et al.*, 1987b; Urano, 1986; see Chapter 2). The potential advantage of hyperthermia in such a schedule will be its cytotoxic rather than its radiosensitizing effect, i.e. the modalities act independently.

If normal-tissue heating can be avoided, hyperthermia in the combined treatment of primary tumours might be given immediately after radiation, as some degree of radiosensitization may add further to the tumour-destructive effect. Otherwise, an interval of approximately 3–4 h is recommended between radiation and hyperthermia to avoid sensitization of the radiation response in the normal tissue. However, as treatment techniques improve, more attention should be given to the possibility of using different portals for the heat and radiation treatments in order to limit this problem, which may also be circumvented by using interstitial implants for one or both modalities.

9.7.2. Recurrent or metastatic tumours

The basis for clinical phase III trials on metastatic or recurrent lesions is somewhat different from that for primary tumours. Recurrent tumours have been used in the past, but in future the planning of such trials should be more rigid, especially with regard to quality assurance and to selection of a single type of tumour in a study. The obvious candidates include recurrent breast carcinoma of the chest wall, recurrent neck nodes in previously irradiated areas and malignant melanoma. The latter is especially well suited for clinical investigations because it is considered by many to respond best to a few high irradiation doses per fraction rather than a conventional treatment schedule. This allows the possibility to combine heat treatment with all radiation fractions.

9.7.2(a) Chestwall recurrences

Recurrent breast carcinoma in the chest wall after mastectomy, either with or without radiotherapy, represents one of the major indications for palliative combined treatment. Although the presence of a local recurrence may have a minor effect on patient survival it is certainly a psychological trauma to the patient and must be treated with much concern. In previously irradiated tumours, it is difficult to control a

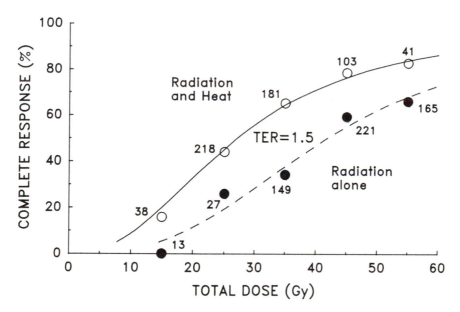

Figure 9.6 Dose–response relationship for the recurrent breast carcinoma on the chest wall treated with radiation alone or combined radiation and hyperthermia. The doses were estimated as described in Figure 9.4. (From Overgaard (1989a).)

subsequent recurrent lesion, but the addition of hyperthermia has markedly improved that situation. As seen in Table 9.2, there are a large number of non-randomized studies which all demonstrate a significant improvement from the combined treatment. When compiling the results from the literature, a well-defined dose–response relationship is seen for both radiation alone and for the combined treatment with a significant improvement in favour of the combined therapy (Figure 9.6). The isoeffect TER at the 50% level is approximately 1.5 (Overgaard, 1988; Perez *et al.*, 1986; van der Zee *et al.*, 1985, 1986, 1988). Recurrent breast lesions are therefore an obvious candidate for combined therapy although careful clarification of various treatment parameters is needed.

9.7.2(b) Malignant melanoma

The other large group of tumours studied is malignant melanoma. This disease has unique radiobiological parameters which allows it to be better controlled by a few large fractions of radiotherapy (Overgaard, 1986). Also, most of the lesions are superficial (either subcutaneous or in lymph nodes). It is therefore obvious that malignant melanoma should be a prime candidate for evaluation of hyperthermia treatment. Table 9.4 shows the isodose results from various published studies. In addition, a dose–response analysis has been performed, taking various fractionation schedules and tumour volume into consideration (Figure 9.7). A more detailed description of this analysis has been presented elsewhere (Overgaard, 1986; Overgaard and Overgaard, 1987). Many of the results have been given as complete response achieved relatively soon after treatment. However, characteristically, response to combined treatment tends to persist and therefore predominantly represents local control. This has been found with several studies which have evaluated the frequency of persistent control as a function of complete response (Emami *et al.*, 1988b; Overgaard, 1986, 1989c; Overgaard and Overgaard, 1987). Typically, there is a strong correlation between these parameters for combined treatment, whereas the tendency for a complete response to recur is higher after radiation alone (Overgaard and Overgaard, 1987; Valdagni *et al.*, 1988b).

Table 9.4   Phase III clinical trials in primary treatment. Multicentre studies.

Protocol	Site	Treatment
ESHO 1–85	Advanced breast	60–70 Gy + 43/60 weekly
ESHO 2–85	Neck nodes	60–70 Gy + 43/60 weekly
ESHO 4–86	Base of tongue	50 Gy + 20 Gy brachytherapy ± interstitial heat × 1
RTOG 84–11	Breast, head and neck, soft-tissue sarcomas	Conventional high-dose radiotherapy ± 42.5–43.5/60 twice weekly
MRC 1	Advanced breast	65 Gy ± heat weekly

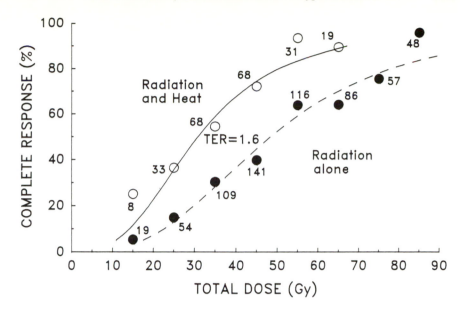

Figure 9.7   Dose–response relationship for malignant melanoma treated with radiation alone and combined radiation and hyperthermia. Total doses were estimated from the assumption of 2 Gy per fraction assuming an α:β ratio of 2.5. (From Overgaard (1989a).)

Biologically, this may be explained by the fact that the most radioresistant cells are in a kind of 'dormant' state, and after some time will give rise to a recurrence if not destroyed. With the combined treatment such cells are likely to be destroyed by heat alone and, therefore, a tumour which disappears completely after combined treatment has a very low probability of containing any viable cells.

9.7.2(c) Strategy for applying hyperthermia with palliative radiotherapy

The treatment schedules in recurrent or metastatic therapy normally do not involve high radiation doses, partly because of previous irradiation of the area and partly because the treatment is normally palliative. Clinical studies with these types of tumours should be aimed at analysing the efficacy of different treatment techniques such as interstitial hyperthermia, quantitative evaluation of different biological factors in human tumours and normal tissues, e.g., detailed studies of the effect of thermotolerance, the study of different fractionation intervals and heat doses, and also a more detailed analysis of the relationship between the hyperthermic cytotoxicity and its radiosensitizing effect by comparing the effect of simultaneous and sequential treatment.

9.8. Clinical trials with heat and radiotherapy

9.8.1. Phase I and II clinical trials in recurrent or metastatic tumours

This constitutes the largest group of trials performed so far and have normally been single-institution studies aimed at defining the feasibility of combined treatment, frequently compared with historical controls (Table 9.2). Such trials have been in progress for many years, and it is unlikely that further such studies will add to our knowledge. Although phase II trials may still be justified for special sites or techniques, patients with the most frequent superficial tumours (neck nodes, recurrent breast carcinoma, etc.) should now be included in randomized phase III studies.

9.8.2. Phase III studies in recurrent or metastatic tumours

There are very few completed studies of this kind. Most have been designed to evaluate the biological rationale of hyperthermia. Thus, studies have analysed the importance of simultaneous compared to sequential heating, or focused on the importance of number of heat treatments (concluding that few heat fractions appear to yield a response equal to that of a higher number). One large multi-institutional study (Perez *et al.*, 1989) on superficial tumours showed no significant benefit of heat in larger tumours, but a major problem was the poor quality of hyperthermia in most cases. This example underlines the importance of good quality control in clinical trials.

　A number of phase III studies are now in progress in order to define the role of hyperthermia in recurrent tumours. Pending the outcome of these trials (Table 9.4), the role of hyperthermia in these tumours will be defined, hopefully within a few years. In addition, such studies can also address more biologically oriented questions.

9.8.3. Phase III studies in primary treatment

Only a few trials with hyperthermia as a part of the primary treatment of locally advanced tumours have been completed. The few small studies performed in, for example, head and neck, and uterine cervix have all indicated a significant improvement in local tumour control. Currently, several multi-institutional studies are underway, organized by ESHO (European Society of Hyperthermia Oncology), RTOG (Radiation Therapy Oncology Group), and MRC (Medical Research Council) (Table 9.5). The outcome of these trials is of utmost importance and participation in them is therefore strongly recommended.

Table 9.5 Phase III clinical trials in recurrent tumours. Multicentre studies.

Protocol	Site	Treatment
ESHO 3–85	Malignant melanoma	8–9 Gy×3±43/60×3
ESHO 5–88	Chest wall (previous radiotherapy)	32 Gy/8 fx twice weekly±43/60 with each fx
RTOG 81–04[1]	Miscellaneous	32 Gy/8 fx twice weekly±43/60 with each fx
RTOG 84–19	Miscellaneous	Brachytherapy±interstitial heat×2
MRC 2a,3a	Chest wall, neck nodes (no previous radiotherapy)	65 Gy±heat weekly
MRC 2b,3b	Chest wall, neck nodes (previous radiotherapy)	28.8 Gy/8 fx/2 weeks±heat weekly

[1] Closed; Perez *et al.* (1989).

9.8.4. Phase I and II trials in deeply located tumours

A last group of clinical trials is the phase I and II evaluation of hyperthermia in deep-seated tumours. These are to a large extent feasibility studies of the technical aspects of the heating capability.

9.9. Concluding remarks

The future role of hyperthermia will depend on the development of better heating equipment. The clinical results so far obtained strongly indicate that there is a place for hyperthermia in oncology, but the final conclusion must await the outcome of controlled clinical trials conducted with a strong emphasis on quality control, including documentation of heating efficiency, and with adequate patient numbers. Until these studies have been completed, hyperthermia must still be considered as an experimental modality and should be limited to institutions where substantial clinical, physical and biological resources are available. In parallel with randomized clinical trials, other studies addressing proper heating schedules must be conducted in order to establish the importance of the number of fractions and their interval, not only with regard to tumour response, but also to clarify the potential risk of increasing normal tissue response. In particular, the possibility of causing late effects needs to be studied.

Providing the technical problems can be overcome so that a hyperthermia can be successfully applied, then the modality must be considered a promising development, especially in treatment of locally advanced, or recurrent, tumours. The results of phase III randomized trials are eagerly awaited.

References

Abe, M., Hiraoka, M., Takahashi, M., Ono, K. and Norhara, H. (1982) Clinical experience with microwave and radiofrequency thermotherapy in the treatment of advanced cancer, *National Cancer Institute Monographs*, **61**, 411–414.

Arcangeli, G., Cividalli, A., Mauro, F., Nervi, C. and Pavin, G. (1979) Enhanced effectiveness of adriamycin and bleomycin combined with local hyperthermia in neck node metastases from head and neck cancers, *Tumori*, **65**, 481–486.

Arcangeli, G., Nervi, C., Cividalli, A. and Lovisolo, G. A. (1984) Problem of sequence and fractionation in the clinical application of combined heat and radiation, *Cancer Research*, **44**, (Suppl.) 4857–4863.

Arcangeli, G., Arcangeli, G., Guerra, A., Lovisolo, G. A., Cividalli, A., Marino, C. and Mauro, F. (1985a) Tumour response to heat and radiation: prognostic variables in the treatment of neck node metastases from head and neck cancer, *International Journal of Hyperthermia*, **1**, 207–217.

Arcangeli, G., Nervi, C., Cividalli, A., Lovisolo, G. A. and Mauro, F. (1985b) The clinical use of experimental parameters to evaluate the response to combined heat (HT) and radiation (RT), in Overgaard, J. (Ed.), *Hyperthermic Oncology 1984*, Vol. 1, London and Philadelphia, Taylor and Francis, pp. 329–332.

Arcangeli, G., Benassi, M., Cividalli, A., Lovisolo, G. A., and Mauro, F. (1987) Radiotherapy and hyperthermia. Analysis of clinical results and identification of prognostic variables. *Cancer*, **60**, 950–956.

Arcangeli, G., Overgaard, J., Gonzalez Gonzalez, D. and Shrivastava, P. N. (1988) Hyperthermia trials, *International Journal of Radiation Oncology, Biology and Physics*, **14**, S93–S109.

Arriagada, R., Mouriesse, H., Sarrazin, D., Clark, R. M. and Deboer, G. (1985) Radiotherapy alone in breast cancer. I. Analysis of tumour parameters, tumour dose and local control: the experience of the Gustave–Roussy Institute and the Princess Margaret Hospital, *International Journal of Radiation Oncology, Biology and Physics*, **11**, 1751–1757.

Astrahan, M., Liggett, P., Petrovich, Z. and Luxton, G., (1987) A 500 KHz localized current field hyperthermia system for use with ophthalmic plaque radiotherapy, *International Journal of Hyperthermia*, **3**, 423–432.

Buyse, M. E., Staquet, M. J. and Sylvester, R. J. (Eds) (1984) *Cancer Clinical Trials. Methods and Practice*, Oxford, New York and Toronto, Oxford University Press.

Calabresi, F., Carlini, P., Arena, M. G., Papaldo, P., Filippo, F. Di., Nallo, A. Di., Begnozzi, L., Colistro, F. and Arcangeli, G. (1989) Superficial hyperthermia combined with cisplatinum chemotherapy: preliminary results, in Sugahara, T. and Saito, M. (Eds), *Hyperthermic Oncology 1988, Vol. 1*, London and Philadelphia, Taylor and Francis, pp. 516–517.

Corry, P. M., Barlogie, B., Tilchem, E. J. and Armour, E. P. (1982) Ultrasound-induced hyperthermia for the treatment of human superficial tumours, *International Journal of Radiation Oncology, Biology, and Physics*, **8**, 1225–1229.

Dahl, O. (1986) Hyperthermia and drugs, in Watmough, D. J. and Ross, W. M. (Eds), *Hyperthermia*, Glasgow and London, Blackie, pp. 121–153.

Datta, N. R., Bose, A. K. and Kapoor, H. K. (1987) Thermoradiotherapy in the management of carcinoma cervix (III B): a contolled clinical study, *Indian Medical Gazette*, **121**, 68–71.

Datta, N. R., Bose, A. K., Kapoor, H. K. and Gupta, S., (1990), Head and neck cancers: results of thermoradiotherapy versus radiotherapy, *International Journal of Hyperthermia*, **6**, 479–486.

Dubois, J. B., Bordure, G., Delauzun, J. P. and Hay, M. (1983) Treatment of superficial tumours by electrontherapy and hyperthermia with 2450 MHz microwaves: thermal dosimetry and preliminary clinical results, *Strahlentherapie*, **159**, 371.

Dunlop, P. R. C., Hand, J. W., Dickinson, R. J. and Field, S. B. (1986) An assessment of local hyperthermia in clinical practice, *International Journal of Hyperthermia*, **2**, 39–50.

Emami, B., Perez, C. A., Bignardi, M., Leybovich, L., Straube, W., vonGerichten, D. and Hederman, M. A. (1988a) A retrospective study of irradiation versus combined irradiation plus hyperthermia in recurrent head and neck cancers. *Abstract Bg-1, Thirty-sixth Annual Meeting of the Radiation Research Society, Phildelphia, Pennsylvania*.

Emami, B., Perez, C.A., Konefal, J., Pilepich, M.V., Leybovich, L., Straube, W., vonGerichten, D. and Hederman, M.A. (1988b) Thermoradiotherapy of malignant melanoma, *International Journal of Hyperthermia*, 4, 373–381.

Fazekas, J. and Nerlinger, T. (1981) Clinical hyperthermia pilot studies, Thomas Jefferson University Hospital, results of 42.5 °C adjuvant to irradiation, *Henry Ford Hospital Medical Journal*, 29, 39–45.

Field, S.B. and Bleehen, N.M. (1979) Hyperthermia in the treatment of cancer, *Cancer Treatment Review*, 6, 63–94.

Field, S.B. and Franconi, C. (Eds) (1987) *Physics and Technology of Hyperthermia*, NATO ASI Series E: Applied Sciences No. 127, Dordrecht, Boston and Lancester, Martinus Nijhoff.

Field, S.B. and Morris, C.C. (1983) The relationship between heating time and temperature: its relevance to clinical hyperthermia, *Radiotherapy and Oncology*, 1, 179–186.

Goldobenko, G.V., Durnov, L.A., Knysh, V.I., Amiraslanov, A.T., Kondratieva, A.P., Matyakin, G.G., Tkachev, S.I., Tseitlin, G. Ya., Ivanov, S.M. and Kozhushkov, A.I. (1987) Experience in the use of thermoradiotherapy of malignant tumors (in Russian), *Medical Radiology*, 32, 36–37.

Goldson, A.L., Smyles, J.M., Ashayeri, E., Dewitty, R., Nibhanupudy, J.R. and King, G. (1987) Simultaneous intraoperative radiation therapy and intraoperative interstitial hyperthermia for unresectable adenocarcinoma of the pancreas, *Endocurietherapy/Hyperthermia Oncology*, 3, 201–208.

Gonzalez Gonzalez, D., van Dijk, J.D.P., Blank, L.E.C.M. and Rümke, Ph. (1986) Combined treatment with radiation and hyperthermia in metastatic malignant melanoma, *Radiotherapy and Oncology*, 6, 105–113.

Gonzalez Gonzalez, D., van Dijk, J.D.P. and Blank, L.E.C.M. (1988) Chestwall recurrences of breast cancer: results of combined treatment with radiation and hyperthermia, *Radiotherapy and Oncology*, 12, 95–103.

Hahn, G.M. (1982) *Hyperthermia and Cancer*, New York and London, Plenum Press.

Hall, R.R., Schade, R.O.K. and Swinney, J. (1974) Effect of hyperthermia on bladder cancer, *British Medical Journal*, 2, 593–594.

Hand, J.W., Lagendijk, J.J.W., Bach Andersen, J. and Bolomey, J.C. (1989) Quality assurance guidelines for ESHO protocols, *International Journal of Hyperthermia*, 5, 421–428.

Herman, T.S., Teicher, B.A., Jochelson, M., Clark, J., Svensson, G. and Coleman, C.N. (1988) Rationale for use of local hyperthermia with radiation therapy and selected anticancer drugs in locally advanced human malignancies, *International Journal of Hyperthermia*, 4, 143–158.

Hiraoka, M., Jo, S., Dodo, Y., Ono, K., Takahashi, M., Nishida, H. and Abe, M. (1984) Clinical results of radiofrequency hyperthermia combined with radiation in the treatment of radioresistant cancers, *Cancer*, 54, 2898–2904.

Hofman, P., Lagendijk, J.J.W. and Schipper, J. (1984) The combination of radiotherapy with hyperthermia in protocolized clinical studies, in Overgaard, J. (Ed.), *Hyperthermic Oncology 1984*, Vol. 1, London and Philadelphia, Taylor and Francis, pp. 379–382.

Hofman, P., Knol, R.G.F., Lagendijk, J.J.W. and Schipper, J. (1989) Thermoradiotherapy of primary breast carcinoma, *International Journal of Hyperthermia*, 5, 1–11.

Hornback, N.B., Shupe, R.E., Shidnia, H., Marshall, C.U. and Lauer, T. (1986) Advanced stage IIIB cancer of the cervix treatment by hyperthermia and radiation, *Gynecology and Oncology*, 23, 160–167.

Hu, Z.-S., Tian, W.-Q., Sun, A.T. and Liao, Z.X. (1989) Intracavitory microwave hyperthermia for cervical cancer in China, in Sugahara, T. and Saito, M. (Eds), *Hyperthermic Oncology 1988*, Vol. 1, London, Taylor and Francis, pp. 586–587.

Israel, L. and Besenval, M. (1980) Local hyperthermia with 13.56 MHz in 46 patients with deep-seated solid tumors, *Abstracts: Radiation Research Society, Orlando, Florida.*

Jaulerry, C., Gaboriaud, G., Chevret, S., Martini, F., Mahé, C., Pontvert, D., Brunin, F., Brugère, J. and Bataini, J.P. (1989) Hyperthermia and radiotherapy or chemotherapy in the retreatment of primary and nodal head and neck carcinomas, in Sugahara, T. and Saito, M. (Eds), *Hyperthermic Oncology 1988*, Vol. 2, Taylor and Francis, pp. 454–457.

Kai, H., Matsufuji, H., Okudaira, Y. and Sugimachi, K. (1988) Heat, drugs, and radiation given in combination is palliative for unresectable esophageal cancer, *International Journal of Radiation Oncology, Biology and Physics*, 14, 1147–1152.

Kapp, D. S. (1986) Site and disease selection for hyperthermic clinical trials, *International Journal of Hyperthermia*, **2**, 139–156.

Kim, J. H. and Hahn, E. W. (1979) Clinical and biological studies of localized hyperthermia, *Cancer Research*, **39**, 2258–2261.

Kim, J. H., Hahn, E. W., Tokita, N. and Nisce, L. Z. (1977) Local tumour hyperthermia in combination with radiation therapy. 1. Malignant cutaneous lesions, *Cancer*, **40**, 161–169.

Kim, J. H., Hahn, E. W. and Tokita, N. (1978) Combination hyperthermia and radiation therapy for cutaneous malignant melanoma, *Cancer*, **41**, 2143–2148.

Kim, J. H., Hahn, E. W. and Ahmed, S. A. (1982a) Combination hyperthermia and radiation therapy for malignant melanoma, *Cancer*, **50**, 478–482.

Kim, J. H., Hahn, E. W. and Antich, P. E. (1982b) Radiofrequency hyperthermia for clinical cancer therapy, *National Cancer Institute Monographs*, **61**, 339–342.

Kim, J. H., Hahn, E. W., Ahmed, S. A. and Kim, Y. S. (1984) Clinical study of the sequence of combined hyperthermia and radiation therapy of malignant melanoma, in Overgaard, J. (Ed.), *Hyperthermic Oncology 1984*, Vol. 1, London and Philadelphia, Taylor and Francis, pp. 387–390.

Kim, Y. H., Ahmad, K., Gottleib, C. F., Byun, Y. S. and Fayos, J. V. (1985) Local hyperthermia and radiation for advanced superficial tumour, *Endocurietherapy/Hyperthermia Oncology*, **1**, 185–191.

Lagendijk, J. J. W., Schipper, J., Hofman, P. and Tan, K. E. W. P. (1989) Hyperthermia of intraocular tumours, in Sugahara, T. and Saito, M. (Eds), *Hyperthermic Oncology 1988*, Vol. 2, London and Philadelphia, Taylor and Francis, pp. 378–381.

Lauche, H. M. El Akoum, H., Fischer, L. and Gautherie, M. (1987) Capacitive radiofrequency hyperthermia: phase-II therapeutic trials and technological advances, *International Journal of Hyperthermia*, **3**, 574–575.

Lee, E. T. (1980) *Statistical Methods for Survival Data Analysis*, Belmont, California, Lifetime Learning Publications.

Lele, P. P. (1983) Physical aspects and clinical studies with ultrasonic hyperthermia, *Textbook of hyperthermia*, Boston, G. K. Hall, pp. 165–187.

Leventhal, B. G. and Wittes, R. E. (1988) *Research Methods in Clinical Oncology*, New York, Raven Press.

Li, R.-Y. Wang, H.-P., Lin, S.-Y. and Zhang, T.-Z. (1985) Clinical evaluation of combined radiotherapy and thermotherapy on carcinoma of the breast (in Chinese), *Clinical Oncology*, **12**, 73–76.

Lindholm, C. E., Kjellen, E. and Nilsson, P. (1989) Low dose radiotherapy with and without hyperthermia in superficial human tumors with evaluation of prognostic factors for tumour response, in Sugahara, T. and Saito, M. (Eds), *Hyperthermic Oncology 1988*, Vol. 2, London and Philadelphia, Taylor and Francis, pp. 618–620.

Luk, K. H., Purser, P. R., Castro, J. R., Meyler, T. S. and Phillips, T. L. (1981) Clinical experiences with local microwave hyperthermia, *International Journal of Radiation Oncology, Biology and Physics*, **7**, 615–619.

Luk, K. H., Francis, M. E., Perez, C. A. and Johnson, R. J. (1983) Radiation therapy and hyperthermia in the treatment of superficial lesions. Preliminary Analysis: treatment efficacy and reactions of skin, tissues subcutaneous, *American Journal of Clinical Oncology*, **6**, 399–406.

Luk, K. H., Pajak, T. F., Perez, C. A., Johnson, R. J., Conner, N. and Dobbins, T. (1984) Prognostic factors for tumour response after hyperthermia and radiation, in Overgaard, J. (Ed.), *Hyperthermic Oncology 1984*, Vol. 1, London and Philadelphia, Taylor and Francis, pp. 817–820.

Manning, M. R., Cetas, T. C., Miller, R. C., Oleson, J. R., Connor, W. G. and Gerner, E. W. (1982) Clinical hyperthermia results of a phase 1 trial employing hyperthermia alone or in combination with external beam or interstitial radiotherapy, *Cancer*, **49**, 205–216.

Marchal, C., Bey, P., Metz, R., Gaulard, M. L. and Robert, J. (1982) Treatment of superficial human cancerous nodules by local ultrasound hyperthermia, *British Journal of Cancer*, **45**(Suppl. V), 243–245.

Marchal, C., Bey, P., Jacomino, J. M., Hoffstetter, S., Gaulard, M. L., and Robert, J. (1985) Preliminary technical, experimental and clinical results of the use of the HPLR 27 system for the treatment of deep-seated tumours by hyperthermia, *International Journal of Hyperthermia*, **1**, 105–116.

Marmor, J. B., Pounds, D. and Hahn, G. M. (1982) Clinical studies with ultrasound-induced hyperthermia, *National Cancer Institute Monographs*, **61**, 333–337.

Meinert, C. L. and Tonascia, S. (1986) *Clinical Trials. Design, Conduct, and Analysis, Monographs in Epidemiology and Biostatistics*, Vol. 8, New York and Oxford, Oxford University Press.

Morita, K., Fuwa, N., Kato, E., Ito, Y., Kimura, C. and Muroka, M. (1988) in Kärcher, K. H. (Ed.), *Progress in Radio-Oncology IV*, Vienna, ICRO, pp. 247–251.

Muratkhodzhaev, N. K., Svetitsky, P. V., Kochegarov, A. A., Alimnazarov, Sh. A., Kuznetsov, V. N. and Shek, B. A. (1987) Hyperthermia in therapy of cancer patients (in Russian), *Medical Radiology*, **32**, 30–36.

Okada, K., Kiyotaki, S., Kawazoe, K., Sato, Y. , Tahara, R. and Kishimoto, T. (1977) Hyperthermic treatment for the bladder tumour (II), *Japanese Journal of Urology*, **68**, 128–135.

Overgaard, J. (1981) Fractionated radiation and hyperthermia. Experimental and clinical studies, *Cancer*, **48**, 1116–1123.

Overgaard, J. (1983a) Histopathological effect of hyperthermia, in Storm, K. (Ed.), *Textbook of hyperthermia*, Boston, G.K. Hall, pp. 165–187.

Overgaard, J. (1983b) Hyperthermic modification of the radiation response in solid tumors, in Fletcher, G. H., Nervi, C. and Withers, H. R. (Eds), *Biological Bases and Clinical Implications of Tumour Radioresistance*, New York, Masson, pp. 337–352.

Overgaard, J. (1985) The design of clinical trials in hyperthermic oncology, in Overgaard, J. (Ed.), *Hyperthermic Oncology 1984*, Vol. 1, London and Philadelphia, Taylor and Francis, pp. 325–338.

Overgaard, J. (1986) The role of radiotherapy in recurrent and metastatic malignant melanoma: a clinical radiobiological study, *International Journal of Radiation Oncolology, Biology and Physics*, **12**, 867–872.

Overgaard, J. (1987a) The design of clinical trials in hyperthermic oncology, in Field, S. B. and Franconi, C. (Eds), *Physics and Technology of Hyperthermia*, NATO ASI Series E: Applied Sciences No. 127, Dordrecht, Boston and Lancester, Martinus Nijhoff, pp. 598–620.

Overgaard, J. (1987b) Some problems related to the clinical use of thermal isoeffect doses, *International Journal of Hyperthermia*, **3**, 329–336.

Overgaard, J. (1987c) Hyperthermia as an adjuvant to radiotherapy. Review of the randomized multicenter studies of the European Society for Hyperthermic Oncology, *Strahlentherapie und Onkologie*, **163**, 453–457.

Overgaard, J. (1988) The role of hyperthermia in radiotherapy, in Kärcher, K. H. (Ed.), *Progress in Radio-Oncology IV*, Vienna, ICRO pp. 247–251.

Overgaard, J. (1989a) The current and potential role of hyperthermia in radiotherapy, *International Journal of Radiation Oncology, Biology and Physics*, **16**, 535–549.

Overgaard, J. (1989b) Clinical trials with hyperthermia and radiotherapy, in Sugahara, T. and Saito, M. (Eds), *Hyperthermic Oncology 1988*, Vol. 2, London and Philadelphia, Taylor and Francis, pp. 57–62.

Overgaard, J. (1989c) Combined hyperthermia and radiation treatment of malignant melanoma, in Sugahara, T. and Saito, M. (Eds), *Hyperthermic Oncology 1988*, Vol. 2, London and Philadelphia, Taylor and Francis, pp. 464–467.

Overgaard, J. (1989d) Hyperthermia in tumour treatment, in Bannasch, P. (Ed.), *Cancer therapy: New Trends*, Berlin, Springer-Verlag, pp. 61–70.

Overgaard, J. and Nielsen, O. S. (1983) The importance of thermotolerance for the clinical treatment with hyperthermia, *Radiotherapy and Oncology*, **1**, 167–178.

Overgaard, J. and Overgaard, M. (1987) Hyperthermia as an adjuvant to radiotherapy in the treatment of malignant melanoma. A clinical study evaluating the effect of simultaneous or sequential radiation and hyperthermia treatment, *International Journal of Hyperthermia*, **3**, 483–502.

Overgaard, J. and Overgaard, M. (1988) Is there a role for hyperthermia in gastrointestinal tract cancer, *Recent Results in Cancer Research*, **110**, 437–454.

Overgaard, J., Matsui, M., Lindegaard, J., Grau, C., Zachariae, C., Johansen, I. M., von der Maase, H. and Nielsen, O. S. (1987a) Relationship between growth delay and modification of local control probability of various treatments given as an adjuvant to irradiation, in Kallman, R. F. (Ed.), *The Use of Rodent Tumors in Experimental Cancer Therapy*, New York, Pergamon Press, pp. 128–132.

Overgaard, J., Nielsen, O. S. and Lindegaard, J. C. (1987b) Biological basis for rational design of clinical treatment with combined hyperthermia and radiation, in Field, S. B. and Franconi, C. (Eds), *Physics and technology of hyperthermia*, NATO ASI Series E: Applied Sciences No. 127, Dordrecht, Boston and Lancester, Martinus Nijhoff, pp. 54–79.

Perez, C. A., Kopecky, W., Baglan, R., Rao, D. V. and Johnson, R. (1981) Local microwave hyperthermia in cancer therapy. Preliminary report, *Henry Ford Hospital Medical Journal*, **29**, 23–37.

Perez, C. A., Kuske, R. R., Emani, B. and Fineberg, B. (1986) Irradiation alone or combined with hyperthermia in the treatment of recurrent carcinoma of the breast in the chest wall. A nonrandomized comparison, *International Journal of Hyperthermia*, **2**, 179–187.

Perez. C. A., Gillespie, B., Pajak, T., Hornback, N. B., Emami, B., and Rubin, P. (1989) Quality assurance problems in clinical hyperthermia and their impact on therapeutic outcome: a report by the radiation therapy oncology group, *International Journal of Radiation Oncology, Biology and Physics*, **16**, 551–558.

Peto, R., Pike, M. C., Armitage, P., Breslow, N. E., Cox, D. R., Howard, S. V., Mantel, N., McPherson, K., Peto, J. and Smith, P. G. (1976a) Design and analysis of randomized clinical trials requiring prolonged observation of each patient. I. Introduction and design, *British Journal of Cancer*, **34**, 585–612.

Peto, R., Pike, M. C., Armitage, P., Breslow, N. E., Cox, D. R., Howard, S. V., Mantel, N., McPherson, K., Peto, J. and Smith, P. G. (1976b) Design and analysis of randomized clinical trials requiring prolonged observation of each patient. II. Analysis and Examples, *British Journal of Cancer*, **35**, 1–39.

Petrovich, Z., Langholz, B., Gibbs, F. A., Sapozink, M. D., Kapp, D. S., Stewart, R. J., Emami, B., Oleson, J., Senzer, N., Slater, J. and Astrahan, M. (1989) Regional hyperthermia for advanced tumors: a clinical study of 353 patients, *International Journal of Radiation Oncology, Biology and Physics*, **16**, 601–607.

Pilepich, M. V., Jones, K. G., Emami, B. N., Perez, C. A., Fields, J. N. and Myerson, R. J. (1989) Interaction of bleomycin and hyperthermia — results of a clinical pilot study, *International Journal of Radiation Oncology, Biology and Physics*, **16**, 211–213.

Riedel, K. G., Svitra, P. P., Seddon, J. M., Albert, D. M., Gragoudas, E. S., Koehler, A. M., Coleman, D. J., Torpey, J., Lizzi, F. L. and Driller, J. (1985) Proton beam irradiation and hyperthermia. Effects on experimental choriodal melanoma, *Archives of Ophthalmology*, **103**, 1862–1869.

Sannazzari, G. L., Gabriele, P., Orecchia, R. and Tseroni, V. (1989) Head and neck tumours: hyperthermia alone, in Sugahara, T. and Saito, M. (Eds), *Hyperthermic Oncology 1988*, Vol. 2, London and Philadelphia, Taylor and Francis, pp. 438–441.

Sapareto, S. A. and Corry, P. M. (1989) A proposed standard data file format for hyperthermia treatments, *International Journal of Radiation Oncology, Biology and Physics*, **16**, 613–627.

Sapareto, S. A. and Dewey, W. C. (1984) Thermal dose determination in cancer therapy, *International Journal of Radiation Oncology, Biology and Physics*, **10**, 787–800.

Sapozink, M. D., Gibbs, F. A., Egger, M. J. and Stewart, J. R. (1986) Regional hyperthermia for clinically advanced deep-seated pelvic malignancy, *American Journal of Clinical Oncology*, **9**, 162–169.

Sapozink, M. D., Astrahan, M. and Boyd, S. (1989) Treatment of benign prostatic hyperplasia with transurethral hyperthermia, *Abstract Ad-9, Thirty-seventh Annual Meeting of the Radiation Research Society, Seattle, Washington, DC*.

Savchenko, N. E., Zhakov, I. G., Fradkin, S. Z. and Zhavrid, E. A. (1987) The use of hyperthermia in oncology (in Russian), *Medical Radiology*, **32**, 19–24.

Scott, R. S., Johnson, R. J. R., Story, K. V. and Clay, L. (1984) Local hyperthermia in combination with definitive radiotherapy: Increased tumour clearance, reduced recurrence rate in extended follow-up, *International Journal of Radiation Oncology, Biology and Physics*, **10**, 2119–2123.

Scott, R., Gillespie, B., Perez, C. A., Hornback, N. B., Johnson, R., Emami, B., Bauer, M. and Pakuris, E. (1988) Hyperthermia in combination with definitive radiation therapy: results of a phase I/II RTOG study, *International Journal of Radiation Oncology, Biology and Physics*, **15**, 711–716.

Shidnia, H., Hornback, N. B., Shupe, R., Shen, R.-N. and Yune, M. (1987) Correlation between hyperthermia and large dose per fraction in treatment of malignant melanoma,

Abstract, 1987 Annual Meeting, International Clinical Hyperthermia Society, Lund, Sweden, p. 26.

Shimm, D. S., Cetas, T. C., Oleson, J. R., Gross, E. R., Buechler, D. N., Fletcher, A. M. and Dean, S. E. (1988) Regional hyperthermia for deep-seated malignancies using the BSD annular array, *International Journal of Hyperthermia*, **4**, 159–170.

Shrivastava, P. N., Saylor, T. K., Matloubieh, A. Y. and Paliwal, B. R. (1988) Hyperthermia thermometry evaluation: criteria and guidelines, *International Journal of Radiation Oncology, Biology and Physics*, **14**, 327–335.

Shrivastava, P. N., Luk, K., Oleson, J., Dewhirst, M., Pajak, T., Paliwal, B., Perez, C., Sapareto, S., Saylor, T. and Steeves, R. (1989) Hyperthermia quality assurance guidelines, *International Journal of Radiation Oncology, Biology and Physics*, **16**, 571–587.

Suit, H. D. (1982) Potential for improving survival rates for the cancer patient by increasing the efficacy of treatment of the primary lesion, *Cancer*, **50**, 1227–1234.

Suit, H. D. and Westgate, S. J. (1986) Impact of improved local control on survival, *International Journal of Radiation Oncology, Biology and Physics*, **12**, 453–458.

Tait, D M. and Carnochan, P. (1987) Thermal enhancement of radiation response: a growth delay study on superficial human tumour metastases, *Radiotherapy and Oncology*, **9**, 231–240.

Tsyb, A. F. and Berdov, B. A. (1987) The use of local hyperthermia for therapy of cancer patients (in Russian), *Medical Radiology*, **32**, 25–29.

U, R., Noell, T. , Woodward, K. T., Worde, B. T., Fishburn, R. I. and Miller, L. S. (1980) Microwave-induced local hyperthermia in combination with radiotherapy of human malignant tumors, *Cancer*, **45**, 638–646.

Urano, M. (1986) Kinetics of thermotolerance in normal and tumour tissues: a review, *Cancer Research*, **46**, 474–482.

Valdagni, R., Kapp, D. S. and Valdagni, C. (1986) N_3 (TNM-UICC) metastatic neck nodes managed by combined radiation therapy and hyperthermia: clinical results and analysis of treatment parameters, *International Journal of Hyperthermia*, **2**, 189–200.

Valdagni, R., Amichetti, M. and Pani, G. (1988a) Radical radiation alone versus radical radiation plus microwave hyperthermia for N_3 (TNM-UICC) neck nodes: a prospective randomized clinical trial, *International Journal of Radiation Oncology, Biology and Physics*, **15**, 13–24.

Valdagni, R., Liu, F.-F. and Kapp, D. S. (1988b) Important prognostic factors influencing outcome of combined radiation and hyperthermia, *International Journal of Radiation Oncology, Biology and Physics*, **15**, 959–972.

van der Zee, J., van Rhoon, G. C., Wike-Hooley, J. L. and Reinhold, H. S. (1985) Clinically derived dose effect relationship for hyperthermia with low dose radiotherapy, *British Journal of Radiology*, **58**, 243–250.

van der Zee, J., Treurniet-Donker, A. D., The, S. K., Helle, P. A., Seldenrath, J. J., Meerwaldt, J. H., Wijnmaalen, A. J., van den Berg, A. P., van Rhoon, G. C., Broekmeyer-Reurink, M. P. and Reinhold, H. S. (1988) Low dose reirradiation in combination with hyperthermia: a palliative treatment for patients with breast cancer recurring in previously irradiated areas, *International Journal of Radiation Oncology, Biology and Physics*, **15**, 1407–1413.

van der Zee, J., van Putten, W. L. J., van den Berg, A. P., van Rhoon, G. C., Hooley, J. L. W., Broekmeyer-Reurink, M. P. and Reinhold, H. S. (1986) Retrospective analysis of the response of tumors in patients treated with a combination of radiotherapy and hyperthermia, *International Journal of Hyperthermia*, **2**, 337–349.

Withers, H. R., Thames, H. D., and Peters, L. J. (1983) A new isoeffect curve for change in dose per fraction, *Radiotherapy and Oncology*, **1**, 187–191.

Yerushalmi, A., Fischelovitz, Y., Singer, D., Reiner, I., Arielly, J., Abramovici, Y., Catsenelson, R., Levy, E. and Shani, A. (1985) Localized deep microwave hyperthermia in the treatment of poor operative risk patients with benign prostatic hyperplasia, *Journal of Urology*, **133**, 873–876.

10. Electromagnetic field propagation and interaction with tissues

C. H. Durney

10.1. Introduction

A good understanding of the basic principles of classical electromagnetics is essential for the design and implementation of hyperthermia generating systems. Although the basic principles and techniques of electromagnetics are well known and are described in many textbooks, their direct application to hyperthermia generating systems is not always fully discussed. The purpose of this chapter is to give a brief summary of those electromagnetic principles and relationships especially pertinent to hyperthermia, with emphasis on physical interpretation and applications, but not including all the detailed derivations of standard electromagnetic theory. The chapter is intended to be useful for clinicians and biologists, as well as for physicists and engineers, especially those new to the field of electromagnetic generation of hyperthermia.

10.2. The basic nature of electric and magnetic fields

It is rather remarkable that all of classical electromagnetics is based on the simple phenomenon that electric charges exert forces on each other. Since it is usually difficult to keep track of all the charges and their interactions in a complicated electrical system, a quantity called electric field is defined and used to account for the forces exerted on charges by each other. The definition of the electric field intensity vector E can be explained in terms of a simple thought experiment. (Vectors are quantities that have both magnitude and direction; vectors are denoted here by bold characters.) The force F exerted on a point test charge q at a given point in space is measured. The electric field intensity E at that point is then defined as

$$E = F/q. \tag{10.1}$$

242

The charge q must be infinitesimally small so that it does not affect the measurement. The units of E are volts per meter (V/m). Although one could, in principle, actually determine whether an electric field existed at a given point in space by placing a small charge at that point and measuring the force on it, it would obviously not be a practical way to detect or measure an electric field intensity. The idealized thought experiment is valuable, though, for understanding the basic nature of electric fields.

The basic law that describes the force between two static charges is Coulomb's force law:

$$F=qQ\hat{R}/4\pi\epsilon_0 R^2. \tag{10.2}$$

where F is the force on the static charge q due to the presence of another static charge Q, and $\hat{R}$ is a unit vector along a straight line from Q to q and pointing toward q. R is the distance between q and Q, and ϵ_0 is the permittivity of free space. In the SI system of units, the units of charge are coulombs (C), and the units of permittivity are farads per meter (F/m). The force in equation (10.2) is repulsive when both q and Q have the same sign and attractive when they have opposite signs. In systems of multiple charges, the force on one charge is the summation of all of the forces acting on it due to each of the other individual charges.

The expression for the force on a charge q placed in an electric field follows from the definition of electric field

$$F=qE. \tag{10.3}$$

The *magnetic flux density vector* B is another force field defined to account for forces in addition to Coulomb's force that can be exerted on a charge when it is moving. The magnitude of B is defined as

$$B=F_m/qv, \tag{10.4}$$

where F_m is the maximum force on the test charge q in any direction, and v is the velocity of q. In the SI system, the units of B are teslas (T). The direction of the force exerted on q by the B field has the special property that it is always perpendicular to both the velocity of the particle and to the B field. This force is given by

$$F=q(v\times B), \tag{10.5}$$

where v is the vector velocity of q and $v\times B$ is a mathematical operation called the vector cross product of v and B. Since the direction of the vector cross product is defined to be perpendicular to both vectors in the cross product, the direction of F is perpendicular to both v and B. A moving charge q in the presence of both an E and a B field will experience a total force given by the sum of the forces of equations (10.3) and (10.5),

$$F = q(E + v \times B), \qquad (10.6)$$

which is called the Lorentz force equation.

Although the basic concepts of electric and magnetic fields are most easily explained and understood in terms of forces on charges, these force relations themselves are not frequently used in designing electromagnetic systems. The more pertinent fundamental laws (which are based on the basic concepts and relationships described above) are discussed in the next section.

10.3. Fundamental laws of electromagnetics

10.3.1. Maxwell's equations in differential form

A famous set of equations called Maxwell's equations are the basis of all classical electromagnetic field theory. The differential forms of these equations for free space are

$$\nabla \times E = -\partial B/\partial t \qquad (10.7)$$

$$\nabla \times B = \mu_0 (J + \epsilon_0 \partial E/\partial t) \qquad (10.8)$$

$$\nabla \cdot E = \rho/\epsilon_0 \qquad (10.9)$$

$$\nabla \cdot B = 0, \qquad (10.10)$$

where μ_0 is the permeability of free space in henrys per meter (H/m), J is current density in amperes per square meter, and ρ is charge density in coulombs per cubic meter (C/m^3). Maxwell's equations, along with some auxiliary relations, describe all of electromagnetic phenomena over a remarkably wide range of frequencies from zero to the extremely high frequencies of ultraviolet light and X-rays.

An important auxiliary relationship that can be derived by taking the divergence of equation (10.8) and combining the result with equation (10.9) is the current continuity equation

$$\nabla \cdot J = -\partial \rho/\partial t. \qquad (10.11)$$

Since any vector field can be completely defined in terms of its curl and its divergence, equations (10.7) through (10.11) can be interpreted in terms of field sources called curl-type and divergence-type sources. In equation

(10.7), $\nabla \times E$ means the curl of E. The curl is a mathematical operation that describes circulation at a point, much like the circulation of water around a drain. Equation (10.7) states that $-\partial B/\partial t$, the negative time rate of change of B is the curl-type source of E. Since curl-type sources produce vector fields that encircle the source, E circulates around B. Equation (10.7) is called Faraday's law. Likewise, in equation (10.8), the current density J and $\partial E/\partial t$, the time rate of change of E, are curl-type sources of B, and B circulates around them.

In equation (10.9), $\nabla \cdot E$ means the divergence of E. Divergence is a mathematical operation that describes the divergence of flux, much like the divergence of smoke from a match. Equation (10.9) states that charge density ρ is a divergence-type source of E. Since field lines begin and end on divergence-type sources, the field lines of E produced by charge begin and end on the charge. Equation (10.10) states that there are no divergence-type sources of B; that is, there are no magnetic charges. Since the divergence of B is zero, the field lines of B form closed loops.

10.3.2. Maxwell's equation in integral form

Equations (10.7) to (10.11) can be converted to integral form by integrating and using Stoke's theorem and the divergence theorem to get

$$\int_l E \cdot dl = -\int_S (\partial B/\partial t) \cdot dS \qquad (10.12)$$

$$\int_l B \cdot dl = \mu_0 \int_S J \cdot dS + \epsilon_0 \mu_0 \int_S (\partial E/\partial t) \cdot dS \qquad (10.13)$$

$$\int_S E \cdot dS = (1/\epsilon_0) \int_V \rho \, dV \qquad (10.14)$$

$$\int_S B \cdot dS = 0 \qquad (10.15)$$

$$\int_S J \cdot dS = -\int_V (\partial \rho/\partial t) \, dV \qquad (10.16)$$

The physical interpretation of equation (10.12) is that the voltage around any closed loop is equal to the time rate of change of the flux cutting that loop. In equation (10.13) the line integral of B around any closed path is equal to the total current passing through any surface bounded by the closed path, which includes both the current due to the motion of charged particles, J, and the displacement current produced by the time rate of change of E. Gauss's law, equation (10.14), states that the total E field flux passing out through any closed surface is equal to the total charge in the volume enclosed by the surface, divided by ϵ_0. As shown by equation (10.15), the total **B** field flux passing through any closed surface is zero. Equation (10.16) is a statement of the conservation of charge: the total change in charge inside any closed surface is equal to the total current passing out through the surface.

This set of equations, either in integral form or in differential form, describes the relations between **E**, **B** and their sources, which are charges and currents. Although any problem in classical electromagnetics can, in principle, be solved by solving these equations (along with some relationships for material interactions that are described later), the solution of these equations is generally very difficult. Consequently, special techniques have been developed for certain classes of functions, and other useful relationships have been derived from these equations. Some of them are described below.

10.3.3. Behaviour with frequency

Although Maxwell's equations are stated in relatively compact form mathematically, they are complex and difficult to solve in the general case. Special techniques have been developed, therefore, for certain ranges of the frequency spectrum. These frequency ranges are most conveniently categorized by the relationship between the wavelength λ and the nominal size of the object or system to which Maxwell's equations are being applied:

$$\lambda > > L \quad \text{circuit theory (Kirchhoff's laws)}$$
$$\lambda \approx L \quad \text{microwave theory}$$
$$\lambda < < L \quad \text{optics.}$$

When $\lambda > > L$, Maxwell's equations may be approximated by circuit theory, principally Kirchhoff's laws, which are much easier to solve than Maxwell's equations. Since the free-space wavelength at 1 MHz is 300 m, any system that will fit in an ordinary room can usually be treated by circuit theory at frequencies below 1 MHz. Historically, circuit theory did not evolve as an approximation to Maxwell's equations, but was developed independently; circuit theory is, nevertheless, an approximation to Maxwell's equations, and a very useful one because it is much simpler.

When the nominal size of the system and the wavelength are of the same order of magnitude, microwave theory (Maxwell's equations solved with a minimum of approximations) must be used. From frequencies of 300 MHz to 300 GHz, the corresponding wavelengths range from 1 m to 1 mm. Hence, in this part of the electromagnetic spectrum, most systems must be treated by microwave theory.

When $\lambda < < L$, which is usually the case at frequencies above 3000 GHz (where the wavelength is smaller then 100 μm), the theory of optics can be used for most systems. Optics is also an approximation to Maxwell's equations, but optical theory did not historically evolve as such an approximation; it was formulated independently from physical observations.

10.3.4. Sinusoidal steady-state field equations

Most electromagnetic systems used for hyperthermia generation involve only sinusoidal steady-state fields. The field equations for the sinusoidal steady-state case can be written in simpler form than those for the general time-varying case because the sinusoidal forcing functions make all the steady-state fields sinusoidal functions of time. For this special case a technique called the *phasor transform* is used. It is defined by

$$\boldsymbol{F}(x, y, z, t) = \text{Re}\,[\tilde{\boldsymbol{F}}(x, y, z, \omega)\,e^{j\omega t}], \tag{10.17}$$

where $\boldsymbol{F}$ represents any field value, (x, y, z) are space co-ordinates, t is time, ω is frequency and j is the square root of minus one. $\tilde{\boldsymbol{F}}$ is the phasor transform of $\boldsymbol{F}$ and is called a phasor. Phasors are functions of space and frequency, but not time. With the definition of equation (10.17), Maxwell's equations can be written in the frequency domain as

$$\nabla \times \tilde{\boldsymbol{E}} = -j\omega \tilde{\boldsymbol{B}} \tag{10.18}$$

$$\nabla \times \tilde{\boldsymbol{B}} = -\mu_0 \tilde{\boldsymbol{J}} + j\omega\mu_0 \tilde{\boldsymbol{E}} \tag{10.19}$$

$$\nabla \cdot \tilde{\boldsymbol{E}} = \tilde{\rho}/\epsilon_0 \tag{10.20}$$

$$\nabla \cdot \tilde{\boldsymbol{B}} = 0 \tag{10.21}$$

$$\nabla \cdot \tilde{\boldsymbol{J}} = -j\omega\tilde{\rho}. \tag{10.22}$$

These equations in the frequency domain are simpler because they do not include time derivatives. Because of the definition in equation (10.17), phasors are complex functions of ω and x, y, z, but not functions of time. Once the phasor is known, the function of time can easily be found from equation (10.17). Often in practice, the symbolic difference between the phasor $\tilde{\boldsymbol{F}}$ and the function of time $\boldsymbol{F}$ is not indicated, but whether a field quantity is a phasor or not is understood from the context. That practice will be followed here, with phasors denoted simply by symbols such as $\boldsymbol{E}$.

10.3.5. Potential functions

In addition to electric and magnetic fields, a number of potential functions have traditionally been used because they are more convenient in many formulations. Two·of the most commonly used potential functions are defined by

$$E = -\nabla\Phi - j\omega A \tag{10.23}$$

$$B = \nabla \times A, \tag{10.24}$$

where Φ is the scalar potential function, and A is the vector potential function. One advantage of the potential functions is that Φ is conveniently related to the source charge density and A is conveniently related to the source current density:

$$\Phi(x, y, z) = (1/4\pi\epsilon_0)\int_V \rho(x', y', z')(e^{-jkR}/R)\,dV' \tag{10.25}$$

$$A(x, y, z) = (\mu_0/4\pi)\int_V J(x', y', z')(e^{-jkR}/R)\,dV', \tag{10.26}$$

where $k = \omega\sqrt{\mu_0\epsilon_0}$, R is the distance from the source point (x', y', z') to the field point (x, y, z) (see Figure 10.1), and the integration is over the volume containing the sources. The integrands of equations (10.25) and (10.26) are the values of the potential functions at (x, y, z) due to the charge and current in one differential volume of V at (x', y', z'). The integrals represent

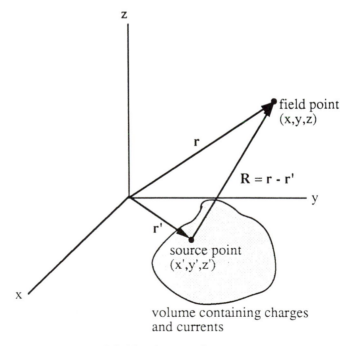

Figure 10.1   Source-point and field-point notation.

the sums of the contributions from all of these differential volumes at all the source points. When the charge density and the current density are known functions, and when they are not affected by the electric and magnetic fields, the field functions can be found by integrating equations (10.25) and (10.26) (by numerical integration, if necessary) and using equations (10.23) and (10.24). Often, however, the charge density and current are affected by the forces exerted on the charges by the electric and magnetic fields, and if so, these effects must be included in the formulation, which results in integral equations with unknowns under the integral signs. These integral equations are much more difficult to solve.

10.4. Electromagnetic fields and materials

10.4.1. Dielectric and magnetic materials

Electric and magnetic fields affect materials by exerting forces on charges in the materials. Two main things happen. First, the electric and magnetic fields exert forces on the charges in the materials and change the charge patterns that originally existed. Then the changed charge patterns in the materials act as sources and produce electric and magnetic fields in addition to the originally applied fields. These two effects result in net electromagnetic fields that are different from the originally applied fields.

Electric fields affect materials in three main ways: (1) polarization of bound charges, (2) orientation of existing electric dipoles, and (3) movement of 'free' charges. Polarization of bound charges is the separation of charges of different signs, which creates dipoles. A dipole is defined as the combination of a positive and a negative charge separated by a small distance. When no electric field is applied, a material is usually electrically neutral; that is, the amount of positive charge is equal to the amount of negative charge and, on the average, the two appear macroscopically to be superimposed and therefore cancel each other. When an electric field is applied, it moves the positive charge in one direction and the negative charge in the other direction, producing electric dipoles and thus polarization. The second effect is the orientation of already existing dipoles; that is, forces exerted on the positive and negative charges of the dipole tend to cause the dipole to line up with E. The third effect, the movement of charges that are not tightly bound (so-called 'free' charges), is called conduction current. All three of these effects are accounted for macroscopically by a property called *electric permittivity*. Biological materials often have high permittivity, which means that the E field has a great effect on the charges in the material. Materials with permittivities significantly greater than that of vacuum are called dielectric materials.

Magnetic fields affect materials mainly by aligning magnetic dipoles in the material. This effect is accounted for macroscopically by a property called *magnetic permeability*. Biological materials are mostly non-magnetic, with permeabilities essentially the same as that of free space.

Expressions for permittivity and permeability can be derived by starting with equation (10.19) and expressing the material current produced by E and B in terms of the measureable properties of the material. For the sinusoidal steady-state case, the current due to charges in the material is

$$J_m = \sigma E + \epsilon_0 x j\omega E + \nabla \times M. \tag{10.27}$$

The term σE represents the drift of conduction ('free') charges in the material; σ is the effective conductivity. The movement of the conduction charges is called 'drift' because forces due to the applied fields only slightly change the random motion caused by thermal excitation. The second term on the right in equation (10.27) is called *polarization current* because it represents the current due to polarization of bound charge and orientation of electric dipoles in the material. χ, the electric susceptibility of the material, accounts for both kinds of polarization. The third term on the right in equation (10.27) is due to alignment of the magnetic dipoles in the material. Before application of a magnetic field, the magnetic dipoles are randomly oriented because of thermal excitation. The slight alignment of each dipole by the applied magnetic field creates an average orientation that is called *magnetization* of the material, M.

$$\nabla \times [(B/\mu_0) - M] = \mu_0 J + j\omega\epsilon_0(1 + \chi + \sigma/j\omega\epsilon_0)E \tag{10.28}$$

which leads to the following definitions:

$$B/\mu_0 - M = H \tag{10.29}$$

$$\epsilon = \epsilon_0(1 + \chi + \sigma/j\omega\epsilon_0) \tag{10.30}$$

$$M = x_m H \tag{10.31}$$

$$\mu = \mu_0(1 + \chi_m) \tag{10.32}$$

$$D = \epsilon E \tag{10.33}$$

and therefore

$$\nabla \times H = J + j\omega D. \tag{10.34}$$

In equation (10.34) J is current due only to free charges not in the material. ϵ is called the *complex permittivity* of the material. χ_m is called *magnetic susceptibility*. The electric and magnetic scalar susceptibilities are defined only for materials called *linear materials*, in which the polarization is proportional to the electric or magnetic field. For non-linear materials the susceptibilities are tensor quantities. D is called *electric flux density*; its units are coulombs per square meter (C/m²). H is called *magnetic field intensity*, with units of amperes per meter (A/m).

The *relative permittivity* or *dielectric constant* is defined by

$$\epsilon = \epsilon_0(\epsilon' - j\epsilon''), \tag{10.35}$$

where ϵ' and ϵ'' are the real and imaginary parts, respectively, of the complex relative permittivity. Complex permittivity is widely used because it accounts macroscopically for all the interactions of the material with electric and magnetic fields in a much simpler way than accounting for forces on individual charges.

10.4.2. Energy absorption in materials

Forces of the applied electric fields give kinetic energy to charges in materials. The *power* transferred to the material is the rate of change of the energy transferred to the material. Usually the energy transferred to the material becomes heat, which is, of course, the object of hyperthermia generation. The heat occurs because of the 'friction' associated with the movement of the induced dipoles, the permanent dipoles and the drifting conduction charges.

Poynting's theorem describes the energy transferred to materials from electric fields. Poynting's theorem for sinusoidal steady-state fields, which is most pertinent to hyperthermia generation, can be derived starting from the phasor form of Maxwell's equations for non-magnetic materials, as follows:

$$\nabla \times E = -j\omega\mu_0 H \tag{10.36}$$

$$\nabla \times H^* = J_m{}^* - j\omega\epsilon_0 E^*, \tag{10.37}$$

where the asterisk means complex conjugate of the complex function. The standard derivation, as given in textbooks on electromagnetic theory, is to take the vector dot product of H^* with equation (10.36), the dot product of E with the conjugate of equation (10.37), subtract the second from the first, and use a vector identity for the divergence of $E \times H^*$ to get

$$\nabla \cdot (E \times H^*) = J_m{}^* \cdot E - j\omega\mu_0 |H|^2 + j\omega\epsilon_0 |E|^2. \tag{10.38}$$

Next, integrate both sides of equation (10.38) over a volume V, use the divergence theorem on the left, and take the real part of both sides to get

$$-\int \text{Re}\,[E \times H^*] \cdot dS = \int \text{Re}\,[J_m{}^* \cdot E]\,dV. \tag{10.39}$$

The term Re ($\boldsymbol{E} \times \boldsymbol{H}^*$), which has units of watts per meter squared (W/m²), is called the *average Poynting vector*. The integral of the Poynting vector over the surface (term on the left) is the total time-averaged power passing in through the closed surface S. The term on the right is the total average power transferred to the charged particles inside V. Equation (10.40) thus states that the total power passing in through S is equal to the total power transferred to particles inside S, which is just a statement of conservation of average power. Poynting's theorem is valid only for relating the total average power passing in through a closed surface to the change in energy of the particle inside the volume the surface encloses. Futhermore, the Poynting vector must include the *total* fields on the surface, both incident and scattered. Applying equation (10.39) incorrectly to only part of a closed surface or not including scattered fields in the Poynting vector gives erroneous and misleading results.

The average power transferred to the material in a differential volume is given by the integrand of the right-hand side of equation (10.39). This time-averaged volume power density is denoted by

$$P = \text{Re} \, [\boldsymbol{J}_m{}^* \cdot \boldsymbol{E}], \tag{10.40}$$

where P has units of watts per cubic meter (W/m³) when $\boldsymbol{J}_m$ and $\boldsymbol{E}$ are root mean square values. A factor of one-half must be included on the right-hand side of equation (10.40) if peak values are used for $\boldsymbol{J}_m$ and $\boldsymbol{E}$.

For non-magnetic materials ($\boldsymbol{M}=0$), $\boldsymbol{J}_m$ can be written in terms of $\boldsymbol{E}$ from equations (10.30) and (10.27) as

$$\boldsymbol{J}_m = j\omega(\epsilon - \epsilon_0)\boldsymbol{E} \tag{10.41}$$

Substituting this into equation (10.40) and using the definition of relative permittivity from equation (10.35) gives

$$P = \text{Re}[-j\omega\epsilon_0(\epsilon' + j\epsilon'') + j\omega\epsilon_0] | \boldsymbol{E} |^2$$

$$P = \omega\epsilon_0\epsilon'' | \boldsymbol{E} |^2. \tag{10.42}$$

The term *specific absorption rate* (SAR) has been adopted by the bioelectromagnetics community to be equivalent to P divided by mass density and expressed in watts per kilogram (W/kg). 'Absorption rate' means the rate of energy absorption; 'specific' means that it is normalized to mass.

Equation (10.42) shows that the energy absorption is directly proportional to ϵ''. Thus, ϵ'' indicates how 'lossy' the material is. Hence, if ϵ'' is zero, no power will be transferred from the applied $\boldsymbol{E}$ to the material. On the other hand, if ϵ'' is large, a relatively large amount of power will be transferred for a given $\boldsymbol{E}$. Values of complex permittivity for materials have been published in various forms. A quantity called the *loss tangent*, defined as

$$\tan \delta = \epsilon''/\epsilon' \tag{10.43}$$

is given in some tables instead of ϵ''. As defined in equation (10.27), σ does not include the losses due to polarization, but in the literature σ is often defined to include them. When it does, the effective conductivity is

$$\sigma = \omega \epsilon_0 \epsilon'' \tag{10.44}$$

and from equation (10.42)

$$P = \sigma |\boldsymbol{E}|^2. \tag{10.45}$$

The 'wetter' a material is, the more lossy it is, and the 'drier' a material is, the less lossy it is. For example, muscle tissue is wetter than bone, and therefore more lossy than bone. Also, a wet piece of paper placed in a microwave oven will be heated while it is wet but, when it dries, it will no longer be heated by the electromagnetic fields of the oven.

10.4.3. Electromagnetic properties of biological materials

In biological tissue the complex permittivity varies greatly with frequency. ϵ' for muscle tissue varies from more than 10^6 at a few Hz to less than a 100 at GHz frequencies. The exceedingly high dielectric constant at the very low frequencies is due to polarization resulting from the charging of membrane interfaces through intracellular and extracellular fluids. The effective conductivity σ ranges from a few millisiemens per centimeter (mS/cm) at low frequencies to nearly a 1000 at the high frequencies.

Curves fitted to published data for measured values of relative permittivity and effective conductivity in animal muscle tissue as a function of frequency are shown in Figures 10.2 and 10.3.

These curves provide a convenient way to get approximate values of the complex permittivity for muscle tissue over a wide range of frequencies. The following two equations (W. D. Hurt, personal communication) that represent a least-squares best fit to the published data are also useful:

$$\epsilon' = \frac{2\,340\,000}{1 + (f/78)^2} + \frac{62\,100}{1 + (f/76 \times 10^3)^2} + \frac{1970}{1 + (f/2.6 \times 10^6)^2}$$

$$+ \frac{30.8}{1 + (f/340 \times 10^6)^2} + \frac{41.3}{1 + (f/23 \times 10^9)^2} + 4.3 \tag{10.46}$$

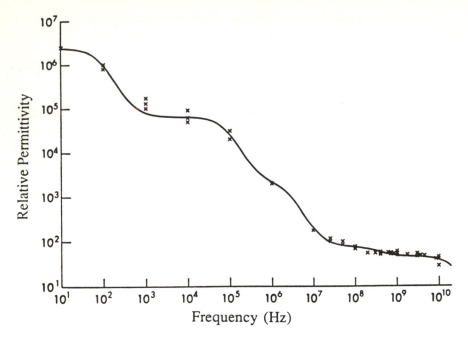

Figure 10.2 Relative permittivity for muscle tissue (W.D. Hurt, personal communication; Durney *et al.*, 1986).

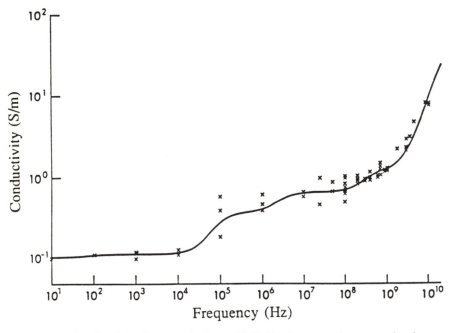

Figure 10.3 Conductivity for muscle tissue (W.D. Hurt, personal communication; Durney *et al.*, 1986).

$$\sigma = \frac{1.67 \times 10^{-6} f^2}{1 + (f/78)^2} + \frac{4.54 \times 10^{-11} f^2}{1 + (f/76 \times 10^3)^2} + \frac{4.21 \times 10^{-14} f^2}{1 + (f/2.6 \times 10^6)^2}$$

$$+ \frac{5.04 \times 10^{-18} f^2}{1 + (f/340 \times 10^6)^2} + \frac{9.99 \times 10^{-20} f^2}{1 + (f/23 \times 10^9)^2} + 0.106 \qquad (10.47)$$

The *Radiofrequency Radiation Dosimetry Handbook* (Durney *et al.*, 1986) contains extensive tabulations from the literature of ϵ' and σ for various kinds of animal tissue. Some representative values from it are given in Table 10.1.

The wide variations of complex permittivity in the various tissues with frequency can be important factors in generating hyperthermia. Even changes in permittivity caused by changes in water due to changes in circulation with temperature can be important. The average value of complex permittivity of the human body has been estimated to be about two-thirds that of muscle tissue. The dependency of power absorption on permittivity as a function of frequency is discussed in a later section.

Table 10.1 Complex relative permittivities of some tissues (Durney *et al.*, 1986).

Frequency (MHz)	Tissue	ϵ'	σ (S/m)
1	Muscle	2000	0.476–0.625
1	Brain	870	0.143–0.233
1	Fat	—	—
10	Muscle	179	0.588–0.667
10	Brain	240	0.222–0.333
10	Fat	—	—
100	Muscle	69–76	0.5–1.04
100	Brain	70–83	0.435–0.625
100	Fat	8–13	0.08–0.085
1000	Muscle	49–61	1.266–1.299
1000	Brain	—	—
1000	Fat	4.3–9.5	0.04–0.91

10.4.4. Boundary conditions

Maxwell's equations require E and H to satisfy certain conditions at the boundary between two materials. These boundary conditions are

$$\epsilon_1 E_{n1} = \epsilon_2 E_{n2} \qquad (10.48)$$

$$E_{t1} = E_{t2} \qquad (10.49)$$

$$\mu_1 H_{n1} = \mu_2 H_{n2} \qquad (10.50)$$

$$H_{t1} = H_{t2}, \qquad (10.51)$$

where subscripts 1 and 2 stand for the two dielectrics and subscripts n and t stand for field components normal and tangential to the boundary, respectively.

As stated in equations (10.48) and (10.49), the normal components of **E** are discontinuous by the ratios of the complex permittivities in the two dielectrics, but the tangential components of **E** are continuous at the boundary. On the other hand, from equations (10.50) and (10.51), the normal components of **H** are discontinuous by the ratios of the complex permeabilities and the tangential components of **H** are continuous.

Although these relations hold only at the boundaries, they can provide valuable insight into the behaviour of electromagnetic fields. For example, Figure 10.4 shows a slab of dielectric material between two parallel metallic plates, with an air gap on each side of the dielectric, and with a potential difference applied between the plates. The boundary conditions at the boundary between the dielectric material and the air require the normal component of **E** to be discontinuous, as indicated in the figure. Thus, when the permittivity of the dielectric material is high, the field inside the material will be much weaker than the field in the air. In this particular geometry, **E** will be nearly uniform (if the plates are large in extent compared to the distance between them so that fringing can be neglected), and so the boundary conditions can be used to infer the relative values of **E** everywhere in both the air and the dielectric. Even if the values of **E** were known not to be uniform, the boundary conditions could give insight into what **E** is like near the boundaries.

Figure 10.5 shows a contrasting example with the dielectric slab rotated from its position in Figure 10.4. With this orientation, the **E** in the dielectric material will be essentially equal to that in air, since the boundary conditions require the **E** in the dielectric to be equal to the **E** in the air at the boundary because they are each tangential to the boundary. These two examples illustrate how the orientation of the fields with respect to the boundaries can significantly affect the fields inside objects.

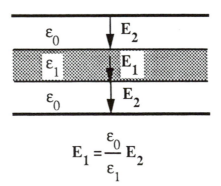

$$E_1 = \frac{\varepsilon_0}{\varepsilon_1} E_2$$

Figure 10.4 Dielectric slab between two metallic plates and the boundary conditions on E.

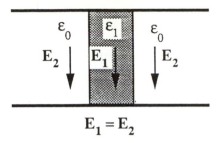

Figure 10.5 Dielectric slab between two metallic plates and the boundary conditions on *E*.

10.5. Electromagnetic field sources

A variety of techniques is used to generate electromagnetic fields for producing hyperthermia, depending on the frequency range in which a device operates. When the frequency is low enough that $kR \ll 1$, $e^{-jkR} \approx 1$ in equations (10.25) and (10.26), which means that propagation time is negligible, circuit theory can often be used along with static field theory in describing electromagnetic field generation. Since $k = 2\pi/\lambda$, $kR \ll 1$ is equivalent to $\lambda \gg L$ (see Section 10.3.3). When kR is not small, microwave techniques must be used to design electromagnetic field sources.

10.5.1. Low-frequency sources

At low frequencies ($kR \ll 1$, $e^{-jkR} \approx 1$ in equations (10.25) and (10.26), the fields are said to be *quasi-static* because the electromagnetic fields vary sinusoidally with time, but their spatial distribution is the same as static fields, since propagation effects are negligible. Quasi-static field sources may be categorized as either divergence-type (electric-type) or curl-type (magnetic-type) sources. For divergence-type sources, equation (10.20) is the main characterizing relation, and for curl-type sources equation (10.19) is the main characterizing relation. The nature of these two kinds of sources can also be explained from equations (10.23)–(10.26).

When a low-frequency source produces mostly charge, with little or no current, the dominating charge is a divergence-type source that produces *E* that begins and ends on charges. In this case, the $\nabla\Phi$ term in equation (10.23) due to the charge dominates, the $j\omega A$ term due to the current is negligible, and the main source of the *E* field is the charge. When a low-frequency source produces mostly current, with little or no charge, the dominating current is a curl-type source that produces primarily *B*, which, if it changes with time, then secondarily produces *E*. Thus in equation (10.23) $j\omega A$ dominates, the $\nabla\Phi$ term is negligible, and the main source of

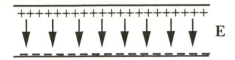

Figure 10.6 *E* field produced by a potential difference between capacitor plates (fringing neglected).

the *B* field is current. The low-frequency sources that produce mostly charge are also called electric-type sources, because they produce *E* directly, and the low-frequency sources that produce mostly current are called magnetic-type sources because they produce primarily *B*, which then produces *E* by its time rate of change.

A familiar example of an *E*-type source is the parallel-plate capacitor (Figure 10.6). A voltage source applied across the plates of the capacitor causes charge to build up on the two plates, and the charge produces *E*. If the voltage is time-varying, the *E* will also change with time, which will act as a curl-type source of *B*, but at low frequencies the *B* so produced will be very weak and will usually be negligible.

Insight into the nature of electric-type sources can often be gained from just two qualitative principles: (1) the electric field lines extend from the conductor of higher potential to the conductor of lower potential, and (2) the field lines must meet the conductors perpendicularly. Corners and sharp edges will concentrate electric fields, which sometimes produces arcing at the corners or edges. Textbooks on static fields give methods for sketching *E* field lines.

A familiar example of a low-frequency magnetic-type source is a loop in which the current is uniform all the way around (Figure 10.7). Magnetic field lines can be visualized from knowing that the *B* field lines will circle

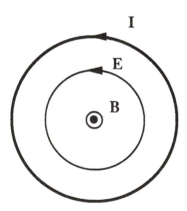

Figure 10.7 Uniform current loop that produces primarily *B* which, as it changes with time, secondarily produces an *E*.

the current according to the right-hand rule: with the thumb in the direction of the current, the fingers of the right hand will indicate the direction of **B**. As indicated by equations (10.26) and (10.24), the net **B** field is the summation of all the fields produced by the differential current elements that make up the current source.

10.5.2. High-frequency sources

At high frequencies for which $kR \approx 1$, the field sources cannot usually be categorized as either electric type or magnetic type. For this case, Maxwell's equations in some form must be solved directly.

Figure 10.8 shows an example of a high-frequency source consisting of a current loop in which the current is not uniform, but changes direction halfway around the loop. According to equation (10.11), this current produces charge at the top and bottom of the loop, and this charge produces **E**. **E** can be found by solving for ρ from equation (10.22), substituting this into equation (10.25), then solving for Φ and **A** from equations (10.25) and (10.26) and substituting into equation (10.23). The result is that the **E** produced by the charge in equation (10.25) is much stronger than the **E** produced by the current in equation (10.26), and the current loop acts more like an electric-type source because the **E** field produced by the charge is much stronger than the **E** produced by the current. As shown in Figure 10.8, the **E** looks more like that produced by capacitor plates (Figure 10.6) than by a loop (Figure 10.7).

Comparison of the field produced by the current loops in Figures 10.7 and 10.8 show how field generation can be described in terms of source charge and current, and how frequency significantly affects the nature of the fields generated. These principles can provide much qualitative under-

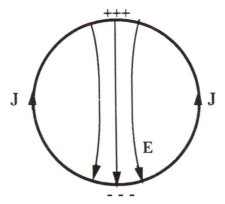

Figure 10.8 A non-uniform current in a loop that produces charge at the top and bottom of the loop. The *E* produced by the charge is much stronger than the *E* produced by the current.

standing that is very valuable in guiding experimental design and interpretation.

Many high-frequency field generating devices consist of some kind of radiating aperture, such as open-ended waveguides, horns and slots or other openings in waveguides. In such devices, the aperture fields are usually considered as the sources of the radiated fields. When such a radiator is placed on or next to a body surface, however, the aperture fields are not usually the same as if the radiator is radiating into free space. Sometimes the aperture fields can be approximated by those of the free-space radiator. When that approximation is not valid, complicated numerical methods are usually needed to calculate the aperture fields as well as the radiated fields.

10.6. *Propagating waves*

Although various mathematical solutions to Maxwell's equations may be formulated, wave solutions have proven to be among the most valuable in many applications, including hyperthermia generation. Wave solutions for the sinusoidal steady-state case are usually obtained by substituting equation (10.33) into equation (10.34), taking the curl of both sides of equation (10.36), and then substituting equation (10.34) into the right-hand side to obtain a wave equation in E. The wave equation is then solved for the particular geometry being considered. The wave equation for a medium in which there is no current due to free charge is

$$\nabla \times \nabla \times E - k^2 E = 0, \tag{10.52}$$

where $k^2 = \omega^2 \mu \epsilon$. k is complex because permittivity and permeability are both complex numbers. With the vector identity

$$\nabla \times \nabla \times E = \nabla (\nabla \cdot E) - \nabla^2 E \tag{10.53}$$

we get

$$\nabla^2 E + k^2 E = 0 \tag{10.54}$$

because $\nabla \cdot E = 0$ since there is no free charge present. Equation (10.53) is a vector point relation that can be written as three simultaneous scalar partial differential equations. These three equations can be solved for regions where k is a constant and boundary conditions used to evaluate the arbitrary constants that arise in the solution to the differential equations. Spherical waves, cylindrical waves and plane waves are the three elementary solutions to equation (10.54).

None of these three kinds of waves actually exist, because they apply only to idealized geometries that cannot be realized physically, but they are sometimes useful approximations to waves that do exist. These three wave solutions are widely used because they are mathematically simple and can provide insight and understanding about electromagnetic field devices, including hyperthermia-generating devices. To illustrate the nature of wave solutions to Maxwell's equation, a brief description of spherical and plane waves is given below.

10.6.1. Spherical waves

The conceptually simplest geometry for which equation (10.54) can be solved is a single spherically symmetric source in free space. For this case, the fields vary only with r in spherical co-ordinates, which we will designate in the standard way by (r, θ, ϕ), and the phasor fields in the frequency domain are

$$E_\theta = E_{01}(e^{-jkr}/r) + E_{02}(e^{jkr}/r), \tag{10.55}$$

where E_{01} and E_{02} are arbitrary constants to be evaluated from the boundary conditions. For the special case where these constants are real, transforming E_θ back to the time domain by multiplying the right hand side by $e^{j\omega t}$ and taking the real part gives

$$E_\theta = E_{01}(e^{-\alpha r}/r) \cos(\omega t - \beta r) + E_{02}(e^{\alpha r}/r) \cos(\omega t + \beta r), \tag{10.56}$$

where α and β are the imaginary and real parts of k, respectively:

$$k = \beta - j\alpha. \tag{10.57}$$

The first term on the right in equation (10.56) is a wave travelling in the $+r$ direction and the second is a wave travelling in the $-r$ direction. β is called the propagation constant.

The H field for this case can be found either by deriving a wave equation in H similar to the derivation of equation (10.54) or using equation (10.55) in equation (10.36). The result is

$$H_\phi = (\mu/\epsilon)^{-1/2} [E_{01}(e^{-jkr}/r) - E_{02}(e^{jkr}/r)] \tag{10.58}$$

in the frequency domain, and

$$H_\phi = (\mu/\epsilon)^{-1/2} [E_{01}(e^{-\alpha r}/r) \cos(\omega t - \beta r) - E_{02}(e^{\alpha r}/r) \cos(\omega t + \beta r)] \tag{10.59}$$

in the time domain. Since E_θ and H_ϕ are constant everywhere on a surface of constant r, which is a spherical surface, the wavefronts are spheres, and the wave is said to be a *spherical wave*. **E** and **H** are perpendicular to each other and to the direction of propagation. The amplitude of the wave decreases with increasing r both because of $e^{-\alpha r}$ and because of $1/r$. Since α represents the loss of a material, the $e^{-\alpha r}$ decrease represents energy transferred from the wave to the material. Even in free space, where there are no losses, the amplitude of the spherical wave decreases with distance as $1/r$.

The *phase velocity* of the wave is defined as the velocity of a point of constant phase on the wave. The phase velocity may be found in terms of β by setting the phase $\omega t - \beta z$ equal to a constant and taking the derivative with respect to time. This gives

$$v_{\mathrm{p}} = \omega/\beta = 1/\mathrm{Re}\,[(\mu\epsilon)^{1/2}]. \tag{10.60}$$

The *wavelength* is defined as the distance between corresponding points on the travelling wave, for example the distance between one trough and the next. The wavelength λ is given by

$$\lambda = 2\pi/\beta. \tag{10.61}$$

The ratio E_θ/H_ϕ for one wave travelling in either the $+r$ or the $-r$ direction is called the *wave impedance*:

$$\eta = (\mu/\epsilon)^{1/2}. \tag{10.62}$$

The wave impedance in free space, where $\epsilon = \epsilon_0$ and $\mu = \mu_0$, is numerically equal to 377 ohms. In biological tissue, the permittivity is relatively high, and the wave impedance is considerably lower than in free space.

The properties of spherical waves can be summarized as:

1. The wavefronts are spherical surfaces.
2. **E** and **H** are perpendicular to each other and to the direction of propagation.
3. The wave impedance is $E/H = (\mu/\epsilon)^{1/2}$.
4. **E** and **H** are constant on any spherical surface.
5. $v_{\mathrm{p}} = 1/\mathrm{Re}\,[\mu\epsilon)^{1/2}]$ is the velocity of propagation and $\lambda = 2\pi/\beta$ is the wavelength. The velocity is lower in material than in free space and the wavelength is shorter in material than in free space.

10.6.2. Plane waves

Another simple solution to equation (10.54) occurs when **E** has only an x

component that varies only with z in rectangular co-ordinates. The fields for this case are

$$E_x = E_{01} e^{-jkz} + E_{02} e^{jkz} \tag{10.63}$$

$$H_y = (\mu/\epsilon)^{-1/2} (E_{01} e^{-jkz} - E_{02} e^{jkz}). \tag{10.64}$$

The e^{-jkz} term represents a wave that propagates in the $+z$ direction, and the e^{jkz} term a wave that propagates in the $-z$ direction. Here again, E and H are perpendicular to each other and to the direction of propagation. Since E and H are constant on surfaces of constant z, which are planes, these waves are called plane waves. Unlike spherical waves, plane waves do not decrease in amplitude with distance in a lossless medium.

A more general plane wave solution in which the waves do not necessarily propagate in the z direction can also be found from equation (10.54). This solution in the frequency domain is

$$E = E_0 e^{\pm jk \cdot r}, \tag{10.65}$$

where

$$k = k_x \hat{x} + k_y \hat{y} + k_z \hat{z} \tag{10.66}$$

is the propagation constant vector and

$$r = x \hat{x} + y \hat{y} + z \hat{z} \tag{10.67}$$

in the co-ordinate vector from the origin to the field point. $\hat{x}$, $\hat{y}$, $\hat{z}$ are unit vectors along the x, y, z axes, respectively. E_0 is an arbitrary constant. The $\pm$ sign indicates two solutions: a wave propagating in the $+k$ direction and one in the $-k$ direction. The components in equation (10.66) must satisfy

$$k_x^2 + k_y^2 + k_z^2 = k^2, \tag{10.68}$$

where k is the magnitude of k.

From equations (10.65) and (10.18),

$$H = (\mu/\epsilon)^{-1/2} n \times E, \tag{10.69}$$

where n is a unit vector in the direction of k. Also, since the divergence of E is zero,

$$k \cdot E = 0. \tag{10.70}$$

The properties of plane waves can be summarized as follows.

1. The wavefronts are planes.
2. **E**, **H** and **k** are all mutually perpendicular. That is, **E** and **H** are perpendicular to each other and to the direction of propagation.
3. The wave impedance is $E/H=(\mu/\epsilon)^{1/2}$.
4. **E** and **H** are constant in any plane perpendicular to **k**.
5. $v_p=1/\mathrm{Re}[(\mu\epsilon)^{1/2}]$ is the velocity of propagation and $\lambda=2\pi/\beta$ is the wavelength. The velocity is lower in material than in free space and the wavelength is shorter in material than in free space.

10.6.3. Plane wave absorption

Equations (10.63) and (10.45) can be used to calculate the absorption of a plane wave as it propagates in a lossy dielectric half-space having the permittivity of muscle tissue. The dielectric half-space is defined to be $z>0$, and the wave propagates in the $+z$ direction. There will be no reflected wave because the half-space extends to $z\rightarrow\infty$, and there is no discontinuity to cause reflections. Therefore E_{02} in equation (10.63) will be zero. Combining equations (10.57), (10.44) and (10.45) then gives

$$P=\omega\epsilon_0\epsilon''E_{01}{}^2e^{-2\alpha z}, \qquad\qquad (10.71)$$

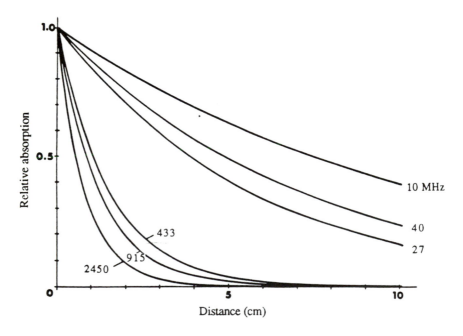

Figure 10.9 Absorption as a function of depth for a plane wave propagating in a dielectric half-space, normalized to values at $z=0$.

where P is the average power absorbed by the dielectric at a given point z. Figure 10.9 shows a graph of P/P_0 versus z for several frequencies. P_0 is the value of P at the surface. Figure 10.9 shows a lot about the nature of power absorption in muscle tissue that is important in understanding electromagnetic hyperthermia generation, most notably that the depth of penetration of the wave decreases drastically with frequency. The data of Figure 10.9 is only qualitatively indicative of what happens in hyperthermia generation because the data is for a plane wave, not the actual waves generated in hyperthermia. The striking decrease in depth of penetration with frequency is characteristic of real devices, though, and it is one of the fundamental obstacles to heating deep-lying tissue.

10.7. Electromagnetic transmission systems

At low frequencies where the wavelength is large compared to the size of the structures being used, electric signals (electric energy) can be conducted from one place to another by pairs of ordinary metallic wires without special considerations about wire size and configuration. However, at high frequencies where the wavelength and the size of the structure have the same order of magnitude, the size and configuration of the conductors can greatly affect how the electric signals are conducted. At high frequencies electromagnetic signals can also sometimes be practically beamed through space without guiding structures. This section describes some transmission systems that are commonly used to conduct electrical signals at higher frequencies.

10.7.1. Transverse electromagnetic transmission lines

Some common systems for transmitting electromagnetic waves are two-wire transmission lines, such as coaxial lines and twin-lead, and waveguides, which usually consist of a hollow metallic pipe of some kind, typically either rectangular or cylindrical. The most commonly used waves on two-wire transmission lines are called transverse electromagnetic (TEM) modes because they have no component of E or H in the direction of propagation. TEM waves cannot exist in hollow waveguides, but they can exist in free space (spherical waves and plane waves, for example, are TEM waves).

One set of equations describes all TEM waves. For example, the equations for a simple two-wire line transmission line (often called twin-lead) are:

$$(\partial V/\partial z) + ZI = 0 \tag{10.72}$$

$$(\partial I/\partial z) + YV = 0, \tag{10.73}$$

where

$$Z = R + j\omega L \qquad (10.74)$$

$$Y = G + j\omega C \qquad (10.75)$$

and R is the equivalent series resistance per unit length of the line, L is the equivalent series inductance per unit length, G is the equivalent shunt conductance per unit length, and C is the equivalent shunt capacitance per unit length.

Solving equations (10.72) and (10.73) simultaneously gives

$$V(z) = V_1 e^{-\Gamma z} + V_2 e^{\Gamma z} \qquad (10.76)$$

$$I(z) = (Y/Z)^{1/2} [V_1 e^{-\Gamma z} - V_2 e^{\Gamma z}], \qquad (10.77)$$

where the complex propagation constant Γ is given by

$$\Gamma = (ZY)^{1/2} \qquad (10.78)$$

and V_1 and V_2 are arbitrary constants that come from solving differential equations. These constants are to be evaluated from the boundary conditions, usually at the input and output of the transmission line. The $e^{-\Gamma z}$ term is a wave travelling in the $+z$ direction and the $e^{+\Gamma z}$ term is a wave travelling in the $-z$ direction. The complex propagation constant Γ has real and imaginary parts defined as

$$\Gamma = \alpha + j\beta, \qquad (10.79)$$

where α is called the attenuation constant and β is called the propagation constant. For lossless transmission lines $(R=G=0)$, $\alpha=0$ and Γ is purely imaginary. Only lossless transmission lines will be considered here.

10.7.1(a) Wave patterns

When a transmission line is infinitely long, as represented schematically in Figure 10.10, a voltage wave and a current wave will travel to the right, but since there is no discontinuity or other source, nothing will cause waves to travel to the left. This corresponds to $V_2=0$ in equations (10.76) and (10.77). The ratio of the voltage to the current for this special case of waves travelling in only one direction is called Z_0, the *characteristic impedance* of the transmission line. From equations (10.76) and (10.77) with $V_2=0$,

$$Z_0 = V(z)/I(z) = (Z/Y)^{1/2}, \tag{10.80}$$

which gives the characteristic impedance in terms of the parameters of the transmission line from equations (10.74) and (10.75). Most coaxial cables are designed to have characteristic impedances of either 50 or 75 ohms. Twin lead of the kind often used for television antennas typically has a characteristic impedance of 300 ohms.

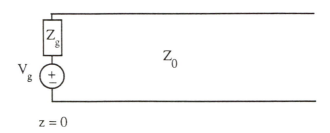

Figure 10.10   Generator connected to an infinitely long transmission line.

When a lumped-element load impedance Z_L is connected across the transmission line at some point $z=d$ (Figure 10.11), waves travelling in only one direction will satisfy the boundary conditions at $z=d$ if $Z_L=Z_0$, in which case the load impedance is said to be matched to the line and consequently does not cause reflections. If Z_L does not equal Z_0, waves travelling in only one direction cannot satisfy the boundary conditions at Z_L and reflected waves are generated. The boundary conditions at Z_L are just Kirchhoff's laws for voltage and current.

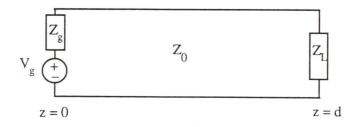

Figure 10.11   Transmission line with a generator and load impedance.

The reflection of waves is accounted for by the reflection coefficient ρ:

$$\rho(z) = V_2 e^{j\beta z}/V_1 e^{-j\beta z}, \tag{10.81}$$

where V_1, V_2 and ρ are complex quantities. With equation (10.81), V and I can be expressed as

$$V(z) = V_1 e^{-j\beta z}(1+\rho) \tag{10.82}$$

$$I(z) = (Y/Z)^{1/2} V_1 e^{-j\beta z}(1-\rho). \tag{10.83}$$

Requiring V and I to satisfy the boundary condition $V/I = Z_L$ at $z = d$ and solving for ρ gives

$$\rho = \rho_L e^{-2j\beta(d-z)}, \tag{10.84}$$

where

$$\rho_L = (Z_L - Z_0)/(Z_L + Z_0) \tag{10.85}$$

is the refection coefficient at the load impedance.

The impedance at any point z along the line with both waves present is defined as

$$Z = V(z)/I(z) = Z_0(1+\rho)/(1-\rho). \tag{10.86}$$

Note that when $\rho = 0$, $Z = Z_0$, since $\rho = 0$ corresponds to no reflected wave.

Substituting for ρ from equation (10.84) into equation (10.86) and using equation (10.85) gives the impedance in terms of Z_0 and Z_L:

$$Z = Z_0 \frac{Z_L + jZ_0 \tan \beta (d-z)}{Z_0 + jZ_L \tan \beta (d-z)} \tag{10.87}$$

which shows that the impedance is a strong function of position along the line. The impedance at $z = 0$ is called the input impedance. The input impedance is equal to the load impedance modified by the effects of propagation along the line. Thus, the load impedance is said to be transformed by the transmission line to the input impedance.

When $Z_L = Z_0$, $Z = Z_0$ everywhere along the line, including the input. Thus, another interpretation is that the characteristic impedance of the line is that impedance which, when connected to the end of the line, produces the same impedance at the input of the line.

Discontinuities of the load impedances produce characteristic voltage and current wave patterns along transmission lines. Figure 10.12 shows these patterns in the time domain for $Z_L \rightarrow \infty$ (an open circuit) at the right

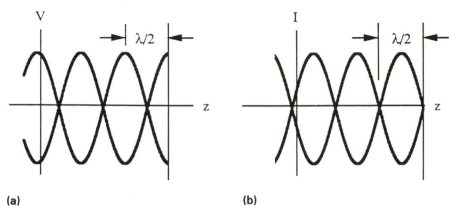

Figure 10.12 **Standing-wave pattern on an open-circuit transmission line. (a) Envelope of the voltage standing wave. (b) Envelope of the current standing wave.**

end of the line. They result from an incident wave travelling in the $+z$ direction and a reflected wave travelling in the $-z$ direction. Since the incident and reflected waves are equal in amplitude but diffrerent in phase, their sum produces nodes at half-wavelength intervals along the line. A node is a point where the voltage or current is zero for all instants of time. At other points on the line, the voltage and current are sinusoidal functions of time with peak values that depend on z. Figure 10.12 shows the pattern at only two instants of time: when all the sinusoids are at their positive peak values and when all the sinusoids are at their negative peak values. This is called the envelope of the wave pattern. Nodes occur only when $Z_L = 0$ or when $Z_L \to \infty$, which corresponds to a short circuit or an open circuit. For other values of Z_L, minima occur, but not nodes. When nodes occur, the wavelength pattern is called a *standing-wave pattern*.

Figure 10.13 shows the top half of the envelope when Z_L is not equal to Z_0, but is not an open or short circuit. The *standing-wave ratio S* is defined

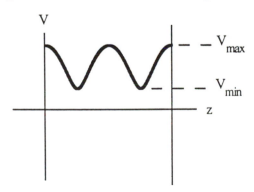

Figure 10.13 **Top half of the voltage envelope when Z_L is not equal to Z_0, but is not an open or short circuit.**

as the ratio of the maximum value to the minimum value in Figure 10.12. It is equal to

$$S=(1+|\rho_L|)/(1-|\rho_L|). \tag{10.88}$$

For a short circuit or an open circuit, $|\rho_L|=1$ and S approaches infinity. When $Z_L=Z_0$, $\rho_L=0$ and $S=1$, which is the minimum value S can have. The standing-wave ratio is commonly used to describe how well the load impedance is matched to the characteristic impedance of the line. For tranferring maximum power to a hyperthermia applicator, for example, S should be equal to 1.

10.7.1(b) Maximum power transfer

In hyperthermia systems, the object is usually to deliver power to an applicator, which is represented by a load impedance. This requires that $Z_L=Z_0$ so that $\rho_L=0$ and $S=1$. When $S=1$, the envelope in Figure 10.13 becomes a straight line and the line is said to be *flat*, and the load impedance is said to be *matched* to the line, since then $Z_L=Z_0$. A voltage standing-wave ratio meter that is used to monitor conditions on the line is an integral part of most hyperthermia systems.

   Some sort of matching network is usually needed to make the equivalent impedance of a hyperthermia applicator equal to the characteristic impedance of the transmission line. An unmatched hyperthermia applicator with an impedance substantially different from Z_0 is nearly useless for heating. Open or shorted stubs, which consist of short sections of transmission line, are sometimes used at higher frequencies to match impedance, but they have the disadvantage that they must be inserted at precise places in the transmission line, and it is not easy to adjust the position of insertion. Double- and triple-stub tuners, which are easier to adjust, are useful when the frequency is high enough that a half wavelength is not too long to be physically manageable, since the tuners contain stubs that must be about a half wavelength long. It is usually desirable to incorporate a matching network right at the load, because otherwise the standing-wave ratio will be high along the unmatched portions of the line, producing excess loss in the line and possibly voltage breakdown.

10.7.1(c) Source and impedance transformation

Any combination of sources and impedances on a transmission line can be transformed into an equivalent transmission line configuration like that shown in Figure 10.11. Expressions for the voltage and current at any point z on this system can be found by requiring equations (10.76) and

(10.77) to satisfy the boundary conditions at the input and output. The result is

$$V = V_g Z_0 [(Z_0 + Z_g)(1 - \rho_L \rho_g e^{-2j\beta d})]^{-1} [e^{-j\beta z} + \rho_0 e^{j\beta z}] \tag{10.89}$$

$$I = V_g [(Z_0 + Z_g)(1 - \rho_L \rho_g e^{-2j\beta d})]^{-1} [e^{-j\beta z} - \rho_0 e^{j\beta z}], \tag{10.90}$$

where

$$\rho_0 = \rho_L e^{-2j\beta d} \tag{10.91}$$

$$\rho_g = (Zg - Z_0)/(Z_g + Z_0). \tag{10.92}$$

These relations allow us to use transformations to make any transmission line system equivalent to the one in Figure 10.11 and then use equations (10.89) and (10.90) to find the voltage and current on the line. The two transformations that can do this are shown in Figure 10.14. The load impedance Z_L is transformed to Z_i by the relation

$$Z_i = Z_0 \frac{Z_L + jZ_0 \tan \beta d}{Z_0 + jZ_L \tan \beta d}. \tag{10.93}$$

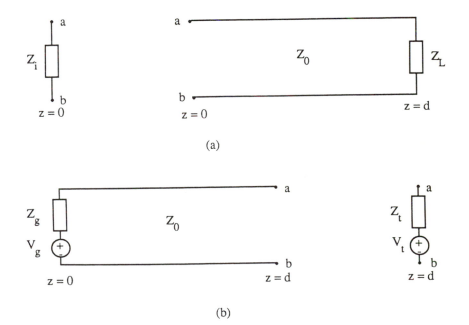

(a)

(b)

Figure 10.14 Impedance and source transformations used in solving transmission line problems.

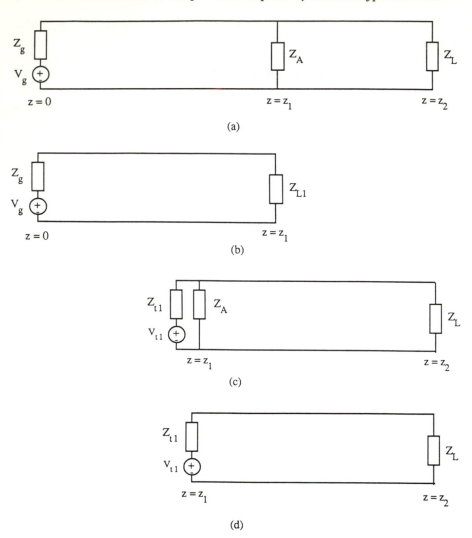

Figure 10.15 Using transformations to solve for voltages and currents on transmission lines. (a) The original problem. (b) Z_L transformed to $z=z_1$ and combined with Z_A. (c) V_g and Z_g transformed to $z=z_1$. (d) V_{t1} and Z_{t1} combined with Z_A using Thevenin's theorem.

The source is transformed by using Thevenin's theorem. V_t is the open-circuit voltage at $z=d$, and Z_t is the impedance at $z=d$ with V_g deactivated (replaced by a short). From equations (10.89) and (10.87),

$$V_t = 2V_g Z_0\, e^{-j\beta d}\, [(Z_0+Z_g)(1-\rho_g\, e^{-2j\beta d})]^{-1} \qquad (10.94)$$

$$Z_t = Z_0\, \frac{Z_g + jZ_0 \tan \beta d}{Z_0 + jZ_L \tan \beta d}. \qquad (10.95)$$

Figure 10.15 gives an example of how these two transformations can be used to solve a transmission line problem. Transforming Z_L to $z=z_1$ and combining it with Z_A gives the equivalent circuit shown in Figure 10.15(b), from which the voltage and current at any point on the left section of the transmission line can be found from equations (10.89) and (10.90). To find the voltages and currents on the right section of the line the source in (a) is transformed to $z=z_1$ as shown in (c), and then another Thevenin's equivalent is constructed to get the circuit shown in (d). Then, a new co-ordinate system with $z=0$ at the left end of the final circuit is constructed, as shown in (d). The voltage and current on the right section of the line can be found from equations (10.89) and (10.90).

The *Smith chart* is a very useful graphical tool for transforming imped-ances, as explained in standard texts on transmission lines. It also provides useful conceptual procedures that can provide increased under-standing and insight. Much transmission line design is now done using computer simulations, although the Smith chart sometimes provides more insight than computer calculations.

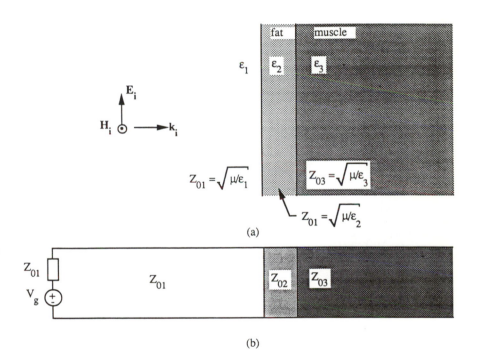

Figure 10.16 (a) Plane wave incident on a fat–muscle planar model. (b) Equivalent transmission line model.

10.7.2. Plane wave transmission

Plane wave transmission problems may also be solved using the powerful transformations described above by constructing a transmission line model equivalent to the plane wave model by changing:

1. V to E.
2. I to H.
3. Transmission line impedances to wave impedances.
4. Generator voltage to twice incident E.

With these changes, the transmission line equations and plane wave equations are exactly equivalent. The incident E is equal to half the generator voltage because half of the generator voltage is dropped across the Z_g, which is taken to be equal to Z_0.

As an example, consider a plane wave incident on a planar fat–muscle model (Figure 10.16). The fields in the fat and muscle layers can be found from an equivalent transmission line model using the techniques illustrated in Figure 10.15.

The equivalent transmission line model is shown in Figure 10.16(b). With the model of Figure 10.16(b), transformations similar to those in Figure 10.15 can be used to find E in the fat and in the muscle. These transformations can easily be represented by computer algorithms. They are also better than the traditional bouncing-wave approach used in some physics texts in which the waves inside the layers are summed in infinite series to calculate the fields. The transformation methods provide both superior calculation procedures and more insight and qualitative understanding of the interactions.

References

Durney, C.H., Massoudi, H. and Iskander, M.F. (1986) *Radiofrequency Radiation Dosimetry Handbook*, 4th ed., Brooks Air Force Base, San Antonio, Texas, US Air Force School of Aerospace Medicine.

11 Electromagnetic applicators

L. Heinzl, S.N. Hornsleth, P. Raskmark and J.B. Andersen

11.1. Introduction

This chapter concentrates on non-invasive electromagnetic applicators, used individually or in phased array systems, while invasive and interstitial devices are described elsewhere in this book. An overview of the basic principles of the known categories of applicators, supplemented with examples of actual applicator designs, is given.

The basic physical mechanism of electromagnetic heating is very simple and can be expressed by a modified form of Ohm's law, which determines the tissue current density J for a given electric field E and a known tissue conductivity σ as $J=\sigma E$. Passing through a lossy medium the current creates a local power deposition having a density

$$P=1/\sigma\,|J|^2=\sigma\,|E|^2 \quad (\text{W/m}^3), \tag{11.1}$$

which is closely related to the specific absorption rate (SAR) by the formula $P=\rho\times\text{SAR}$ where ρ is the tissue density.

The tissue conductivity σ depends to a large extent on the water content of the tissue and on the choice of frequency.

11.2. Effect of electric field orientation

The purpose of the applicator is, for a given distribution of tissue, to provide a sufficient E field to obtain therapeutic temperatures in the tissue volume to be treated and avoid excessive heating of healthy tissue.

Considering inhomogeneous and layered tissue, it turns out that the orientation of the electric field relative to the interfaces has a great influence on the power deposition pattern obtained. Figure 11.1 shows the dramatic differences in the power patterns when electric fields directed, respectively, normal and parallel to the interfaces are applied to a typical multilayer configuration.

(a)   **(b)**

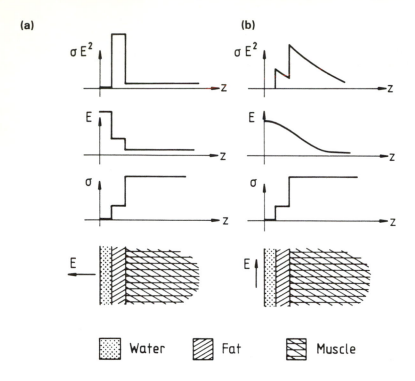

Figure 11.1 Conductivity, electric field and power density distributions in water bolus, fat, and muscle layers with electric field (a) normal to and (b) parallel to interfaces.

In the cases where the E field is normal to the interfaces, the continuity of the total current, including both conductive and displacement currents, leads to the following variation of the power density $P(z)$:

$$P(z)=\sigma(z)\,|E(z)|^2=\sigma(z)\frac{|D|^2}{|\epsilon(z)|^2} = \frac{\omega\,\epsilon_0\,\epsilon''(z)}{[\epsilon'(z)]^2+[\epsilon''(z)]^2}|D|^2, \qquad (11.2)$$

where D, the electric displacement, is continuous over the boundaries, ϵ' is the relative dielectric constant and ϵ'' is related to conductivity by $\epsilon''=\sigma/\omega\epsilon_0$. The power deposition in the water is small because $\epsilon''\approx 0$ and in the muscle tissue it is small because ϵ'' (and ϵ') is large (Figure 11.1(a)), which explains the excessive heating of the fat layer.

In the case where the E field is tangential to the interfaces, the E field remains continuous across the boundaries between the layers directly leading to jumps in power density which corresponds to the jumps in conductivity according to equation (11.1). This explains why the power dissipation in the muscle region now exceeds the power in the fat layer, while the power losses in the water bolus still remain low (Figure 11.2(a)). If the source of the E field is placed outside the tissue, one must take into account the attenuation of the E field with increasing depth as shown in the figure. This matter has been treated in more detail in Chapter 10.

11.3. Applicator designs

Rather than classifying the applicators by the choice of frequency it is more meaningful from an engineering point of view to classify the different applicator types in accordance with the physical mechanism involved in establishing the E field. All known applicators fit into at least one of the following three categories:

- Capacitive applicators.
- Inductive applicators.
- Aperture applicators.

Some applicators can be considered to contain elements from more than one category.

The following sections describe the basic theory of each category followed by examples of actual applicator designs.

11.3.1. Capacitive applicators

The capacitive applicator is a quasistatic device normally used at low frequencies where the wavelength in tissue is large compared to tissue dimensions. This means that no radiation takes place and the heating effect obtained is only due to dielectric losses in the tissue which can be considered as part of a capacitive electric circuit. The applicator can be considered as the electrodes of a capacitor and is simply excited by applying an electric potential across it.

For the capacitive applicator, the distributions of the E field and the related currents are fully determined by Laplace's equation. The field distribution is frequency independent and is only determined by the geometry of the applicator set up and of tissue dimension and composition. The design of the applicator electrodes can be adapted to the treatment situation by choosing the shape and size in accordance with the treated volume. In many arrangements a metallic couch is used as the second electrode, while the primary electrode is used to concentrate the field in the volume to be treated.

A simplified configuration is shown in Figure 11.2(a), where the primary electrode is a circular disc. The Laplace equation is solved for homogeneous muscle tissue of infinite thickness (in practice this means several disc diameters). The power-deposition pattern is shown in Figure 11.2(b).

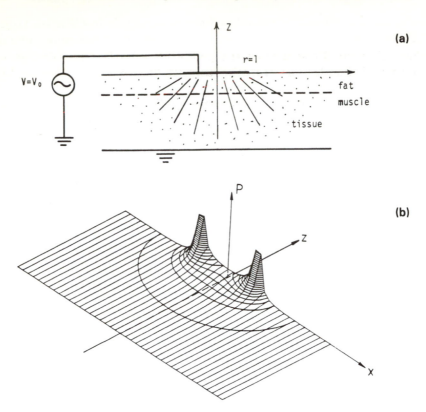

Figure 11.2 (a) A schematic diagram of a capacitive disc applicator with a distant pick-up electrode.
(b) Calculated power distribution in a centre plane below the disc. (Andersen *et al.*, 1984.)

The result is general and can be referred to any size of disc diameter. Two features are apparent, the edge singularities and the rapid decay away from the plate. The peaking of the E field near the edges is a general phenomenon where two different media meet. For a sharp edge the field behaves as

$$E \approx r^{-\alpha}, \tag{11.3}$$

where r is the distance from the edge and α depends on medium contrast and wedge angle. From the graph it is seen that apart from the singularities the maximum density is obtained at and is limited to the tissue surface below the primary electrode. At some depth the primary electrode can be considered as a charged monopole leading to a decay of the E field in any direction:

$$E \approx r^{-2} \text{ or } P \approx R^{-4} \tag{11.4}$$

where R is the distance from the primary electrode. The arrangement shown will therefore be useful mainly for the treatment of superficial tumours.

The major drawback of the capacitive applicator is that the direction of the E field is normal to the tissue surface. This means excessive heating of the fat layers placed between the skin and the deeper muscle tissue, as described above. However, the drawback can to some extent be eliminated by placing a water bolus between the electrodes and the skin of the patient. This can, at the same time, separate the edges of the electrodes from tissue (so that the high peaks remain in the water bolus) and provide cooling of the skin and fat layers.

The advantages of capacitive applicators are the low complexity of instrumentation due to the low frequency (<30 MHz) and simple applicator design.

11.3.1(a) Examples

The Thermotron RF8 (manufactured by Yamamoto Vinyter Co., Osaka, Japan) is an 8 MHz capacitive system, which is based upon a number of circular disc electrodes of different sizes (10–25 cm diameter). Because of the relatively large electrodes, the system is designed for treatment of deep-seated as well as superficial tumours.

An alternative design is described by van Rhoon *et al.* (1988) using two ring electrodes surrounding the tissue as shown at Figure 11.3. The applicators operate in the frequency range 10–40 MHz. Apart from the areas below the electrodes, this set-up will provide an electric field mainly parallel to the skin surface avoiding the excessive heating of fat layers which usually is characteristic for capacitive applicators. The capacitive ring applicator has similarities to the gap applicator, which is described in Section 11.3.3.

In a three-electrode configuration suggested by Morand and Bolomey (1987), individual phases of the electrode potentials are used to control the power deposition.

11.3.2. Inductive applicators

While the source in a capacitive device is the electrical potential applied across the tissue, the source of the inductive applicator is a current distribution (linear or two dimensional) placed outside and parallel to the tissue surface.

The mechanism of the inductive applicator can be understood by considering the electromagnetic fields around a small alternating current

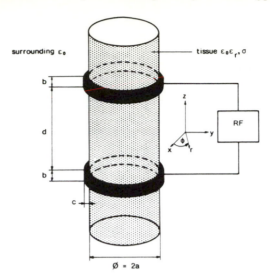

Figure 11.3 A schematic representation of the ring capacitor plate applicator. The significant parameters are indicated: (a) tissue cylinder radius, (b) electrode width, and (c) thickness, (d) gap width and the operating frequency (van Rhoon *et al.*, 1988).

element I (also known as an electric dipole) placed parallel to the tissue surface. The current element is surrounded by closed magnetic field lines which, when induced scattering from tissue is ignored, pass the tissue interface undisturbed due to the non-magnetic performance of the body. At any point the varying magnetic field induces a local electric field E. In the direction normal to the current element the E field is, as shown in Figure 11.4(a), normal to the H field and parallel to the current of the element and consequently parallel to the tissue surface, which is in accordance with the preferred field orientation shown in Figure 11.1(b).

The E field and the alternating H field are coupled by the Maxwell equation:

$$\nabla \times E = -\mu_0 \frac{\partial H}{\partial t}. \tag{11.5}$$

It should be noted that the induced E field does not depend on the dielectric constant ϵ of tissue and that it is proportional to frequency (the time derivation of the H field in the frequency domain can be written as $j2\pi fH$). Equation (11.5) is solved for an electric dipole in most elementary antenna textbooks:

$$E \approx I \left(\frac{1}{\gamma r} - \frac{j}{(\gamma r)^2} - \frac{1}{(\gamma r)^3} \right) e^{-j\gamma r}, \tag{11.6}$$

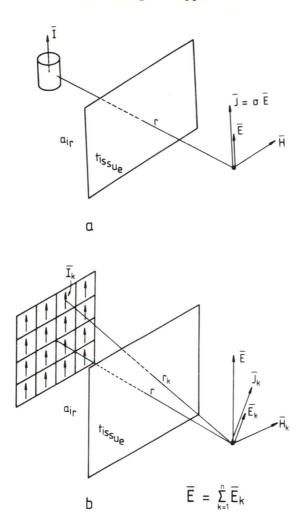

$$\bar{E} = \sum_{k=1}^{n} \bar{E}_k$$

Figure 11.4 (a) Field and current orientations in tissue when exposed by an electric dipole parallel to the tissue surface. (b) Field and current orientations from a plane applicator, equivalized by an array on *n* electric dipoles. The figure shows the contributions from the *k*th dipole and the total *E* field at a position in front of the applicator.

where $\gamma = k_0 \sqrt{(\epsilon' - j\epsilon'')}$. $k_0 = \omega \sqrt{\mu_0 \epsilon_0}$ is the free-space propagation constant and r is the distance from the current element. The formula consists of three parts. The first can be identified as a wave propagating away from the current element, the second represents the inductive field around the element and the last the quasi-stationary field component. From the formula it can be seen that at high frequencies, dependent on the product

γr, the quasi-stationary field will be vanishing even close to the applicator and that the inductive component will dominate in the outer parts of tissue, while the radiating component will take over at depth. In inductive applicators the inductive field will be the main contributor in the tissue volumes to be treated.

The exponential factor $e^{-j\gamma r}=e^{-\alpha r}e^{-j\beta r}$ expresses the exponential decay and the time retardation of the propagating field. A plane wave decays in the same way as the exponential factor. It is important to realize that the penetration into tissue is frequency dependent. Penetration depth δ is defined as the distance where the power density has decreased by a factor e^{-2}. The half-power distance $\delta_{\frac{1}{2}}$ is the distance over which the power density is reduced by a factor of two. Figure 11.5 shows half-power distance as a function of frequency in muscle tissue. It is apparent that the operating frequency of inductive applicators should be kept in the interval 40–150 MHz. If a higher frequency is chosen, penetration into depth will be poor, whilst if the frequency is lower, the applicator becomes a capacitive device.

Practical inductive applicators can be considered as an array of current elements; alternatively, electric or magnetic dipoles and the related E field and power deposition can be determined by integrating the contributions of all individual elements taking into account the effect of phase shifts and field orientations [see Figure 11.4(b)] (Andersen, 1985).

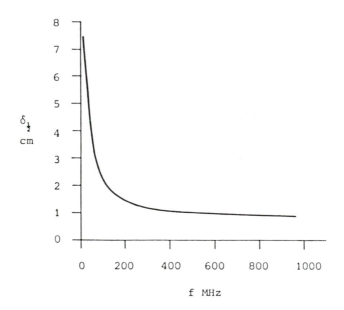

Figure 11.5 Half-power penetration depth in muscle tissue as a function of frequency (Andersen, 1987).

The features of the inductive applicators can be deduced from the performance of the above-described current element. The capacitive effect will only be present at the edges of the applicators, as the charges built up in the current elements will cancel when put together in the array. Furthermore, the potentials involved will be relatively small as the inductive applicator turns out to be a low-impedance device. Except in areas close to the edges of the applicator, the induced electric field will remain parallel to tissue surface, thus avoiding the excessive power deposition in the fat layers [see Figure 11.1(b)] which characterized the capacitive applicator. The penetration-depth curve in Figure 11.5 can still be used as a guideline for choice of frequency even though applicator size plays an important role (see Figure 11.7).

11.3.2(a) Examples

The single-turn planar coil has been a common applicator, used both as a pancake applicator oriented parallel to the tissue and as a concentric coil surrounding the tissue. Both applicators suffer from a zero in power deposition at the central axis of the coil when used at frequencies where the circumference of the coil is small compared to the wavelength. In this case, opposed current elements will have opposite polarization and cancel each other at the centre of the coil. From this it can be seen that the concentric coil is not an ideal applicator as a deep-heating device. However, it still has the advantage of low fat heating as the induced currents flow parallel to tissue surface and, because of the nature of the design, there will be no unwanted hot spots due to edge effects. To increase the treatment volume the coil can be constructed from a wide strip of metal which may overlap at the ends to provide the tuning capacity for the device (Storm *et al.*, 1982).

In the applicator design suggested by Andersen *et al.* (1984), the current loop is turned so that the plane of the loop is normal to the tissue surface (see Figure 11.6(a)). Due to differences in the phases of the field contributions from front and return currents, the nulling effect of the above-mentioned applicators is avoided and a beam-like distribution of power into tissue is obtained. In the actual design the return current is replaced by a reflector surface (see Figure 11.6(b)). The loop current is turned to resonance by a capacitor placed behind the reflector plate. The position of the feeding point is chosen to give best impedance matching to the radio-frequency generator. As above, the treatment volume can be increased by widening the conductors and an even current distribution on the applicator can be ensured splitting the conductor into parallel strips [Figure 11.6(c)].

Improvements of the above applicator have been suggested by Tiberio *et al.* (1984), who have created a balanced feeding arrangement which minimizes the effect of the stray capacitances at the ends of the applicator,

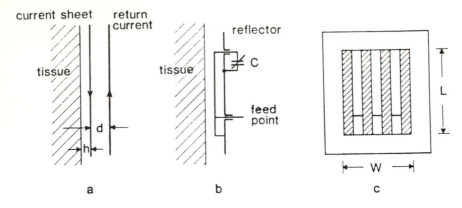

Figure 11.6 Inductive applicator (Andersen *et al.*, 1984). (a) Current sheet parallel to the tissue surface and return current behind the current sheet. (b) Schematic side view of the applicator where the return currents are replaced by the reflector surface; the loop current is turned to resonance by the capacitor behind the reflector. The current is tapped at a suitable impedance point. (c) Front view of the applicator consisting of four parallel conductors.

and by Johnson *et al.* (1987), who have created a flexible applicator which can conform to a curved tissue surface.

The concentric helical applicator (Ruggera and Kantor, 1984) is an improvement of the concentric coil applicator since it contains a component of current in the axial direction.

11.3.3. Aperture applicators

In the aperture applicator the source of the tissue currents is an electro-magnetic wave-front which is known in a limited area (the aperture) in front of the tissue surface. The wave will continue to propagate into the tissue in front of the aperture. From Love's equivalence principle (Balanis, 1982) it is known that the electric and magnetic fields in the aperture can be replaced by equivalent sheets of electric and magnetic currents.

The current sheets can again be divided into a collection of magnetic and electric dipoles which individually contribute to the field in the tissue. From this it can be understood that the basic performance of the aperture applicators is the same as for inductive applicators. Penetration depth will depend upon frequency and upon aperture size related to wavelength in tissue.

When dimensions of the aperture are large compared to wavelength the electromagnetic fields in the tissue below the aperture will propagate as a plane wave with the direction of the *E* field parallel to the tissue surface, as was the case for inductive applicators. Turner and Kumar (1982) have shown that electrically small apertures have smaller penetration depth

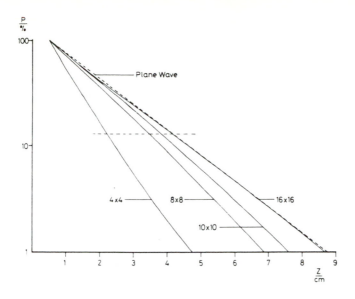

Figure 11.7 Dependence of the aperture size on penetration into muscle tissue related to plane wave penetration (434 MHz) (Hand and Hind, 1986).

and relatively higher surface heating than large apertures, as shown in Figure 11.7. However, the aperture dimensions should not exceed more than one wavelength to avoid higher-order modes in the aperture.

11.3.3(a) Examples

The most commercially used aperture applicator is the horn antenna, which can be fed from a rectangular or circular waveguide. Both are characterized by a cut-off frequency f_c below which no propagation can take place:

$$f_{c,\,rec}=\frac{c}{2a\sqrt{\epsilon_r}}; \quad f_{c,\,cir}=0.29\,\frac{c}{r\sqrt{\epsilon_r}}, \tag{11.7}$$

where a is the width of the rectangular waveguide and r the radius of the circular waveguide (Ramo *et al.*, 1965).

For horn antennas with a small flare angle, it is a good approximation to assume that the distribution of the aperture field for the open-ended horn is the same as for the incident field in the waveguide. In the case of a rectangular horn this means that using a co-ordinate system, with the x axis along the wide wall and the y axis along the narrow wall, the fundamental mode will have an aperture distribution given by

$$E_y=E_0\sin\frac{\pi}{a}x, \quad E_x=0, \quad H_x=H_0\sin\frac{\pi}{a}x, \tag{11.8}$$

where a is the width of the horn. Applying Love's principle, this is equivalent to a tapered array of x-directed magnetic dipoles and a tapered array of y-directed electric dipoles, where the tapering is in accordance with the field variations across the aperture. The tapering is due to boundary conditions in the waveguide and reduces the effective treatment volume compared to a current-driven applicator of the same dimensions. For horn applicators having a wide flare angle a quadratic phase shift is superimposed on the tapering given by equation (11.8) due to differences in the propagation distances within the horn. This will provide a tendency to defocus the wave penetrating into tissue so decreasing the penetration depth.

To enable a horn applicator of practical size to operate at a frequency appropriate for hyperthermia applicators, the cut-off frequency may be decreased by filling the horn with de-ionized water sealed at the aperture. This water bolus will simultaneously provide skin cooling, protect against excessive heating from singularities near the sharp edges of the horn and reduce the impedance-matching problems of the applicator as a water–skin interface gives much less reflection than an air–skin interface.

Normally, horn and waveguide applicators are operated in the frequency range 200–2500 MHz and their use is restricted to the treatment of superficial tumours, although Paglione *et al.* (1981) have used a large water-filled ridged waveguide applicator for deep heating at frequencies as low as 27 MHz.

Even though a real focus cannot be obtained from a plane aperture applicator (Andersen, 1987) some modifications of the power deposition can be achieved by correcting the phase shifts from all the individual elements of the aperture so that all contributions for the destination arrive in phase at the target independent of the actual path lengths. The focusing can be obtained by tapering the field distributions in the aperture with respect to phase (Nikawa *et al.*, 1986) by inserting a microwave lens in front of a normal rectangular waveguide. The lens also provides a better coupling to the tissue as it conforms to the shape of the body.

Another microwave antenna, which has been turned into an applicator, is the microstrip applicator. This consists of a metallic groundplane, a thin dielectric sheet of relatively high dielectric constant ($\epsilon_r \approx 10$–30) on top of which is a metallic plate (circular or rectangular), which is in turn covered by another sheet of dielectric material. The source of propagation is the electric field in the narrow aperture between the plate and the groundplane. As the circuit is a high Q-factor resonant structure, the H field in the aperture can be considered equal to zero. This means that the radiation from the applicator is equivalent to that from a magnetic ring current placed in the aperture around the plate. In contrast to the inductive planar coil applicator discussed earlier, there is, by proper choice of dimensions related to frequency, a phase reversal around the magnetic

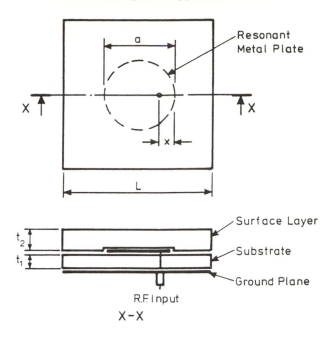

Figure 11.8 Microstrip applicator with a circular resonator patch. The position of the feeding point x is chosen to optimize impedance matching (Johnson *et al.*, 1984).

current loop which will give broadside radiation. As with other applicators, the microstrip applicator requires a water bolus to protect tissues against near-field effects.

A microstrip applicator with a circular resonator plate, designed by Johnson *et al.* (1984), is shown in Figure 11.8. The feeding of the applicator is simply done from the back of the applicator. The distance of the feeding point from the centre of the plate determines the impedance level of the loaded applicator.

The microstrip applicators are operated in the frequency range 200–1000 MHz. Since the operating frequency and size of applicator dimensions are mutually related by the resonance conditions, a large treatment volume requires large dimensions and low frequency, and vice versa.

The applicators mentioned above are used mainly for superficial treatment. To obtain deep heating some means of focusing the field will be necessary. A ring aperture placed around the patient and with a uniform electric field in the longitudinal direction of the patient will create a local interference maximum in the central region.

Lagendijk (1983) has designed a coaxial transverse electromagnetic applicator. The patient is placed inside in a hollow inner conductor in a large coaxial structure. The inner conductor is removed above the patient volume to be heated and the electric field between the two parts of the inner conductor creates the desired ring aperture.

A similar arrangement is used in the gap applicator designed by Raskmark and Andersen (1984a) (Danish Hyperthermia Foundation, 1987). A schematic drawing is shown in Figure 11.9(a). This applicator also has a coaxial structure, but in this case the patient acts as a lossy inner conductor and the ring aperture is established by making a gap in the outer conductor. The applicator is excited by applying a voltage across the gap. The structure has the advantage that modes propagating in the longi-

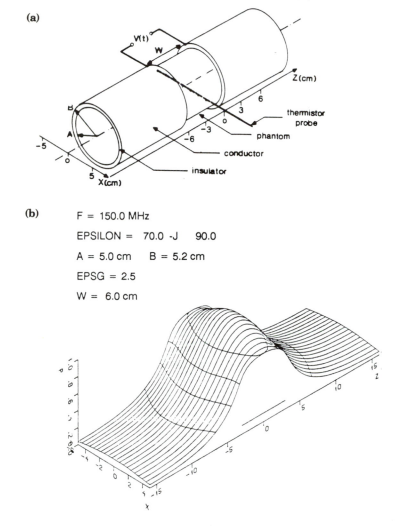

Figure 11.9 Gap applicator design (Raskmark and Andersen, 1984a) (a) Schematic drawing of the gap applicator. (b) Power deposition distribution in a longitudinal section containing the centre-line. Significant parameters are shown in the figure.

tudinal direction of the patient decay very quickly, resulting in a tight coupling of the applicator to the treatment volume. The theoretical power distribution in a 10 cm diameter muscle cylinder is shown in Figure 11.9(b).

Raskmark and Andersen (1984b) 'split' the cylindrical gap applicator into four individual gap applicators. This allows the possibility of individual control of phase and amplitude which is used as described for phased array systems (Section 11.4). In addition, this arrangement is more flexible in that it allows the set-up to conform to the body providing a tighter coupling to the tissue. The split gap applicator can also be used as a single applicator for superficial treatment. These gap applicators are operated in the frequency range 40–150 MHz. At lower frequencies they will behave as the capacitive ring applicator mentioned above.

The two gap applicators discussed are examples of cylindrically focused apertures, where the focusing effect is due to the cylindrical shape of the applicator giving the same path length from all elements of the applicator to the centre-line, where it creates constructive interference.

11.4. Phased array systems

A number of systems which create some focusing of the electromagnetic power have been reported. All these use phase shifting as a means of obtaining the desired focusing. Three different set-ups are normally considered: co-planar arrays, annular arrays and movable applicator arrays.

All phased array systems rely on the effect known as constructive interference, simply meaning that by proper phasing of the individual applicators their power contributions will interact in a positive manner at a certain point.

11.4.1. Constructive interference

The power deposited in a small volume of constant conductivity σ from one applicator is $P = \sigma |E_1|^2$. The addition of an extra applicator producing a field strength $|E_2|$ with the same polarization will result in a total power deposition:

$$P = \sigma(|E_1|^2 + |E_2|^2) \tag{11.9}$$

The applicators are non-coherent when there is no common phase reference in the system. This would be the case when two independent radio-frequency generators are used to supply the applicators.

If the signals to the two applicators are coherent then

$$P(\alpha) = \sigma |E_{\text{tot}}|^2 = \sigma[(|E_1| + \cos(\alpha)|E_2|)^2/(sin(\alpha)|E_2|)^2\langle, \tag{11.10}$$

where α is the phase difference between the E fields at the point of interference.

If constructive interference, i.e. $\alpha=0$, is achieved, the total power will be $P(\alpha=0)=\sigma \mid E_1+E_2\mid^2$, equal to twice the power of the non-coherent system. It can be seen that the non-coherent case corresponds to $\alpha=\pi/2$. If $\alpha=\pi$, the result is destructive interference and the total power is $P(\alpha=\pi)=\sigma \mid E_1-E_2\mid^2$. This indicates that controlling the phase difference can give some control of the power deposition pattern and explains why this type of system is referred to as a phased array.

Continuing this line of thought one might suggest applying N coherent applicators and thus obtaining a total power $P'=\sigma \mid E_1+E_2+\ldots E_N\mid^2$. Unfortunately, the penetration depth of the individual applicator is a function of size as shown by Turner and Kumar (1982), and different studies (Arcangeli *et al.*, 1984; Wait, 1985; Jouvie *et al.*, 1986) suggest that 4–8 applicators are a good compromise.

This, however, does not mean that choosing N to be large, i.e. $N\to\infty$, is a bad choice, because this will be equivalent to a continuous source completely surrounding the body, which is actually the case for a number of applicator systems (Raskmark and Andersen, 1984a; Lagendijk, 1983; Tiberio *et al.*, 1988). In these situations the system is no longer referred to as a phased array because the only way to control the power deposition pattern is by moving the body around inside the applicator.

Of course, changing the amplitudes of each individual applicator will also affect the power deposition pattern. For a four-applicator system seven parameters can be adjusted in this manner, namely four amplitudes and three phase differences.

11.4.2. Choice of frequency

As shown by Andersen (1984), the frequency will affect the power deposition pattern in two ways. Since the penetration depth is determined by losses in the tissue it is a decreasing function of frequency. On the other hand the focal spot size will also be a decreasing function of frequency. Figure 11.10 shows the relation between half-power penetration depth and frequency for a plane wave and the relation between focal spot size and frequency in the cyclindrical case (two dimensional).

Figure 11.10 clearly shows that the choice of frequency will depend on the size of the volume to be treated. If the volume to be treated is large, e.g. the abdominal region, a low frequency must be chosen ($f<70$ MHz) in order to assure reasonable penetration. In this way some of the power-steering capabilities will be lost due to the large focal spot size, and hence these systems can also be classified as regional heating devices.

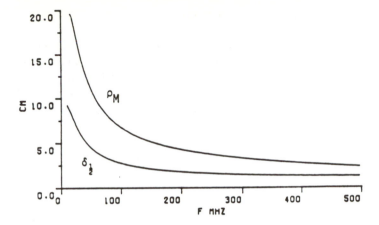

Figure 11.10 Half-power penetration depth $\delta_{\frac{1}{2}}$ for a plane wave, and focal spot size ρ_M (two-dimensional case) in muscle tissue as a function of frequency (Andersen, 1984).

11.4.3. Effects of changing amplitude and phase

In order to evaluate the effects of amplitude and phase changes in the phased array systems, a simple model describing each applicator by an array of magnetic dipoles (Turner and Kumar, 1982; Andersen, 1985) is used. The model is only valid in a homogeneous region of space.

The model for each applicator has been fitted to actual E field measurements (Gross and Raskmark, 1989) performed in the elliptical Centre for Devises and Radiological Health (CDRH) phantom (Allen *et al.*, 1988) filled with muscle equivalent saline solution ($\sigma = 1.1$ S/m, $\epsilon_r = 75.5$) (Raskmark, 1989). Several four-applicator phased array simulations (Hornsleth and Heinzl, 1989) and measurements have been performed with this set-up (Figure 11.11) which show good agreement.

Since temperature simulations are more meaningful from a clinical point of view, the simulated SAR patterns were used to predict steady-state temperature distributions using the three-dimensional inhomogeneous temperature simulator developed at the University of Arizona, Tucson (Chen *et al.*, 1990).

The following assumptions were made. In electrical terms the region to be treated was considered homogeneous. In thermal terms, the region consisted of a large volume of muscle with resting blood flow with a solid block of tumour at the centre having no blood perfusion. The limiting temperature in the healthy tissue was chosen as 42.5°C. This is clearly a simplified situation but nevertheless gives valuable insight into the effects of constructive interference on the resulting temperature distribution (Hornsleth *et al.*, 1989).

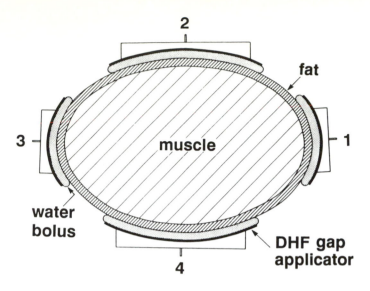

Figure 11.11 The set-up used in phased array measurements.

Figure 11.12 shows the steady-state temperature distribution for the case of constant phase and constant amplitude. It is obvious that the two hot spots will prevent therapeutic heating of the tumour.

In Figure 11.13 the phase of each applicator was changed in such a way that constructive interference would take place at the centre of the ellipse (conjugate phase and constant amplitude), and the effect on the temperature distribution is evident since the hot spot is now located in the tumour.

11.4.4. Control strategies

Because of the large number of parameters that can be used to change the heating pattern of a phased array system, it is necessary to have a standard method by which a good ('optimum') choice can be made. Normally three approaches are considered:

- Computerized treatment planning.
- Conjugate field matching.
- Feedback temperature control.

11.4.4(a) Computerized treatment planning

Recently, Nguyen and Strohbehn (1989) have reported numerical methods to optimize the position, orientation, amplitudes and phases in a movable

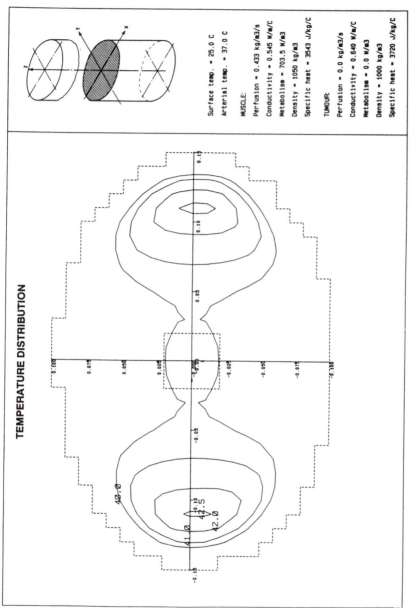

Figure 11.12 Temperature distribution in an electrically homogeneous elliptic muscle cylinder, with a block tumour at the centre having no blood perfusion, when using four applicators with the same amplitude and phase.

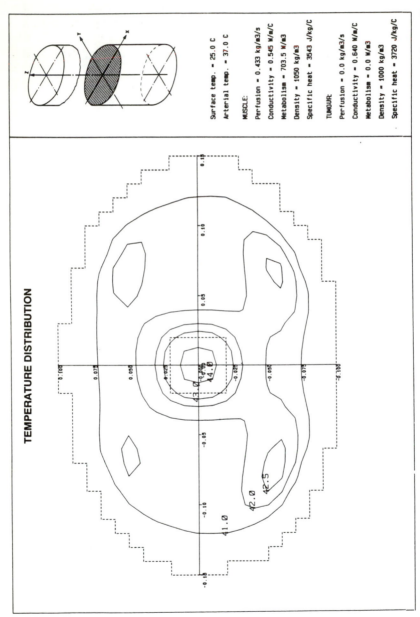

Figure 11.13 Temperature distribution in the same set-up as shown in Figure 11.12 but with the phases of the applicator changed to give 'optimum' power deposition at the centre.

multi-applicator phased array system. This provides clinicians with a tool by which they can determine the 'optimum' treatment parameters.

The problems with such methods is of course that an error in the model will lead to temperature distributions that are different from those predicted. Since modelling of blood flow variations during treatment is extremely difficult, some temperature measurements should be performed during the treatment in order to check the temperature distribution.

11.4.4.(b) Conjugate field matching

One way of obtaining information on the 'optimum' choice of amplitude and phase is by using conjugate field matching (Gee *et al.*, 1984; Andersen, 1985; Loane *et al.*, 1986). This method uses a radiating probe that is inserted into the body at the location of the desired focus. The applicators are then used as receiving antennas, and for each applicator the complex voltage is measured.

These complex voltages are used to determine the feeding amplitude and phase when the applicators are used as transmitters. To give constructive interference at the test point, the exciting phases must be chosen as the conjugate of the test voltages. However, this does not give us a unique answer since three different amplitude strategies can be adopted:

- *Conjugate amplitude.* The exciting amplitude is also chosen as the conjugate; this is known as maximum gain.
- *Inverse amplitude.* The exciting amplitude is chosen as the inverse of the received voltages. This corresponds to the individual waves arriving at the focal point with same amplitude and phase, giving maximum resolution.
- *Constant amplitude.* From a practical point of view this is the easiest since all exciting amplitudes are equal and only phase control is required.

It should be noted that inhomogeneities are accounted for. Since the radiating test probe will radiate through the actual medium. The method requires the use of a vector voltmeter, which may not be available in all institutions. In practice the radiating probe would be inserted into the patient in the correct place and with the correct polarization. The method will give good power deposition patterns but, since there is no guarantee that this will lead to the desired temperature distributions, temperature measurements should still be performed during the treatment.

11.4.4.(c) Feedback temperature control

This method is based on temperature measurements performed during the actual treatment. These measurements are used as input to a control algorithm that computes new values of amplitude and phase parameters concurrently. When using a four-applicator system, seven parameters can be controlled, and therefore seven temperatures must be measured. Unfortunately, the temperature can, at the most, only be controlled in the feedback points, and hot spots may occur in other areas (Knudsen and Heinzl, 1989).

11.4.5. Applicator coupling

Because of the close coupling nature of the applicators used in phased arrays some mutual coupling will take place. The result of this is that changing amplitude and/or phase on one applicator will effect the other applicators. In practice this will make amplitude and/or phase adjustments difficult. To overcome this problem an iterative adjustment procedure in which the amplitudes and phases of all applicators are constantly monitored can be used. Alternatively, feedback loops can be implemented for continuous control of amplitudes and phases.

11.4.6. Examples of phased arrays

11.4.6(a) Annular phased array

The commercially available BSD-2000 system (manufactured by the BSD Medical Co., Salt Lake City, Utah, USA) (Turner, 1982; Turner and Kumar, 1982) consists of eight dipoles paired together giving four power input channels. The Sigma 60 applicator (60–120 MHz) includes dipoles which are arranged concentrically surrounding the patient's torso (Figure 11.14). The gap between the patient and applicators is filled with a large water bolus that provides electromagnetic coupling and surface temperature control. The system provides radial power steering by a four-channel independent control power and phase. Axial positioning is possible by positioning the applicator along the support couch. The rigid transparent dipole support provides reproducible dipole positioning for both treatment and numerical modelling. The electronic phase and power steering permit either deep/regional or local heating.

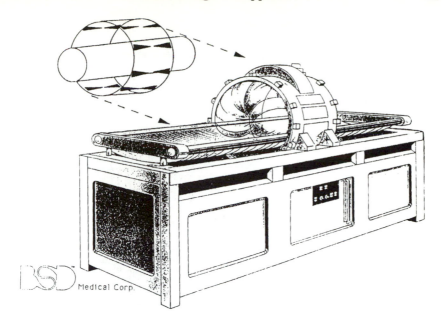

Figure 11.14 The BSD-2000 phased array hyperthermia system. The figure shows the patient coach and applicator system. (Figure by courtesy of Dr Turner, BSD Medical Co.)

11.4.6(b) Movable multi-applicator phased array

The movable multi-applicator system (Raskmark and Andersen, 1984b) uses four freely movable applicators, mounted with ball joints on four electronically steered arms. This ensures a high degree of freedom in both positioning and orientation of the individual applicators. Each applicator is equipped with a water bolus which provides individual temperature control and electromagnetic coupling. The applicators used are the gap applicators previously described. The system also provides control of the four power channels to give amplitude and phase control of the radial power deposition pattern. Axial control is obtained by the position of the applicators. Because of the close coupling structure of the applicators and the flexibility in the number of applicators used, this system provides the capabilities of both local and deep/regional heating (Figure 11.15).

11.4.6(c) Coplanar phased arrays

The coplanar phased arrays (Gee *et al.*, 1984; Hand *et al.*, 1986) are not designed for deep regional heating but some degree of focusing may be used to achieve better penetration to facilitate heating of subcutaneous

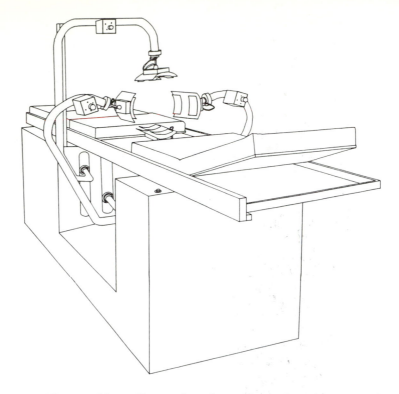

Figure 11.15 The movable applicator phased array hyperthermia system. The figure shows the patient coach, and four applicators and the supporting arms. (Figure by courtesy of the Danish Hyperthermia Foundation.)

tumours. When using this type of array it is necessary that the effective field size of the individual applicator is equal to the physical dimensions. Coplanar arrays require compact applicators and should be, to some extent, conformal.

11.5. Applicator feeding

11.5.1. Hyperthermia heating system

In the previous text the discussion has been concentrated on the inter-action of the applicators with the tissue in single- and multi-applicator systems. In a typical hyperthermia treatment situation, as shown on Figure 11.16, the power feeding must also be considered.

A number of elements such as power meters, baluns, impedance matching networks, the water boluses and finally the applicators are

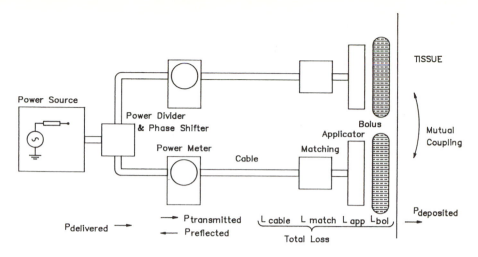

Figure 11.16 A typical hyperthermia heating set-up.

included in the feeding network. A brief introduction will be given, but a detailed description of design procedures may be found in relevant radio-frequency literature (e.g. Brown *et al.*, 1973).

The aim is to deposit sufficient power in the tissue. An ideal power source having a constant voltage source and a real internal impedance (typically 50 ohms) is assumed and modelled as shown in Figure 11.16. The maximum power transfer to a load will take place when the load impedance at the source is equal to the source impedance. The applicator is connected to the source by a cable and since voltages propagate with a limited velocity, it is appropriate to describe the total voltage on the cable as consisting of a transmitted and a reflected voltage. Assuming equal cable and source impedances, the total power delivered to the system from the source can be expressed as the power transmitted reduced by the power reflected from the load. This corresponds to a measurement of transmitted and reflected power separately with a through-line power meter. The difference is the net power passing through the meter.

The power deposited in the tissue will be reduced by the total losses and the following relations for the set-up can be deduced from Figure 11.16. The efficency of the system is defined as

$$\text{Efficiency} = P_{\text{deposited, tissue}} / P_{\text{delivered, source}} \tag{11.11}$$

$$P_{\text{delivered, source}} = P_{\text{transmitted}} - P_{\text{reflected}} \tag{11.12}$$

$$P_{\text{deposited, tissue}} = P_{\text{delivered, source}} - P_{\text{loss}} \tag{11.13}$$

where

$$P_{loss} = P_{cable\ loss} + P_{match\ loss} + P_{applicator\ loss} + P_{bolus\ loss}. \qquad (11.14)$$

Since power sources have a limited capability, it is usually necessary to minimize the losses and to keep the reflections at the load at a low level.

It should also be noticed that the reflected power contributes to unwanted heating of cables, etc., and in severe cases damage to the source or cables may occur. The reflected power can be removed from the power source by a circulator, but the deposited power will still be the difference between transmitted and reflected power at the applicator. A circulator will only protect the power source, not improve efficiency.

11.5.2. Matching network

The impedance matching network should transform the applicator impedance to the nominal (cable) impedance and it is usually necessary to consider the variations in the applicator impedance created by changing coupling, positioning of the patient, etc. An additional source for multi-applicators is change in mutual couplings.

In principle, the technique is the same for different types of applicators since they all can be described as a complex load having a real and an imaginary part. Thus, two parameters must be adjusted. Using a transmission line approach the line stretcher and stub tuner are well suited (Brown *et al.*, 1973). A line stretcher adds phase shift until the real part of the impedance is matched and the stub (open or shorted) then outbalances the imaginary part. Although there is an infinite number of solutions, only cable lengths within half a wavelength are of interest. Considering the ability to adjust the matching circuit it is not always the shortest cable length that gives the best-suited matching. At low frequencies line lengths may be too long and therefore can be replaced by capacitors or inductors. At high frequencies waveguides may be used.

It is important to integrate the matching in the applicator whenever possible. If part of the matching circuit is connected via a cable, this cable length must not be changed. The matching is often made adjustable, and in a good applicator design only small adjustments are necessary. Adjustments are made to maximize the power delivered. Occasionally, it is necessary to have a small amount of reflected power (rather than zero as expected), an effect caused by having a source impedance different from the cable impedance.

11.5.3. System losses

The total losses must be as small as possible in order to get a high efficiency. The losses in the matching network and cables may be reduced by using high-quality components such as Teflon-filled cables and vacuum capacitors. At the same time, these types of components secure stable high-power operation with no drift in matching circuits. Adjustable coils should be avoided.

It is important to notice that losses in the bolus will reduce the power deposited in the tissue. Consequently, the conductivity of the bolus liquid must be kept as low as possible. De-ionized bolus water can be used, but it must be replaced regularly.

The power delivered in the tissue cannot be determined from the power measurements if the total losses are not known. The overall efficiency in a typical set-up may be as low as 50%.

11.5.4. Stray radiation and balance

If the applicator is designed to eliminate stray radiation, the applicator housing and the tissue outside the treatment area will be at zero radio-frequency potential. The tissue impedance can be thought of as a floating impedance excited by a balanced source. The applicator creates a symmetric field balanced around the housing's zero radio-frequency potential. On the other hand, power generators, cables, etc., are unbalanced having the ground and cable screen at zero radio-frequency potential. A network which transforms the mode from balanced to unbalanced is called a balun and can conveniently be incorporated in the applicator feeding or matching network (Brown *et al.*, 1973).

If the applicator balance is violated by tilting or by loading with a very inhomogeneous tissue, an unbalanced applicator loading will be created and the balun tends to create a similar balanced component in the cable screen. This results in radio-frequency currents that will off-set the zero radio-frequency levels in the cable and applicator causing stray radiation.

11.5.5. Power dividers and phase shifters

In multi-applicator systems with only one power amplifier available, the power is divided into the applicators as shown in Figure 11.16. A power divider with a 1:1 dividing ratio is made by two quarter-wavelength transformers consisting of cables having an impedance $\sqrt{2}$ times the nominal (source) impedance. The transformers are inserted in each branch transforming the matched impedance to the double of the nominal impedance. The parallel connection reproduces the nominal impedance on the common input.

If necessary, a small phase shift can be introduced in one branch by inserting a small length of cable (having nominal impedance) between the output of the power divider and the power meter. Different cable lengths may be switched electronically to change phase relations, and the amplitude can be controlled to some extent by adjusting the matching circuit. The circuit is simple but it can only be recommended for set-ups with no more than two applicators. In multi-applicator arrays it is preferable to use a separate power amplifier for each applicator and to have the signals generated from an electronically controlled common reference.

References

Allen, S., Kantor, G., Bassen, H. and Ruggera, P. (1988) CDRH RF phantom for hyperthermia system evaluation, *International Journal of Hyperthermia*, **4**(1), 17–37.

Andersen, J. B. (1984) Electromagnetics of Hyperthermia, in Bajser, Z., Baxa, P. and Franconi, C. (Eds) *Applications of Physics to Medicine and Biology*, Singapore, World Scientific, pp. 337–353.

Andersen, J. B. (1985) Theoretical limitations on radiation into muscle tissue, *International Journal of Hyperthermia*, **1**(1), 45–55.

Andersen, J. B. (1987) Electromagnetic power deposition: inhomogeneous media, applicator and phased arrays, in Field, B. and Franconi, C. (Eds), *Physics and Technology of Hyperthermia*, Dordrecht, Boston and Lancaster, Martinus Nijhoff, pp. 159–188.

Andersen, J. B., Baun, A. A., Harmark, K., Heinzl, L., Raskmark, P. and Overgaard, J. (1984) A hyperthermia system using a new type of inductive applicator, *IEEE Transactions on Biomedical Engineering*, **31**(1), 21–27.

Arcangeli, G., Lombardini, P. P., Lovisolo, G. A., Marsiglia, G. and Piatelli, M. (1984) Focusing of 915 MHz electromagnetic power on deep human tissues: a mathematical model study, *IEEE Transactions of Biomedical Engineering*, **31**(1), 47–52.

Balanis, C. A. (1982) Aperture antennas, and ground plane edge effects, *Antenna Theory, Analysis and Design*, New York and Chichester, Wiley, Chapter 11, pp. 446–531.

Brown, R. G., Sharpe, R. A., Hughes, W. L. and Post, R. E. (1973) Transmission-line matching, *Lines, Waves and Antennas: The Transmission of Electric Energy*, New York, The Ronald Press, Chapter 4, pp. 76–107.

Chen, Z. P., Miller, W. H., Roemer, R. B. and Cetas, T. C. (1990) Errors between two- and three-dimensional thermal model predictions of hyperthermic treatments, *International Journal of Hyperthermia*, **6**(1), 175–191.

Danish Hyperthermia Foundation (1987) Electromagnetic applicator for localizing hypethermia heating in a system, United States Patent 4 702 262.

Gee, W., Lee, S. W., Bong, N. K., Cain, C. A., Mittra, R. and Magin, R. L. (1984) Focused array hyperthermia applicator: theory and experiment, *IEEE Transactions on Biomedical Engineering*, **31**(1), 38–46.

Gross, E. J. and Raskmark, P. (1989) Phased Array Hyperthermia: An Experimental Investigation, in Sugahara, T. and Saito, M. (Eds), Vol. 1, London, Taylor and Francis, pp. 724–725.

Hand, J. W. and Hind, A. J. (1986) A review of microwave and RF applicators for localized hyperthermia, in Hand, J. W. and James, J. R. (Eds) *Physical Techniques in Clinical Hyperthermia*, New York and Chichester, Wiley, pp. 98–148.

Hand, J. W., Cheetham, J. L. and Hind, A. J. (1986) Absorbed power distributions from coherent microwave arrays for localized hyperthermia, *IEEE Transactions on Microwave Theory and Techniques*, **35**(12), 1317–1321.

Hornsleth, S. N. and Heinzl, L. (1989) Simulation and discussion of power deposition from a four-applicator phased array hyperthermia system in Sugahara, T. and Saito, M. (Eds), *Hyperthermic Oncology 1988*, Vol. 1, London, Taylor and Francis, pp. 698–699.

Hornsleth, S. N., Andersen, J. B., Chen, Z. P., Roemer, R. B. and Cetas, T. C. (1989) 3D temperature distributions from electromagnetic phased arrays, in Sugahara, T. and Saito, M. (Eds) *Hyperthermic Oncology 1988*, Vol. 1, London, Taylor and Francis, pp. 684–685.

Johnson, R. H., James, J. R., Hand, J. W., Hopewell, J. W., Dunlop, P. R. C. and Dickinson, R. J. (1984) New low profile applicators for local heating of tissues, *IEEE Transactions on Biomedical Engineering*, 31(1), 28–37.

Johnson, R. H., Preece, A. W., Hand, J. W. and James, J. R. (1987) A new type of lightweight low-frequency electromagnetic hyperthermia applicator, *IEEE Transactions on Microwave Theory and Techniques*, 35(12), 1317–1321.

Jouvie, F., Bolomey, J.-C. and Gaboriaud, G. (1986) Discussion of capabilities of microwave phased arrays for hyperthermia treatment of neck tumours, *IEEE Transactions on Microwave Theory and Techniques*, 34(5), 495–501.

Knudsen, M. and Heinzl, L. (1989) Self-tuning temperature control for multi-applicator hyperthermia systems, in Sugahara, T. and Saito, M. (Eds) *Hyperthermic Oncology 1988*, Vol. 1, London, Taylor and Francis, pp. 751–752.

Lagendijk, J. J. W. (1983) A new coaxial TEM radiofrequency/microwave applicator for non-invasive deep-body hyperthermia, *Journal of Microwave Power*, 18, 367–376.

Loane, J., Ling, H., Wang, B. F. and Lee, S. W. (1986) Experimental investigation of retro-focusing microwave hyperthermia applicator: conjugate field matching scheme, *IEEE Transactions on Microwave Theory and Techniques*, 34(5), 490–494.

Morand, A. and Bolomey, J. C. (1987) A Model for impedance determinations and power deposition characterization in three-electrode configurations for capacitive radio frequency hyperthermia, *IEEE Transactions on Biomedical Engineering*, 34(3), 217–232.

Nguyen, D. B. and Strohbehn, J. W. (1989) Optimal positions, orientations, amplitudes and phases of movable microwave applicator in hyperthermia treatment, in Sugahara, T. and Saito, M. (Eds) *Hyperthermic Oncology 1988*, Vol. 1, London, Taylor and Francis, pp. 738–739.

Nikawa, Y., Watanabe, H., Kikuchi, M. and Mori, S. (1986) A direct-contact microwave lens applicator with a microcomputer-controlled heating system for local hyperthermia, *IEEE Transactions on Microwave Theory and Techniques*, 34(5), 626–630.

Paglione, R., Sterzer, F., Mendecki, J., Friedental, E. and Botstein, C. (1981) 27 MHz ridged waveguide applicators for localized hyperthermia treatment of deep seated malignant tumors, *Journal of Microwave Power*, 16, 71–80.

Ramo, S., Whinnery, J. R. and van Duzer, T. (1965) Characteristics of common waveguides and transmission lines, *Fields and Waves in Communication Electronics*, New York and London, Wiley, Chapter 8, pp. 420–485.

Raskmark, P. (1989) Permittivity measurements, in Sugahara, T. and Saito, M. (Eds) *Hyperthermic Oncology 1988*, Vol. 1, London, Taylor and Francis, pp. 726–727.

Raskmark, P. and Andersen, J. B. (1984a) Focused electromagnetic heating of muscle tissue, *IEEE Transactions on Microwave Theory and Techniques*, 32(8), 887–888.

Raskmark, P. and Andersen, J. B. (1984b) Electronically steered heating of a cylinder, in Overgaard, J. (Ed.) *Hyperthermic Oncology 1984*, Vol. 1, London, Taylor and Francis, pp. 617–620.

Ruggera, P. S. and Kantor, G. (1984) Development of a family of RF helical coil applicators which produce transversely uniform axially distributed heating in cylindrical fat–muscle phantoms, *IEEE Transactions on Biomedical Engineering*, 31(1), 98–106.

Storm, F. K., Elliot, R. S., Harrison, W. H. and Morton, D. L. (1982) Clinical RF hyperthermia by magnetic loop induction: a new approach to human cancer therapy, *IEEE Transactions on Microwave Theory Techniques*, 30, 1149–1158.

Tiberio, C. A., Raganella, L. and Franconi, C. (1984) Symmetric pure induction 27 MHz applicator, in Bajzer, Z., Baxa, P. and Franconi, C. (Eds) *Applications of Physics to Medicine and Biology*, Singapore, World Scientific, pp. 585–586.

Tiberio, C. A., Raganella, L., Banci, G. and Franconi, C. (1988) The RF toroidal transformer as a heat delivery system for regional and focused hyperthermia, *IEEE Transactions on Biomedical Engineering*, 35(12), 1077–1085.

Turner, P. F. (1982) United States Patents 4 589 423 and 4 672 980.

Turner, P. F. and Kumar, L. (1982) Computer for applicator heating patterns, *National Cancer Institute Monograph*, 61, 521–523.

van Rhoon, G. C., Visser, A. G., van den Berg, P. M. and Reinhold, H. S. (1988) Evaluation of ring capacitor plates for regional deep heating, *International Journal of Hyperthermia*, 4(2), 133–142.

Wait, J. R. (1985) Focused heating in cylindrical targets: part 1, *Transactions on Microwave Theory and Techniques*, 33(7), 647–649.

12 Power deposition models for hyperthermia applicators

K. D. Paulsen

12.1. Introduction

This chapter focuses on the development of computational models for characterizing power deposition patterns produced by hyperthermia applicators. The emphasis will be on the formulation and implementation of the mathematical framework required to make the desired calculations. The majority of the discussion will centre on non-invasive electromagnetic hyperthermia applicators, though sizeable portions will also be applicable to interstitial and ultrasonic techniques. It will be assumed that the reader is familiar with the contents of Chapters 10 and 11, although a certain amount of repetition of this material is unavoidable in order to facilitate the discussion. Familiarization with Chapters 13 and 14 will also enable the reader to make the appropriate analogies and extensions of the material covered in the present chapter to those specific situations.

Before embarking on a discussion of how to develop a computational model, it is instructive to consider why one would want to develop a computational model of the hyperthermia treatment process in the first place. The motivation for using computational models arises in both clinical and engineering settings. As a clinical tool, computational models are potentially important primarily because of the present impracticality (or inability) of making detailed temperature or SAR measurements throughout the treatment region, particularly for deep-seated sites. Furthermore, the degree of variability/complexity among individual patients in terms of parameters (such as tumour size, shape, location, blood flow, etc.) which can significantly influence the treatment outcome, makes intuition about the probability of the success or failure of a particular heating difficult to gain. Computational models may provide a way to obtain some insight into the expected outcome of particular treatments or to compare the effectiveness of various applicators on general classes of tumour sites. From an engineering standpoint, computational models may be a valuable tool in the design and development of

305

hyperthermia applicators. This value is readily recognized when one considers the time, effort and expense needed for the carefully planned laboratory experiments that are required to extract the detailed measurement maps needed for analysis. Optimization and parameter sensitivity studies are often well suited to computational models and can be performed in a cost-effective manner. In fact, one can envision the applicator development process as consisting of an appropriate computational model coupled with a few key laboratory experiments (which most likely the simulation studies would have suggested).

The extensive potential of computational models in hyperthermia has been best summarized by the identification of four areas where modelling would play a role in improving power delivery and evaluating resulting dosimetry (Roemer and Cetas, 1984): (1) *comparative*, (2) *prospective*, (3) *concurrent* and (4) *retrospective* hyperthermic dosimetry. The purpose of comparative dosimetry is to evaluate the performance of different devices under the same clinical situations in order to establish guidelines for the use of one system over another. Prospective dosimetry incorporates all available information about a specific patient and considers the array of possible heating devices that may be used in order to plan the treatment course most suitable for that particular patient. In concurrent dosimetry one attempts to infer temperature distributions during the treatment based on a few temperature measurements and then modifies the power delivery (if possible) to improve the therapeutic temperature rise in the tumour. In retrospective dosimetry the complete temperature profile is also inferred from measured data, but this inference occurs after the treatment in order to correlate temperature distributions with observed therapeutic benefit and to evaluate the ability of a device to elevate the entire tumour volume above some predefined therapeutic threshold.

Even with the brief discourse above, it is evident that computational models could have a significant impact on understanding and improving hyperthermia therapy. Realizing this potential is contingent on the development of an appropriate computational model, and two issues must be considered: (1) how well does the mathematical description represent the underlying physical mechanisms and (2) how well does the computer model solve the mathematical description as posed. Unfortunately, determining and implementing the appropriate computational model is a difficult task often consisting of numerous complex issues and trade-offs (which makes the identification of a single 'best' model for all situations not really possible). An in-depth discussion of such trade-offs relevant to hyperthermia power deposition calculations is beyond the intended scope of this chapter. Alternatively, some broader generalizations will be given to provide some intuition into the model-selection process. Within this general scope, efforts will be made to highlight the strengths and weaknesses of competing methods and to determine the 'best' choices (when possible) under generalized circumstances. To accomplish this general treatment of the complex process of developing and implementing an appropriate numerical model, the following outline will be followed. First, the required mathematical framework or governing equations will be discussed for the electromagnetic power deposition problem. Secondly, several problem formulation issues will be addressed which directly impact the structure of any numerical implementation. Consideration of these issues will provide the basis for making generalizations about the nature of the numerical methods which are currently competing for superiority in the hyperthermia power deposition calculation context. This discussion will lead to an examination of the required data input for computational models in hyperthermia — an issue which, to date, has remained largely unaddressed.

Finally, the chapter will conclude with a section devoted to the current status and future directions of electromagnetic modelling in hyperthermia research. The evolution of computational models in hyperthermia will be briefly traced over the last decade to their present form. Here the emphasis will centre on what present models can and cannot do, where they have been successful and where they have failed, and what some of the pressing questions and issues are that must be addressed (and solved) if computational models are to improve and realize their potential in hyperthermia research.

12.2. Mathematical formulation

The modelling of hyperthermia treatments had traditionally been a two-step process: (1) to compute the heating rate or power deposition patterns produced in the body by the heating source, and (2) to compute the redistribution of energy due to thermal conduction and blood flow. To date these two steps have been regarded as decoupled such that solution of the second step does not alter that of the first. Note, however, that the second step cannot be undertaken until the first has been completed because the output of the first step provides input to the second step.

Although the process of hyperthermia treatment simulation has been computationally decoupled, it is important that analyses based on hyperthermia simulations remain coupled. The computational decoupling conveniently allows the two facets of treatment simulation — power absorption and heat transfer — to be discussed separately, but care must be exercised in interpreting any results based solely on one of these aspects. This is especially true of power deposition calculations which, if interpreted in isolation, can result in misleading conclusions due to the profound effects of blood flow on heat transfer. Given that similar power absorptions could easily result in completely different thermal profiles due to varying blood perfusions (e.g. see Paulsen *et al.*, 1984), the direct correlation of power deposition patterns with successful treatment outcome would be difficult to establish. Thus, power deposition patterns will be viewed herein as intermediary quantities and not as end results. However, while they involve secondary quantities which alone are of less interest in clinical hyperthermia, power deposition calculations are essential ingredients in all thermal simulations of hyperthermia treatments.

The two heating sources that have been used most often to induce hyperthermia are electromagnetic (EM) and ultrasonic (US) radiation. The EM energy deposited at a point in tissue is proportional to the square of the magnitude of the electric field at that point. Mathematically, the instantaneous power absorbed per unit volume of tissue can be written as

$$P_e = \sigma \, |\boldsymbol{E}|^2, \tag{12.1}$$

where σ is the electrical conductivity and $\boldsymbol{E}$ is the electric field. Similarly, acoustic power is dissipated through the conversion of mechanical energy to internal energy as US waves propagate through tissue. Mathematically, the instantaneous power absorbed per unit volume of tissue from US radiation can be written as

$$P_u = \alpha I, \tag{12.2}$$

where α is the absorption coefficient and I is the acoustic intensity. Thus, knowledge of the EM or US field produced by a particular therapy system is needed in order to calculate its power deposition pattern. Making such a calculation obviously requires a mathematical framework which represents the underlying physics. Unfortunately, the propagation of acoustic waves is highly non-linear, but under certain circumstances a linear theory can be developed which is accurate enough to be useful (Swindell, 1986). In contrast, the propagation of EM waves is linear in most instances and a well-established linear mathematical framework is firmly in place. Since this is the case, the discussion in this section will centre around the EM power deposition problem. Further, where the ultrasound equations have been linearized, their forms mimic the EM equations and discussions concerning their solution and interpretation will be similar to the EM case.

The foundation of EM theory lies in the four first-order partial differential equations known collectively as the Maxwell equations. It is possible to show that only two of the four Maxwell equations are needed to uniquely define the electric and magnetic fields (provided that conservation of charge is maintained) (Shen and Kong, 1987). These two equations can be written as

$$\nabla \times \boldsymbol{E} = -\mu \frac{\partial \boldsymbol{H}}{\partial t} \tag{12.3a}$$

$$\nabla \times \boldsymbol{H} = \epsilon \frac{\partial \boldsymbol{E}}{\partial t} + \sigma \, \boldsymbol{E}, \tag{12.3b}$$

where E is the time-varying electric field, H is the time-varying magnetic field, μ is the magnetic permeability, ϵ is the permittivity and σ is the electrical conductivity. Equation set (12.3) will be referred to as the time-domain form of the Maxwell equations. Most EM hyperthermia systems have sources which vary harmonically in time such that the time variation of any wave can be represented by the complex exponential factor $e^{-j\omega t}$ where ω is the angular frequency of the sinusoidal source producing the wave and j is $\sqrt{-1}$. Under these conditions the propagation problem can be solved in a simplified form where the time variation is removed and only the spatial variations are involved. Computations of this type will be referred to as occurring in the frequency domain. The frequency domain equivalents of equation (12.3) that will be considered here are

$$\nabla \times \boldsymbol{E} = j\omega\mu\boldsymbol{H} \tag{12.4a}$$

$$\nabla \times \boldsymbol{H} = -j\omega\epsilon^{*}\boldsymbol{E}, \tag{12.4b}$$

where E is the complex amplitude of the electric field, H is the complex amplitude of the magnetic field, $\epsilon^{*} = \epsilon + j\sigma/\omega$ is the complex permittivity and ω is the radian frequency. Selecting either the time or frequency domain for carrying out the relevant computations will have certain ramifications for the solution techniques and will be discussed in more detail later in the chapter. For the sake of brevity, the rest of the development in this section will be given in terms of the frequency domain. The equivalent time-domain forms of any equations presented will be discussed later as the need arises.

Equations (12.4) must be supplemented with certain boundary conditions at interfaces between different tissues since, formally, the differential operators do not exist at such boundaries. Inside the body, the conditions on the electric and magnetic fields at a tissue interface are

$$\hat{\boldsymbol{n}} \cdot (\epsilon_{1}^{*}\boldsymbol{E}_{1} - \epsilon_{2}^{*}\boldsymbol{E}_{2}) = 0 \tag{12.5a}$$

$$\hat{\boldsymbol{n}} \times (\boldsymbol{E}_{1} - \boldsymbol{E}_{2}) = 0 \tag{12.5b}$$

$$\hat{\boldsymbol{n}} \times (\boldsymbol{H}_{1} - \boldsymbol{H}_{2}) = 0 \tag{12.5c}$$

$$\hat{\boldsymbol{n}} \cdot (\mu_{1}\boldsymbol{H}_{1} - \mu_{2}\boldsymbol{H}_{2}) = 0, \tag{12.5d}$$

where the subscripts distinguish the two tissues forming the boundary and $\hat{\boldsymbol{n}}$ is a unit vector normal to the interface. When sources are present on a boundary, equations (12.5a) and (12.5c) must be modified such that

$$\hat{\boldsymbol{n}} \cdot (\epsilon_{1}^{*}\boldsymbol{E}_{1} - \epsilon_{2}^{*}\boldsymbol{E}_{2}) = \rho \tag{12.6a}$$

$$\hat{n} \times (H_1 - H_2) = J \qquad (12.6b)$$

where J and ρ are the surface current and charge densities, respectively, residing on the boundary. The introduction of J and ρ requires that conservation of charge be maintained through the continuity equation

$$\nabla \cdot J = j\omega\rho. \qquad (12.7)$$

In the context of electromagnetically-induced hyperthermia, μ is effectively constant, but σ and ϵ vary with tissue type as well as frequency. Hence equations (12.5) will have to be satisfied when making power deposition calculations in tissue.

Since the task is to calculate power deposition patterns, and power absorption is directly related to the electric field, two choices seem clear: (1) to compute the electric field directly or (2) to compute the magnetic field directly and then take its curl (i.e. apply equation (12.4b) to obtain E). To isolate E, divide equation (12.4a) by $j\omega\mu$, take the curl of equation (12.4a) and substitute equation (12.4b). The magnetic field can be isolated in a similar manner by dividing equation (12.4b) by $j\omega\epsilon^*$, taking the curl of equation (12.4b) and substituting equation (12.4a). Performing these operations results in the following second-order partial differential equations in the quantity of interest — either E or H,

$$\nabla \times \frac{1}{j\omega\mu} \ \nabla \times E + j\omega\epsilon^* E = 0 \qquad (12.8a)$$

$$\nabla \times \frac{1}{j\omega\epsilon^*} \ \nabla \times H + j\omega\mu H = 0. \qquad (12.8\beta)$$

Since μ is constant, equations (12.8) can be rewritten as

$$\nabla \times \nabla \times E - k^2 E = 0 \qquad (12.9a)$$

$$\nabla \times \frac{1}{k^2} \ \nabla \times H - H = 0, \qquad (12.9b)$$

where $k^2 = \omega^2\mu\epsilon^*$. One of these two equations, in conjunction with any required boundary conditions, needs to be solved in order to calculate power deposition.

When acoustic wave theory is linearized, a wave equation can be written in terms of acoustic pressure:

$$\nabla^2 P + k^2 P = 0, \qquad (12.10)$$

where k is a complex wavenumber whose real part is approximately $2\pi/\lambda$ (λ is the wavelength in tissue) at the frequencies of interest in hyperthermia and whose imaginary part is the attenuation coefficient in the medium (Swindell, 1986). Equation (12.10) is a scalar version of equation (12.9a) thus, the same solution techniques that apply to equations (12.9) will also be applicable to the linearized US theory expressed by equation (12.10). The acoustic intensity, which is needed to calculate the US power deposition, is proportional to the magnitude squared of the acoustic pressure. The relationship between the instantaneous acoustic intensity and the acoustic pressure at a point in tissue can be expressed as

$$I = \frac{|P|^2}{Z} \,, \tag{12.11}$$

where Z is the acoustic impedance. The acoustic impedance is the product of the medium density and velocity of propagation of sound in the medium (Swindell, 1986). The propagation velocity is relatively constant in many tissues (with a few notable exceptions which are very important) and likewise the densities of many tissues are also quite similar. Hence, reflections (i.e. boundary conditions) can be ignored for many types of interfaces and the problem becomes a calculation in homogeneous media. However, in the presence of interfaces between muscle and bone or muscle and air, the impedance mismatches cannot be ignored and suitable boundary conditions must be applied. Under such circumstances the mismatch is so great that almost all of the energy is reflected which results in additional and potentially excessive heating. Care must be exercised to account for these boundaries, and knowledge of their location can be especially important.

12.3. Solution methods

Now that the basic equations that need to be solved have been discussed, the task of arriving at solutions to these equations must be considered. At this point, it is important to keep in mind the purpose of any calculations to be made. If the power deposition calculations are intended to fall within the four categories of hyperthermic dosimetry — comparative, prospective, concurrent or retrospective — then there must be some accounting for irregular boundaries and inhomogeneous tissue properties and numerical techniques are in order. If the purpose of the power deposition calculations is to illuminate some basic physical principles, then analytical or quasi-analytical solutions under simplified, but representative, circumstances may be the best way to gain insight into the problem. Since this chapter is primarily concerned with calculations having the former purpose in mind, the emphasis here will be on numerical methods.

Numerical computation is a valuable form of present day analysis because high-speed computers are capable of carrying out approximations to such a high degree of precision that they are essentially as accurate as exact solutions. This capability is especially important when considering problems that are not solvable with analytical methods such as those routinely faced in clinical hyperthermia. The basic strategy is to reduce a functional equation into an algebraic equation or matrix of algebraic equations. Insight into characterizing numerical methods can be gained by making the distinction between the form of a functional equation and the process by which it is converted into an algebraic one. Such a distinction is not always 'clear cut' because certain choices concerning the form of the functional (or operator) equation often dictate the type of algebraic conversion that can take place and vice versa. Nonetheless, considering the factors which influence the form of the functional equation allows one to make some general observations about the structure of various numerical techniques without knowing a great deal about the details of their specific algebraic conversion processes. Thus, before highlighting some of the specifics of particular techniques for generating an algebraic representation of a given functional equation, several factors which impact the form of the functional equation will be discussed. These factors include choices of (1) whether an integral or differential operator is used, (2) whether calculations are carried out in the frequency or time domains, and (3) whether the total field or its incident and scattered components are computed.

12.3.1. Integral versus differential operators

If an integral operator equation is used to find the solution to equations (12.3) (or equations (12.4)), then the operator is considered to have global influence because the integrands contain functions which are non-zero over the entire limits of integration. This results in an algebraic representation in which each algebraic equation involves many (typically all) variables in the problem. Further, the integrands of the integral equation are typically singular at certain points within the problem space and extra effort must be spent on the conversion process at such points. The conversion of an integral operator into a matrix equation would then produce a matrix that is 'full' or 'dense' (i.e. in general it has all non-zero entries though it may have some zero entries in special cases). The mathematical details of integral equation solutions to power deposition problems of interest in hyperthermia have recently been discussed by Bardati (1986) and Paulsen (1989).

For unbounded problems (i.e. problems which include boundaries at infinity — a typical situation in EM wave calculations), integral equation operators are formulated so that they exactly satisfy the radiation boundary conditions at infinity. As a consequence, for hyperthermic situations the problem domain need not extend outside the body (except perhaps to account for external applicators). While boundary conditions at infinity are automatically satisfied, boundary conditions at tissue interfaces can be problematic for some integral equation formulations (Borup *et al.*, 1987). This is not a fundamental limitation but more a consequence of the inconvenience of imposing local constraints with globally influential functions. The enforcement of the tissue boundary conditions does not occur naturally for integral equation operators especially as the degree of electrical heterogeneity increases within the problem domain. In many instances these tissue boundary conditions have been ignored because of the high cost associated with their implementation.

In contrast to the global nature of integral operators, differential operator equations have local influence. Their conversion to an algebraic representation results in each equation containing only a few of the variables in the problem. Hence, the conversion of a differential operator equation into an algebraic matrix equation produces a matrix which contains only a few non-zero entries. Such a matrix is usually referred to as being 'sparse'. In addition, the conversion process for differential operators typically only involves non-singular functions, so there are not any points within the problem domain which require special attention. Because of the compatibility of the local influence of the differential operator with the local nature of the tissue boundary conditions, these interface requirements (equations (12.5)) are easily enforced as part of the solution process, thus making differential operator equations a natural choice when the problem domain consists of a high degree of electrical heterogeneity. However, for unbounded problems, additional assumptions are needed for applying the required boundary conditions at infinity. These requirements are not met as a natural consequence of the operator (as in the integral operator case) and approximations must be made at finite boundaries which are intended to mimic the behaviour of the boundary as if it were at infinity.

Based only on the knowledge of a numerical method as having either an integral or differential operator, important information concerning the structure and capabilities of the method can be inferred without knowing much about the particular details of the actual conversion of the functional equation into an algebraic one. Specifically, one can make some general observations about the nature of any matrix equation which might arise and the ability of the resulting numerical method to enforce both local (tissue boundary conditions) and global (boundary conditions at infinity) constraints on the quantities being calculated. In fact, one can in general assert that the strong points of an integral operator are the weak points of a differential operator and vice versa. There are, of course, exceptions to these generalizations, but from the point of view of understanding the basic structure of various numerical methods and their advantages and disadvantages, such generalizations are a valuable tool.

12.3.2. Frequency versus time domains

Another choice which impacts the form of the functional equation is whether the problem is solved in the frequency or time domain. For the EM hyperthermia problem both integral and differential operator equations can be cast in either of these domains. As mentioned earlier, in frequency domain calculations an explicit time dependence is assumed ($e^{-j\omega t}$) such that the time variable is removed from the problem. Frequency domain calculations involve matrix equations as a result of the conversion process of the functional equation. Hence, a major concern for frequency domain calculations is the computer memory requirements that are needed to assemble the matrix equation and the computer run-time that is needed to solve the algebraic system.

For dense matrices, the storage requirement is N^2 and the run-time is of the order N^3 for a linear system consisting of N equations in N unknowns. These computer resources can be drastically reduced for sparce matrices, but quantification of the reduction tends to be problem specific. Sparse matrices are often categorized as being 'banded' (Golub and Van Loan, 1983), in which case one can estimate the storage requirements as proportional to $N^{3/2}$ in two dimensions and $N^{5/3}$ in three dimensions with concomitant run-times of order N^2 and $N^{7/3}$, respectively (for N equations in N unknowns). It is worth noting that in the frequency domain, there is typically very little extra computational effort needed for a change in the source excitation when direct matrix solution techniques are used. This is important in hyperthermia simulations where one may be interested in comparing several different EM source excitations. Further, as long as the matrix is well conditioned (i.e. non-singular), direct methods do not have any convergence issues (though round-off errors can be a consideration for very large N).

Other sparse matrix solution techniques that are commonly available have storage requirements which are proportional to N (Golub and Van Loan, 1983; Axelsson, 1977). These methods require an iterative procedure whereby an initial guess to the solution is made and some rule is used to improve on the initial guess until the change in the solution between two successive iterations is less than some criterion. For these types of matrix solvers, convergence and rate of convergence are often issues. In some instances, convergence can be guaranteed in, at most, N steps for a matrix consisting of N equations and N unknowns (Hestenes and Stiefel, 1952). For sparse matrices the run-time is usually proportional to N for each iteration but the number of iterations needed tends to be problem specific and can sometimes be significantly altered through preconditioning techniques (Axelsson, 1985). Thus, run-time results can vary a great deal depending on the problem. Typically, if one can store the matrix (without requiring too much memory) the run-time will be shorter for the direct solution techniques relative to iterative procedures. This has certainly been the case for two-dimensional problems in hyperthermia power deposition calculations, but remains to be seen in three-dimensional computations. In fact, iterative solutions for frequency domain calculations have become popular in three dimensions because of the extremely large matrix storage demands that can be imposed by direct methods. Thus in three dimensions, iterative solution techniques are making possible calculations that would otherwise be impossible due to lack of computer memory. Note that iterative techniques can also be used to solve dense matrices, but they provide no advantages over direct methods from the matrix storage requirement point of view (unless one is willing to continuously reassemble the matrix in piecewise fashion as it is needed during each iteration cycle). In some special circumstances integral equation approaches, which normally produce dense matrices, can be cast in a form where the matrix algebra inherits the convolutional nature of the functional integral equation (Borup *et al.*, 1987; Borup and Gandhi, 1984). As a result, matrix multiples can be performed with a fast Fourier transform. This efficient computation of a matrix product is then coupled with an iterative solution technique such that the run-time is of order $N\log_2 N$ per iteration and the memory requirements scale proportionally to N. Finally, it must be kept in mind that when iterative techniques are used, a change in the source excitation forces the entire solution procedure to be re-executed.

While frequency-domain calculations are an elegant extension of classical analysis, one also has the option of computing in the time domain. Time-domain formulations remove the need for matrix storage and inversion. The quantity of interest can be represented as a weighted combination of values at previous time instances. In this way the calculation can 'march-on-in-time' without ever needing to invert a matrix since the quantities at the new time level are completely dependent on their previously calculated values. However, for this kind of time-stepping scheme there are usually restrictions on the size of the time step that can be taken to advance the solution. If the time step is too large, the algorithm can become unstable. This potential instability will not be discussed in depth here (see Ames, 1977; Tijhius, 1984), but one should realize that the time domain imposes such restriction. Since no matrix storage is required, time domain techniques conserve memory, but run-times can be long due to the integration through time. Quantifying the run-time is difficult since the size of the time step tends to be problem dependent especially for heterogeneous media. For the time harmonic excitations that occur in EM hyperthermia, typically only three or four periods are needed to reach the dynamic steady state (which is the primary concern), but the number of time steps per period can range from a few hundred to many thousands. Further, a change in the electrical properties within the computational domain alters the propagation speed of the waves to be computed which directly affects the maximum allowable size of the time step. Also note that in the time domain, the entire 'march-on-in-time' procedure must be repeated from time zero to the dynamic steady state whenever the source excitation is changed. For these reasons, time domain solutions generally have longer run-times than their frequency domain counterparts, but because of the memory conservation, time domain approaches appear promising in three-dimensional hyperthermia simulation. At the present time, memory requirements seem to be the limiting factor in three-dimensional EM calculations in hyperthermia, thus approaches which conserve memory, even at the expense of run-time, are highly desirable.

It is worth noting that differential operators exercise local influence in time while their integral counterparts have global time dependence. Thus, differential operator equations have simple time domain representations which take maximum advantage of the explicit nature of the calculations. Integral operator equations on the other hand have more complicated time domain representations (Bolomey *et al.*, 1978; Mansur and Brebbia, 1982). Because of their global nature, the solution at the new time level does not depend just on the previous one or two time steps (as in the differential operator case), but on a more complete time history. This requires the solution at many past time instances to be saved (and stored), thus eradicating one of the most attractive features of time domain calculations — memory conservation.

As with differential and integral operators, one can conclude that the frequency and time domains are contrasting in that the strengths of frequency domain calculations are the weaknesses of time domain calculations, and vice versa. Hence, one sees that general observations about numerical methods can again be made based only on categorizing the form of the functional equation and not the algebraic conversion process.

12.3.3. Scattering versus boundary value formulations

Another issue of particular importance in hyperthermia modelling is the treatment of the EM source. To date, two approaches have been used most often: (1) the scattering formulation and (2) the boundary value formulation. Historically, the scattering formulation has been associated with integral equation operators, but it has been used in conjunction with differential equation operators as well (Bardati, 1986; Rine *et al.*, 1987; Sullivan *et al.*, 1988). The scattering formulation has also appeared in the hyperthermia literature in both its frequency and time domain forms (Rine *et al.*, 1987; Sullivan *et al.*, 1988). The boundary value formulation, on the other hand, has historically been most often associated with differential equation operators but has also appeared with integral equation operators (Paulsen *et al.*, 1984, 1988a). Like the scattering formulation, it too has been implemented in both its frequency and time domain forms (Lynch *et al.*, 1988; Paulsen *et al.*, 1984).

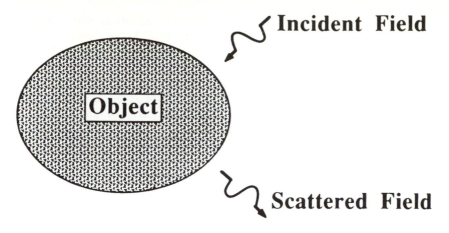

Figure 12.1 Conceptualization of the incident and scattered fields in the scattering formulation.

The scattering formulation sets up a basic mathematical framework for the treatment of a particular source by considering the electric (and/or magnetic) field as the sum of an incident and scattered component:

$$E = E^i + E^s \qquad (12.12)$$

The incident field E^i, is defined to be the field radiated by the sources in the absence of the body. The scattered field, E^s, can then be viewed as the difference between the total field and the incident field; that is, the field that results as a consequence of the presence of the body (or scatterer). Figure 12.1 shows a conceptualization of the incident and scattered fields. In this figure, an object (the body in the hyperthermia context) is impinged upon by the incident field which in turn gives rise to the scattered field. Incident and scattered fields do not exist physically (i.e. only the left side of equation (12.12) can be directly measured in the laboratory; the right side can only be inferred from the context of the particular experiment). However, the decomposition of field quantities into these components is a construct which facilitates the mathematics and its interpretation.

The scattering formulation has been used extensively in computational electromagnetics and is particularly well suited for a classical type of electromagnetics calculation where the source is located far away from the scattering object. For such problems the scattering formulation affords two conveniences because of this large distance between the source and interacting body: (1) the incident field E^i can be assumed to be a plane wave, giving E^i a very simple form computationally, and (2) the interaction of the scattered field with the source can be ignored. Clearly, the body and source are in close proximity in hyperthermia so the incident field must be taken as the near field of the source (in general a significantly more complex calculation) and the interaction of the scattered field with the source must be included. This second requirement also significantly complicates the calculations that need to be made and is often ignored (i.e. the scattered field is allowed to not 'see' the source). In some two-dimensional situations of interest in hyperthermia the boundary conditions that must be applied on the scattered field at the source boundary are quite simple (e.g. $E^s=0$) but in the general three-dimensional problem, this interaction of the scattered field with the source has yet to be included in the near-field situation required of hyperthermia power deposition calculations.

An alternative to the scattering formulation is the boundary value approach where the total field is computed directly rather than through its incident and scattered components. As an example of the issues involved in this strategy, consider the requirements for boundary value solution of the electric field (similar statements can be made for the magnetic field). In this case, the boundary value scheme requires that either E or $\hat{n}\times H$ be specified on the source with the loading in place. Then its counterpart is computed as part of the solution process. Basically, the boundary value formulation represents the inner structure of complicated EM sources by imposing surface fields which produce equivalent effects within the body. In this way the details of a particular source need not be accounted for directly, provided a suitable equivalent field representation can be found. Figure 12.2 illustrates the replacement of a complex source structure with an equivalent surface field representation.

The boundary value formulation can be used as a design tool such that if the imposed E, for example, results in an unrealistic requirement for $\hat{n}\times H$ (due to the presence of the load), a new E distribution can be tried until the resulting $\hat{n}\times H$ is more physically realizable. While certainly not ideal in that an initial guess or measurement of the boundary conditions with the load in place is required, this strategy may be better than using the scattering formulation for hyperthermia applications because one cannot readily account for the interactions of the scattered field with the nearby source. Thus, the sum of the incident and scattered field at the location of the source will not accurately reflect the value of the total E nor the total $\hat{n}\times H$ when the source is near the scatterer. By using the boundary value formulation on the total fields, a self-consistent, although not necessarily practical (i.e. clinically relevant) set of quantities (i.e. E and $\hat{n}\times H$) will result from the solution process.

To date, the boundary value formulation has been applied in hyperthermia power deposition calculations by imposing an ideal distribution for a particular source without any loading. In this situation, the boundary value formulation implicitly assumes that the presence of the body will not alter the quantity imposed. For hyperthermia sources which are known to be susceptible to loading effects, this type of boundary value model may not be advisable and experimental measurements will be

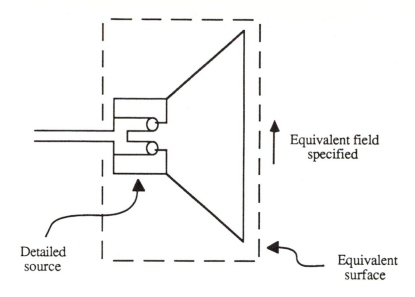

Equivalent field
specified

Detailed
source

Equivalent
surface

Figure 12.2 Representation of an EM source by an equivalent surface field.

needed to provide the required values under typical loading conditions. If the source is particularly sensitive to changes in the loading (e.g. large changes from patient to patient) then the possibility of supplementing the computational model with a few measurements becomes an impracticality and other types of source representations for computational EM hyperthermia models must be sought.

Success has been obtained in this area for two-dimensional models consisting of time harmonic sources which can be represented by one or more perfectly conducting strips (Lynch *et al.*, 1986). To date, this approach has been used to study capacitive heating devices with various numbers, sizes and shapes of electrodes. The source formulation requires that the continuity equation (equation 12.7) be directly enforced along with the governing Maxwell equations. The only knowns that are needed for the solution method are the location and amount of current that is injected onto a particular conducting surface. The resulting distribution of current and charge on the source is determined as part of the solution process with the body in place, thus accounting for the effects of the body on these distributions. This formulation appears to be extendable to three-dimensional problems, as well as to problems having sources with finite conductivity, but the details have yet to be worked out.

12.3.4. Data input: Mesh generation

The mechanics of converting the operator equation into an algebraic representation can also be discussed in general terms. One aspect of this conversion process which is common to all of the numerical techniques under consideration here and which is particularly relevant for hyperthermia power deposition calculations is the generation of the required data input. For problems that involve complicated domains (which hyperthermia modelling certainly does!), a common strategy is to expand the

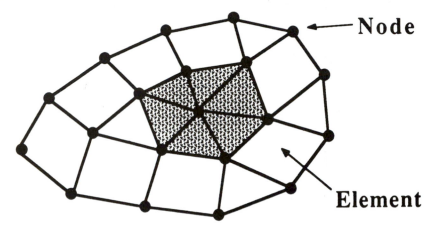

Figure 12.3 A mesh consisting of nodes and elements.

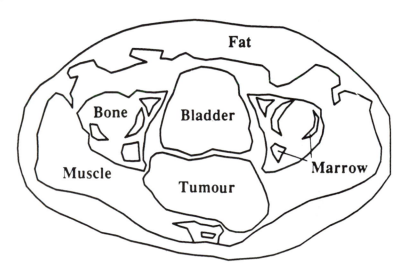

Figure 12.4 Boundaries of the major anatomical features of a computed-tomography cross-section.

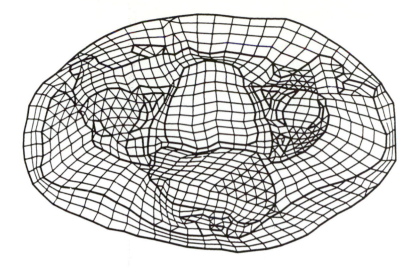

Figure 12.5 A computational domain based on the geometry in Figure 12.4.

unknown solution, or approximate the operator itself, in terms of simple functions that only exist over a subregion of the overall problem domain. The use of such a strategy necessitates the adoption of a scheme for dividing up the domain and prescribing the form of these functions over a piece of local geometry. A domain that has been divided into a set of subregions is called a *grid* or *mesh*. Typically, the subregions within the grid are referred to as elements (or cells) while the points of interconnection of the elements are called nodes. In many cases, the governing equations are enforced at the nodes and the unknowns, once computed, represent the numerical solution at these nodal points. In other situations the element centre serves as the position where each equation is assembled and the computed results are considered to be the value of the solution at these locations. Figure 12.3 shows an example of the basic concept of a mesh and the associated nodes and elements. To illustrate a mesh in the context of calculating power deposition patterns in hyperthermia, Figures 12.4 and 12.5 have been included. Figure 12.4 shows a drawing of a typical patient geometry where the boundaries of the major tissues and organs have been delineated. Figure 12.5 shows a grid that has been constructed from the anatomy of the cancer patient displayed in Figure 12.4.

Two types of elements — (1) domain elements or (2) boundary elements — can comprise a mesh depending on the numerical techniques being implemented. As the names suggest, domain elements are used to subdivide the entire problem space whereas boundary elements are used to subdivide only the boundaries of the problem. Hence, in three dimensions domain elements represent small volumes and in two dimensions they consist of incremental areas. Likewise, boundary elements correspond to pieces of surface area on a three-dimensional object and degenerate into contours in the two-dimensional case. In many situations, the same one- and two-dimensional domain elements can be used as two- and three-dimensional boundary elements, respectively. These savings of both numbers of elements and their ease of construction resulting from the reduction in problem dimension, makes boundary elements attractive for grid generation. Figure 12.6 shows four views of a three-dimensional boundary element grid which consists of a collection of two-dimensional linear area elements composing the body surface. Figure 12.7 illustrates some simple area and volume elements that are commonly used.

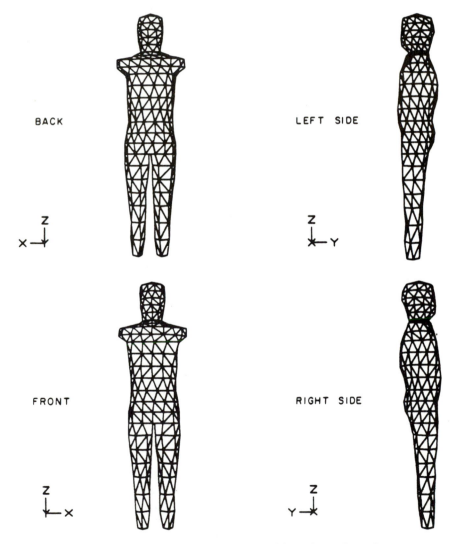

Figure 12.6 A three-dimensional grid constructed from boundary elements.

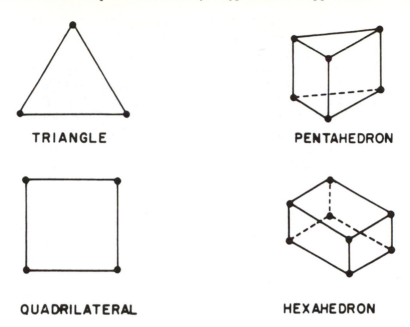

Figure 12.7 Simple area and volume elements.

The advantage of using a numerical method which requires the problem domain to be divided into small subregions is that the irregular boundaries and homogeneous tissue properties of clinical problems can be easily accommodated. Clearly, where the elements are large, the interpolation will be coarse and where the elements are small, the interpolation will be fine. The placement of the elements for a given problem in terms of their size and shape is governed by issues which centre around the desired accuracy to be achieved. The ability of the elements to resolve the problem geometry and to account for changes in the solution (where the solution varies rapidly more elements are often needed; where it varies slowly fewer elements may be required) are important considerations. The correspondence between the approximated solution and the unknown exact solution and the rate at which they converge as the elements are made progressively smaller, is beyond the intended scope of this chapter. Research is being done in this area as it pertains to the hyperthermia problem in order to establish some guidelines for mesh generation. One useful rule of thumb for typical patient geometries and frequencies of interest is that if the elements are small enough to resolve the anatomical features, then acceptable (better than 5%) accuracy should result (Lynch *et al.*, 1985).

If numerical techniques that are based on dividing the problem domain into an interconnecting set of subregions or elements are to be incorporated into routine clinical use for purposes of predicting hyperthermia dosimetry, then the grid construction process must be automated. This

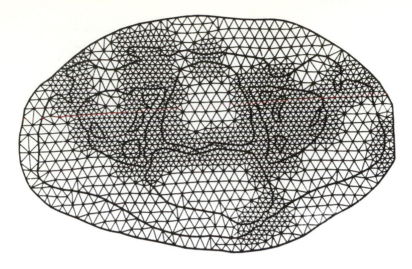

Figure 12.8 A mesh automatically generated from the geometry in Figure 12.4.

automation should be complete enough so that an individual with little understanding of the underlying numerical method can construct an acceptable grid with anatomical detail derived from computed-tomography (CT) scans. Considering the geometric detail shown in Figure 12.4, this task, while essential, is complex. Further, since the size of the elements within the grid directly affects the accuracy of the resulting solution, rules of thumb about nodal spacings like the one described above must be developed. Figure 12.8 shows a grid that has been constructed with an automatic mesh generation scheme where the user has control over the maximum element size (Lynch *et al.*, 1988). This grid is based on the same set of boundaries from which the mesh in Figure 12.5 was developed. The difference is in the time and effort required to arrive at the mesh of Figure 12.8 as compared to that of Figure 12.5. With the automatic grid generation algorithm, Figure 12.8 was constructed using less than 10 min of CPU time on a MicroVax II workstation and with little user intervention. Figure 12.5 on the other hand, took many hours of interactive graphical input from a user skilled in grid generation concepts. Such automated algorithms are now becoming available and will bring the routine use of numerical techniques closer to the clinic.

Although the use of variable element sizes in mesh construction provides the flexibility needed to resolve detailed geometries, the complexity of generating this kind of grid by hand or of developing automated procedures, has been problematic. As a result, numerical approaches which rely on grids with uniform spacing throughout the problem domain have a certain appeal in terms of the simplified data input. In a uniform grid, an indexing system can easily be used such that

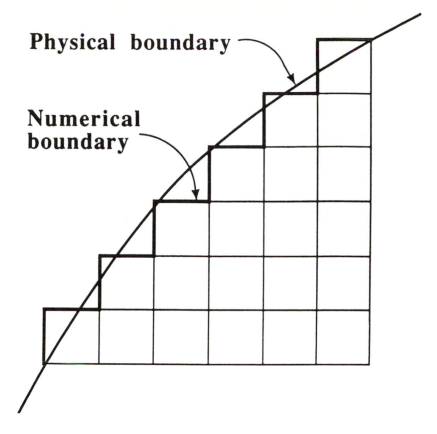

Physical boundary

Numerical boundary

Figure 12.9 A curved boundary discretized with uniform elements.

the indices of each node designate its co-ordinates. Further, for some numerical approaches, the uniform mesh spacing provides some computationally attractive features that would be lost if non-uniform spacing were used. The price paid for the inherent simplicity of a uniformly spaced grid is in terms of the inability of such a mesh to resolve curved boundaries. Figure 12.9 shows an example of the 'stair-stepping' effect that results when a uniform grid is used to discretize a curved boundary. Of course, the effect of this 'stair-stepping' can be reduced by decreasing the grid spacing; however, in a uniform grid this requires decreasing the nodal spacing throughout the entire problem domain and, hence, can cause excessive detail in regions of less importance. This meshing inefficiency is the primary argument against the use of uniform grids. The availability of non-uniform automatic mesh generation, at least in two dimensions, makes the variable-sized-element mesh the grid of choice for clinical hyperthermia problems. Hence, numerical methods which can readily accommodate variable nodal spacing within a grid are to be preferred in this context. Note that the computational ecomonies associated with some methods, when uniform grids are used, are significant and must also be considered in the choice of one numerical technique over another (Borup *et al.*, 1987).

In three-dimensions, the choice of a non-uniform mesh of domain elements should also be favoured over a uniformly spaced grid. However, the current lack of automated grid generation algorithms which can handle the complex three-dimensional shapes of tissues and organs makes the uniform mesh very practical. Further, the same indexing approach can still be used in three dimensions with a uniform grid and the simplicity in the data input it affords is even more magnified in three dimensions. This is partly responsible for the early success of three-dimensional numerical calculations based on methods which exploit uniform grid spacing (Sullivan *et al.*, 1987, 1988). A potentially competing three-dimensional grid which utilizes non-uniform nodal spacing is a boundary element grid. Such a grid (e.g. see Figure 12.5) can be constructed by connecting a series of two-dimensional contours in the remaining third dimension. The relative ease of constructing this type of grid, which has the advantages of non-uniform spacing for resolving complex geometrical shapes, will make methods that use boundary elements attractive options in three dimensions, at least as far as data input is concerned.

Three-dimensional automatic grid generation is an active research area in its own right and currently represents a significant obstacle in all of three-dimensional numerical analysis. This is particularly true in hyperthermia because the problem domains are extremely complex. This inability to automatically generate complicated three-dimensional body models (i.e. to produce the needed data input) raises some interesting questions for near-term hyperthermia modelling in three dimensions: What level of accuracy is needed from computational models and what level of anatomical detail will be required to result in such accuracy? For example, is the spatial averaging of tissue properties and the stair-step approximations of uniform grids adequate or even desired (since one only knows the tissue properties and boundaries approximately anyway)? What about tissue interface boundary conditions: Do they play critical roles in affecting the solution? Is it essential that they are accounted for exactly? Answers to these types of questions must be forthcoming if meaningful results are to be obtained from three-dimensional calculations. If the anatomical detail is critical enough to require the use of irregular variable-spaced meshes then the hyperthermia research community must be willing to devote significantly more effort and resources than are presently allocated towards researching techniques for generating meshes of complicated domains. Otherwise, regardless of whether the techniques to make the necessary calculations exist, they will be of little use if the data input cannot be produced.

Having discussed the major factors which influence the form of the governing functional equation and having seen that these factors are directly related to the structure of the underlying algebraic represent- ation (which is what is actually implemented on the computer), attention is now turned to several specific techniques for generating the algebraic equivalent (or presumed equivalent) form. For most methods, the starting point is generally the same: a second-order partial differential equation in the quantity of interest (i.e. equations (12.9)). With this starting point, five approaches have been used most often in hyperthermia power deposition calculations. Below is a brief overview of these methods. These methods have been discussed in much greater detail by Paulsen (1990).

12.3.5. Domain integral equation method

In the domain integral equation (DIE) approach the EM sources are treated as inhomogeneous terms in the governing second-order differ- ential equation. An analytical form of the solution to the inhomogeneous partial differential equation, which is based on an unbounded-space Green's function, is then numerically implemented. The approach is attractive for problem formulation because the EM sources are easily handled as inhomogeneous terms in the partial differential equation, and the Green's function solution is an exact solution firmly embedded in classical theory. In effect, these calculations represent the superposition of point source solutions where the boundaries are located at infinity. The approach is unattractive numerically because the singular Green's function can be difficult to evaluate, and the kernels of the domain integrals have global influence. This global influence in turn leads to a full matrix which requires inversion. The use of domain integral approaches in electromagnetics has been well described in the hyperthermia literature (Bardati, 1986; Borup and Gandhi, 1984; Hill *et al.*, 1983). Historically, the DIE method has been referred to as the method of moments technique. However, the method of moments can be viewed as a general weighted residual approach (Finlayson, 1972), and as such is inde- pendent of whether an integral or differential operator is used. Hence, the terminology 'domain integral equation' is used here to highlight the use of an integral functional equation even though in the literature 'method of moments' has evolved to be synonymous with the use of integral operator equations.

12.3.6. Finite element method

In the finite element method (FEM), the solution to the governing second- order partial differential equations is approximated by simple functions upon which the differential operations are carried out exactly. Typically, the EM sources are handled through boundary integrals which arise from the integration-by-parts of the second-order differentiation. The FEM requires domain integration, but in contrast to the DIE approach, the boundaries are closed and finite. Also, unlike the DIE approach, the exact solution is approximated by an expansion of an incomplete basis. This feature, however, leads to the numerical attractiveness of finite elements since the integrations that must be performed are carried out on simple, regular functions which lead to a sparse matrix that is effectively inverted by LU (lower-upper) decomposition (Golub and van Loan, 1983). Further, geometrical irregularity as well as regional heterogeneity are easily handled with the FEM. One drawback is that knowledge of the source is required around the boundary of the entire domain of interest. Since many EM problems are fundamentally unbounded, this requirement can be limiting and explains, in part, why integral equations have dominated numerical electromagnetics. Examples of FEM application in the hyperthermia context can be found in Lynch *et al.* (1986) and Strohbehn *et al.* (1986).

12.3.7. Finite differences

In finite differences (FD) an exact solution is obtained for operators which approximate those contained in the governing partial differential equation. Typically, the first-order differential Maxwell equations relating the electric and magnetic fields have been used in the time domain. With this formulation, an explicit time-stepping scheme can be employed which leads to the desirable feature that no matrix solutions are needed. This makes the finite difference time-domain (FDTD) approach particularly attractive for large scale EM problems (three-dimensional), but can be counterbalanced by the required iteration procedure through time. As with the FEM, the application of FD is limited in unbounded problems and requires that additional assumptions be made in such cases to terminate the problem domain. Further complications occur when the problem domain has highly irregular geometrically detailed boundaries which, if represented accurately, require the use of variable grid spacings throughout the mesh or excessive detail in regions of less importance (for a uniform grid). The use of a variable-sized mesh detracts from one of the main advantages of the FD method — its algorithmic simplicity. Examples of the use of FD in EM hyperthermia problems can be found in Lau and Sheppard (1986) and Sullivan *et al.* (1987).

12.3.8. Boundary element method

Boundary integral statements are derived by employing Green's theorem on the governing second-order partial differential equation. Integrations are performed only around the boundaries of homogeneous, source-free domains; hence, all EM sources must reside on these boundaries. After the set of boundary equations is assembled and solved, calculation of interior fields can be made as the appropriate weighting of the known boundary values. In direct contrast to the FEM, the BEM is ideal for homogeneous, unbounded problems. For inhomogeneous media, this approach suffers from the additional complexity of either the inclusion of domain integration or the creation of multiple interior boundaries enclosing homogeneous subdomains. One of the most appealing features of the BEM is the reduction in problem dimension because only boundary integration is required. Like the DIE approach, boundary integrals are exact expressions for the problem solution based in classical theory. Unfortunately, also like the DIE approach, the BEM leads to a full matrix and has singular kernels which can be difficult to evaluate numerically. BEM methods are described in Paulsen *et al.* (1988a, b) for hyperthermia power deposition problems.

12.3.9. Hybrid element method

A fifth approach that has been used, but which does not formally introduce any new numerical techniques, is the hybrid element method (HEM). Hybrid formulations are numerical methods which combine two or more of the preceding approaches. For example, a hybrid element technique has been established to couple the advantages of finite and boundary element methods while avoiding the inherent weaknesses of each method when used alone (Lynch *et al.*, 1986; Paulsen *et al.*, 1988b). As discussed above, the FEM is best suited for problems with geometric irregularity and region heterogeneity which are confined by a finite boundary that completely contains the problem domain and on which known boundary conditions can be applied. The major weakness of the FEM is that, in general, it cannot be used in cases where portions of the problem domain extend to infinity without additional assumptions and approximations. On the other hand, the BEM is ideally suited to problems containing large homogeneous regions which can extend to infinity; however, it is weakened for problems with domains having severe heterogeneity by its globally influential, singular kernels which must be integrated over the increased number of boundaries. Hence, the HEM employs the FEM in regions of heterogeneity (i.e. inside the patient) and the BEM in regions of unbounded homogeneity (such as the background medium and/or bolus which contains the patient as well as the EM source). The FEM effectively treats the variation of the tissue electrical properties with the use of simple piecewise linear basis functions which can be integrated easily and which lead to sparse sets of algebraic equations. The BEM is used to couple the finite element equations to the remote source and the surrounding infinite domain by removing the requirement that known boundary conditions be applied at all finite element boundaries. In this manner, the

HEM provides a formal approach to the problem of a heterogeneous body irradiated by a detached source.

12.4. Comparison of methods

Of the five methods that have been discussed for numerically solving the hyperthermia EM problem, it is difficult, if not impossible to rank one of these techniques as definitively superior to the others because of the many trade-offs that exist. Nonetheless, each method does have inherent strengths and weaknesses, and some kind of general ranking or ordering of these numerical techniques in terms of their preferred use in hyperthermia power deposition modelling is needed, especially for those who wish to use computational models in hyperthermia research, but do not plan to enter the research areas devoted to the development of the computational models, themselves. Presented here is one analysis which provides such a ranking of the numerical methods discussed in this chapter. It is by no means the only framework or criterion upon which a comparative ranking could be made, but rather one strategy for providing guidelines for the selection of one numerical technique over another.

The basic premise on which the comparative rankings are made is that all the varied trade-offs, nuances, and idiosyncracies that exist between these numerical methods can be distilled into one overriding consideration — *computational integrity versus computational efficiency*. The computational integrity of a method is a measure of how well the numerical technique solves the governing equations as posed. This is to be distinguished from the question of how well the mathematical framework represents the physical problem, which is certainly germane to hyperthermia, but is not considered to be part of the computational integrity of a given numerical method. Computational efficiency, on the other hand, asks the question of how powerful a computer is needed and how long will it take to make the desired calculations. Basically, computational efficiency assesses the computing power or resources required. Thus, by assigning each method a relative ranking in terms of its computational integrity and its computational efficiency, one can gain a general sense of the suitability of the various methods for hyperthermia EM modelling.

The ideal method would be ranked at the top of both categories and this is the basic challenge facing researchers in the model development arena — how to devise a method that can simultaneously rise to the top of both categories. At the present time, such a method does exist and the individual model user must decide on the relative emphasis to be placed on computational integrity and computational efficiency. Clearly, a numerical method which is computationally efficient but which is based on such extensive approximations that computed results are questionable, is of little use in hyperthermia modelling. Likewise, a numerical technique, which is mathematically rigorous, yet so complex or computationally demanding that a solution cannot even be computed, is of little value to hyperthermia research.

Fundamentally, computational integrity should have top priority. In three-dimensional work, however, a curious situation exists (at least for the moment) where insisting on computational integrity can make the problem so inefficient that either it cannot be solved or it is severely limited in size. Thus, for three-dimensional power deposition calculations, computational efficiency can be the sole factor determining whether the problem can be attempted. This is in contrast to two-dimensions where one can insist on computational integrity and not worry so much about efficiency for most problems in hyperthermia. Of course, even in two dimensions one desires methods which are as efficient as possible, but there is no reason to gain efficiency at the expense of integrity because of the availability of low-cost computer workstations capable of running very complex two-dimensional models of power deposition. Hence, if the model user has access to moderate computer power, he/she need only be concerned with the relative computational integrity of the numerical methods under consideration.

While this basic trade-off between computational integrity and computational efficiency provides a clear 'bottom line' approach to favouring one numerical technique over another, the criterion for actually evaluating the integrity and efficiency for a particular numerical method is somewhat more subjective. A variety of criteria are possible, and what now follows is one such evaluation which will provide a means for arriving at relative rankings of the computational integrity and efficiency of the five numerical methods discussed in this chapter.

To establish a criterion for the computational integrity of a particular numerical technique it is instructive to review the distinguishing features of the general hyperthermia EM problem. These features are: (1) the body (computational domain) is irregularly shaped, heterogeneous and lossy, (2) typically the electrical properties are such that in the frequency ranges of interest, neither conduction nor displacement currents can be ignored, (3) the body is in the near-field of the source, and (4) the EM fields are unbounded. Based on this list one concludes that the numerical method which maintains a high level of computational integrity must employ a computational mesh which accommodates geometric irregularities and tissue property heterogeneities. Further, it must satisfy the appropriate boundary conditions at interfaces where these electrical properties change abruptly. Since the body is in the near-field of the heating source, one must account for the appropriate boundary conditions or interactions with the source. In addition, no energy or waves may be allowed to propagate into the problem domain from any boundary which contains (surrounds) the body and all EM sources. That is, all outward travelling waves must satisfy the radiation boundary condition at infinity.

Table 12.1 Computational integrity.

Method	Grid	Tissue boundary conditions	Radiation condition
DIE	U,V	N,Y	Y
FEM	V	Y	N
FDTD	U	N	N
BEM	V	Y	Y
HEM	V	Y	Y

U, uniform grid; V, variable grid; N, does not satisfy; Y, does satisfy.
Indicators are given for the conventional implementations of the methods found in the hyperthermia literature. In cases of notable exceptions, two indicators are given — the first being the convention, the second being the exception.

Table 12.1 indicates the ability of the five numerical techniques under consideration to satisfy these constraints on computational integrity. Several comments about Table 12.1 are in order. First, the treatment of the hyperthermia source has not been included since both the scattering and boundary value formulations have shortcomings (see pp. 316–319) and most of the numerical methods can accommodate either approach. Second, in some instances, limitations on the computational integrity of a particular method may be due to the implementation of the technique rather than to the method, itself. In these cases, two indicators are given: the first for the way in which the computational integrity is effected for the implementations that have typically been used in hyperthermia applications and the second for any restrictions on the computational integrity due to fundamental limitations in the method itself.

Table 12.2, which has been constructed from the data in Table 12.1, provides the computational integrity rankings. Based on typical implementations of the methods in hyperthermia power deposition problems, the hybrid element method and the boundary element method come out on top with essentially identical computational integrity rankings. Next are the finite element method, the domain integral equation method and the FDTD technique, respectively. If one considers limitations in computational integrity imposed by the techniques themselves, then the DIE joins the HEM and BEM at the top since there are no

Table 12.2 Computational integrity rankings.

	Convention		Exception
1.	HEM	1.	HEM
	BEM		BEM
			DIE
3.	FEM		
		4.	FEM
4.	DIE		
		5.	FDTD
5.	FDTD		

'Convention' rankings based on typical implementations in the hyperthermia literature.
'Exception' rankings based on notable exceptions in the hyperthermia literature.

fundamental reasons why the DIE cannot satisfy the internal tissue boundary conditions and be used with irregular grids (although typically the extra effort required to preserve these features has not been exercised in the hyperthermia context).

Unfortunately, there have not been careful studies performed to determine whether the accuracy required for hyperthermia simulations is greatly affected if one uses a method at the bottom of the computational integrity list or one at the top. Even more fundamentally, the level of accuracy required for meaningful hyperthermia treatment simulations is not well understood. However, one can make some qualitative arguments suggesting that the tissue interface conditions are important in shaping the field patterns and resultant power absorption. For example, the excessive heating in the fat layer that has been observed clinically with some EM devices has been attributed to the jump discontinuity in the normal component of the electric field (12.5a). Such clinical observations suggest that modelling the interface boundary conditions accurately is an important step toward predicting these kinds of clinical problems.

Since, in two-dimensional work, low-cost computer workstations have evolved to the point where the computing power is sufficient to demand that computational integrity not be sacrificed for computational efficiency, the limitations imposed by computational efficiency need only be considered in the three-dimensional realm. Further, at the moment, the memory requirements for three-dimensional computations are more restrictive than the length of run-time. If the needed storage arrays cannot be 'fitted' into the computer, the run-time is of secondary concern because the computations cannot be started. An additional important consideration that is often overlooked is the effort required to generate the computational mesh (i.e. the problem geometry). Clearly, if the data input cannot be produced, the computations cannot be carried out regardless of how powerful a computer is available.

Thus, the computational efficiency of the numerical methods used for hyperthermia power deposition calculations will be measured in terms of computer memory requirement and ease of data generation (grid con-

Table 12.3 Computational efficiency.

Method	Memory	Grid
DIE	$N^2, 0\,(N)$	U
FEM	$0\,(N^{5/3}), 0\,(N)$	V
FDTD	$0\,(N)$	U
BEM	$0\,(N^{4/3})$	V
HEM	$0\,(N^{5/3})$	V

U, uniform grid; V, variable grid; $0\,(x)$, Order x.
Memory estimates are based on a cube of N total nodes and assume one unknown per node (i.e. N equations in N unknowns). If two values are given, the first represents the conventional computer storage strategy while the second represents more specialized schemes.

Table 12.4 Computational efficiency rankings.

Convention	Exception
1. FDTD	1. FDTD
2. BEM	DIE
3. FEM	3. FEM
4. HEM	4. BEM
5. DIE	5. HEM

'Convention' rankings based on typical computer storage
strategies and grid constructions.
'Exception' rankings based on specialized storage schemes
along with typical grid constructions.

struction). Table 12.3 shows these measures of computational efficiency
for the five numerical methods. Under the storage requirements category,
more than one value is given for several of the methods. Again, the first
value indicates the memory requirement for the conventional implementation of the particular method. However, because of the significant restrictions that the storage burden can place on a given numerical method
(particularly for three-dimensional vector problems), specialized techniques have been developed in many instances and the storage requirements for these cases are indicated as well. Table 12.4 shows the resulting
computational efficency rankings considering the conventional and
specialized storage strategies. In both cases, the FDTD method can be considered to be the most efficient. Conventional rankings has the boundary
element method second since in this approach only the boundaries need to
be discretized. Thus, while the resulting matrix is full, the total number of
unknowns is less than for the other methods which require domain
(volume) elements. Further, grid construction is significantly easier, even
though it reflects the actual geometry of the problem. Rounding out this
list are the FEM, HEM and DIE. If some of the more specialized
storage/solution techniques are to be considered, one finds that the DIE is
essentially indistinguishable from the FDTD. Both use uniform grids and
both have storage requirements proportional to N. The constants of pro-
portionality which affect the actual storage requirements are different for
the two methods, but these subtleties are secondary to this simplified
analysis and for the purposes of discussion the two methods can be considered equivalent from a computational efficiency viewpoint. Close
behind these two methods is the FEM which can also have storage requirements proportional to N (again with some subtle differences in the pro-
portionality constants), but which is regarded as somewhat less efficient
according to the criteria used here because finite element methods
typically involve grids composed of irregular variable-spacing meshes
which are more difficult and time consuming to construct. These improved
efficiencies of the DIE and FEM allow them to rise above that of the BEM.

As a result, the HEM is found at the bottom of the special rankings.

The basic trade-off between computational integrity and computational efficiency is quite clear. The FDTD is at the top of both rankings in Table 12.4 which rate computational efficiency, but is at the bottom of the list in Table 12.2 where computational integrity is being considered. Another illustration of this trade-off is observed with the DIE. If one permits specialized efficiencies to be taken advantage of with the DIE, it appears essentially at the top of the 'exception' computational efficiency list of Table 12.4. However, by doing so, one forces the computational integrity ranking to be taken from the 'convention' list in Table 12.2 where the DIE is next to last. Conversely, when the computational integrity of the DIE is taken from the 'exception' list in Table 12.2 where it is at the top (essentially indistinguishable from the BEM and HEM), then its concomitant computational efficiency is found in the 'convention' list of Table 12.4 where it is ranked at the bottom.

As a final remark, one must bear in mind that in actual practice there are many subtleties which blur the distinctions or rankings that have been given in Tables 12.1–12.4. Tables 12.1–12.4 should be regarded as a first-order analysis which is useful because it reveals some general guidelines for evaluating the relative merits of various numerical methods. However, it does not truly reflect the complexity and interplay of issues that must be considered in order to definitively establish one numerical method as superior to another. In fact, at this point in hyperthermia research, some questions of a more fundamental nature need to be answered before a strong case can be made for one particular numerical technique as the 'method of choice' for power deposition calculations.

12.5. *Electromagnetic modelling: current status and future directions*

The ability to simulate hyperthermic treatments is recognized as an essential tool for understanding and improving hyperthermia therapy. The role of modelling in hyperthermia can only be expected to become more important as the clinical problems become more complex and as the heat delivery technology becomes more sophisticated. Even though the field of simulating hyperthermia treatments is still in its incipient stages, models have evolved from simple analytical solutions on idealized geometries to complex two- and three-dimensional numerical calculations on realistic CT-based anatomies. This section will discuss the progress that has been made in calculating power deposition patterns produced inside the body by electromagnetic means.

The 1980s have seen numerical simulation of hyperthermia treatments become one of the major areas of interest to hyperthermia researchers. Numerous papers have appeared on this topic; however, due to space limitations no attempt will be made to provide an extensive list of such references here. Several more comprehensive reviews of this literature are available in other sources (Paulsen, 1990; Strohbehn and Roemer, 1984). The success of recent numerical models in predicting the performance of certain therapy devices prior to the accumulation of significant quantities of clinical data has demonstrated the impact that numerical simulation can have on the evaluation of these systems. Qualitative comparisons as well as limited quantitative comparisons with clinical data suggest that numerical models can accurately predict the dominant capabilities and limitations of hyperthermia systems such as annular array applicators and magnetic induction coils. Specifically, the modelling can provide insight into the types of thermal distributions that might be seen in the clinic, show regions where systems may have difficulty in achieving good temperature distributions, and show the importance of blood flow data.

In general, two emphases have emerged in the power absorption studies: (1) initial analysis of complicated situations based on simplified models, and (2) detailed power deposition computations using complex models where as much information as possible (e.g. patient anatomy, source type and location, etc.) is included. There has been a natural evolution from the first type of emphasis to the second as the fundamental understanding of power deposition in hyperthermia has grown and as the need to solve clinical problems where irregular boundaries and tissue heterogeneities are the norm has increased.

While modelling efforts in hyperthermia have evolved from a simple analytical solution to one-, two-, and even three-dimensional numerical simulations, all models have important roles to fulfil in hyperthermia research. Analytical solutions provide the crucial checks on numerical algorithms and the avenue for initial analyses of complicated problems that are particularly useful for illuminating basic principles. One- and two-dimensional models can also provide these kinds of simplified results which are generally easier to interpret, but which often lead to greater physical intuition. Two-dimensional models may perform a screening function such that three-dimensional computations can be avoided in cases which show little promise in two-dimensions. They may also be of use in routine clinical practice if they consistently provide qualitative or semi-quantitative treatment information which can be used to produce more effective heat delivery. Three-dimensional models can provide information about all 'end effects', thus they will be the most realistic and accurate prediction tools. Three-dimensional calculations can also act as a standard against which the validity of any simplifying assumptions, including those that lead to models of lower dimensionality, can be compared.

Increases in model complexity in terms of the irregular boundaries and tissue heterogeneities in clinical problems has likewise progressed from homogeneous cylinders to layered cylinders and ellipses to heterogeneous body models based on CT scans of actual patients. Simple homogeneous and layered tissue models play a similar role to that discussed above for analytical solutions. In general, they are useful in initial stages of analysis where the goal is to understand the basic mechanisms of complicated processes. They provide a foundation upon which more-complex body models can be implemented. Further, this level of model complexity (i.e. homogeneous or layered) allows for comparisons of theory and experiment under controlled and interpretable circumstances which can readily be constructed in the laboratory. An important, but often ignored consideration, is how well the theoretical predictions compare with phantom experiments. However, as with increases in model dimensionality, increases in anatomical detail will be needed to account for the effects caused by tissue heterogeneities and will be required in modelling where this level of structure is essential to the accuracy or interpretation of the computed results.

The majority of numerically calculated power absorption patterns that have been reported to date are based on two-dimensional approximations. In terms of the five, numerical methods outlined in the previous section — domain integrals, finite elements, finite differences, boundary integrals and hybrid techniques — the domain integral approach has been used most often to calculate fields in biological tissue.

While posed in three dimensions, a number of investigators have made simplifying assumptions to reduce the three-dimensional formulation to a two-dimensional problem in order to obtain the kind of computational resolution that is needed for making detailed power deposition calculations. If the z direction is taken to be the long axis of the patient, the incident field is assumed to be along this axis, and no variation in geometry, electrical properties or source characteristics is allowed in the z dimension, then the calculation of the electric field in the plane transverse to the z direction is a scalar problem. These assumptions have been used to calculate electric fields in two-dimensional body cross-sections where the geometry and electrical properties have been preserved via CT-scan information (e.g. see Iskander *et al.*, 1982; van Den Berg *et al.*, 1983). Domain integrals have also been used to calculate power deposition in two-dimensional body cross-sections where the electric field components are assumed to be in the plane of analysis (Hill *et al.*, 1983).

The use of finite elements in hyperthermia power deposition problems, while not as prevalent as the use of the domain integral approach, has increased substantially over the last several years. A Galerkin finite element formulation of the Maxwell equations for calculating power deposition patterns has recently been reported (Lynch *et al.*, 1985). In this work, several clinical devices have been modelled as boundary value problems in which case the FEM with simple linear and bilinear elements has been used to solve the governing equations. The details of the patient interior have been preserved by generating finite grids from CT scans of

actual cancer patients. The use of the FEM in the simulation of the electromagnetically induced hyperthermia has recently been reviewed
(Strohbehn *et al.*, 1986). While these computations are based on frequency
domain solutions to the Maxwell equations, the FEM has also been used in
the time domain to calculate power deposition patterns for hyperthermia
(Lynch *et al.*, 1988).

The FD method has been used in the simulation of hyperthermia
treatments largely to make the thermal calculations involved in the
modelling process. In the frequency domain, there has been little use of FD
in order to make power deposition calculations. An impedance formulation which is essentially an FD treatment of the Maxwell equations
where displacement currents have been assumed negligible has been
reported (Armitage *et al.*, 1983; Gandhi *et al.*, 1984). The FD approach has
been implemented in the time domain and this FDTD approach has
recently been used to make three-dimensional power deposition calculations in models containing very large numbers of elements. The application of the FDTD method to three-dimensional hyperthermia problems
will be discussed further below.

Boundary integral equations (or boundary elements) have appeared in
simulations of power deposition patterns through hybrid formulations
(Lynch *et al.*, 1986) and in three-dimensional computations (Paulsen *et al.*,
1988a). A number of clinical settings of interest result in EM problems
which are fundamentally unbounded. As illustrative examples of these
situations, two-dimensional models of a capacitive heating system and an
array of radiating applicators which do not completely surround the
patient have been posed, and sample calculations have been shown (Lynch
et al., 1986).

While the two-dimensional models described above have led to studies
which have been informative and agree qualitatively with clinical obser-
ations, three-dimensional numerical models of hyperthermia therapy
systems are needed. In some cases, the two-dimensional models are too
crude to be used for patient-specific treatment planning and have shown
poor correlation with clinical results.

Studies have been reported where electric fields induced in three-
dimensional block models of man have been calculated in the context of
potential health hazards due to EM radiation. These works have relied on
the DIE formulation in order to make the necessary computations. More
recently, this same three-dimensional approach has been used in hyperthermia applications to analyse the heating of tissue between parallel
capacitive plates (Hessary and Chen, 1984) and the absorption of energy in
the body outside the area of direct treatment exposure (Hagmann and
Levin, 1986). While these computations are based on three-dimensional
formulations which were pioneering in nature, the models of the body that
were used were intended to study gross effects rather than detailed energy
deposition patterns.

However, studies are beginning to emerge where detailed energy deposition patterns are being calculated inside inhomogeneous three-dimensional models of the body (Sullivan *et al.*, 1988). This work is based on the FDTD method which requires a large number of elements for good resolution, but is reasonably straightforward to implement. As discussed earlier, this technique has an advantage over frequency domain methods because large matrices do not need to be stored and inverted. While the stepping through time to steady state tends to increase computation time, the decreased storage at least makes detailed three-dimensional problems solvable whereas the corresponding frequency domain solution often becomes impossible due to the massive matrix storage requirements.

An alternative formulation to the Maxwell equations which has been applied to hyperthermia problems is based on boundary elements. This approach is expected to require considerably fewer unknowns than the FDTD method, but it leads to a more computationally complex solution (Paulsen *et al.*, 1988a). Three-dimensional frequency domain formulations of the FEM and HEM have also been developed and implemented for application to problems of interest in hyperthermia modelling (Paulsen *et al.*, 1988b). To date, only three-dimensional calculations on simple homogeneous and heterogeneous cylindrical geometries have been attempted in this work. The drawback, at the moment, seems to be the extremely large amounts of memory required for large-scale problems. A finite element time-domain method, which should possess the same economies in storage that accompany the FDTD, but which can readily be implemented on non-uniform meshes, has recently been discussed (Lynch *et al.*, 1988). To date, only two-dimensional calculations have been attempted with this formulation. Further work will be required to explore the advantages and disadvantages of these and other approaches in three dimensions. It can be expected that these techniques will expand and improve in the near future as more investigators begin to solve increasingly realistic three-dimensional problems.

While advancements have been made in the field of computer simulations of hyperthermia, several gaps in the current knowledge exist: (1) the effects of detailed three-dimensional calculations on the evaluation, design and development of hyperthermia therapy devices and (2) the extent to which numerical models can predict specific power deposition patterns and resultant temperature distributions.

Clearly, the general problem to be solved is three-dimensional and most of the numerical formulations discussed herein have been posed as such. The major limitation of two-dimensional models is precisely their two-dimensional nature. Some of the ramifications of this shortcoming have been discussed in the literature (Hagmann, 1987). Cases where two-dimensional models are inadequate have been shown both numerically (Hagmann and Levin, 1986) as well as experimentally (Turner, 1984). Typically, the cross-sectional dimensions in a clinical setting are on the order of the longitudinal dimension and the extent of the errors introduced in two-dimensional models by ignoring the third dimension has not been carefully quantified. While three-dimensional calculations which have the resolution necessary to account for the details of the body anatomy are beginning to appear, their use in routine analysis problems may take many years of further research. This leaves one with the questions of what is the role of two-dimensional analysis in simulating hyperthermia treatments, and should two-dimensional models be regarded as worthless due to flaws in their basic assumptions or do they have some intrinsic merit in predicting hyperthermia dosimetry which should be exploited?

One view would be that the simpler plane calculation represents a 'best case' analysis since treatments which fail to produce satisfactory heating patterns in two dimensions would not be expected to perform better in three dimensions. At a minimum, then, the plane analysis could be used as a screening mechanism to reduce (or even avoid) the computing in the third dimension where solutions are not only expensive, but also more difficult to display and interpret. The extent to which three-dimensional performance correlates with two-dimensional performance remains an open question that needs to be addressed with more research. A main thrust of future two-dimensional modelling should be directed toward establishing the propriety of the two-dimensional assumptions.

This kind of study can take place in two forms: (1) comparisons of two-dimensional numerical results with phantom and clinical measurements, and (2) comparisons of two-dimensional numerical calculations with planar results generated from full three-dimensional computations. Generally, the comparisons of numerical versus experimental results that have been published have been intended to lend some further credibility to the numerical models that were being developed, and as such were not the main focus of the study. In contrast to this approach, studies are needed that will make these comparisons of primary interest in order to investigate under what conditions the numerical results are, or are not, in agreement with measured data and, when agreement is lacking, to determine the cause of such disagreement (i.e. boundary conditions, inappropriate equations or assumptions, etc.). If a range of validity of a model can be established, then the degree of this validity and how often the conditions under which the model is valid occur in clinical problems will ultimately decide its utility. This same sort of protocol which has an emphasis on rigorously testing a model for conditions of both validity and invalidity is also needed for comparing three-dimensional predictions with experiments.

The foregoing discussion on the role of two-dimensional models is in no way intended to suggest that two-dimensional models are adequate or to diminish the importance of three-dimensional calculations. The fact that three-dimensional computations are more difficult both to develop as well as to interpret provides no scientific basis to prefer two-dimensional models over three-dimensional ones. Three-dimensional models are essential to the future success of simulating hyperthermia treatments, and the major emphasis in hyperthermia modelling should be directed toward such three-dimensional calculations. Further, their impact on the evaluation, design, and development of hyperthermia therapy devices is still relatively unknown and may provide new insights that will improve hyperthermia delivery systems. It seems clear that as future applications call for more detailed and accurate calculation of power deposition patterns, numerical modelling techniques will be required that account for the full-dimensional effects within realistic body anatomies.

Hence, the emphasis of future work in the development of algorithms and methods for simulating hyperthermia treatments should be on three-dimensional numerical models which account for tissue heterogeneity and irregular geometry. Coupled with this effort, special attention should be paid to comparisons between experiment and theory. These comparisons are appropriate in both two and three dimensions and must be performed with the intent of thoroughly examining the range and extent of model validity in representative clinical settings rather than presenting isolated situations of agreement. Numerical simulations have had a positive impact on clinical hyperthermia to date and can continue to do so provided the future studies which are needed support a scientific rationale for the clinical applicability of numerical simulations in hyperthermia dosimetry problems.

References

Ames, W. F. (1977) *Numerical Methods for Partial Differential Equations*, 2nd edn, New York, Academic Press.

Armitage, D. W., Le Veen, H. H. and Pethig, R. (1983) Radiofrequency-induced hyperthermia, computer simulation of specific absorption rate distributions using realistic anatomical models, *Physics Medicine Biology*, **28**, 31–42.

Axelsson, O. (1977) Solution of linear systems of equations, iterative methods, in *Sparse Matrix Techniques*, Berlin, Springer-Verlag.

Axelsson, O. (1985) A survey of preconditioned interative methods for linear systems of algebraic equations, *BIT*, **25**, 166–187.

Bardati, F. (1986) Models of electromagnetic heating and radiometric microwave sensing, in Hand, J. W. and James, J. R. (Eds), *Physical Techniques in Clinical Hyperthermia*, Letchworth, Research Studies Press, pp. 327–382.

Bolomey, J. C., Durix, C. H. and Lesselier, D. (1978) Time-domain integral equation approach for inhomogeneous and dispersive slab problems, *IEEE Transactions on Antennas Propagation*, **26**, 658–667.

Borup, D. T. and Gandhi, O. P. (1984) Fast-Fourier-transform method for the calculation of SAR distributions in finely discretized models of biological bodies, *IEEE Transactions on Microwave Techniques*, **32**, 355–360.

Borup, D. T., Sullivan, D. M. and Gandhi, O. P. (1987) Comparison of the FFT conjugate gradient method and the finite-difference time-domain method for the 2-D absorption problem, *IEEE Transactions on Microwave Theory Techniques*, **35**, 383–395.

Finlayson, B. A. (1972) *Method of Weighted Residuals and Variational Principles*, New York, Academic Press.

Gandhi, O. P., Deford, J. F. and Kanai, H. (1984) Impedance method for calculation of power deposition patterns in magnetically-induced hyperthermia, *IEEE Transactions on Biomedical Engineering*, **31**, 644–651.

Golub, G. H. and van Loan, C. F. (1983) *Matrix Computations*, Baltimore, Johns Hopkins University Press.

Hagmann, M. J. (1987) Difficulty in using two-dimensional models for calculating the energy deposition in tissues during hyperthermia, *International Journal of Hyperthermia*, **3**, 465–476.

Hagmann, M. J. and Levin, R. L. (1986) Aberrant heating, a problem in regional hyperthermia, *IEEE Transactions on Biomedical Engineering*, **33**, 405–411.

Hessary, M. K. and Chen, K. M. (1984) EM local heating with HF electric fields, *IEEE Transactions on Microwave Theory Techniques*, **32**, 569–576.

Hestenes, M. and Stiefel, E. (1952) Method of conjugate gradients for solving linear systems, *Journal of Research of the National Bureau of Standards*, **49**, 409–436.

Hill, S. C., Christensen, D. A. and Durney, C. H. (1983) Power deposition patterns in magnetically-induced hyperthermia, a two-dimensional low-frequency numerical analysis, *International Journal of Radiation Oncology, Biology and Physics*, **9**, 893–904.

Iskander, M. F., Turner, P. F., DuBow, J. B. and Kao, J. (1982) Two-dimensional technique to calculate the EM power deposition in the human body, *Journal of Microwave Power*, **17**, 175–185.

Lau, R. W. M. and Sheppard, R. J. (1986) The modelling of biological systems in three dimensions using the time-domain finite-difference method, I. the implementation of the model, *Physics Medicine Biology*, **31**, 1247–1256.

Lynch, D. R., Paulsen, K. D. and Strohbehn, J. W. (1985) Finite element solution of Maxwell's equations for hyperthermia treatment planning, *Journal of Computational Physics*, **58**, 246–269.

Lynch, D. R., Paulsen, K. D. and Strohbehn, J. W. (1986) Hybrid element method for unbounded electromagnetic problems in hyperthermia, *International Journal for Numerical Methods in Engineering*, **23**, 1915–1937.

Lynch, D. R., Paulsen, K. D., Sullivan, J. M. and Strohbehn, J. W. (1988) Hyperthermia analysis on finite elements, *Proceedings of the Ninth Annual Conference of the IEEE Engineering in Medicine and Biology Society*, pp. 1293–1295.

Mansur, W. J. and Brebbia, C. A. (1982) Formulation of the boundary element method for transient problems governed by the scalar wave equation, *Applied Mathematics Modelling*, **6**, 307–311.

Paulsen, K. D. (1990) Calculation of power deposition patterns in hyperthermia, in Gautherie, M. (Ed.), *Clinical Thermology*, Vol. 5, Thermal modelling and thermal dosimetry, Berlin, Springer-Verlag.

Paulsen, K. D., Strohbehn, J. W. and Lynch, D. R. (1984) Theoretical temperature distributions produced by an annular phased array type system in CT-based patient models, *Radiation Research*, **100**, 536–552.

Paulsen, K. D., Strohbehn, J. W. and Lynch, D. R. (1988a) Theoretical electric field distributions produced by three types of regional hyperthermia devices in a three-dimensional homogeneous model of man, *IEEE Transactions on Biomedical Engineering*, **35**, 35–45.

Paulsen, K. D., Lynch, D. R. and Strohbehn, J. W. (1988b) Three-dimensional finite boundary and hybrid element solutions of the Maxwell equations for lossy dielectric media, *IEEE Transactions on Microwave Theory Techniques*, **36**, 682–693.

Rine, G., Dewhirst, M. W., Samulski, T. V. and Wallen, A. (1987) Modelling of SAR values in tissue due to slab loaded waveguide applicators, *Proceedings of the Ninth Annual Conference of the IEEE Engineering in Medicine and Biology Society*, pp. 1000–1001.

Roemer, R. B. and Cetas, T. C. (1984) Applications of bioheat transfer simulations in hyperthermia, *Cancer Research*, **44**, 4788s–4798s.

Shen, L. C. and Kong, J. A. (1987) *Applied Electromagnetism*, 2nd edn, Boston PWS.

Strohbehn, J. W. and Roemer, R. B. (1984) A survey of computer simulations of hyperthermia treatments, *IEEE Transactions on Biomedical Engineering*, **31**, 136–149.

Strohbehn, J. W., Paulsen, K. D. and Lynch, D. R. (1986) Use of finite element methods in computerized thermal dosimetry, in Hand, J. W. and James, J. R. (Eds), *Physical Techniques in Clinical Hyperthermia*, Letchworth, Research Studies Press, pp. 383–451.

Sullivan, D. M., Borup, D. T. and Gandhi, O. P. (1987) Use of the finite-difference time-domain method in calculating EM absorption in human tissues, *IEEE Transactions on Biomedical Engineering*, **34**, 148–157.

Sullivan, D. M., Gandhi, O. P. and Taflove, A. (1988) Use of the finite-difference time-domain method for calculating EM absorption in man models, *IEEE Transactions on Biomedical Engineering*, **35**, 179–185.

Swindell, W. (1986) Ultrasonic hyperthermia in Hand, J. W. and James, J. R. (Eds), *Physical Techniques in Clinical Hyperthermia*, Letchworth, Research Studies Press, pp. 288–326.

Tijhius, A. G. (1984) Toward a stable marching-on-in-time method for two-dimensional transient electromagnetic scattering problems, *Radio Science*, **19**, 1311–1317.

Turner, P. F. (1984) Hyperthermia and inhomogeneous tissue effects using an annuar phased array, *IEEE Transactions on Microwave Theory Techniques*, **32**, 874–882.

van Den Berg, P. M., DeHoop, A. T., Segal, A. and Praagman, N. (1983) A computational model of electromagnetic heating of biological tissue with application to hyperthermia cancer therapy, *IEEE Transactions on Biomedical Engineering*, **30**, 797–805.

13 Techniques for interstitial hyperthermia

P. R. Stauffer

13.1. Introduction

In response to the increasing demand for instrumentation to produce well-localized heating of deep-seated tumours in the pelvis, abdomen, thorax and brain, several different interstitial heating technologies have evolved. Although technically impractical for many patients, invasive heating approaches offer significant advantages over externally applied heat for the required combination of tissue penetration and localization of heat into a tumour at depth.

There are three general methods used to produce well-localized heating of tumour volumes located at depth in the body. Radio frequency localized current field (RF-LCF) equipment has been used successfully to produce ohmic tissue heating in the frequency range 0.5–27 MHz, normally between rigid needle electrode pairs implanted through templates to ensure parallel orientation. Implanted microwave antennas (400–2450 MHz) also have been used due to their compatibility with plastic catheter brachytherapy procedures as well as their capability of dielectric tissue heating at distance from the antenna surface. Clinical application of several 'hot source' techniques for producing thermal conduction heating of tissue has just begun. These include inductively heated ferromagnetic 'thermoseeds', hot-water tubes, resistance wires and optical-fibre-coupled laser illumination. An overview of the basic operating principles for all interstitial heating modalities will be given in Section 13.3. For more detailed descriptions of the underlying physical principles, the interested reader is referred to the primary references at the end of this chapter, or to reviews of interstitial heating techniques by Strohbehn and Douple (1984), Strohbehn (1987) and Hand *et al.* (1990). Subsequent sections briefly describe some practical implant and thermometry considerations for clinical use of the techniques. Finally, the advantages and disadvantages of the current interstitial applicators are summarized and used to form guidelines for the most appropriate clinical sites of application for each modality.

13.2. Techniques for interstitial hyperthermia

13.2.1 Implantable microwave antennas

Implantable microwave antennas are under investigation at many institutions. Although existing antennas have worked successfully in certain deep-seated tumour volumes (Emami *et al.*, 1987; Salcman and Samaras, 1983; Coughlin *et al.*, 1985), *in vivo* thermal dosimetry studies (Lyons *et al.*, 1984; Matsumoto *et al.*, 1986; Sneed *et al.*, 1986; Samaras, 1984; Satoh *et al.*, 1988b) show that tissue heterogeneity significantly complicates the induced temperature distributions over those predicted by computer simulations (de Sieyes *et al.*, 1981; Mechling and Strohbehn *et al.*, 1986; Strohbehn *et al.*, 1982; Zhang *et al.*, 1988). The isotherm contours form a roughly ellipsoidal pattern along the long axis of typically used implantable dipole antennas, with little heating near the ends (Iskander and Tumeh, 1989; Strohbehn *et al.*, 1979; Satoh *et al.*, 1988a; Stauffer *et al.*, 1987). New antenna designs incorporating multiple active sections, 'sleeves', 'chokes', helical coils or other radiating structures (Satoh and Stauffer, 1988; Lee *et al.*, 1986; Wu *et al.*, 1987; Turner, 1986; Lin and Wang, 1987) have been developed to increase the uniformity, controllability or distribution of heating along the antenna axis. While improved field-shaping capabilities have resulted, substantial changes in the effective heating length of microwave antennas for a given source frequency and insertion depth have been difficult.

13.2.2. RF-LCF electrodes

Temperature uniformity within arrays of RF-LCF electrodes depends largely on the homogeneity of tissue electrical properties along the conductive electrode length and on parallel orientation of the implants. Since the RF electrodes deposit energy in tissue between electrodes in proportion to the square of the current density, significant temperature variability along the electrode length can result. Numerical modelling (Strohbehn, 1983; Zhu and Gandhi, 1988) and engineering details (Doss and McCabe, 1976; Astrahan and Norman, 1982; Stauffer, 1984) of RF-LCF heating systems have been presented previously, and the heating performance of LCF arrays contrasted with other interstitial modalities both theoretically (Mechling and Strohbehn, 1986; Strohbehn, 1987) and experimentally (Stauffer *et al.*, 1989). Clinical RF-LCF treatments are in progress at several institutions (Emami *et al.*, 1987; Puthawala *et al.*, 1985; Vora *et al.*, 1982; Linares *et al.*, 1986; Manning and Gerner, 1983; Cosset *et al.*, 1985) for applications allowing implantation of multiple metal needles. Although it is technically simpler to control electric current distributions around RF electrodes than around shorter-wavelength microwave antennas, electrodes which can shape the heating pattern along the electrode length as well as provide compatibility with flexible catheter brachytherapy procedures are still under development (Kapp *et al.*, 1988; Prionas *et al.*, 1989).

13.2.3. Thermal conduction 'hot source' techniques

Thermal conduction 'hot source' techniques contrast sharply with the other interstitial approaches in several respects. Tissue electrical properties do not affect the heating dynamics since no power is deposited directly in tissue. Consequently, blood flow, heat source temperature and implant spacing are the major determinants of the induced temperature distribution. Along with this beneficial elimination of one major heating variable comes the requirement for larger numbers of higher-temperature heat sources, since the tissue is heated solely by thermal redistribution of heat from the induced local heat peaks. Following a lengthy development, clinical trials with inductively heated, curie point, ferromagnetic seeds have been initiated (Brezovich *et al.*, 1984; Shimm *et al.*, 1989). More recently, a preliminary report was given of clinical trials with implanted resistance wire heaters. Thermal dosimetry studies have begun on other 'hot source' techniques, such as the use of hot-water tubes (Brezovich *et al.*, 1989; Hand *et al.*, 1990) and optical-fibre-coupled laser irradiation (Daikusono *et al.*, 1987).

13.3. Basic principles of operation

The 'hot source' techniques are the simplest form of interstitial heating in that no power is deposited directly in tissue. Instead, tissue is heated by thermal redistribution of heat within implanted arrays of essentially equal temperature hot sources. Although each of the techniques rely upon simple transfer of energy away from heated surfaces, there are significant differences in both the methodology of implant heating and the resulting distribution of temperature along the implant length.

13.3.1. Hot-water tubes

Perhaps the most basic form of interstitial hyperthermia is accomplished with an array of needles or catheters through which hot water is circulated. The temperature distribution will come to a 'pseudo' thermal equilibrium (slowly varying in time along with changes in local perfusion) as the additive superposition of individual patterns of temperature rise from each implant (Matloubieh *et al.*, 1984) since there is no direct power deposition or focusing of heat in tissue between sources. In addition to the radially directed temperature gradient between sources, the surface temp-erture of each implant falls off gradually in the direction of water flow as heat energy is taken from the water and conducted or convected (via blood perfusion) into surrounding tissue. Other than this linear gradient, there is little chance of controllably modifying the temperature distribution along each implant except via minor increases in the thermal resistance around portions of the tubing with thicker layers of (catheter) insulation.

Due to the mechanics of water flow through a tube, there is a velocity profile across the lumen such that water flows faster in the centre and slowest near the tubing wall. For the case of water flowing through a tube which is losing heat energy to surrounding tissue at a lower temperature, there is a corresponding temperature profile across the lumen with the highest temperature at the centre falling off toward the tubing–tissue interface. Thus, temperature gradients are established both radially out from the tubing centre and axially along the tubing length which are dependent on flow velocity through the tubing. In practical use of the tech-nique, water flow should be increased until a rapid, turbulent flow is obtained through each tube of the array to thoroughly mix the contents and to minimize transit time of the water through tissue. For a tube with an internal diameter of 1.5 mm implanted a length of 10 cm in tissue, this corresponds to a water flow rate of roughly 2.5 ml/s to limit the temp-erature drop to approximately 0.3–0.5 °C (Hand *et al.*, 1990).

In order to heat an array of water tubes, there are three basic configur-ations: (1) connecting tubes together in series ensures the same water flow rate through each tube, but requires an impractically high flow rate to ensure that there is no temperature drop from the first to the last tube; (2) connecting tubes in parallel to a single hot-water source reduces the flow rate required for minimal temperature drop, but introduces the desir-ability of flow indicators in each tube in order to guarantee that appropriate flow is not lost in any of the tubes due to inadvertant con-striction; and (3) for maximum adjustability of heating in response to vari-ations in internal array tissue temperature, a separate adjustable temp-erature hot-water source can be connected to each individual tube. Systems to date have used a single temperature source feeding multiple implanted tubes in parallel, with sufficient water pressure to assume appropriate flow in each tube (Brezovich *et al.*, 1989).

13.3.2. Resistance wire heaters

As with hot-water tube arrays, tissue heating via high-resistance implant needles is accomplished via thermal conduction and convection of heat energy radially away from cylindrical heat sources. Similar consider-ations regarding the tissue temperature profiles parallel to hot sources apply as for the hot-water tubes, except that the resistance needles are less likely to have a uniform surface temperature. This is because the basic heating mechanism of the needles is an ohmic loss from the electric current (I) flowing through the resistive implant material. From Ohm's law, $I = V/R$ where V is the DC (or low frequency) voltage applied across the total series resistance R from end to end along the needle. The total heating power deposited in each needle is $P = I^2 R$. If the resistance is distributed uniformly along the implant length (as in the case of a homogeneous implant needle), for any induced needle current there will be a correspondingly uniform heat generation along the needle length. Thus, if heat removal via tissue cooling is non-uniform along the implant, then the surface temperature must be non-uniform as well. The variation of temperature along the needle surface obtained for tissues of various cooling capacities has yet to be determined.

By constructing the needles from more than one resistivity material, heating can be localized at depth. Since a single 'series' current must flow from end to end in each needle, power deposition per unit length will be greatest in those portions of the needle with the highest resistance. Thus, with appropriate treatment pre-planning and needle construction, it should be possible to localize heating into a tumour region at depth with just a single electrical power supply for all implants. Alternatively, adjustable, real time control over the axial heating distribution may be possible by varying the relative voltage of power supplies connected individually to multiple, insulated segments along each 'needle'.

Proper isolation of all low-frequency implant needles and electrical wire connections from the body is essential to avoid electrical shock hazard. While a thin dielectric coating should provide adequate electrical insulation, extreme care must be taken to maintain the integrity of the dielectric or catheter wall and to avoid any patient contact with the electrical wire connections.

13.3.3 Fibre-optic coupled laser irradiation

Nd/YAG lasers have been used for a variety of medical applications. One involves the use of optical fibre cables to deliver laser energy directly into tumours (Brown, 1983). Because of the high power density at the distal end of the fibre, only a limited volume of tissue can be heated from a single fibre without damaging the tip. To extend the heated region, special synthetic frosted sapphire crystals have been developed which mount on the ends of optical fibre cables and serve to diffuse the laser energy into the surrounding medium. Due to the extremely short wavelength of radiation, essentially no penetration of the laser light is obtained in tissue. Rather, it is dissipated as heat energy at the tissue–crystal interface. The sapphire 'probes' are available in a number of sizes and shapes though thermal dosimetry has been reported only for a conical tip probe 5 mm long and 2.2

mm in diameter (Daikuzono *et al.*, 1987). Tissue hyperthermia results from thermal conduction redistribution of heat energy away from the implanted array of sapphire crystals. Although the present system appears suitable for heating only small volumes, the principles of heating are similar to other thermal conduction techniques which require a single connecting cable per heated source.

13.3.4. Inductively heated ferromagnetic seeds

As in the previous three techniques, ferromagnetic seeds heat tissue by thermal conduction and convection of heat radially away from the implanted sources. Unlike the other techniques, the ferromagnetic seeds can be constructed of any desired length and do not require external connections. Along with this advantage of easily localized heating at depth comes the complication that near the ends of finite length linear heat sources, heat is lost axially off the ends as well as radially. Thus, the distribution of tissue temperature along the implant length is not uniform, even for equal temperature sources. At increasing radial distance from the source, the axial direction temperature profiles become increasingly peaked near the middle of the implant, showing little heating near the ends at larger distances (Figure 13.1). This complication can be accommodated in one of two ways. The tumour volume can be over-implanted by the seeds such that the centrally heated zone corresponds with the tumour boundaries. Preferably, the 'ferroseeds' can be broken into short segments and inserted as multiseed strings. With proper seed length and interseed spacing, and/or using higher temperature seeds near the ends, the overall axial heating characteristic of the seed string can be adjusted as necessary to achieve uniform or even peripherally enhanced heating patterns.

The ferromagnetic material is heated by inducing large eddy currents on the surface of the metal via an externally generated magnetic field (Stauffer *et al.*, 1982, 1984a). The heating power absorbed in a cylindrical metallic implant of length (L) is dependent primarily on the seed characteristics: radius (a), magnetic permeability (μ), electrical conductivity (σ) and the frequency (ω) and strength of the magnetic field (H_0), assuming seed strings are implanted 1 cm apart and neglecting end effects, each 1 cm long ferromagnetic seed must heat approximately 1 cm^3 of tissue. The absorbed power density in each 1 cm long cylindrical seed available to heat the surrounding 1 cm^3 tissue volume is

$$P_I = \pi a \, (\omega\mu/2\sigma)^{1/2} H_o{}^2 \times 10^{-2} \text{W/cm}^3. \tag{13.1}$$

A low frequency RF field (≤ 500 KHz) is generally used since significant direct tissue heating results at higher frequencies from magnetic-field-induced eddy currents in body tissue. The power deposited directly in 1 cm^3

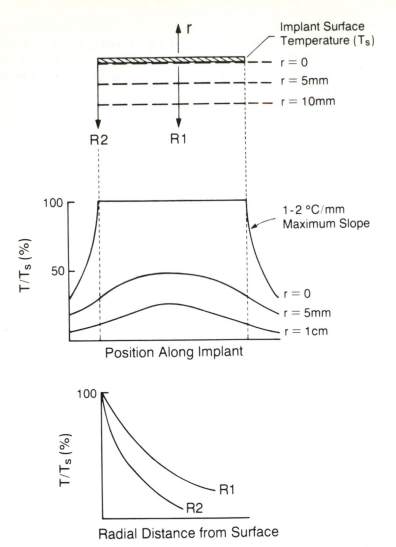

Figure 13.1 Distribution of tissue temperature adjacent to an equal-temperature heat source. Note the rapid shortening of the uniformly heated length with increasing radial distance from the source.

of tissue of electrical conductivity σ_t located at distance r from the centre of the body by the same magnetic field but without implants is

$$P_T = \omega^2 \mu_r^2 r^2 \sigma_t H_o^2 \times 10^{-6} \, \text{W/cm}^3. \tag{13.2}$$

For practical valued parameters, heating around the implant predominates at frequencies below 2 MHz and direct tissue power absorption predominates above 10 MHz. Since there is a practical lower limit on

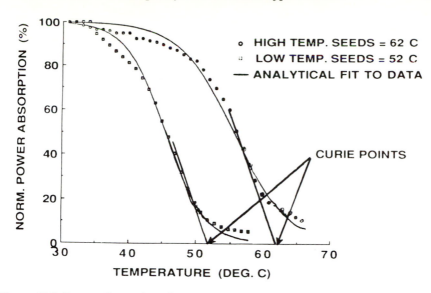

Figure 13.2 Power absorption characteristics of thermally regulating ferromagnetic seeds. Note the rapid drop in heating efficiency as seed temperatures approach the Curie points of the two different alloys. Adapted from T. C. Cetas *et al.* (private communication).

frequency to limit the risk of direct nerve stimulation, a frequency around 50–100 KHz appears to provide a good combination of safety and adequate differential heating of interstitial implant-sized ferromagnetic seeds in the human torso (Atkinson *et al.*, 1984; Stauffer *et al.*, 1984b).

While most of the parameters in the above two equations are not easily varied during treatment, the relative seed heating can be affected by changes in the permeability and/or electrical conductivity of the material. One such change occurs naturally as a consequence of heating the seed. The permeability of any ferromagnetic material drops rapidly as the material approaches a specific temperature — its 'Curie point'. As this temperature is reached, heating efficiency drops, providing a leveling of the temperature rise and, in fact, proportional temperature control (Burton *et al.*, 1971; Demer *et al., 1986; Chen et al.,* 1988). By alloying together a highly permeable material (iron, nickel or cobalt) with another metal having a lower Curie point temperature, alloys can be manufactured that have a high permeability suitable for efficient induction heating at low (body) temperatures which become non-magnetic with very poor heating capabilities as the material approaches its Curie point (Figure 13.2; Deshmukh *et al.*, 1984; Brezovich and Liu, 1987). These implants are thus thermally self-regulating in that little change in seed temperature occurs for variations in external magnetic field strength (Figure 13.3), or similarly for variations in tissue cooling rates within the array. Furthermore, the thermal regulation occurs at all points along the seed so that generally the entire seed surface remains at one Curie point

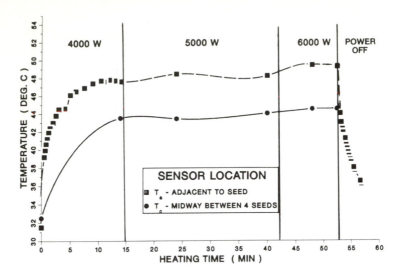

Figure 13.3 Thermoregulating characteristics of a 1×1×6 cm array of thermoregulating ferromagnetic seeds (c/o Dr T. C. Cetas) in porcine thigh muscle. Note the 43°C minimum tissue temperature obtained using seeds with surface temperatures of 48–49°C, and the minimal 1°C change in seed temperature for a 50% increase in applied power.

temperature, regardless of variations in the cooling rate of surrounding tissue. As with the previous techniques, the tissue temperature distribution within an array of inductively heated ferromagnetic seeds reaches a 'pseudo' thermal equilibrium which is dependent on ferromagnetic seed spacing and the pattern of blood perfusion (Matloubieh *et al.*, 1984).

13.3.5. RF-LCF

By connecting implanted metal 'electrodes' in pairs to a power generator that oscillates in the low RF range instead of individually to a constant DC or powerline frequency source, movement of electrons (current) can be forced through the primarily resistive tissue medium. Typical impedance between two electrodes spaced 1–2 cm apart and implanted 5 cm deep into tissue is in the range of $Z=20$ ohms$-20°$ to 15 ohms$+5°$ for 0.5 to 15 MHz operating frequencies. Ohmic heating of tissue between needles results, as well as thermal conduction and convective redistribution of the heat. Each pair of needles forms a closed circuit with the tissue and RF generator, so that the same electric current must flow through each needle surface and spread out into the intervening tissue (Figure 13.4). Since there is an unlimited number of parallel pathways in tissue for current to flow, the tissue current density (J) is highest at the point of smallest cross-sectional area through which the current must flow (i.e. the needle–tissue interface). Power is deposited directly in tissue according to the simple relation

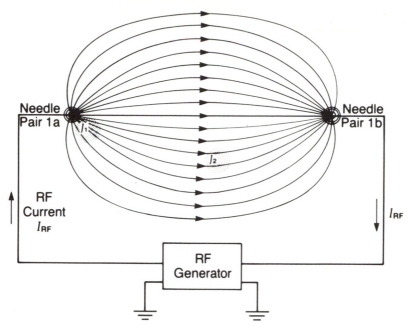

Current Density $J_1 \gg J_2 (A/m^2)$

Power Density $P_D = \rho J^2/2 = \sigma E^2/2$ (W/m³)

Figure 13.4 Representation of RF current spreading out into tissue from implanted needle electrodes. Note the highest current density is in the vicinity of the electrode surface.

$P_D = \rho J^2/2$, where ρ is the tissue resistivity and P_D is the average absorbed power density in tissue. Thus, the power absorption is concentrated in the immediate vicinity of the electrodes. As the current spreads out in the surrounding medium, correspondingly lower heating rates are obtained. If part of the tissue volume between electrodes has lower resistance to the flow of current, the current will preferentially flow through that region and result in increased power deposition. Conversely, current will tend to avoid regions of higher resistance (such as the fat tissue layer) leaving them relatively cool. Finally, if portions of the needle implants are insulated with a thin dielectric coating, the RF current can be concentrated in the region between bare metal sections. This results from the very large capacitive impedance to the flow of current through the dielectric insulation relative to the low-impedance pathway between the bare metal sections. Figure 13.5 illustrates the effectiveness of this approach by displaying the temperature distribution along the length of a pair of partially insulated electrodes implanted in muscle tissue *in vivo*.

At present there are four basic approaches to the distribution of heating between multiple needles of a large LCF implant array. The simplest

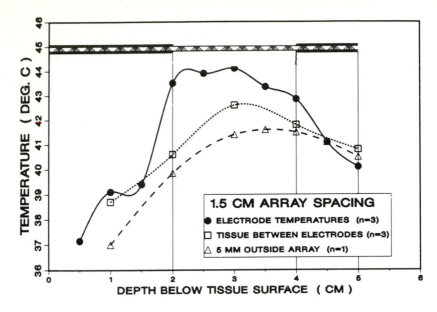

Figure 13.5 Tissue temperature distributions with depth into dog thigh muscle inside an array of four flexible metal braid electrodes. A section of the polyethylene outer coating was stripped away as indicated on the figure to localize heat into the tissue region between metal sections.

approach is to connect together all needles into two (or more) concentric rows, semicircular rings, or other geometric pattern, and then to apply power from a single amplifier to two adjacent formations (Cetas *et al.*, 1980; Manning and Gerner, 1983; Stauffer, 1984; Milligan and Dobelbower, 1984). Since both the tissue heating and tissue cooling tend to be non-uniform between multiple needles (in correspondence with the surrounding electrical and thermal tissue properties), control over the induced temperature distribution is extremely limited for a single power source. A second approach is to have a single power amplifier with a large number of computer-controlled output relays (switches) which are each hard-wire connected to pre-planned needle pairs (Astrahan and Norman, 1982). During operation of the system, the computer opens and closes the relays such that power is applied sequentially to each electrode pair for a short period of time (0.1–0.5 s). Following an initial ramping of power to an appropriate level, the power output is held constant while the time interval (dwell) that each electrode pair is connected to the amplifier is adjusted by the system computer in accordance with a single control temperature feedback signal for each needle pair. Since power is applied to a given pair for only a brief time and then interrupted for a much longer time (sufficient to heat all other pairs) this system has an inherent temperature variation (ripple). In treatments with 15 or less needle pairs spaced 1.5 cm apart in moderate blood flow tissue, the ripple is generally less than 2°C at the needle surface and less than 0.4°C in tissue midway between needles, provided the control temperature feedback is from within the implant needles for fastest response time of the controller. A third approach to LCF heating differs in that a separate well-isolated power amplifier is hard-wire dedicated to each pre-planned pair of needles. With this type of system, RF power is applied to all needle pairs simultaneously. The power to each amplifier is controlled by duty cycle variation in response to the 'average' computer-calculated temperature of all sensors in the vicinity of that specific needle pair. In all three of the above hardware implementations, the LCF needles are hard-wire paired with adjacent needles 1–2.5 cm apart, with either parallel or diagonally crossing current paths. For this reason, it is normally beneficial to pre-plan electrical connection of wire pairs to needles in zones of similar blood perfusion cooling. In clinical application of the technique, it may be necessary to adjust the electrode pair wiring configuration during treatment to accommodate significantly different needle temperatures, especially for larger electrode spacings (Figure 13.6)

Although not commercially available as yet, development continues on a fourth power distribution scheme (Prionas *et al.*, 1989) which should provide increased adjustability of the heating patterns. To obtain more control over the distribution of heat along the length of implanted electrodes, the needles can be broken into short, electrically insulated segments which are connected separately to the power source(s). Normally

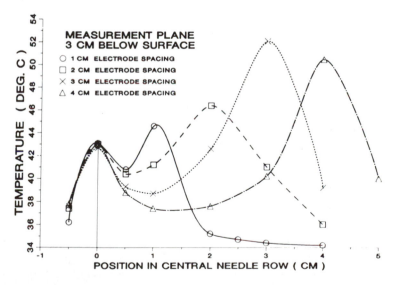

Figure 13.6 Temperature profiles between electrodes at various separations in dog thigh muscle. Steady-state temperatures were measured mid-depth in the central row of three parallel rows of No. 17 needles; each needle was inserted to a depth of 5 cm. Note the high temperatures between electrodes for spacings as large as 2 cm due to thermal conduction smearing of the tissue power deposition pattern.

a time-sequenced power generator would be used to distribute power between the many segments due to the large number of possible electrode pair heating combinations. Control routines can be configured to apply power between the coolest two electrode segments at each moment, between the coolest electrode segment and all other segments connected together, or between time-sequenced adjacent pairs. Obviously, combinations of the above techniques are possible, such as using separate power amplifiers for each needle pair in addition to time sequencing of relative power to individual segments of each needle.

13.3.6. Implantable microwave antennas

In the interstitial heating modalities described above, heat energy is transferred into tissue either via conduction and convection from heated surfaces, or deposited directly in tissue by ohmic (joule) heating from electric currents through resistive tissue. As the frequency is increased, dielectric as well as ohmic losses must be considered within the tissue volume and the radiation characteristics of the implanted electromagnetic sources become important. At sufficiently high frequencies (above approximately 300 MHz) the radiative mode of electromagnetic propagation and dielectric losses in tissue are dominant. While multiple needles can still be placed in an array and activated with high frequency RF (microwave) power, the current is no longer constrained to flow directly between metal 'electrodes'. Instead, each implant becomes a radiating antenna with a complex electric field generated in the surrounding tissue. With only a single 'active' element, the electric current may return to the generator through alternative pathways such as the referred earth ground connection which is capacitively coupled to the body surface. With multiple implanted elements, there may be cross-coupling between antennas as well as to the referred ground. In order to limit the radiated heating pattern to a tumour-sized volume, several antenna designs have been developed using miniature (1–2 mm outside diameter) coaxial cable feedlines implanted directly in tissue so that a preferred return path for the current is inserted with each powered element. Generally, microwave antennas have been designed for efficient radiation into lossy dielectrics at the allowed medical research (ISM band) frequencies of 915 or 2450 MHz. At these frequencies, the wavelength is approximately matched to tumour dimensions and the radial penetration into tissue is limited to 1.5 cm or less. For 915 MHz, the wavelength in free space ($\lambda_o = c/f$) is 33 cm while in soft tissue it is shortened to $\lambda = \lambda_0/\epsilon_r^{1/2} \simeq$ 4–5 cm due to the high dielectric constant (ϵ_r) of soft tissue. At 2450 MHz, the wavelength in tissue is about 1.5 cm. The use of lower-frequency sources from 300 to 500 MHz has been suggested for heating larger tumours (Trembly *et al.*, 1988; Iskander and Tumeh, 1989).

The power deposition rate at any point in tissue around the antennas is determined from $P_D = \sigma_t E_T^2/2$ where $\sigma_t = (1/\rho)$ is the electrical conductivity and $E_T = \Sigma E_i$ is the electric field (magnitude and direction) at that point summed from all sources. For an array of coherently driven (equal phase) radiators, there is significant interaction between antennas in terms of constructive and destructive interference of the individual electric fields (E_i's). Because the square of the total field superimposed from multiple antennas (E_T^2) can be significantly greater than the sum of the individual fields squared ($E_1^2 + E_2^2 + ...$), tissue power deposition can be significantly enhanced in regions between antennas where the electric fields are in phase. For predominantly linearly polarized antennas, this occurs near the array centre (both laterally and longitudinally within the array). For increasing axial distance from the 'junction' plane of dipole-type radiators, the electric field orientation becomes progressively skewed with respect to the antenna axis, so that significant phase cancellation occurs near the tip and back along the antenna feedline (King *et al.*, 1983). For an array of coherently driven circularly polarized (helical coil) radiators, significant phase cancellation of the individual electric fields occurs throughout the region between antennas of the array.

Significant interaction of antenna electric fields of any polarization will also occur using coherently driven but physically unmatched antennas, unmatched-length coaxial feedline cables, or antennas with staggered insertion depths due to irregularly shaped tumours. This may lead to unexpected high and low SAR regions within the array if the antennas are driven by a single generator–power splitter combination, or by phase coherent generators. Additionally, due to the short distance between antennas relative to the radiation wavelength, the entire array volume is in the near field of the radiating apertures. Thus, the electric fields surrounding the antennas have a complex structure which leads to further unpredictability of the set regions of constructive and destructive interference. Until individual phase shifters are available to adjust the relative phasing of antennas either before (Trembly *et al.*, 1986), or continuously during, treatment (Trembly *et al.*, 1988), improved volume heating uniformity is most likely to occur using phase incoherent sources to eliminate the complications of a set constructive–destructive interference pattern from the 'coherent' antenna array. *In vivo* thermal dosimetry experiments in our laboratory have shown an improvement of approx. 10% in volume heating uniformity within 2 cm arrays of phase incoherent antennas compared to the same, equal depth antennas driven coherently (see Table 13.1). This was due primarily to a reduction of peak temperatures in the central array region and the consequent spreading of the effective heating axially along the antennas and somewhat peripherally. While the central 'overheated' zone could be reduced alternatively by moving the antennas further apart, this reduces the length of effectively heated tissue along the

Table 13.1 Coherent versus incoherent antenna array heating uniformity. Percentage of 64 measured points achieving greater than 50% of the maximum temperature rise within a 2 cm square by 7 cm deep implant array in canine muscle *in vivo*.

Animal No.	3.5 cm dipole		3.5 cm helical coil		3.5 cm dipole/coil tip	
	Coherent (%)	Incoherent (%)	Coherent (%)	Incoherent (%)	Coherent (%)	Incoherent (%)
1	59	67	72	88	69	81
2	61	64	61	75	66	86
3	45	63	64	67	—	—
$x \pm SD$	55 ± 9	65 ± 2	66 ± 6	77 ± 10	67.5	83.5

x, mean; SD, standard deviation.

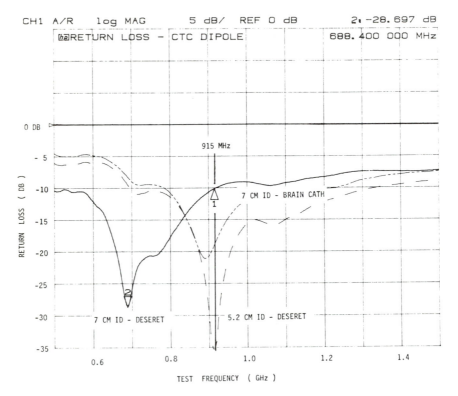

Figure 13.7 Electrical matching characteristics of a commercial dipole microwave antenna in phantom material. The antenna is well matched (< -30 db return loss) to the generator and surrounding media at 915 MHz when implanted 5.2 cm in a polyurethane catheter (1.64 mm outside diameter). Antenna resonance shifts to 688 MHz for 7 cm insertion in the same catheter, and to 889 MHz in a thicker-walled silicon catheter (2.52 mm outside diameter). Though exhibiting different antenna radiation efficiencies, all configurations can heat tissue at 915 MHz.

central array axis due to increased heat loss off the ends of implants spaced further apart (Figure 13.1).

The first implantable antenna design was constructed by stripping off the outer conductor of a coaxial cable for a distance equal to $\lambda/4$ in the catheter–tumour antenna load. Several variations on the design followed, such as leaving the end section of inner conductor bare or insulated, leaving the terminal portion of the outer conductor attached to the antenna but separated from the feedline at a 'junction' point, etc. (de Sieyes *et al.*, 1981). These radiating structures have had many labels, including monopole, dipole, electromagnetic syringe and slot radiator among others. When this type of antenna is implanted to a depth equal to twice the active element length ($\lambda/2$ total), then the structure appears to operate like a symmetric half-wave dipole with current minima at the tip and skin entrance points, and a maximum near mid-depth at the level of the junction. This produces a circularly symmetric Gaussian (football)-shaped heating pattern in tissue around each dipole-type radiator. The longitudinal extent of heating is primarily dependent on the frequency and electrical properties of the surrounding medium. Since there is a large difference between air ($\epsilon_r=1$), catheter ($\epsilon_r=2$), and tumour tissue ($\epsilon_r=49$), the heating is largely dependent on the relative diameter of the antenna and implant catheter. Significant alterations of the axial direction heating profile of the antennas can be seen for tight-fitting

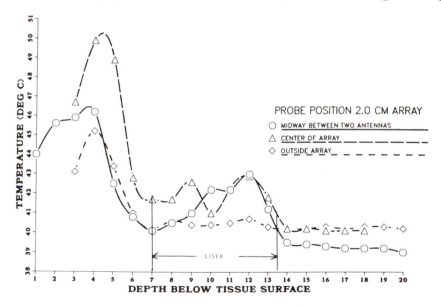

Figure 13.8 Temperature distribution within a 2.0 cm square array of four commercial interstitial dipole microwave antennas implanted percutaneously to a depth of 14 cm in pig liver tissue. Unacceptable heating of superficial fat and skin tissues was observed at the power levels necessary to heat highly perfused normal liver.

antennas in thin or thick wall catheters, loosely fitting antennas in air- or oil-filled catheters, or antennas inserted directly into the tissue (Figure 13.7). Also, due to the extreme difference in electrical properties of air and tissue, the depth of insertion of antennas can have a major effect on the radiation pattern if the electric field is oriented along the axis of implantation (Figure 13.8; Denman *et al.*, 1988; Satoh and Stauffer, 1988). Newer antenna designs, with chokes to limit radiation back along the antenna feedline or electric field orientations parallel instead of normal to the tissue surface, are becoming available for improved localization of heating into small tissue regions at depth (Figure 13.9).

13.4. Thermometry considerations

Due to the sharp localization of heating possible with the invasive modalities, steep thermal gradients (as high as 1–2°C/mm) often occur within the treatment volume which place increased demands on thermo-

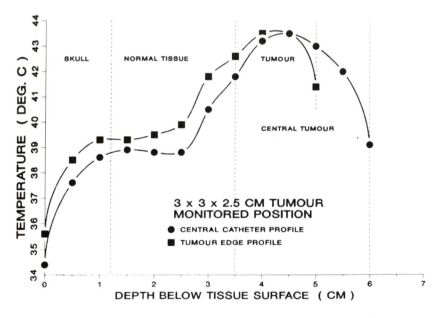

Figure 13.9 Temperature distribution within a 2.0 cm square array of four 2.5 cm long helical coil antennas in brain tissue. The vertical lines represent tumour and normal tissue boundaries. Note the excellent localization of heating into the tumour at depth and sparing of overlying normal brain, skull and scalp tissues. The treatment prescription was 42.0°C minimum tumour temperature.

metry. The available techniques for invasive thermometry have been described numerous times (Cetas, 1982; Samulski, 1988). Guidelines describing the minumum acceptable thermometry for documenting clinical treatments are still evolving (Luk and Shrivastava, 1987). The present discussion will be limited to an identification of the most appropriate thermometry equipment and dosimetry procedures for each interstitial heating modality.

The most common technique used for characterizing and controlling interstitial heat treatments is the use of stationary multiple-sensor probes. While this approach normally produces only a small number of measured points due to probe hardware costs and system data acquisition times, it provides a minimum level of temperature information for real time feedback control of the heat sources and integration of thermal dose. Monitoring points of particular interest include maximum and minimum temperatures of the heat source surface, tissue midway between two heat sources, centrally between four heat sources, at the deep and lateral tumour periphery, and at major tissue interfaces (Figure 13.10). For critical tissue regions such as brain or liver, the number of extra temperature probes specified in Figure 13.10 may not be feasible.

Depending on the hardware configuration, additional important temperature information may be obtained without added surgical trauma. At least one manufacturer supplies implantable microwave antennas with miniature thermistor sensors embedded inside the coaxial cable dielectric near the hottest section of the antenna. Another is developing special implant catheters which have as many as 4–7 fibre optic sensors embedded in the walls of the hollow implant catheters for measuring tissue temperature as close as possible to the heat source without the need for

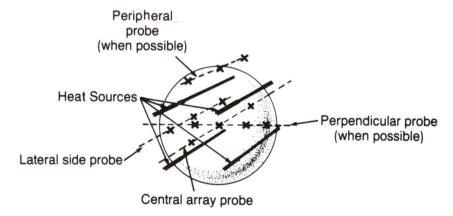

Figure 13.10 Recommended tissue sites for documentation of temperature in interstitial heating applications, in addition to source temperature measurements. Mapping of temperatures at other positions along the catheters is helpful when possible.

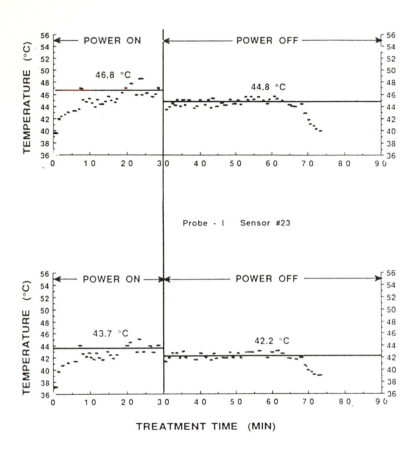

Figure 13.11 Time-temperature recordings at two points between an RF electrode that was de-energized at *t*=30 min to monitor the effect of fewer heating electrodes on tumour temperature. Although this increased the active electrode spacing to 2.5 cm in this region, the tumour was adequately heated without this pair. If left energized, the automatic control routine would have limited the average power to this pair in order to maintain the desired set-point temperature. The original Data Acquisition System temperature data have been replotted in entirety on redrawn axes for improved clarity.

additional catheters. While temperatures measured in these positions may be useful for balancing power to each element of the array, the data should be interpreted with extreme care to account for possible errors from self-heating artifact and thermal conduction smearing of high thermal gradients along the implant surface (Astrahan *et al.*, 1988). As with the LCF-electrode-mounted thermometry, the heat sources should be cycled off for 5–10 min periods in order to read true tissue temperature at distance from all heat sources (Figure 13.11) or (preferably) additional probes should be placed in tissue to measure minimum thermal doses in tissue.

Even with recent improvements in data acquisition capabilities, the need for increased thermometry information during the treatment continues to outpace hardware implementation at most sites. While automated computer-controlled mapping systems have been introduced (Gibbs, 1984; Engler *et al.*, 1987), most users have been forced to perform manual thermal mapping, consisting of repeated incremental movements of temperature sensors along implanted 'dosimetry' catheters, to record multiple temperatures.

While any of the temperature monitoring systems can be configured to operate with any of the interstitial heating modalities, the conventional use of thermometry systems is as follows. For monitoring tissue temperatures within *microwave antenna arrays*, fibre optic probes have provided the most reliable and artifact-free thermometry. Shortcomings of the fragile, expensive fibre optic probes with marginally acceptable accuracy and stability characteristics are offset by the inherent advantages of minimal interactions between the small diameter, low mass, non-metallic fibres with electromagnetic and thermal fields (Chan *et al.*, 1988). Alternatively, high-resistance lead thermometers have been used successfully to improve temperature accuracy and stability while still minimizing electrical artifacts. Limitations of these probes are related to their larger physical dimensions, thermal mass and cost. RF shielded and filtered thermocouples or thermistors have also been used, normally by sampling during brief interruptions of microwave power to determine and subtract out the electrical artifact. Control temperature feedback from sensors located closest to the hottest portion of each antenna will yield the best temporal power regulation and balance between individual antennas (i.e. the least ripple in tissue temperature) but by itself cannot be expected to control the tissue temperature distribution to meet the prescribed minimum thermal dose. Thus control schemes should use measurements of minimum and maximum tumour temperatures midway between antennas to control the overall power level to each geometric grouping of 2–5 antennas and the individual antenna-tissue interface temperatures for frequent adjustment of relative power balance between antennas.

Monitoring temperature in lower frequency *RF-LCF applications* is somewhat easier due to the small size of temperature probes relative to the longer-wavelength fields and the capability of placing temperature sensors inside hollow RF 'needle' electrodes. As source frequencies drop below 10 MHz ($\lambda_0 = 30\,\mathrm{m}$), commercially available high-precision RF-shielded thermocouple systems become the system of choice (Milligan and Panjehpour, 1983) due to their accuracy, stability, low cost per channel and availability in multiple-sensor 'linear array' probe format. The use of more expensive and finicky optical fibre or thermistor thermometry systems with LCF is allowable, but normally not justified.

For clinical LCF treatments, an absolute minimum of one temperature sensor per needle pair is necessary to control and balance heating between pairs; a minimum of one tumour temperature measured in each needle allows reconfiguring of wire connections during treatment if needed to improve the balance between electrodes. Multiple-sensor probes should be placed in as many needles as possible for continuous monitoring of temperatures at various depths in the tumour and overlying normal tissue. Feedback control temperature sensors should be located inside the heating needles, as opposed to in the tissue, in order to decrease the feedback response time and corresponding temperature ripple from excessive thermal lag time in the control loop. In order to do this, temperatures measured inside the heating needles must be correlated with tissue temperatures at distance from the heat sources by turning power off to selected needle pairs temporarily during treatment (for approximately 5 min) and recording the new equilibrated temperatures (Figure 13.11).

Multisensor thermocouple probes appear to provide the most cost-effective tissue temperature monitoring for the *thermal conduction heating modalities* as well. While no problems from heating field perturbation or field induced temperature artifacts are expected with resistance wire, laser or hot-water tube heating techniques, RF filtering and shielding of the thermocouple readout electronics is required to eliminate electrical interference from the magnetic induction equipment during heating with ferromagnetic seeds.

13.5. Summary: modality selection criteria

Analogous to the determination of optimum external heating modality, there is no single invasive heating procedure which is best suited for all implant cases. Rather, choice of interstitial technique for a particular patient must be based on a number of factors including tumour dimensions and location, tissue properties, and possible implant configurations (array spacing, template or pull-through design, flexible or rigid implants, etc). While technological development in this area is still progressing rapidly, the current status of clinical utilization of interstitial hyperthermia may be summarized as follows:

RF local current field electrodes appear most useful for heating tumours which can be implanted with an array of parallel, equal length metal needles. Applications in the gynaecologic region are most appropriate with possibilities for certain chest wall, neck and extremity locations, as well as intra-operative abdominal or pelvic masses. Using current commercial systems, the maximum control possible is power regulation to each pair of electrodes using one feedback temperature sensor per pair, or one software-generated feedback control signal from multiple sensors per pair. The use of bare metal needle arrays in tissues with high perfusion or

Table 13.2 General comparison of interstitial heating modalities.

Implantable microwave antennas

Advantages
1. Compatible with plastic catheter brachytherapy techniques
2. Deposit power in tissue at distance from antennas, can heat outside implant array boundaries
3. Fewest sources needed to heat a region, best technique for high-perfusion tissues
4. Individual power control possible in 100% of sources
5. Commercial availability of equipment

Disadvantages
1. Non-uniform and difficult to adjust depth-heating pattern
2. Insertion depth and heating length restrictions
3. Complicated thermometry, extra tissue probes necessary to control heating pattern
4. Electrical matching and isolation of antennas necessary for balanced array heating
5. Cost increases rapidly with number of probes, power channels
6. Usually control only 25% of antennas (groups of 2–4)

RF current heating

Advantages
1. Simplest power generation and distribution of electromagnetic techniques
2. Best thermometry, measurement of 100% implants possible
3. Power deposition steerable around tumour via computer
4. Segmented electrode arrays possible for improved three-dimensional control

Disadvantages
1. Requires parallel electrodes for uniform electric field
2. Many electrical wires and thermometry probes required to control heating – awkward set-up
3. No control of depth localization with bare needle electrodes
4. Present electrodes not compatible with plastic catheter brachytherapy techniques
5. Temperature control limited to ≤ 50% of implants (i.e. power applied to electrode needle pairs)

Inductively heated Curie point seeds

Advantages
1. Maintains constant seed temperature in variable perfusion tissue
2. No external connections, permanent implants possible
3. Heat localization any depth, uniform temperature any length
4. Cylindrically, axially symmetric heating independent of σ, ϵ
5. Alternated heat and radiation sources possible
6. Simplest thermometry – no computer required
7. Can use higher temperature seeds at implant ends to counter end-losses and increase peripheral heating
8. Temperature control in 100% of sources

Disadvantages
1. No power deposited between seeds, closer spacing required
2. Pre-treatment plan critical, difficult adjustment of source temperatures during treatment
3. Higher initial expense, electromagnetic-impulse-shielded room required
4. Patient restrictions, no existing metal implants allowed
5. Quick reduction of source temperature not possible

Hot-water/laser/resistance heaters

Advantages
1. Cylindrically symmetric heating, independent of σ, ϵ
2. Heat sources safe, no electromagnetic leakage radiation hazard
3. Positive identification of highest temperature in tissue
4. Least complicated, least expensive systems
5. Temperature control possible in 100% of sources

Disadvantages
1. No power deposited between sources, closer spacing required
2. Set heating power per unit length for resistance wires, no adaptation to variable tissue cooling along sources.
3. No localization at depth, overheated low-perfusion tissues
4. Many heat source and temperature probe connections necessary to apply treatment
5. Awkward – many adjustable temperature baths, insulated lines
6. Additional set-up time from large number of source connections

significant heterogeneity of thermal or electrical properties is likely to produce unacceptable temperature differences between needles and temperature variability along the needle length. With appropriate temperature monitoring techniques, all implants can be energized in order to maximize control of heating within the array.

Implantable microwave antennas are most useful for heating applications requiring non-parallel or flexible catheter implants, and in tissues with high blood perfusion. This includes large head and neck lesions, brain tumours, and cases requiring heat localization at depth (for those antennas which provide surface tissue sparing). Minimum thermometry to take advantage of available control flexibility is one sensor per antenna near the antenna surface and one sensor near the centre of each geometric array building block.

Curie point ferromagnetic seeds now have sufficient thermal self-regulation to allow clinical treatments with relatively few measurements of source temperatures. They should be useful for cases requiring a high degree of localization with few externalized connections. Future use in

Table 13.3 Future directions of research for interstitial heating.

All interstitial systems

1. Improved temperature measurement of heat sources
2. Scanning movement of short heat sources
3. Pre-determination of relative tumour blood flow
4. Combined three-dimensional hyperthermia-brachytherapy treatment planning

Implantable microwave antennas	**Inductively heated Curie point seeds**
1. Improved uniformity and control of axial heating pattern	1. Multiple temperature ferromagnetic seeds within array
2. Families of different length, insertion depth independent heating patterns	2. Rapid insertion–exchange techniques for real time variation of seed temperatures
3. Routines for power control of each antenna	3. Improved thermoregulation ratio of Curie point seeds
4. Practical phase modulation system for scanning-phase coherent 'hot' spot around array for improved heating volume	4. Interspersed radiation and ferromagnetic seed sources
	5. Permanent implants

RF current heating	**Hot-water/laser/ resistance heaters**
1. Flexible electrodes to increase use with brachytherapy	1. Multiple, variable temperature water source systems
2. Dielectric-coated adjustable heating length electrodes	2. Partially insulated water tube implants
3. Hardware and software for controlling power to each electrode	3. Single-ended implant techniques
4. Segmented electrode hardware and software	4. Heterogeneous material resistance wire implants for variable axial heating
5. Partially coated metal stylets in conductive plastic catheters	5. Segmented resistance wire implants with individual power supplies
	6. Longer laser diffuser crystals, integral temperature sensors

connectorless, permanent implant applications appears feasible. While closer seed spacing may be necessary than for the RF or microwave techniques, this may lead to decreased sensitivity of array heating uniformity to variable tissue perfusion due to the independent autoregulation of temperature in each seed. Techniques for rapid seed exchange during treatment, and for pre-treatment planning using measured relative blood perfusion data to guide placement of variable temperature seeds must be developed to provide sufficient adaptation of heating to individual tumour requirements. Potential sites for other *hot source techniques* should be similar to the ferromagnetic seed applications except for possible restrictions from external connection requirements and the localizability of heating at depth. Except for the combination of laser-induced hyperthermia heating with photodynamic therapy, exclusive clinical applications for these recently introduced thermal conduction heating techniques have not been demonstrated as yet.

Advantages and disadvantages of the interstitial heating techniques are summarized in Table 13.2. Due to rapid evolution of the technologies, future adjustments to this table will be necessary. In fact, solutions to many of the problems which restrict the clinical use of interstitial heating are currently under investigation, as summarized in Table 13.3. With these ongoing developments, we anticipate an expanding role for interstitial thermoradiotherapy in the treatment of well-localized deep-seated disease.

References

Astrahan, M. A. and Norman, A. (1982) A localized current field hyperthermia system for use with 192-iridium interstitial implants, *Medical Physics*, **9**(3), 419–424.

Astrahan, M. A., Luxton, G., Sapozink, M. D. and Petrovich, Z. (1988) The accuracy of temperature measurement from within an interstitial microwave antenna, *International Journal of Hyperthermia*, **4**(6), 593–608.

Atkinson, W. J., Brezovich, I. A. and Chakraborty, D. P. (1984) Useable frequencies in hyperthermia with thermal seeds, *IEEE Transactions on Biomedical Engineering*, **31**(1), 70–75.

Brezovich, I. A. and Liu, I. S. (1987) Effect of catheters on the performance of self-regulating thermoseeds, *Proceedings IEEE EMBS Meeting*, Piscataway, pp. 1629–1630.

Brezovich, I. A., Atkinson, W. J. and Chakraborty, D. P. (1984) Temperature distributions in tumour models heated by self-regulating nickel-copper alloy thermoseeds, *Medical Physics*, **11**, 145–152.

Brezovich, I. A., Meredith, R. F., Henderson, R. A., Brawner, W. R., Weppelmann, B. and Salter, M. M. (1989) Hyperthermia with water-perfused catheters, in Sugahara, T. and Saito, M. (Eds), *Hyperthermic Oncology, 1988*, Vol. 1, London, Taylor and Francis, pp. 809–810.

Brown, S. G. (1983) Tumour therapy with the Nd:YAG laser, in Joffe, S., Muckerheide, M. and Goldman, L. (Eds), *Neodymium-YAG Lasers in Medicine and Surgery*, New York, Elsevier, pp. 51–59.

Burton, C. V., Hill, M. and Walker, A. E. (1971) The RF thermoseed — A thermally self-regulating implant for the production of brain lesions, *IEEE Transactions on Biomedical Engineering*, **18**, 104–109.

Cetas, T. C. (1982) Invasive thermometry, in Nussbaum, G. (Ed.), *Physical Aspects of Hyperthermia*, New York, American Institute of Physics, pp. 231–265.

Cetas, T. C., Connor, W. G. and Manning, M. R. (1980) Monitoring of tissue temperature during hyperthermia, *Annals of the New York Academy of Science*, **335**, 281–297.

Chan, K. W., Chou, C. K., McDougall, M. A., and Luk, K. H. (1988), Perturbation due to the use of catheters with non-perturbing thermometry probes, *International Journal of Hyperthermia*, **4**(6), 699–702.

Chen, J. S., Poirier, D. R., Damento, M. A., Demer, L. J., Biencaniello, F. and Cetas, T. C. (1988) Development of NI-4 Wt. Prct. Si thermoseeds for hyperthermia cancer treatment. *Journal of Biomaterials Research*, **22**, 303–319.

Cosset, J. M., Dutreix, J., Haie, C., Gerbaulet, A., Janoray, P. and Dewar, J. A. (1985) Interstitial thermoradiotherapy: A technical and clinical study of 29 implantations performed at the Institut Gustave–Roussy, *International Journal of Hyperthermia*, **1**(1), 3–13.

Coughlin, C. T., Wong, T. Z., Strohbehn, J. W., Colacchhio, T. A., Sutton, J. E., Belch, R. Z. and Douple, E. B. (1985), Intraoperative interstitial microwave-induced hyperthermia and brachytherapy, *International Journal of Radiation Oncology, Biology and Physics*, **11**, 1673–1678.

Daikuzono, N., Joffe, S. N., Tajiri, H., Suzuki, S., Tsunekawa, H. and Ohyama, M. (1987) Laserthermia: a computer-controlled contact Nd:YAG system for interstitial local hyperthermia, *Medical Instrumentation*, **21**(5), 275–277.

Demer, L. J., Chen, J. S., Buechler, D. N., Damento, M. A., Poirier, D. R. and Cetas, T. C. (1986) Ferromagnetic thermoseed materials for tumour hyperthermia, *Proceedings IEEE EMBS Meeting*, Piscataway, pp. 1148–1453.

Denman, D. L., Foster, A. E., Lewis, G. C., Redmond, K. P., Elson, H. R., Breneman, J. C., Kereiakes, J. G. and Aron, B. S. (1988) The distribution of power and heat produced by interstitial microwave antenna arrays: II. The role of antenna spacing and insertion depth, *International Journal of Radiation Oncology, Biology and Physics*, **14**(30), 537–545.

Deskmukh, R., Damento, M., Demer, L., Forsyth, K., DeYoung, D., Dewhirst, M. and Cetas, T. C. (1984) Ferromagnetic alloys with curie temperatures near 50°C for use in hyperthermic therapy, in Overgaard, J. (Ed.), *Hyperthermic Oncology, Vol. 1*, London, Taylor and Francis, pp. 599–602.

de Sieyes, D. C., Douple, E. B., Strohbehn, J. W. and Trembly, B. S. (1981) Some aspects of optimization of an invasive microwave antenna for local hyperthermia treatment of cancer, *Medical Physics*, **8**, 174–183.

Doss, J. D. and McCabe, C. W. (1976) A technique for localized heating in tissue: an adjunct to tumor therapy, *Medical Instrumentation*, **10**, 16–21.

Emami, B., Perez, C. A., Leybovich, L. Straube, W. and Vongerichten, D. (1987) Interstitial thermoradiotherapy in treatment of malignant tumours, *International Journal of Hyperthermia*, **3**(2), 107–118.

Engler, M. S., Dewhirst, M. W., Winget, J. W. and Oleson, J. R. (1987) Automated temperature scanning for hyperthermia treatment planning, *International Journal of Radiation Oncology, Biology and Physics*, **13**, 1377–1382.

Gibbs, F. A. (1984) Thermal mapping in experimental cancer treatment with hyperthermia: Description and use of semiautomatic system, *International Journal of Radiation Oncology, Biology and Physics*, **9**, 1057–1063.

Hand, J. W., Trembly, B. S. and Prior, M. V. (1990) Physics of interstitial hyperthermia: radiofrequency and hot water techniques, in Urano, M. and Douple, E. (Eds), *Hyperthermia and Oncology, Vol. 3, Interstitial Hyperthermia*, Zeist, VSP in press.

Iskander, M. F. and Tumeh, A. M. (1989) Design optimization of interstitial antennas, *IEEE Transactions on Biomedical Engineering*, **36**(2), 238–246.

Kapp, D. S., Fessenden, P., Samulski, T. V., Bagshaw, M. A., Cox, R. S., Lee, E. R., Lohrback, A. W., Meyer, J. L. and Prionas, S. D. (1988) Stanford University Institutional Report. Phase I evaluation of equipment for hyperthermic treatment of cancer, *International Journal of Hyperthermia*, **4**(1), 75–116.

King, R. W. P., Trembly, B. S. and Strohbehn, J. W. (1983) The electromagnetic field of an insulated antenna in a conducting or dielectric medium, *IEEE Transactions on Microwave Theory and Techniques*, **31**, 574–583.

Lee, D., O'Neill, M. J., Lam, K., Rostock, R. and Lam, W. (1986) A new design of microwave interstitial applicator for hyperthermia with improved treatment volume, *International Journal of Radiation Oncology, Biology and Physics*, **12**, 2003–2008.

Lin, J. C., and Wang, Y. J. (1987) Interstitial microwave antennas for thermal therapy, *International Journal of Hyperthermia*, **3**, 37–47.

Linares, L. L., Nori, D., Brenner, H., Shieu, M., Ballon, D., Anderson, L., Alfieri, A., Brennan, M., Fucs, Z. and Hilaris, B. (1986) Interstitial hyperthermia and brachytherapy: a preliminary report, *Endocurie Therapy/Hyperthermic Oncology*, **2**, S39–S44.

Luk, K. and Shrivastava, P. (Eds) (1987) *Hyperthermia Quality Assurance Guidelines*, Philadelphia, American College of Radiology.

Lyons, B. E., Britt, R. H. and Strohbehn, J. W. (1984) Localized hyperthermia in the treatment of malignant tumours using an interstitial microwave antenna array, *IEEE Transactions on Biomedical Engineering*, **31**(1), 53–62.

Manning, M. R. and Gerner, E. W. (1983) Interstitial thermoradiotherapy, in Storm, F. K. (Ed.), *Hyperthermia in Cancer Therapy*, Boston, G.K. Hall, pp. 467–477.

Matloubieh, A. Y., Roemer, R. B. and Cetas, T. C. (1984) Numerical simulation of magnetic induction heating of tumors with ferromagnetic seed implants, *IEEE Transactions on Biomedical Engineering*, **31**(2), 227–235.

Matsumoto, K., Stauffer, P. R., Fike, J. R. and Gutin, P. H. (1986) Hyperthermia for malignant brain tumours, *Neurological Surgery*, **14**(8), 965–972.

Mechling, J. A. and Strohbehn, J. W. (1986) A theoretical comparison of the temperature distributions produced by three interstitial hyperthermia systems, *International Journal of Radiation Oncology, Biology and Physics*, **12**, 2137–2149.

Milligan, A. J. and Dobelbower, R. R. (1984) Interstitial hyperthermia, *Medical Instrumentation*, **18**(3), 175–180.

Milligan, A. J. and Panjehpour, M. (1983) The relationship of temperature profiles to frequency during interstitial hyperthermia, *Medical Instrumentation*, **17**(4), 303–306.

Prionas, S. D., Fessenden, P., Kapp, D. S., Goffinet, D. R. and Hahn, G. M. (1989) Interstitial electrodes allowing longitudinal control of SAR distributions, in Sugahara, T. and Saito M. (Eds), *Hyperthermic Oncology, 1988*, Vol. 2, London, Taylor and Francis, pp. 707–710.

Puthawala, A. A., Syed, N., Sheikh, A. M., Khalid, M. A., Rafie, S. amd McNamara, C. S. (1985) Interstitial hyperthermia for recurrent malignancies, *Endocurie Therapy/Hyperthermia Oncology*, **1**, 125–131.

Salcman, M. and Samaras, G. M. (1983) Interstitial microwave hyperthermia for brain tumours: results of a phase-I clinical trial, *Journal of Neuro-Oncology*, **1**, 225–236.

Samaras, G. M. (1984) Intracranial microwave hyperthermia: Heat induction and temperature control, *IEEE Transactions on Biomedical Engineering*, **31**, 63–69.

Samulski, T. V. (1988) Current technologies for invasive thermometry, in *Biological, Physical and Clinical Aspects of Hyperthermia*, edited by B. Paliwal, F. Hetzel, and M. Dewhirst (New York: American Institute of Physics) pp. 168–181.

Satoh, T. and Stauffer, P. R. (1988) Implantable helical coil microwave antenna for interstitial hyperthermia, *International Journal of Hyperthermia*, **4**(5), 497–512.

Satoh, T., Stauffer, P. R. and Fike, J. R. (1988a) Thermal distribution studies of helical coil microwave antennas for interstitial hyperthermia, *International Journal of Radiation Oncology, Biology and Physics*, **15**, 1209–1218.

Satoh, T., Seilhan, T. M., Stauffer, P. R., Gutin, P. H., Sneed, P. K. and Fike, J. R. (1988b), Implantable helical coil microwave antenna for experimental brain hyperthermia, *Neurosurgery*, **23**, 564–569.

Shimm, D., Stea, B., Cetas, T. C., Buechler, D., Carter, L. P., Chen, J., Dean, S., Fletcher, A., Guthkelch, A. N., Haider, S., Hodak, J., Iacono, R., Lutz, W., Obbens, E., Rossman, K., Sinno, R., Spétzler, R. and Cassady, J. (1989) Clinical results of interstitial hyperthermia using thermally regulating ferromagnetic seeds, in Sugahara, T. and Saito, M. (Eds) *Hyperthermic Oncology, 1988*, Vol. 2, London, Taylor and Francis, pp. 536–539.

Sneed, P. K., Matsumoto, K., Stauffer, P. R., Fike, J. R., Smith, V. and Gutin, P. H. (1986) Interstitial microwave hyperthermia in a canine brain model, *International Journal of Radiation Oncology, Biology and Physics*, **12**, 1887–1897.

Stauffer, P. R. (1984) Simple RF matching network for conversion of electrosurgical units or laboratory amplifiers to hyperthermia treatment devices, *Medical Instrumentation*, **18**(6), 326–328.

Stauffer, P. R., Cetas, T. C. and Jones, R. C. (1982) A system for producing localized hyperthermia in tumors through magnetic induction heating of ferromagnetic implants, *National Cancer Institute Monograph*, **61**, 483–484.

Stauffer, P. R., Cetas, T. C. and Jones, R. C. (1984a) Magnetic induction heating of ferromagnetic implants for inducing localized hyperthermia in deep seated tumors, *IEEE Transactions on Biomedical Engineering*, **31**(2), 235–251.

Stauffer, P. R., Fletcher, A. M., DeYoung, D. W., Dewhirst, M. W., Oleson, J. R. and Cetas, T. C. (1984b) Observations on the use of ferromagnetic implants for inducing hyperthermia, *IEEE Transactions on Biomedical Engineering*, **31**(1), 76–90.

Stauffer, P. R., Satoh, T., Suen, S. A. and Fike, J. R. (1987) Thermal dosimetry characterization of implantable helical coil microwave antennas, *Proceedings IEEE EMBS Meeting, Pascataway*, pp. 1633–1635.

Stauffer, P. R., Sneed, P K., Suen, S. A., Satoh, T., Fike J. R., Matsumoto, K. and Phillips, T. L. (1989) Comparative thermal dosimetry of interstitial microwave and radiofrequency-LCF hyperthermia, *International Journal of Hyperthermia*, **5**(3), 307–318.

Strohbehn, J. W. (1983) Temperature distributions from interstitial RF electrode hyperthermia systems: theoretical predictions, *International Journal of Radiation Oncology, Biology and Physics*, **9**, 1655–1667.

Strohbehn, J. W. (1987), Interstitial techniques for hyperthermia, in Field, S. B. and Franconi, C. (Eds), *Physics and Technology of Hyperthermia*, Dordrecht, Martinus Nijhoff, pp. 211–240.

Strohbehn, J. W. and Douple, E. B. (1984), Hyperthermia and cancer therapy: a review of biomedical engineering contributions and challenges. *IEEE Transactions on Biomedical Engineering*, **31**(12), 779–787.

Strohbehn, J. W., Bowers, E. W., Walsh, J. E. and Double, E. B. (1979) An invasive antenna for locally induced hyperthermia for cancer therapy, *Journal of Microwave Power*, **14**, 339–350.

Strohbehn, J. W., Trembly, B. S. and Douple, E. B. (1982) Blood flow effects on the temperature distributions from an invasive microwave antenna array used in cancer therapy, *IEEE Transactions on Biomedical Engineering*, **29** 649–666.

Trembly, B. S., Wilson, A. H., Sullivan, M. J., Stein, A. D., Wang, T. Z. and Strohbehn, J. W. (1986) Control of SAR pattern with an interstitial microwave antenna array through variation of antenna driving phase, *IEEE Transactions on Microwave Theory and Techniques*, **34**(5), 568–571.

Trembly, B. S., Wilson, A. H., Havard, J. M., Sabatakakis, K. and Strohbehn, J. W. (1988) Comparison of power deposition by inhase 433 and phase-modulated 915 MHz interstitial antenna array hyperthermia systems, *IEEE Transactions on Microwave Theory and Techniques*, **36**(5), 908–916.

Turner, P. F. (1986) Interstitial EM applicator/temperature probes, *Proceedings IEEE EMBS Meeting*, Piscataway, pp. 1454–1457.

Vora, N., Forell, B., Joseph, C., Lipsett, J. and Archambeau, J. (1982) Interstitial implant with interstitial hyperthermia, *Cancer*, **50**, 2518–2523.

Wu, A., Watson, M. L., Sternick, E. S., Bielawa, R. J. and Carr, K. C. (1987) Performance characteristics of a helical microwave interstitial antenna for local hyperthermia, *Medical Physics*, **14**(2), 235–237.

Zhang, Y., Dubal, N. V., Hambleton, R. T. and Joines, W. T. (1988) The determination of the electromagnetic field and SAR pattern of an interstitial applicator in a dissipative dielectric medium, *IEEE Transactions on Microwave Theory and Techniques*, **36**(10), 1438–1444.

Zhu, X. L. and Gandhi, O. P. (1988) Design of RF needle applicators for optimum SAR distributions in irregularly shaped tumours, *IEEE Transactions on Biomedical Engineering*, **35**(5), 382–388.

14 Principles of ultrasound used for generating localized hyperthermia

J. W. Hunt

14.1. Introduction

One difficult and demanding challenge in the field of hyperthermia is the necessity of developing sources that will produce acceptable temperature distributions throughout the treatment field and lower temperature outside this region. The importance of reaching the temperature target should not be overemphasized, since cold spots in the treatment field produced have the potential of producing a regrowth of the tumour. Thus, the clinical response requires either a flat profile distribution throughout the treatment field, or must allow much higher treatment in certain regions of the treatment field so that the cooler regions will be adequately treated.

An ultrasound beam has considerable physical advantages over most other heating agents. The wavelength is short and the beam edges can be better defined. In addition, its wavelength allows the beam to be effectively focused, so that beam intensity can be greater at the tumour site than at the point of contact on the patient's skin.

Table 14.1 compares the wavelength for various microwave and ultrasound beams that have similar penetrations. For example, a 1000 MHz microwave beam has a wavelength of 3.9 cm in muscle, while a 3 MHz ultrasound source has a wavelength of only 0.05 cm. Most ultrasound treatments have been carried out in the frequency range between 0.1 and 5 MHz; the lower frequencies having better penetration, while the higher frequencies have better focusing properties. No extra protection is needed for the patient or for the technical personnel, and there is no requirement for special electrical shielding. Furthermore, there is no limitation to applicator size or shape, so different fields can be generated. Highly focused beams can be rapidly scanned throughout the field to allow a better thermal dose distribution. This flexibility is of particular importance, especially when we want to consider ways of accounting for the different acoustic properties in the body and the efficient and variable

371

Table 14.1 Penetration of microwave and ultrasound into tissue for planar beams.

Tissue	Microwave						Ultrasound					
	100 MHz			915 MHz			0.5 MHz			3.0 MHz		
	d_{80} (cm)	d_{50} (cm)	λ (cm)	d_{80} (cm)	d_{50} (cm)	λ (cm)	d_{80} (cm)	d_{50} (cm)	λ (cm)	d_{80} (cm)	d_{50} (cm)	λ (cm)
Fat	6.7	43	106	1.6	12.8	13.7	4.5	30	0.3	1.2	7.9	0.05
Muscle	0.8	4.8	27	0.3	2.9	4.5	2–3	~19	0.3	~0.5	3.6	0.05
Bone	~7.0	~40	100	~1.6	13.0	~14	0.3	2.7	0.7	~0.02	~0.15	0.12

Penetration d is the depth at which the power drops to 80% (d_{80}) and 50% (d_{50}) of the power at the skin surface.
Wavelength λ assuming an average velocity in tissue of 1500 m/s.

cooling by the blood passing through different structures in the tumour and its neighbouring tissues.

Although these advantages noted above seem to indicate that ultrasound should be the technique of choice, there are two serious limitations to its use in specific regions of the body. Firstly, ultrasound cannot be transmitted into the body through gas, and so cannot be used in the lung or other regions of the body with interfaces containing gas. Secondly, serious problems are found near tissue to bone interfaces due to the poorer transmission and the high absorption of ultrasonic waves near the bone interface. This absorption produces high temperatures and severe patient pain. Therefore, ultrasound is usually limited to specific regions of the body where air and bone are not found along the beam's path.

14.1.1. Historical background

The use of therapeutic ultrasound spans over 45 years. The initial experience was one of considerable enthusiasm, then a period of pessimism, and now new revival. Kremkau (1979) has produced an excellent historical review of cancer therapy using therapeutic ultrasound.

The initial studies of ultrasound were made possible by the early experiments of Jaques and Pierre Curie (1880, reprinted 1973), who showed that when a force is produced across a piezoelectric crystal, electrical charges are produced. Later, they demonstrated the opposite effect; that electrical fields produce distortions of the crystal. Langevin in 1920, was the first to demonstrate the generation of ultrasound waves from a quartz plate (Biquard, 1972). The potential use of ultrasound heating was first suggested by Freundlich *et al.* (1932), but the first clinical demonstration of cancer control was by Pohlman *et al.* (1939) and Horvath (1944, 1948).

At the same time, several investigators in Japan, Germany and Russia (1933–1947) began to study the effects of ultrasound on tumours in mice and other animals. The serious clinical use of ultrasound began in 1944, in which Horvath (1944, 1948) and Woeber (1949, 1956) were the most active in the field. The most effective medical responses occurred when the ultrasound treatments were combined with radiotherapy. During the period between 1960 and 1975, some groups were continuing their experiments on the biological effects of ultrasound, but there appears to have been little activity concerning the therapeutic aspects of ultrasound.

A revival in the field was instigated by the careful experiments of Dewey *et al.* (1977) and Sapareto *et al.* (1978) who demonstrated the close relationship between cellular killing at different excess temperatures and treatment times. At the present time, most of the cellular killing by ultrasound appears linked to hyperthermia (Lele, 1979), although both cavitation and shear forces produced by these intense beams must be considered for special conditions (Dunn and Frizzel, 1982; NCRP Report 74, 1983).

14.2. Basic physics of ultrasound

14.2.1. Introduction to ultrasound

Ultrasound waves have the same fundamental properties as found at audible sound frequencies. The general principles of waves passing through media are described in Figure 14.1. These longitudinal wavefronts, produced by the cyclic compression or rarefaction of particles found at the surface, generate the acoustic waves which move away from this source.

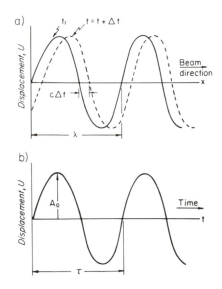

Figure 14.1 Curves describing sinusoidal waves: (a) moving through the medium for times t_1 and t_2; (b) the amplitude of the disturbance in time at a particular position in the field.

In Figure 14.1(a) the instantaneous spatial distribution is shown for a wave propagating in a medium. The motion in the x-direction is the product of the temporal separation Δt and the velocity c of the wave moving in the medium (i.e. $c\Delta t$). Figure 14.1(b) illustrates how the amplitude of the wave changes sinusoidally in time at a specific position. The period of the wave τ is the inverse of frequency ν. The distance between crests of the wave in Figure 14.1(a) is the wavelength λ, as defined by the relationship

$$\lambda = c\tau = c/\nu. \tag{14.1}$$

Thus, this sinusoidal wave travelling in the x-direction is expressed by the equation

$$u = u_0 \sin \omega (t - x/c), \tag{14.2}$$

where u_0 is related to the maximum amplitude of the disturbance at a specific time and ω is the angular frequency of the wave (i.e. $\omega = 2\pi\nu$). This equation neglects attenuation. The velocity of the wavefront depends upon the density ρ_0 of the undisturbed medium. The bulk modulus K_{ad} under adiabatic conditions is related to the strain of the sample under a specific stress. The velocity can be calculated from the simple relationship

$$c = \sqrt{(K_{ad}/\rho_0)}. \tag{14.3}$$

The value of c is independent of frequencies for a small disturbance. Note, however, that this value of c is not valid for large wave amplitudes where Hooke's law breaks down and the wavefronts are distorted (i.e. shock waves are formed).

14.2.2 Propagation of ultrasound beams

The formulations used for the propagation of acoustic waves are based on the well-known wave equations (Wells, 1977). These equations show that a wave with amplitude u_0 moving through a medium (see equation (14.2)) can be described by a sinusoidal wave with a fixed phase shift,

$$u = u_0 \sin (\omega t - \phi), \tag{14.4}$$

where $\phi = kx$ and $k = 2\pi/\lambda$. The velocity of the particles is calculated from the differential of equation (14.4)

$$v = \frac{du}{dt} = u_0 \omega \cos (\omega t - \phi). \tag{14.5}$$

From the formulation of the wave equation, the instantaneous particle pressure is given by

$$p = \rho_0 c\, v = \rho_0\, c\, u_0\, \omega\, (\cos (\omega t - \phi)). \tag{14.6}$$

14.2.3. Acoustic properties

Understanding the acoustic properties is crucial when making decisions for the most-effective hyperthermia sources to be used for a specific site of the body. There are three general properties of ultrasound that must be considered before using an ultrasound source: attenuation and absorption of the beam that produces beam penetration and local heating, the velocity differences at interfaces that produce beam steering, and the impedance

Table 14.2 **Acoustic properties in selected materials.**

Material	Density ρ_0	Velocity c	Characteristic impedance	Absorption coefficients at approx. 1 MHz (dB/cm)	Approximate frequency dependency of α
	$(10^3 \, \text{kg/m}^3)$	(m/s)	$(10^6 \, \text{Kg/m}^2/\text{s})$		
(a) Non-biological materials					
Water	1.0	1480 (~20°C)	1.5	0.0022	ν^2
Air	0.0012	330	0.0004	12	ν^2
Castor oil	0.95	1500	1.4	0.95	ν
Mercury	13.6	1450	20	0.0005	ν^2
Aluminium	2.7	6400	17	0.018	ν
Polymethylmethacrylate	1.19	2680	3.2	2.0	ν
(b) Biological samples					
Muscle	1.04	1532	1.62	1.2	ν
Fat	.92	1478	1.36	0.61	ν
Blood	1.04	1556	1.63	0.12	$\nu^{1.5}$
Bone	1.82	3445	6.27	13.9	ν^2

$\alpha \simeq a\nu^b$ where a is the attenuation parameter and b the frequency dependency.

mismatch that can produce serious reflection. The acoustic properties for selected non-biological and biological samples are summarized in Table 14.2.

14.2.3(a) Attenuation and absorption

When considering a plane pressure wave passing through an attenuating medium, the equation becomes

$$p = p_0 \, e^{-\alpha x} \cos \omega \, (t - x/c), \tag{14.7}$$

where α (nepers/cm) is the amplitude attenuation coefficient per unit path length at the observation point, and p_0 and p are the amplitude of the pressure wave at the surface and at distance of x into the medium respectively. The amplitude of the wave decreases exponentially as it propagates into the medium. This coefficient consists of two parts, $\alpha = \alpha_{scatt} + \alpha_{abs}$, where α_{scatt} is the scattering coefficient, while α_{abs} is the true absorption coefficient. The latter coefficient is usually assumed to have the largest value, although accurate values of α_{abs} are often difficult to obtain (Goss *et al.*, 1979).

14.2.3(b) Intensity

The intensity of the beam, for most practical aspects, is related to the square of the particle pressure. This can be calculated from the total mass of particles with a maximum velocity v per unit volume. The total energy density E_0 is therefore given by

$$E_0 = \rho_0 v_0^2/2 \, (\text{J/m}^3) \tag{14.8}$$

The energy travels through the medium with the wave velocity c, thus the intensity I_0 is determined by

$$I_0 = c \, E_0, \tag{14.9}$$

or

$$I_0 = \rho_0 \, c \, v_0^2/2. \tag{14.10}$$

In the SI system, I_0 has units of W/m². For practical and historical reasons, however, most of the published data are in units of W/cm² or mW/cm².

Table 14.3 shows the parameters found for a plane wave with an intensity I propagating through water. The absorption losses are the

Table 14.3 Field parameters for plane waves in water: values are for an intensity of 1 W/cm² at 1 MHz.

Parameter	Value	Dependence
Wavelength λ	0.15 cm	ν^{-1}
Velocity, c	1500 m/s	
Peak particle displacement	1.8×10^{-8} m	$I^{1/2}, \nu^{-1}$
Peak particle velocity, v	0.122 m/s	$I^{1/2}$
Peak particle pressure, p_0	1.8 atm [$\simeq 1.8 \times 10^5$ Pa (N/m²)]	$I^{1/2}$
Radiation pressure, F	0.69 Kg/m²	I
Heat equivalent (Joule's constant)	0.24 cal/s/cm²	I

Data from Wells (1969).

square of the value of the pressure amplitude for the absorption factor of the attenuation α_{abs}. Thus, the reduction of the intensity is calculated from 2α, as given by

$$I = I_0\, e^{-2\alpha x}, \tag{14.11}$$

while the deposition rate of the energy D in depth of the body is given by

$$D = (dI/dx)_{abs} = 2\alpha_{abs}\, e^{-2\alpha x}. \tag{14.12}$$

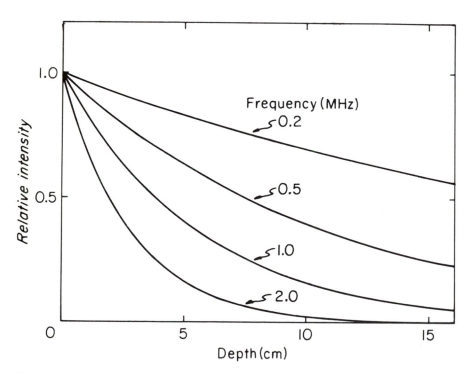

Figure 14.2 The attenuation of beam intensities for various frequencies from a plane wavefront, at an attenuation of 0.8 dB/cm-MHz.

Typical depth–dose curves for typical attenuation coefficients are shown in Figure 14.2. Thus, for superficial hyperthermia, frequencies of 2–1 MHz are used, while for deeper heating, frequencies of approximately 0.5 MHz are used. A general 'rule of thumb' is that at least 50% of the energy reaches the tumour.

The absorption is the conversion of the acoustic energy to mechanical energy: this is a type of friction, in which the motion of the particles is damped by the viscous forces in the medium. The heating in the tissues is, therefore, directly related to this absorption coefficient α_{abs}.

14.2.3(c) Radiation pressure

The acoustic waves also produce steady forces at any interfaces. The forces are related directly to the incident energy reflected at this interface. For the example in which a planar beam is absorbed perfectly on a test object, the radiation pressure is expressed by

$$F = W/c, \tag{14.13}$$

where F is the force pressing against the object in the direction of the propagation of the wave and W is the ultrasonic power (watts). For a perfect reflector, the force is twice as great. By careful design, this force can be measured accurately, and thus this is often used to calibrate the beam power. These aspects will be discussed in more detail in a later section.

14.2.4. Reflection and refraction of acoustic beams

When a wave encounters interfaces between two media, a certain part of the energy is reflected back. In addition, if the beam passes through from one medium to another, in which the acoustic velocities are different, the beam can be steered by refraction. Both concepts are similar to those found for optical waves. Reflection and refraction are crucial in understanding the strengths and weaknesses of ultrasound heating.

14.2.4(a) Acoustic impedance

The amplitude of these reflections from the interface between two media is linked to their acoustic characteristic impedance Z. This is calculated by

$$Z = p/v, \tag{14.14}$$

where v and p are defined in equation (14.5) and (14.6). In the case of planar

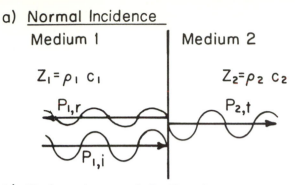

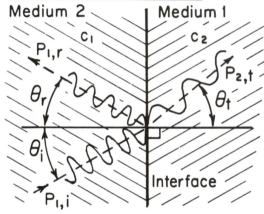

Figure 14.3 (a) Reflection at normal incidence between media with different acoustic characteristic impedances. (b) Refraction (Snell's Law) and reflection between two non-viscous media.

waves in a non-absorbing medium, mechanical or acoustic impedance is pure resistance, thus

$$Z = \rho_0 c. \tag{14.15}$$

Reflection at normal incidence is shown in Figure 14.3(a) with the waves of $P_{1,r}$ and $P_{2,t}$ going forward and backward from the interface. The ratio of the reflected and incident wave amplitude of the pressure at normal incidence R is given by

$$R = \frac{Z_2 - Z_1}{Z_2 + Z_1}, \tag{14.16}$$

while the ratio of the amplitude of the transmitted wave divided by the amplitude of the incident wave T is given by

$$T = 2\frac{Z_2}{Z_2 + Z_1}.$$ (14.17)

Similarly, the reflected and transmitted beam intensities, R_I and T_I, respectively, are determined by

$$R_I \simeq R^2 = [(Z_2 - Z_1)/(Z_2 + Z_1)]^2,$$ (14.18)

and

$$T_I = (4 Z_2 Z_1)/(Z_2 + Z_1)^2.$$ (14.19)

Table 14.2 summarizes selected values of the characteristic acoustic impedance for non-biological and biological materials. There is little energy reflection from water to fat or muscle interfaces, thus excellent transmission into the body can be achieved in most soft tissues in the body. There is almost total reflection of the acoustic energy at water to air interfaces, so the transmission through the lung is small. Also, large reflections can be predicted from the tissue to bone interfaces, as calculated from the values summarized in Table 14.2 and equation (14.17)

14.2.4(b) *Refraction of acoustic beams*

When an acoustic beam passes at an acute angle from one medium to a second with a different velocity, the beam is refracted and reflected. A simple example of this is shown in Figure 14.3(b) with the waves going forward and backward from the interface. The angle of the reflected wave θ_i, determined from the angle normal to the interface, has the same angle as the incident wave θ_i. The direction of the transmitted wave θ_t is calculated from Snell's law:

$$\sin \theta_i / \sin \theta_t = c_1/c_2.$$ (14.20)

In this particular example, if the acoustic velocity in medium 2 is greater than in medium 1, the transmitted beam is refracted closer to the interface. If the value of θ_i, or c_2 is increased until the value of θ_t reaches an angle of 90°, then perfect reflection occurs. The critical angle is given by $\theta_1 = \sin^{-1}(c_1/c_2)$. While total reflection is expected from a tissue to bone interface in the beam, it is also important to note that total reflection may occur in soft tissues. For example, if the beam encounters a fat ($c_1 = 1480$ m/s) to muscle ($c = 1590$ m/s) interface, perfect reflection occurs if the incident angle θ_1 is $>67°$. This phenomenon could greatly alter the temperature distribution. Also, the refraction by fat layers can significantly change the beam profile.

From the simple wave theory summarized by Wells (1977), the intensity of the reflected I_r/I_i and refracted I_t/I_i beams at the interface between medium 1 and medium 2, respectively, are given by

$$\frac{I_r}{I_i} = \frac{Z_2 \cos \theta_i - Z_1 \cos \theta_t}{Z_2 \cos \theta_i + Z_1 \cos \theta_t} \tag{14.21}$$

and

$$\frac{I_t}{I_i} = \frac{4 Z_2 Z_1 \cos \theta_i}{(Z_2 \cos \theta_i + Z_1 \cos \theta_t)^2}. \tag{14.22}$$

From the data in Table 14.2 and from equation (14.21), we can see that for most soft tissues, the transmission is excellent. Even for bone at normal incidence, about 60% of the acoustic energy still passes through the interface.

At this point in this chapter, the analysis only describes the compression or longitudinal waves found in liquids and non-viscous media. In general, this approximation appears to be reasonable for most soft tissues. When stiffer tissues such as bone are considered, the effects of shear waves become important. If the incident longitudinal wave is at an angle $\theta_i > 0$ at the interface, the waves can be split into two components through mode conversion: (i) longitudinal and (ii) shear waves, which have a velocity about half that of the velocity for longitudinal waves and have very high absorption coefficients. These shear waves are of considerable concern in evaluating the excess heating at bone interfaces (Dunn and Frizzell, 1982).

14.3. Beam profiles

In the previous sections, the beam was assumed to consist of a plane wavefront that moves in the x direction. This is only correct for a beam at a great distance away from the source. For hyperthermia studies, however, large ultrasound sources tightly coupled to the body are used. Thus, complex diffraction patterns are usually found in practice. In the design of transducers used for hyperthermia studies, the ability to predict the beam characteristics is of crucial importance so that properties of the actual beams can be evaluated.

In contrast to traditional planar sources used extensively for local surface heaters, new systems use strongly focusing transducers that allow better temperature distributions at depth in the body. This section will review the general methods for calculating the beam profiles for a disc-shaped transducer, a spherically shaped transducer and, finally, the theoretical assessment of more novel ultrasound systems which have been tested in different laboratories.

14.3.1. Diffraction theories

Many factors affect the calculation of the transmitted beam profiles. The more important ones are source geometry, type of excitation (for most studies, continuous wave modes are used), non-linear aspects of the beam, etc. Most of the theoretical treatments described in the literature are based on Huygen's principle, in which the radiating source is divided into infinitesimally small elements, with each element radiating hemispherical wavelets. In general, each element must have a size much smaller than wavelength of the acoustic beam. The three-dimensional acoustic fields can then be calculated from the evaluation of a diffraction integral established by Rayleigh (Strutt, 1945).

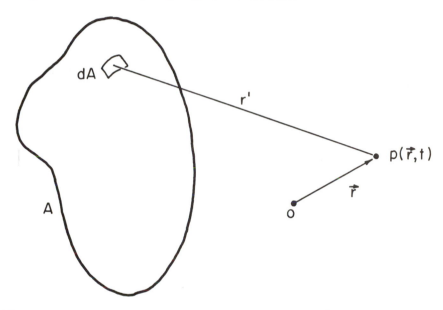

Figure 14.4 Geometry for calculating the pressure wave, $p(\vec{r}, t)$ at a distance r generated from a small element dA from the piezoelectric source of area A.

There are many ways of calculating the ultrasound distribution. The particular analysis described here can be used for both pulsed and continuous waves. The pressure $p(\vec{r}, t)$ at a distance r' from a point $\vec{r}$ relative to an arbitrary origin 0 for a generalized source is shown in Figure 14.4 (for greater detail see Arditi *et al.*, 1981; Hunt *et al.*, 1983). The acoustic field of most ultrasound imaging systems obeys the classical theory of linear acoustics. With Rayleigh's formulation (Strutt, 1945), the value of $p(\vec{r}, t)$ in the field of an ultrasonic radiator is given by

$$p(\vec{r},\ t) = -\rho_0 \frac{\partial \psi}{\partial t}(\vec{r}, t),$$

(14.23)

where ρ_0 is the density of the medium and $\psi(\vec{r}, t)$ is the velocity potential. Using the velocity of sound c, and the parameters given in Table 14.2, the velocity potential of a uniformly excited source of surface area A is given by

$$\psi(\vec{r}, t) = \frac{1}{2\pi} \int_A \frac{v_0\,(t - r'/c)}{r'}\, dA, \tag{14.24}$$

where $v_0(t)$ is the instantaneous normal particle velocity at the face of the source as described in equations (14.5) and (14.6). This analysis assumes the wave propagation is for a linear, non-attenuating, homogeneous medium, and if the source is not flat, the curvature of the transducer is much smaller than the wavelengths.

14.3.1(a) Plane disc aperture

The acoustic field of plane pistons has been studied in great detail by various investigators (Hueter and Bolt, 1955; Kinsler and Frey, 1962; Wells, 1977; Zemanek, 1971). As explained by Zemanek (1971), there are two regions (Figure 14.5). The first region is called the near field (Fresnel

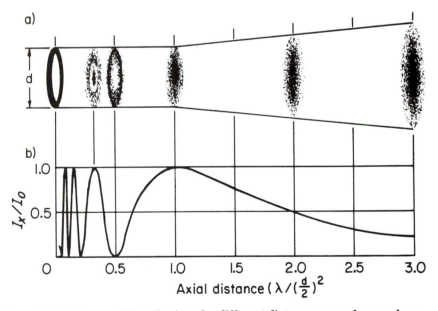

Figure 14.5 Ultrasound distributions for different distances away from a plane disc, (a) three-dimensional distributions of the field and (b) the intensities along the beam axis. The normalized value 1.0 defined in the text is the transition between the Fresnel and Fraunhofer zones.

region) where the acoustic pressure distribution shows complex interference patterns in which the phase differences from different parts from the source are large enough for reinforcement or destruction interference to occur. The second region is where the phase differences are small (Fraunhofer region). Here, the beam has no interference patterns, and the beam slowly expands at depth. Thus, for many experiments, the Fraunhofer region can be approximated to a plane wavefront.

The common transducer is in the form of a disc which behaves like an expanding piston that generates a beam in the direction normal to the plane surface. From derivations based on Rayleigh's equation in three-dimensional analysis noted earlier, two distinct beam zones are observed. From Huygen's principle, the hemispherical wavelets from each of the small elements dA (see Figure 14.4) make up the disc. The phases generated by the wavelets at point $p(\vec{r}, t)$ are determined from the wave equation (14.2) and pressure amplitudes related to the distance from the element, $1/r'$. Thus a series of constructive and destructive patterns (Fresnel zones) are found near the disc. Figure 14.5(a) depicts the predicted constructive and destructive pressure intensities along the beam axis. To a first approximation, each Fresnel zone has the same area.

Along the central axis of the beam, the relative intensity is given by

$$\frac{I_x}{I_0} = \sin^2 \frac{\pi}{\lambda} \{[(d/2)^2 + x^2]^{1/2} - x\}, \tag{14.25}$$

where I_0 is the maximum intensity, I_x is the intensity at a distance x from the transducer, λ is the wavelength in the propagating medium and d is the diameter of the disc.

Figure 14.5(b) describes the central intensity distribution. The scale on the ordinate was normalized so that the value at the transition between the Fresnel or near zone and the Fraunhofer or far zone was unity (beyond this point, no rapidly changing diffraction distributions are found). From the solution of equation (14.25), the maximum along the axis is given by

$$x_{max} = \frac{d^2 - \lambda^2 (2m+1)^2}{4\lambda (2m+1)} \tag{14.26}$$

and the minimum by

$$x_{min} = [(d/2)^2 - \lambda^2 n^2]/2n\lambda, \tag{14.27}$$

in which $m=0, 1, 2, \ldots$ and $n=1, 2, 3, \ldots$ Thus, from equation (14.26), if $d^2 >> \lambda^2$,

$$x'_{max} = d^2/4\lambda. \tag{14.28}$$

where x'_{max} is the last axial maximum (i.e. $n=1$), then, in the Fraunhofer zone, where $x > > x'_{max}$, then the intensity follows the inverse square law, i.e. proportional to $1/x^2$. For uniform excitation over the disc, the pressure amplitude by Kinsler and Frey (1962), is described by

$$p(x) = \frac{\rho_0}{x} \frac{J_1(\frac{\pi d}{\lambda} \sin \phi)}{\frac{\pi d}{\lambda} \sin \phi},$$ (14.29)

where J is the Bessel function of the first kind. From this equation, $\phi = \sin^{-1}(1.22\lambda/d)$, is the diverging angle of the beam.

14.3.1(b) More complex source geometries

Many solutions have been developed to determine the three-dimensional acoustic fields from different transducer designs such as the plane circular piston (Oberhettinger, 1961), the plane rectangular piston (Lockwood and Willette, 1973), the spherically focused piston (Penttinen and Luukkala, 1976) and a thin annulus (Arditi *et al.*, 1981). From the Rayleigh formula described in equations (14.23) and (14.24), the velocity potential, $\psi(\vec{r}, t)$ is separated by two parts,

$$\psi(\vec{r}, t) = v_0(t) * h(\vec{r}, t),$$ (14.30)

where $h(\vec{r}, t)$ is the impulse response of the transducer and $v(t)$ is as defined in equation (14.5). Thus, the two functions enable us to separate the two components: the vibrating transducer's surface velocity v_0, (i.e. related to the excitation voltage), and $h(\vec{r}, t)$ the impulse response that describes the geometric factors of the transducer (i.e. related to the time for each wavelet to reach the observation point). The function is the convolution product. The impulse response is defined by $h(\vec{r}, t)$

$$h(\vec{r}, t) = \frac{1}{2\pi} \int_A \frac{\delta(t - r'/c)}{r'} \, dA,$$ (14.31)

where δ is the Dirac impulse function, A is the transducer's surface, and the other variables are as defined in Figure 14.4. and equations (14.23) and (14.24). Combining equations (14.23) and (14.30), the instantaneous pressure distribution for a particular excitation function can be calculated by

$$p(\vec{r}, t) = \frac{-\rho_0 \partial v_0(t)}{\partial t} * h(\vec{r}, t).$$ (14.32)

Thus, the beam profiles may be evaluated by the specific transducer shape at different points in the field. Analytical solutions are available to calculate the predicted beam distribution for simple transducer geometries.

However, new transducer designs are often complex, and so a numerical computation is best. Powerful programs were developed by Arditi *et al.* (1981) at the University of Toronto. Based on these programs, more-complex source shapes, such as multi-element annular arrays (that enable one to generate a variable focal length), and conically shaped transducers (that project a line focus onto the body), can be calculated by the super-position of the standard solutions or by the numerical computation tech-nique. For any educational or medical groups, these programs are available at nominal cost from our laboratory. Some of these computer simulations described in later parts of the paper are based on these programs.

14.3.1(c) Sinusoidal excitation

In the continuous wave mode, the excitation velocity $v_0(t)$ can be expressed by the Euler formula in the complex form

$$v(t)=u_0\,e^{j\omega_0 t}=u_0\,(\cos \omega_0 t+\sin \omega_0 t), \tag{14.33}$$

where ω_0 is the driving angular frequency and u_0 is the velocity amplitude of the face of the source. From equations (14.31) and (14.32), the instant-aneous pressure becomes

$$p(\vec{r}, t)=j\rho_0 u_0 \omega_0\,e^{j\omega_0 t} \int_{t_1}^{t_2} e^{-j\omega_0 t'} h\,(\vec{r}, t')\,\mathrm{d}t'. \tag{14.34}$$

Details of the analytical solutions of the Fourier transform of the impulse response are covered by Arditi *et al.* (1981) and Hunt *et al.* (1983).

14.3.1(d) Spherically shaped aperture

At the focal plane of a spherically focused aperture of diameter d and radius R of the curved transducer (i.e. the focal length), the off-axis pressure amplitude approximate to the Airy distribution,

$$p(\xi)=\rho_0\,c\,\frac{u_0}{\lambda}\,\frac{\pi d^2}{4R}\,\frac{2J(\xi)}{\xi}, \tag{14.35}$$

where λ is the wavelength, J is the first-order Bessel function, and the distribution is normalized by the factor $\xi=\rho'/(\lambda R/\pi d)$ with ρ' equal to the radial distance off-axis.

Figure 14.6 summarizes the focusing properties of a typical transducer. The full-width, half-maximum (FWHM) of the lateral amplitude profile may be derived from equation (14.32), and yields a value of

$$\mathrm{FWHM}=1.41\lambda\,(f\mathrm{number}), \tag{14.36}$$

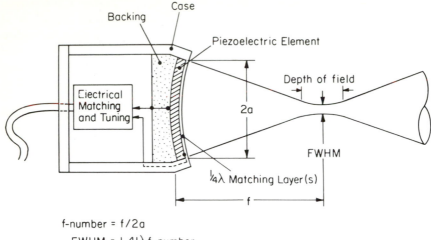

f-number = f/2a
FWHM = 1.41λ f-number

Figure 14.6 The important properties of a spherically shaped focused transducer. In most power sources, the transducers are air-backed, and often have a 1/4λ matching layer for maximum acoustic efficiency.

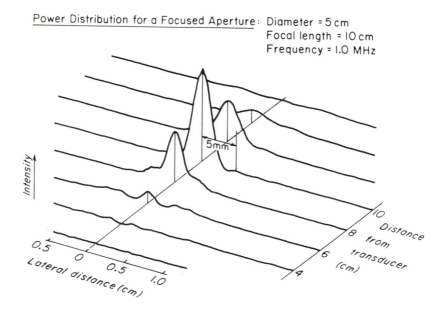

Power Distribution for a Focused Aperture: Diameter = 5 cm
Focal length = 10 cm
Frequency = 1.0 MHz

Figure 14.7 The theoretical intensity profiles for a focused aperture with an *f* number of 2.0.

where f number$=f/d=f/2a$ (i.e. the focal length divided by the aperture diameter). For strongly focused apertures, the depth of field (DOF) of the beam intensity distribution along the axial direction is given by (Kossoff, 1979)

$$DOF = 7.1\,\lambda\,(f\text{number})^2\,(cm). \qquad (14.37)$$

Thus, for a 5.0 cm diameter aperture with a focal length of 10 cm and a frequency of 1.0 MHz in water, the FWHM$=0.4$ mm and the DOF$=4.2$ cm: a cigar-shaped hot spot is produced in an absorbing medium.

Figure 14.7 shows the intensity profiles closer and farther away from the transducer. The off-axis components are not large but still must be considered for any analysis when the beam is scanned across the field.

14.4. Practical ultrasound sources

14.4.1. Transducer materials

The ultrasound beams used for hyperthermia are usually generated by piezoelectric effects. If one compresses the piezoelectric material, an electrical potential can be detected across the sample. In the same way, if an electric voltage is passed through to opposite sides of the material, the sample is distorted. Many crystals exhibit the piezoelectric effect but, for practical reasons, ferroelectric ceramics and special plastic films are now the most common materials.

The principle of the piezoelectric effect for a quartz plate is shown in Figure 14.8. The conductive plates are placed across the crystal with the

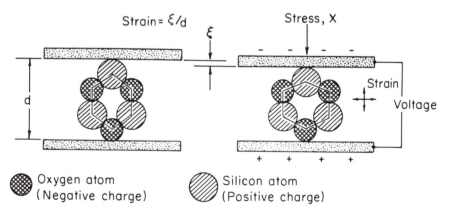

Figure 14.8 The principle of the piezoelectric effect used to generate ultrasound beams for a quartz crystal in which the stress changes the position of the negatively charged oxygen atom and the positively charged silicon atom, thus producing an electrical voltage across them.

specific orientation to the crystal lattice shown here, so that unbalanced charges from the silicon (+) and oxygen (−) molecules are found on opposite sides of the highly polar crystalline structure. If a voltage is passed across this plate with the polarization shown in Figure 14.8, there is a change of crystal thickness ξ. When this crystal is coupled to a medium, a pressure wave moves away from this surface.

Although there is a wide range of piezoelectric materials available which can be used to generate ultrasound waves, there are few materials that are satisfactory for high-powered sources. The original crystal plates were often of quartz or other suitable crystals, but these were often difficult to produce or were expensive for large transducers. Thus, most of the transducers today are made from the class of piezoelectric ceramics which have ferroelectric properties. In particular, the family consists of a sintered mixture made of lead zirconate–titanate ($PbZrO_3PbTiO_3$) (often called 'PZT') with small amounts of impurities added to improve the properties. When these microcrystals are subjected to an external, very high electric field at an elevated temperature (100—200°C), the domains tend to line up along the direction of the field. Thus, the domains are polarized or 'poled' so that excellent piezoelectric properties are obtained. Also, the Curie point or transition temperature (the temperature at which the piezoelectric properties are lost) is high for these ceramics. Thus, most systems use piezoelectric ceramics to generate high-power ultrasound beams.

The properties of some common piezoelectric materials used for power sources are shown in Table 14.4. Certain properties should be noted. The transducers must have good electrical-to-mechanical conversion of the energy in the transducer, as described by the piezoelectric factor k_{33} for the thickness mode as summarized in Table 14.4. Also, mechanical losses in the transducer should be low, so unwanted heating in the transducer will be as low as possible. Sometimes air cooling is needed to obtain maximum output. Under optimum conditions, the overall energy efficiency will be very high, i.e. 99% for quartz and approximately 80% for ferroelectric ceramics (Berlincourt, 1971), if the mechanical Q factors are high (i.e. low losses in the material). Most of the materials have extremely high dielectric constants (i.e. high permittivities). At the transducer's resonance, the electrical reactive component should be much less than the resistive component. Also, a good electrical impedance match must be obtained between the radio-frequency driving generator and the transducer. However, a high dielectric constant is often a disadvantage for large transducers since the electrical impedance of these are low. Thus, special electrical circuits are often needed to match the radio-frequency generator (usual 50 ohms) with the impedance of the transducer.

The most common transducer material for high-power ultrasound generation is the piezoelectric ceramic PZT4, or its equivalent, although other materials are sometimes used for special needs.

Table 14.4 Selected properties of piezoelectric materials.

Property	Parameters	Units	Quartz	Piezoelectric material	
				PZT4[1]	PC–11[2]
Velocity of sound	c_T	m/s	5970	4600	5400
Density	ρ	10^3 kg/m³	2.20	7.5	7.6
Acoustic impedance	Z_0	10^6 Rayleigh	12.1	34.5	41.0
Dielectric constant	$\epsilon_r = \epsilon/\epsilon_0*$	Unitless	3.75 at 20°C	1300	170
Piezoelectric constants	h_{33}	10^8 V/m		26.8	—
	e_{33}	C/m²		15.1	—
	d_{33}	10^{-2} C/N		289	52.8
	g_{33}	10^{-3} Vm/N		25.1	35.1
Piezoelectric coupling factor	k_{33}	Unitless	0.10	0.7	0.5
	$k_{31}{}^1$				0.03
(Mechanical losses)$^{-1}$	Q	Unitless	>25 000	>500	1500
Curie point	T	°C	573	328	355

ϵ_0 = Permitivity of free space $(8.854 \times 10^{-12}$ F/m).
[1] PZT4, lead zirconate–titanate ferroelectric ceramic.
[2] PC–11, similar properties to lead titanate (PT) and manufactured by Hitachi Metals Ltd, Tokyo, Japan.

Recently, a newer type of material called PT (lead titanate, $PbTiO_3$), which has a very low value of k_{31} (radial modes) and low electrical losses, shows considerable promise. This is particularly useful for arrays, in which cross-talk from element to element is greatly reduced.

Because of the high velocities and high density of the transducers, there is also a large acoustic impedance mismatch between the transducer and the coupling medium (usually water). Since most of the energy reverberates inside the transducer, most transducers are air-backed or of a low acoustic loss material so that the maximum output efficiency can be achieved.

14.4.2. Basic transducer design

Transducers have a fundamental resonance given by

$$\nu_0 = c_t/2d_t, \tag{14.38}$$

where c_t is the acoustic velocity in the transducer across the transducer material and d_t is the thickness of the transducer's electrodes. In addition, many harmonics can be excited near the resonance as defined by:

$$\nu = n_h c_t/2d_t \tag{14.39}$$

for $n_h = 1, 3, 5, 7 \ldots$. Different frequencies can be obtained from a single transducer, although the electrical impedance will have considerable differences for these harmonics.

14.4.2(a) Transducer frequency response

The electrical impedance response of a typical transducer can be modelled by the various equivalent circuits first proposed by Mason (1948, 1950). Near its resonance, the response of the transducer can be approximated by the improved equivalent circuit shown in Figure 14.9(a), developed by Krimholz *et al.* (1970) and with an accurate assessment of the frequency response. Note that there are two resonances: one is the parallel resonance whose impedance is large at a frequency slightly greater than the fundamental frequency defined by equation (14.38), while the second is the series resonance whose value of the impedance is small and with a frequency slightly smaller than the fundamental frequency.

It is of considerable importance to design the system such that there is an excellent match between the radio-frequency power source and the transducer at the resonance. This is closely linked to the electrical properties shown in Figure 14.9(b). In general, the phase shift between the

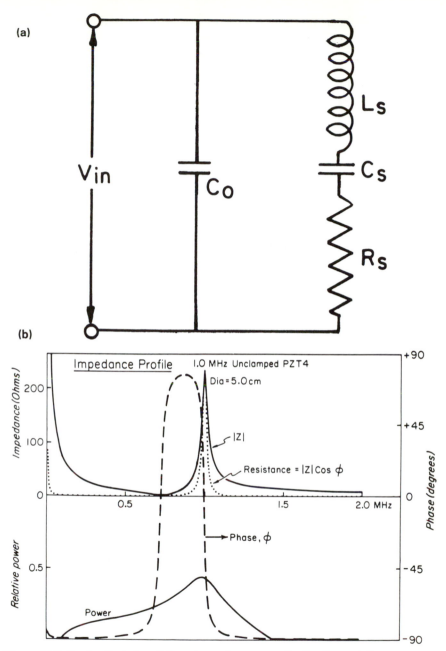

Figure 14.9 Equivalent circuit for a transducer near the fundamental resonance, consisting of series inductive L_s, resistive R_s and capacitive C_s components in parallel with the bulk capacitive C_0 element of the transducer. (b) Impedance profiles for a PZT4 transducer with a diameter of 5 cm and a resonance frequency of 10 MHz for various frequencies. The solid line is the resultant impedance $|z|$, the dotted line is the resistivity component related to actual acoustic output and the dashed line represents rapid phase changes. The output from a 50 ohm source is shown.

excitation and the transducer should be as small as possible and the resistance component of the transducer should match that of the source to obtain the maximum transfer of energy from the source. Thus, different quarter-wave layers with intermediate impedances (Desilets *et al.*, 1978) and/or passive electrical matching circuits are used to optimize the power output (Augustine and Andersen, 1979).

14.2(b) Basic transducer design

Most high-power transducers are based on the general design shown in Figure 14.6. The transducer is usually air-backed, without any matching layers between the transducer and the coupling medium. Some recent designs used for hyperthermia use quarter-wave layers (Umemura and Cain, 1987) to improve the acoustic performance.

Another problem is the type and quality of the piezoelectric material. Due to the narrow bandpass nature of air-backed transducers, these are prone for phase shifts in localized spatial regions. These are found particularly in the large piezoelectric sources where production of irregular profiles across the source are often observed. When the transducer consists of multiple elements, interactions due to radial modes can also yield serious cross-talk from element to element, which can only be eliminated by isolating the elements acoustically and electrically, or by using special piezoelectric ceramics such as PT (see Table 14.4) in which the radial modes are greatly reduced.

Ultrasound beams can be focused effectively, and various transducer designs have been developed to make use of this special property. The common focused source is the spherically shaped aperture shown in Figure 14.6 or, for practical reasons, a planar disc combined with a plastic lens made of polystyrene or methacrylate. The predicted distribution of the acoustic profiles produced by a 5 cm diameter, 1 MHz source with a focal length of 10 cm is shown in Figure 14.7. At the focus, a single peak with small side-lobes is observed, while closer or farther from the focus, complex diffraction patterns are found.

Because the lateral FWHM of the beam at the focus is small (in this example FWHM=4.2 mm), the beam is usually scanned throughout the irradiated field to produce an improved temperature distribution. More-complex geometries are used to shape the beam distribution, phased arrays are used to adjust the beam positions, and special lenses and acoustic mirrors are used to produce unusual beam properties. These will be described in more detail in a later section of the Chapter.

14.4.3. Ultrasound measurements

In using any ultrasonic hyperthermia system, several tests are needed to evaluate its performance. These include transducer resonances, beam profiles and power output for different frequencies.

14.4.3(a) Transducer impedance

In Section 14.4.2(a), the theoretical response of transducer versus frequency was described. The actual response is often different: some of the differences are linked to unwanted resonances in the transducer. For example, the unwanted radial waves generated in transducers often distort the distribution of the beam (these are not considered in the present theory). These are observed in the impedance versus frequency profile. The most convenient way of reading the impedance profiles is to use commercial instruments, such as a 'vector impedance meter', although if such an instrument is unavailable, simpler systems can be used. For example, if one excites the transducer with a swept-frequency generator, in series with a high-impedance resistor (=10 000 ohms), the observed output across the transducer can be observed by an oscilloscope, giving insight into the impedance and phase to assess the resonance performance.

14.4.3.(b) Detection of beam profiles

Two general methods are used for observing the beam profiles: piezoelectric hydrophones and transient thermosensor probes.

Piezoelectric hydrophones are used to test the properties of the beam by transforming the mechanical acoustic amplitude of the beam into an electrical signal. Piezoelectric sensors of a ceramic material, such as PZT5, have been used for many years. These are very sensitive but have a narrow frequency response.

More recently, hydrophones using a special piezoelectric plastic film called polyvinylidene difluoride (PVDF or $PVDF_2$), has been used because of its superior properties such as its wide frequency response (Swartz and Plummer, 1980), its linear response up to very high powers (Meeks and Ting, 1983) and its ability to be fabricated with sensitive regions as small as 0.5 mm. Some aspects of the PVDF material are summarized by Hunt *et al.* (1983). Unfortunately, the signal from the PVDF hydrophone is smaller than from one using PZT5, hence a preamplifier is often placed close to the piezoelectric receiver to produce an adequate signal.

The hydrophone signals detect the amplitude of the pressure wave so that the signal is usually squared to calculate the output (there are errors found in this approach near rapidly changing fields (Hutchins and Archer-Hall, 1983). It appears that the stability of the PVDF detector is good, so standard laboratories in some countries are able to check these outputs. Membrane hydrophones developed by Bacon (1982), and a commercial

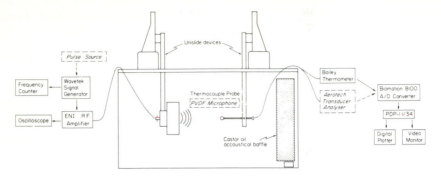

Figure 14.10 A scanning thermometry systems developed by Munro *et al.* (1982) to observe the acoustic intensities across the beam.

array of sensors available by GEC-Marconi, Chelmsford, UK, allow one to rapidly obtain details of the beam profiles.

Another convenient way of checking the beam profiles is to use a transient thermosensor probe. A small, highly absorbing bead is moulded on the junction of a thin thermocouple. The acoustic beam is absorbed on this bead and an equilibrium is quickly reached between the beam absorption and the cooling water bath. This system measures the beam intensity. Such a system is shown in Figure 14.10 (Munro *et al.*, 1982) in which an automatic scanner is used to obtain intensity profiles of the beam. Unfortunately, this technique is not particularly useful for obtaining accurate values of the absolute intensity because the beam absorption (which is related to temperature) is dependent on frequency and bead diameter. A more-detailed analysis for different absorbers and temperature–time responses in ultrasound fields is given by Martin and Law (1980, 1983).

A more-sensitive test of the transducer quality is to use a small hydrophone, such as a PVDF probe, to measure phase changes across the transducer. The larger transducers show peaks and valleys in the beam since these are more sensitive to small imperfections of the transducer flatness, or inhomogeneities in the transducer material. One test is to scan across the transducer with a small hydrophone and, using an oscilloscope, compare phases of the excitation and received signals. Because of the narrow pass band which is characteristic of an air-backed transducer, and large phase changes, serious irregular distributions are observed in SAR contour maps. Better quality control, quarter-wave layers that broaden the frequency pass band, and frequency modulation will all improve the beam profiles.

14.4.3(c) Radiation force power meters

As noted in Section 14.2.3(c) and equation (14.13), the radiation force power detector simply measures the force exerted on a reflector by the sound beam. The force F per unit area on a target or reflector can be given as

$$p = KE = K(I/c) \tag{14.40}$$

and

$$p = F/A, \tag{14.41}$$

where E is the energy density, I the intensity, K is a parameter whose value depends upon the target type and physical configuration of the reflector, and A is the cross-section of the transducer. If a perfect absorber or perfect reflector at normal incidence to the plate is considered, the value of K is 1 or 2, respectively. If a beam directed onto a perfect reflector consisting of a circular cone at an incident angle θ normal of the surface, $K = 1 + \cos 2\theta$ (Borgnis, 1953). Figure 14.11(a) illustrates the principle of a simple radiation force meter. The extra force produced by the beam is simply balanced by the extra weights on the left side of the scale. Many designs based on this principle are used to determine beam power and transducer efficiency.

Another technique to calculate the radiation force, called the ball radiometer, is shown in Figure 14.11(b). A small rigid sphere (usually made of quartz) is suspended in the acoustic field and allows simple calibration of

Ultrasound Intensity Measurements

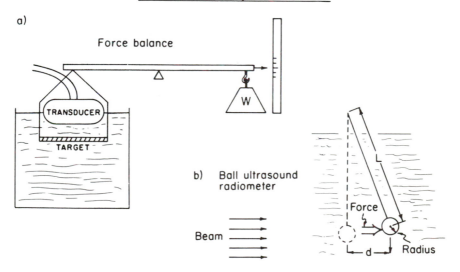

Figure 14.11 Two general radiation force meter designs used to calibrate ultrasound beam intensities. (a) Force meter using a sensitive balance. Different arrangements have been developed to calibrate the output from large plane sources. (b) Ball ultrasonic radiometer used to test the acoustic intensities across a beam.

the absolute average intensity over this weighted cross-section. According to Hasegawa and Yosioka (1969, 1975) the intensity of the acoustic beam is determined from the deflection d of the sphere with a radius a supported by a thin thread of length L. The balance between the acoustic force F and the mechanical static force is given by

$$F = m_b\, g\, d/(L^2 - d^2)^{1/2}. \qquad (14.42)$$

The intensity is calculated by

$$I = F\, c/\pi a^2 Y, \qquad (14.43)$$

where m_b is the effective mass of the sphere (i.e. corrected for buoyancy), and g is the acceleration due to gravity. The factor Y is a small correction for the material and also related to the ratio of the radius of the sphere and the ultrasound wavelength, a/λ.

The ball radiometer is a way of obtaining information on the local intensity. However, if the ball is comparable with the size of the beam peaks, only an average value of the power is obtained.

14.5. *Practical ultrasound systems and their performances*

14.5.1. Common ultrasound systems

There are two general types of ultrasound designs that are effective for hyperthermia; one includes the planar transducers designed for surface heating and the second includes the focused or multiple sources that are developed for deep heating. While these systems have quite different geometries, the general approaches are similar. All of the transducers are air-backed for maximum efficiency. The acoustic beam generated at the front face of the transducer must propagate through a low-loss acoustic media and match the acoustic impedance of the body. For most hyperthermia systems water is used.

In addition, the temperature of the media must be well controlled so that the skin temperature is lower than the deeper-seated tissues. Even at a depth of 2 cm in the body, temperature changes are observed that are related to the temperature fluctuation of the bath. In general, the temperature of the water bath should be controlled in an accurate and stable manner (to within $\pm 0.2\,^{\circ}$C).

Another common problem for these high-powered sources is the accumulation of air bubbles which coat the surface of the transducer and other parts of the system due to cavitation in the water. This cavitation is greatly reduced by degassing the coupling water. One way of achieving this is to boil or evacuate the water sample. A second method is to use a special positive-displacement pump that evacuates the air from the circu-

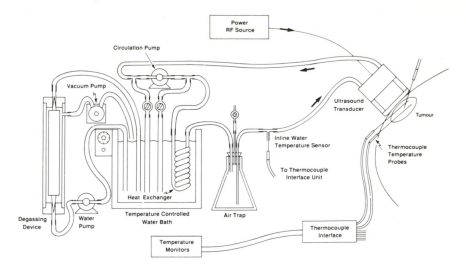

Figure 14.12 A typical ultrasound system with a degassing arrangement as used in clinical hyperthermia protocols. (Adapted from the system used by Kapp *et al.*, 1988).

lation system, the accumulated gas being removed by a degassing system such as that developed by Pounds and Britt (1984), and a newer complete system shown in Figure 14.12, is used for their clinical hyperthermia system developed (Kapp *et al.*, 1988). A degassing device uses a vacuum pump and air trap to reduce the potential of cavitation near the transducer. The third approach to reducing the trapped gas is to run the ultrasound system at maximum power for about an hour which drives out the air in the water circulation system.

14.5.2. Shallow heating

14.5.2(a) Circular sources

The principle of a surface ultrasound heater is shown in Figure 14.13. The piezoelectric disc is excited by a high-powered radio-frequency source with output power of up to 200 W. The circulating water inside the applicator is covered by a thin membrane such as silicone rubber. In addition, a thin layer of acoustic gel, designed for use in diagnostic ultrasound, is placed between the membrane and the skin to obtain a good acoustic coupling. Any air bubbles trapped between these layers must be eliminated.

As shown in Figure 14.2, penetration of the beam is dependent upon the frequency and attenuation coefficients found in the body. This beam has features that are similar to microwave and inductive surface applicators. As shown in Figure 14.14, suitable surface cooling is needed to compensate for the higher power absorption near the surface of the body. The

Ultrasound Surface Heater

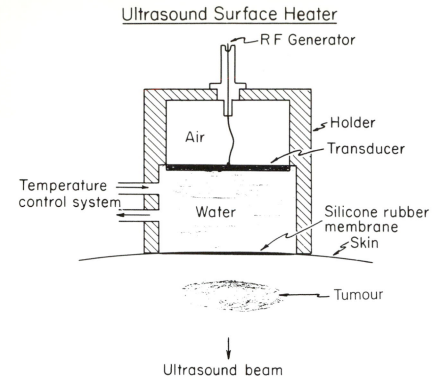

Figure 14.13 The general principles of a superficial ultrasound heater, with a rapidly flowing, temperature-controlled water cooling system. A coupling medium such as an acoustic gel must be used between the membrane and the skin to eliminate beam loss due to trapped air. (Adapted from Hunt, 1982.)

maximum temperature reached along the axis of the beam can be altered by adjusting the surface temperature of the body, or by selecting the optimum transducer frequency. As a general principle, the frequency should be selected so that at least 50% of the acoustic intensity reaches the tumour, and with a SAR distribution that produces the most homogeneous temperature distribution throughout the tumour. As shown in Figures 14.2 and 14.14, a selection of suitable frequencies and surface cooling is needed to obtain a useful depth–dose distribution into the body produced by a plane ultrasound source. While there is still considerable controversy regarding criteria for the acceptable temperature distribution, my personal view is that we should be striving for thermal doses that are compatible with the tough, well-known radiation therapy criteria. Therefore, temperature contours at 80% of the maximum found in the treatment field should be the goal.

Details of two typical ultrasound applicators are shown in Figure 14.15 (Munro *et al.*, 1982). One persistent problem is the fabrication of a trans-

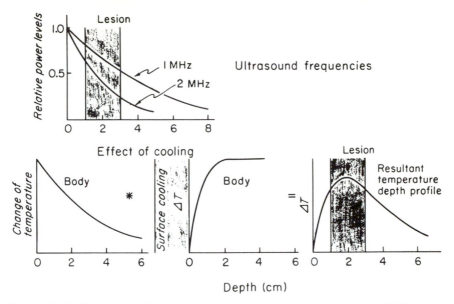

Figure 14.14 The use of suitable plane source ultrasound frequencies with the correct attenuation, combined with convolutions of the surface cooling, to obtain a useful depth–dose distribution in the body.

ducer system that produces an acoustic beam comparable to that predicted from the theoretical assessments described in Section 14.3. Various methods are used to obtain a water seal between the transducer and the rest of the holder; one is shown in Figure 14.15(a). The transducer is compressed between two conductive O-rings (Chromerics, Woburen, Mass., USA) which provide both a water seal and good electrical connection between the radio-frequency amplifier and the front and back electrodes. An alternative arrangement is shown in Figure 14.15(b), in which the disc is gently supported by a polystyrene foam ring and a silicon ring. The electrical contacts are established with standard solder junctions on both sides of the transducer.

Figure 14.16 summarizes some of the simulation and experimental acoustic distributions for a 5.0 cm diameter, 2.22 MHz piezoelectric transducer. Figure 14.16(a) is the simulation and shows a fairly flat intensity distribution across the transducer (except for apparent sharp peaks and valleys along the axis might be probably due to a small computation artifact).

Figure 14.16(b) demonstrates the compression of the O-rings across the transducer. The wide, intense peak found along the axis is not acceptable. Similar effects were described by Pounds and Britt (1984), who suggested that these unexplained results were due to the radial and flexional modes of the resonating transducer. However, they were able to improve the distributions, either by selecting the optimum frequencies or by sweeping

a)

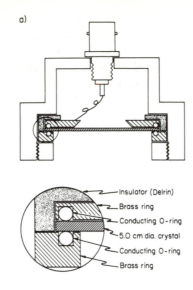

b)

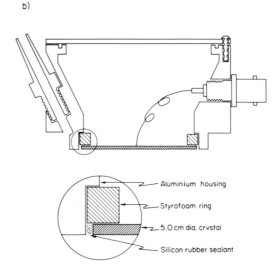

Figure 14.15 Two transducer designs used to produce superficial heating using: (a) a silver-loaded conductive O-ring to seal the transducer from the circulating water 'cuff' to obtain the optimum temperature distribution in the field and (b) a transducer supported by a polystyrene foam, and sealed with silicon rubber. (From Munro *et al.*, 1982.)

the frequency near the transducer's fundamental frequency resonance.

A considerable improvement in the beam distribution is achieved by using a polystyrene foam support, as shown in Figure 14.16(c) (Munro *et al.*, 1982). This intensity distribution is much closer than expected from the theoretical simulation, Figure 14.16(a). Although the average distribution is much flatter, blood cooling and thermal conductivity will reduce the temperatures near the periphery of the field. To compensate for the cooling, the transducer is cut into a series of annular rings. By applying more excitation to the outside rings of the transducer, the resultant intensity distribution observed is as shown in Figure 14.16(d). Thus, some flexibility is available by using this design.

Ocheltree and Frizzell (1987, 1988) have made an elegant assessment of the cooling near the periphery of the treatment field using the bioheat transfer equation. They suggest that a defined value of the energy deposition rate, biased toward the periphery of the treatment field, should produce a flat temperature distribution throughout the tumour, while the temperature distribution should drop rapidly outside this field. This approach will work when blood flow is either homogeneous, or complex but symmetric within the tumour.

An improved transducer design by Keilman (1989) appears to have a superior beam profile, in that the sharp and often irregular peaks have been eliminated in their contour maps. By adding a rotating diffusing plate, intense peaks which produce unwanted hot spots in the field that may be painful for the patient (particularly near bone) are reduced. Also, quarter-wave layers which broaden the frequency band pass of the system, enabling a large range of beam frequencies (1.5–2.5 MHz) to be used from the same source are incorporated. No clinical experience is yet available on this system, but it has considerable potential.

14.5.2(b) Novel planar transducer arrays

Several interesting ultrasound sources are based on an array of elements. For example, simple surface heaters may consist of square arrays, such as those described by Benkeser *et al.* (1984), Dickinson (1984) and Underwood *et al.* (1987). These arrays have the potential of producing an acoustic field that would better match the treatment field, and could improve the thermal profiles by compensating for the extra cooling at the periphery of the treatment field and/or correcting for different blood flow cooling in the treatment field. Figure 14.17 (Underwood *et al.*, 1987) shows the profiles for a 4×4 array in which each 2 cm square element was excited independently. The figure also shows close agreement between the theoretical and experimental profiles for one, two and four elements. In Figure 14.18 the changes of the distribution when power was reduced to 50% on the right-hand element are shown. Thus, considerable flexibility can be obtained with this system.

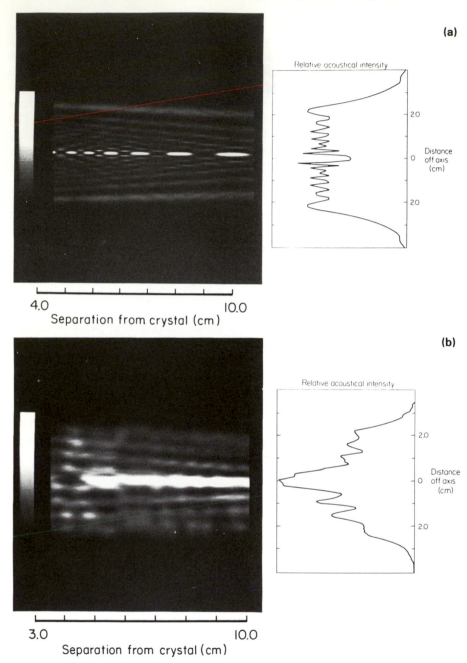

Figure 14.16 Beam intensity distributions produced by a 5.0 cm, approximately 2.0 MHz piezoelectric transducer divided into four annular elements. The profiles were calculated at 6 cm from the source and the distributions show different levels of absorbed radiation:

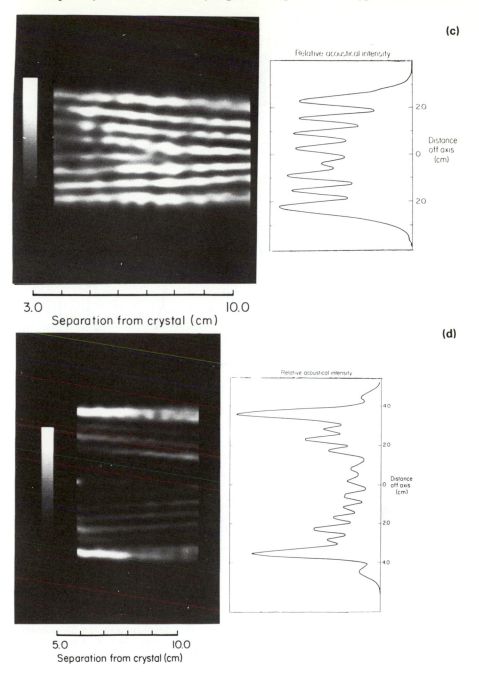

(a) theoretical; (b) beam produced by the system shown in Figure 14.16(a) with O-rings; (c) system shown in Figure 14.16(b); (d) excess power used on the outside elements. (Munro *et al.*, 1982.)

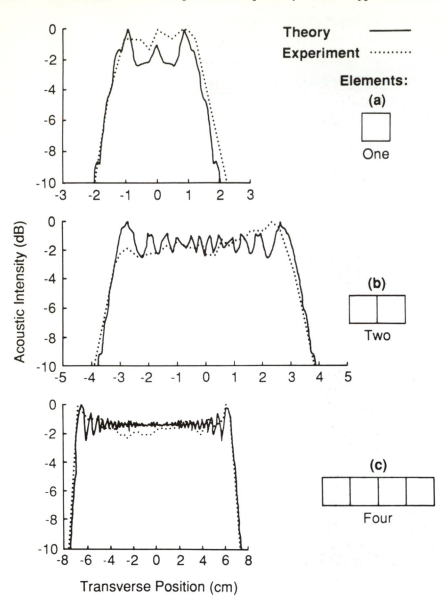

Figure 14.17 The multi-element ultrasound applicator developed by the Bioacoustic Research Laboratory, University of Illinois, USA, and consisting of 4×4 array, each with a width of 2 cm. (a) to (c) are the acoustic intensity profiles for different number of elements. (Adapted from Underwood *et al.*, 1987.)

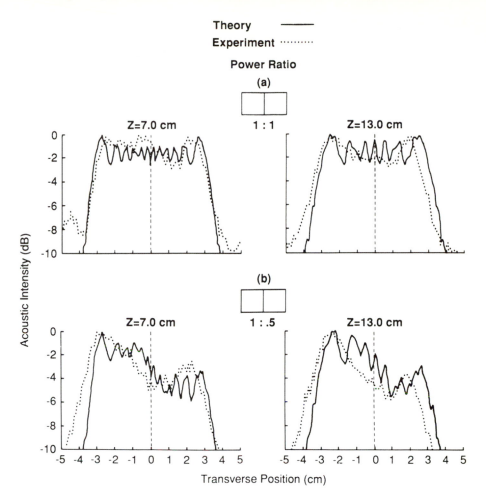

Figure 14.18 The same applicator described in Figure 14.17, except in (b) the power level in the right has been reduced to 50%. This would allow the compensation of different acoustic absorptions, or for different rates of cooling in each section of the array. (Adapted from Underwood *et al.*, 1987.)

14.5.3. Deep-heating ultrasound devices

Because of the special properties described in Section 14.1, ultrasound is particularly attractive for deep heating. There are two general approaches being tested in the laboratory and in clinical trials. In one, single or multiple acoustic beams are directed or focused onto the tumour. In the other, a well-focused beam is scanned rapidly across the tumour to generate an improved temperature distribution.

14.5.3(a) Fixed focused sources

Because of the sharp focus produced by a spherically shaped aperture, special transducer arrangements are used to obtain an improved temperature distribution at depth in the body.

One interesting concept has been tested by Coleman and coworkers (1986) [Figure 14.19(a)] by using the special properties of the eye. In this system, the transducer forms a beam with a focus in the vitreous liquid (where there is little absorption) and which expands beyond this region. The malignant growths found in the back wall of the eye can be effectively heated by this expanding beam.

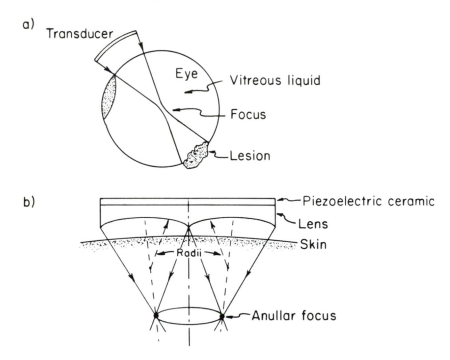

Figure 14.19 (a) A technique for heating the back wall of the eye (adapted from Coleman *et al.*, 1986). (b) The piezoelectric ceramic transducer coupled to a special lens to form an annular focus inside the body (adapted from Beard *et al.*, 1982; Higgins *et al.*, 1984).

Another focusing design is based on the unusual lens shown in Figure 14.19(b) which generates an annular focus some distance inside the body. This lens can be made of a plastic (Lele, 1981; Beard *et al.*, 1982) or of the high-velocity material, aluminium (Higgins *et al.*, 1984). Although the SAR of the annular focus will be large, the blood flow in the tissue has the effect of smoothing out the changes, thus producing a flatter temperature distribution at the focus (see Section 14.5.2(a) which refers to a discussion

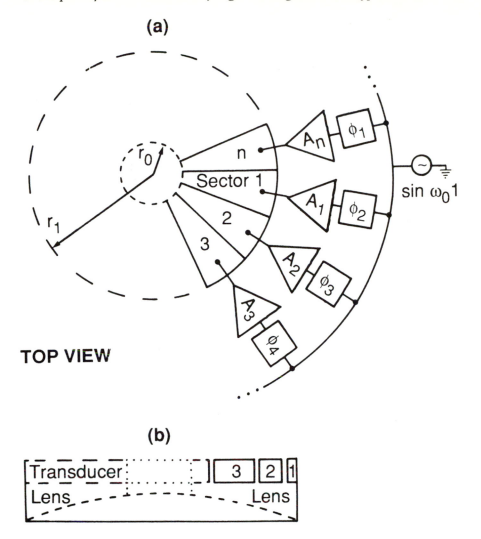

Figure 14.20 The 'vortex phased array' developed by Cain *et al.* (1986) and Umemura *et al.* (1987) in which combined pie-shaped elements are excited by independent sources with defined phases combined with a spherically-shaped lens to produce an annular acoustic beam. (Adapted from Cain *et al.*, 1986.)

of this problem by Munro *et al.* (1982) and Ocheltree and Frizzell (1987). While these general ideas have considerable merit, both experimental (J. W. Hunt and A. Worthington, personal communication) and theoretical phased annular arrays (Cain and Umemura, 1986) showed a strong, central focus which occurred several centimeters beyond the annular focus. This peak would make the beam unacceptable for most hyperthermia treatments.

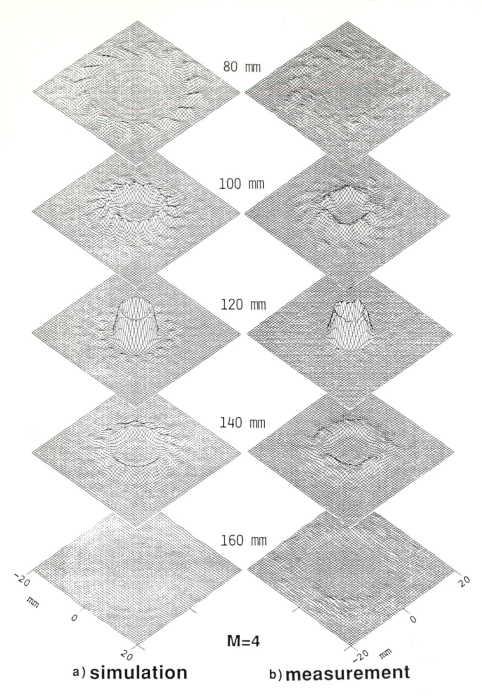

80 mm

100 mm

120 mm

140 mm

160 mm

M=4

a) simulation   b) measurement

Figure 14.21 The three-dimensional field distribution produced by the vortex phased array: (a) theoretical field distribution and (b) the measured distribution. Note the very close agreement. (Adapted from Umemura *et al.*, 1987.)

14.5.3(b) Novel focused transducer designs

Novel designs have been developed by Cain and Umemura (1986) and Umemura and Cain (1987) that improve focused beams with annular focus, but without the unwanted extra foci. One method of achieving this is based on a 'Sector Vortex Array' (see Figure 14.20, in which the transducer(s) is (are) divided into pie-shaped elements (often in 32 elements in two annular rings)), and by suitable phasing of the power amplifiers, the diameter of the annular focus can be altered electronically. As shown by Figure 14.21(a) and (b), the theoretical and experimental studies by Umemura and Cain (1987) show that the actual designs are feasible, and that there is no unwanted central peak closer to or farther from the focus. Other new designs include suitable focusing and steering for rectangular arrays (Ibbini *et al.*, 1987).

14.5.3(c) Multiple plane transducers

As described in Section 14.3.1(a), there is a gentle focusing of the beam at the transition between the Fresnel and near zones produced with a plane transducer, i.e. $x'_{max} = d^2/4\lambda$. Fessenden *et al.* (1984) and Hynynen *et al.* (1983) have made use of this by directing six large transducers (≈ 6 cm in diameter) offset laterally from the focus. In the Stanford system (Fessenden *et al.*, 1984) each transducer is mounted on a spherical shell at an angle of approximately 25° from the axis of the system. These can be tilted slightly for improved intensity distributions. By using a low-frequency beam (0.3 MHz), the system was designed to obtain well-defined regions in phantoms at depths as great as 15 cm.

While the principles are reasonable, the effectiveness of this design is limited. This is due to the extremely large size of the transducer assembly that must be coupled to the body, with the result that most regions of the body cannot be heated effectively. In addition, there is little flexibility in adapting this design for different acoustic absorption or variable tissue cooling in the irradiated field.

14.5.3(d) Scanned focused transducers

The ability to focus the beam is usually an advantage as tissues outside the field can be spared. However, focused beams have diameters smaller than tumours, so mechanical or electronic scanning is needed to steer the beam in such a particular pattern that an improved temperature distribution can be obtained.

The most extensive studies, using scanning techniques, have been carried out by Lele (1982, 1984). In one of his designs, a 0.9 MHz focused ultrasound source is scanned rapidly in single or multiple trajectories. His

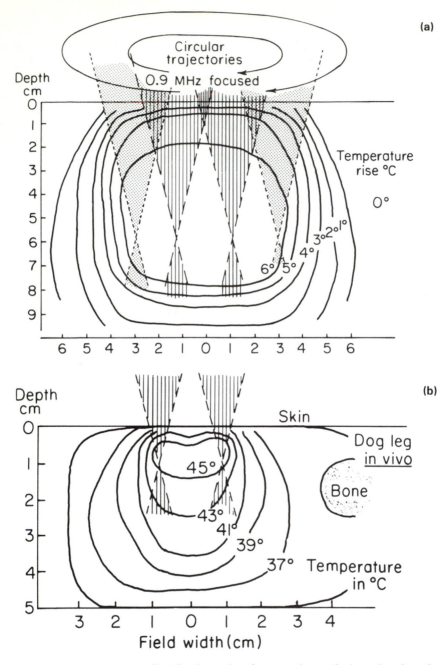

Figure 14.22 The temperature distribution using the scanning technique developed by Lele and coworkers (adapted from Lele, 1982). (a) Temperature distribution for two scanning beams at 0.9 MHz and focused at a 6 cm depth in beef muscle *in vitro*. (b) Temperature distribution by a single circular trajectory in the leg of a dog *in vivo* at a frequency of 1.8 MHz.

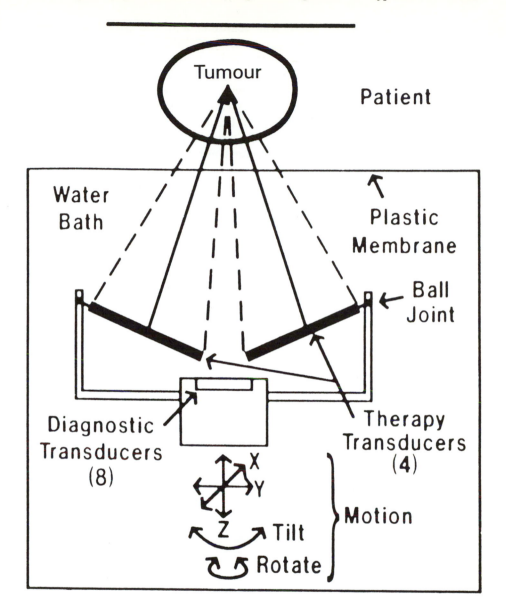

Figure 14.23 The multiple transducer scanning device based on the Octoson diagnostic ultrasound scanner (Ausonic Inc., Australia) modified with four, therapy inducers (diameter=13 cm, radius=34 cm, f=1 MHz). This system is immersed in a water bath and covered with a thin plastic membrane. The gantry can be scanned under computer control with five degrees of freedom. (From Hynynen *et al.*, 1987.)

premise was that if the beam was scanned near the periphery of the tumour, blood flow and thermal conductivity would produce a more homogeneous temperature distribution throughout the whole tumour volume. Figure 14.22(a) shows the temperature distribution from a beef muscle sample *in vitro* using two circular trajectories at radii of 1 and 3 cm from the axis. An excellent temperature distribution was found at a depth of 2–7.5 cm and at a width of 7 cm. Figure 14.22(b) shows reasonable heating distribution in a leg muscle mass of a dog *in vivo*. Again, a reasonable distribution was found with no obvious excess temperatures near the bone.

Hynynen *et al.* (1987) have expanded the concept proposed by Lele by using ultrasound systems adapted from a diagnostic scanner. As shown in Figure 14.23, the prototype system uses four 7 cm diameter air-backed transducers, two on each side of an Octoson diagnostic transducer array. They explored the hyperthermia distributions using 0.5 and 1.0 MHz beams, and focal lengths of 35 cm and 25 cm, respectively. The mechanical assembly allows the beam to be moved in three dimensions with a speed of up to 5 cm/s controlled by a microcomputer. Preliminary studies using this system are very promising and will be covered in more detail in another section.

In the high intensities near the focus of the beam, non-linear effects are expected (Swindell, 1985), where the beam has a saw-tooth shape (see Figure 14.24). Hynynen (1987) has demonstrated these non-linear effects by comparing the temperatures reached using a high-intensity pulsed beam with a continuous wave mode at the same average power levels. He has shown a maximum temperature gain of up to 2°C at a temperature elevation of 5°C in a dog's thigh *in vivo* using the pulsed beam. This could give considerable advantage compared to traditional treatments, but the

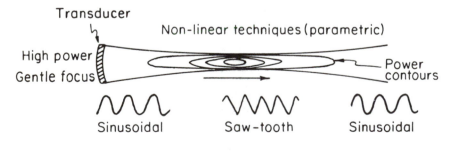

Figure 14.24 The distortion of an acoustic wave transmitted into water near the focus.

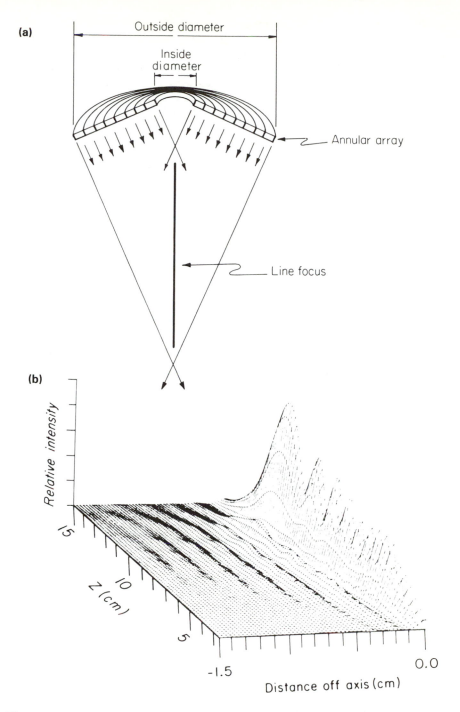

Figure 14.25 (a) A conically shaped transducer designed to generate a line focus (sectional view). (b) The three-dimensional distribution produced by this distribution. Note that the average intensity increases away from the source. If the

(c)

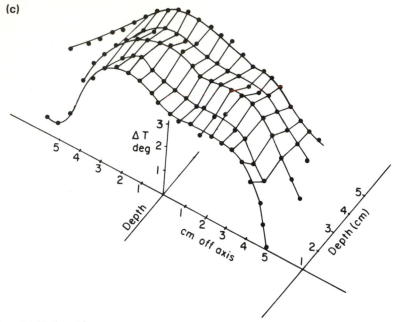

Figure 14.25 (cont.)
frequency is chosen correctly, the intensity can compensate for the attenuation. (c)
The temperature distribution produced by a 24° converging conical beam with a
circular trajectory of 3 cm radius.

non-linear aspects of the beam will be more difficult to monitor and control.

A different concept is being developed by Hunt *et al.* (1990). The general concept is based on a strongly focused conically shaped source, as shown in Figure 14.25(a). The converging ultrasound beam produces a line focus that projects into the body (Foster *et al.*, 1981). The beam can be scanned rapidly to cover the field (i.e. ≈ 10 s for a complete cycle). The particular strength of this technique is its flexibility. If the transducer is cut into a series of annular elements, different regions of the array can be excited so that tumours with various depths and thicknesses can be heated effectively. The simulated profiles of the beam are produced by the conical beam along the axis of the cone (see Figure 14.25(b)). This design compensates for beam attenuation but a ripple is predicted along the axis from diffraction. Figure 14.25(c) shows improved temperature distribution produced by scanning the transducer in a conical lens at a radius from the axis of 3 cm.

14.5.4. Clinical experience using ultrasound systems

While therapeutic ultrasound has been used for many years, critical clinical studies are still needed to demonstrate that ultrasound is an effective hyperthermia moiety. As part of a special report which evaluated hyperthermia systems, Corry *et al.* (1988) compared the thermal distribution of several sites of the body: head and neck, breast, axilla, chest wall, abdomen and pelvis, and other extemities for treatments of 202 tumours. For these tumours, the mean maximum temperature in the treatment fields was 46.1°C, the mean average temperature was 42.6°C and the overall average temperature was 43.1°C. In most sites, these distributions were better than those for magnetic induction and interstitial radiofrequency implants. The success of obtaining a complete regression of the tumour (Dewhirst *et al.*, 1984) is closely related to the lowest temperatures found in the treatment field. Unfortunately, the ability to attain a high thermal dose is usually limited by patient pain and potential treatment complications due to unwanted higher temperatures in the field and in neighbouring tissues.

Kapp *et al.* (1988) evaluated 37 treatment fields using their ultrasound superficial heating systems in comparison with their extensive clinical experience with microwave devices. In general, only good therapeutic average temperatures (≈ 43°C) could be attained for smaller transducers (4 cm diameter). For larger transducers (6–8 cm diameter), average treatment temperatures were 42°C and 41.1°C, respectively, with patient pain limiting the maximum treatment temperatures reached.

Shimm *et al.* (1988) described their preliminary experiences of a multiple transducer scanning system (Hynynen *et al.*, 1987), shown in Figure 14.24, while a summary of their results are given in Table 14.6. Good hyperthermic temperatures are found in the treatment fields for selected sites of the body (extremities, head and neck, brain and pelvis). Again, however, certain regions of the treatment fields are not adequately heated. Localized cooling by the blood appears to be the culprit, with large differences between predicted and observed temperature distributions in the

Table 14.5 Treatment investigations by Shimm *et al.*, (1988) using scanned focused ultrasound relating tumour site and temperature.

Site	Maximum temperature during treatment		
	Highest sensor[1] (°C)	*Lowest sensor[2]* (°C)	*Sensors >42.5°C[3]*
Extremity	44.7	41.7	35/78
(3 patients, 4 tumours)	(41.5–49.2)	(38.3–43.8)	(45%)
Head and neck	42.0	37.8	0/68
(1 patient, 2 tumours)	(41.6–42.3)	(37.5–38.1)	
Brain	44.8	41.8	42/59
(2 patients, 2 tumours)	(43.9–45.9)	(41.0–42.9)	(71%)
Pelvis	44.2	40.6	110/199
(8 patients, 9 tumours)	(38.0–46.3)	(38.0–43.1)	(55%)

[1] Mean highest-measured maximum temperature for each site.
[2] Mean lowest-measured maximum temperature for each site.
[3] Time-averaged mean temperature. In this column the values are the number of sensors reaching 42.5°C divided by the total number of sensors.

treatment field (Langendijk, 1987). In an attempt to reduce this problem of localized cooling, rapid heating techniques using intense ultrasound beams are being explored (Hunt *et al.*, 1987) where the rate of heating is considerably greater than the rate of cooling. This approach has the potential to reduce the temperature gradients found in treatment fields. Until these cooling problems are solved, the potential of ultrasound (and most other techniques) as a standard therapeutic agent for controlling cancer may be limited.

14.6. Conclusions

The physical properties of ultrasound have many attractive character-istics that make it the technique of choice for both shallow and deep heating in selected regions of the body. A particular advantage is the short wavelength of the ultrasound beam which allows more flexibility in con-trolling localized power deposition. This can be used to account, to some extent, for blood flow cooling, allowing sharper beam edges and enables one to tailor the beam shape to match the treatment field. Because of small acoustic absorption in the subcutaneous fat layers in the skin, ultrasound does not demonstrate the excess heating in this region plagued by some microwave and radio-frequency devices. Also, intense focused acoustic beams can be used in a rapid heating technique which has the potential to reduce thermal gradients, such as those due to heterogeneous blood cooling in different regions of a tumour. Multiple-element arrays and scanning techniques have the potential to adjust the power deposition so that a more homogeneous temperature distribution is obtained. However, to obtain this distribution, data from many thermal detectors must be available to make the necessary corrections to the power distributions. Another advantage of ultrasound is that by using lower frequencies (<1 MHz) the beam can be focused at depths of 10 cm or greater, while still producing small beam widths.

There are also drawbacks to ultrasound hyperthermia. In deeper lesions, the sites for effective treatments are limited since, while only a small acoustic window is available, a large focused beam must be directed towards the tumour — at the same time avoiding neighbouring air and bone structures. This prevents the use of ultrasound hyperthermia. Furthermore, the beam has a cigar-like shape, and thus the boundary of the treatment field along the direction of the beam is not distinct. However, when the limitations of ultrasound are clearly delineated, it is likely that new designs will be fabricated which may produce acceptable beam distributions throughout the tumour. Thus, ultrasound should have a promising future in the field of hyperthermia.

Acknowledgements

I would like to acknowledge financial support from the Medical Research Council of Canada, the National Cancer Institute of Canada and the Ontario Cancer Treatment and Research Foundation. I would like to thank Katherine Wilton, Leif Heinzl, Stephen Brown, Geoff Lockwood and Andrew Kerr for their critical suggestions in the various drafts of this paper.

References

Arditi, M., Foster, F. S. and Hunt, J. W. (1981) Transient fields of concave annular arrays, *Ultrason. Imaging*, **3**, 37–61.

Augustine, L. J. and Andersen, J. (1979) An algorithm for the design of transformerless broadband equalizers of ultrasonic transducers, *J. Acoust. Soc. Am.*, **66**, 629–635.

Bacon, D. R. (1982) Characteristics of a PVDF membrane hydrophone for use in the range 1–100 MHz, *IEEE Transactions of Sonics and Ultrasonics*, **29**, 18–25.

Beard, R. E., Magin, R. L., Frizzell, L. A. and Cain, C. A. (1982) An annular focus ultrasonic lens for local hyperthermia treatment of small tumours, *Ultrasound Med. Biol.*, **8**, 177–184.

Berlincourt, D. (1971) Piezoelectric crystals and ceramics, in Mattiat, O. E. (Ed.), *Ultrasonic Transducers and Materials*, New York, Plenum, pp. 63–124.

Biquard, P. (1972) Paul Langevin, *Ultrasonics*, **10**, 213.

Benkeser, P. J., Frizzell, L. A., Holmes, K. R., Ryan, W., Cain, C. A. and Goss, S. A., (1984) Heating of a perfused tissue phantom using a multi-element ultrasonic hyperthermia applicator, *IEEE Ultrasonics Symposium*, pp. 685–688.

Borgnis, F. (1953) Acoustic radiation pressure of plane compressional wave, *Rev. Mod. Phys.*, **25**, 653–664.

Cain, C. A. and Umemura, S.-I. (1986) Concentric-ring and sector-vortex phased-array applicator for ultrasound hyperthermia, *IEEE Trans. Microwave Theor. and Tech.*, **34**, 542–551.

Coleman, D. I., Lizzie, F. L., Burgess, S. E. P., Silverman, R. E., Smith, M. E., Driller, J., Rosado, A., Ellisworth, R. M., Hai, K. B. B., Abramson, D. H. and McCormick, B. (1986) Ultrasonic hyperthermia and radiation in the management of intraocular malignant melanoma, *Amer. J. Ophthalmology*, **101**, 635–642.

Corry, P. M., Jobboury, K., Kong, J. S., Ammour, E. P., McCraw, F. J. and Le Duc, T. (1988) Evaluation of equipment for hyperthermia treatment of cancer, *Int. J. Hyperthermia*, **4**, 53–74.

Curie, P. J. and Curie, P. (1880) Crystal physics, development by pressure of polar electricity in hemihedral crystals with inclined faces, *Comptes Rendus Hebdomadaires des Seances de L'Academia des Sciences, Paris*, **91**, 294. Translated by Lindsay, R. B. and reprinted in Lindsay, R. B. (Ed.) (1973) *Acoustics: Historical and Philosophical Development*, Stroudsburg, P. A. Dowden, Hutchinson and Ross, pp. 372–374.

Desilets, C. S., Fraser, J. D. and Kino, G. S. (1978) The design of efficient broad-band piezoelectric transducers, *IEEE Trans. Sonics Ultrason.*, **25**, 115–125.

Dewey, W. C., Hopwood, L. E., Sapareto, S. A. and Gerweck, L. E. (1977) Cellular responses to combinations of hyperthermia and radiation, *Radiology*, **123**, 463–474.

Dewhirst, M. W., Sim, D. A., Sapareto, S. A. and Connor, W. G. (1984) Importance of minimum tumor temperature in determining early and long-term responses of spontaneous canine and feline tumours to heat and radiation, *Cancer Research*, **44**, 43–50.

Dickinson, R. J. (1984) A non-rigid mosaic applicator for local ultrasound hyperthermia, in Overgaard, J. (Ed.), *Hyperthermic Oncology 1984*, Vol. 1, London, Taylor and Francis, pp. 671–674.

Dunn, F. and Frizzell, L. A. (1982) Bioeffects of ultrasound, in Lehmann, J. F. (Ed.), *Therapeutic Heat and Cold*, 3rd edn, Williams and Wilkins, pp. 386–403.

Fessenden, P., Lee, E. R., Anderson, T. L., Strohbehn, J. W., Meyer, J. L., Samulski, T. V. and Marmor, J. B. (1984) Experience with a multitransducer ultrasound system for localized hyperthermia of deep tissues, *IEEE Trans. Biomed. Eng.*, **31**, 126–135.

Foster, F. S., Patterson, M. S., Arditi, M. and Hunt, J. W. (1981) The conical scanner: a two transducer ultrasound scatter imaging technique, *Ultrasonic Imaging*, **3**, 62–82.

Freundlich, H., Söllher, K. and Fogowski, F. (1932) Klinische biologische wircungen von ultraschallwellen, *Klinische Wochenschrift*, **11**, 1512–1513.

Goss, S. A., Frizzell, L. A. and Dunn, F. (1979) Ultrasonic Absorption and Attenuation in Mammalian Tissues, *Ultrasound in Med. and Biol.*, **5**, 181–186.

Hasegawa, T. and Yosioka, K. (1969) Acoustic-radiation force on a solid elastic sphere, *J. Acoust. Soc. Am.*, **46**, 1139.

Hasegawa, T. and Yosioka, K. (1975) Acoustic radiation force on fused-silica spheres and intensity determination, *J. Acoust. Soc. Am.*, **58**, 581–585.

Higgins, P. D., Zeng, X.-W., Zagzebski, J. A., Paliwal, B. R. and Steeves, R. A. (1984) Versatility of distributed focus ultrasound in treatment of superficial lesions, *Int. J. Radiat. Oncol. Biol. Phys.*, **10**, 1923–1931.

Horvath, J. (1944) Ultraschallwirkung Bein Menschlichen Sarkom, *Strahlentherapie*, **75**, 119.

Horvath, J. (1948) Morphologische Untersuchungen uber die wirkung der Ultrasahallwellen auf das karzinomgewebe, *Strahlentherapie*, **77**, 279.

Hueter, F. F. and Bolt, R. H. (1955) *Sonics: Techniques for the use of Sound and Ultrasound in Engineering and Science*, New York, Wiley.

Hunt, J. W. (1982) Applications of microwave, ultrasound and radio-frequency heating in Third International Symposium: cancer by hyperthermia, drugs and radiation, *Natl. Inst. Health Monograph*, **61**, 447–456.

Hunt, J. W., Arditi, M. and Foster, F. S. (1983) Ultrasound transducer for Pulse-echo medical imaging, *IEEE Transactions Biomed. Eng.*, **31**, 453–481.

Hunt, J. W., Worthington, A., Urchuk, S. and Bernstein, M. (1987) *IEEE/9th Annual Conference of the Engineering in Medicine and Biology Society*, pp. 1941–1942.

Hunt, J. W., Lockwood, G., Worthington, A., Hutchins, D., Mare, H. and Taylor, R. (1990) Design of conical transducers to produce strongly focused ultrasound beams for deep heating, *Int. J. Hyperthermia*, submitted.

Hutchins, D. A. and Archer-Hall, J. A. (1983) Particle velocity near-fields of disk radiators with variable drive, *Acustica*, **53**, 123–131.

Hynynen, K. (1987) Demonstration of enhanced temperature elevation due to nonlinear propagation of focused ultrasound in dog's thigh, *In Vivo, Ultrasound in Med. and Biol.*, **13**, 85–91.

Hynynen, K., Watmough, D. J. and Mallard, J. R. (1983) Local hyperthermia induced by focused and overlapping ultrasonic fields — an *in vivo* demonstration, *Ultrasound in Med. and Biol.*, **9**, 621–627.

Hynynen, K., Roemer, R., Anholt, D., Johnson, C., Xu, A., Swindell, W. and Cetas, T. (1987) A scanned, focussed, multiple transducer ultrasonic system for localized hyperthermia treatments, *Int. J. Hyperthermia*, **3**, 21–35.

Ibbini, M., Ebbini, E., Umemura, S.-I. and Cain, C. A. (1987) Ultrasonic phased arrays for hyperthermia: new techniques based on field pattern synthesis using a field conjugation method, *IEEE Ultrasonics Symposium Proceedings*, **2**, 863–866.

Kapp, D. S., Fessenden, P., Samulski, T. V., Bagshaw, M. A. and Keilman, G. (1988) Stanford University Institutional Report: phase I evaluation of equipment for hyperthermia treatment of cancer, *Int. J. Hyperthermia*, **4**(1), 75–115.

Keilman, G. (1989) New ultrasound hyperthermia applicators with improved bandwidth and spatial uniformity, *Abstract, Thirty-seventh Annual Meeting of the Radiation Society and the Ninth Annual Meeting of the North American Hyperthermia Group*, Ak-6, pp. 32.

Kinsler, L. E., and Frey, A. R. (1962) *Fundamentals of Acoustics*, 2nd edn, New York, Wiley.

Kossoff, G. (1979) Analysis of focusing action of spherically curved transducer, *Ultrasound in Med. and Biol.*, **5**, 359–365.

Kremkau, F. W. (1979) Cancer therapy with ultrasound: a historical review, *J. of Clin. Ultrasound*, **7**, 287–300.

Krimholtz, R., Leedom, D. A. and Matthaei, G. L. (1970) New equivalent circuits for elementary piezoelectric transducers, *Electron. Lett.*, **6**, 398–399.

Langevin, P. (1920) British Patent Specifications IYS, 457 No. 146, 691.

Lagendijk, J. J. W. (1987) Heat transfer in tissue, in Field, S. B. and Franconi, C. (Eds), *Physics and Technology of Hyperthermia*, NATO ASI Series, Series E: Applied Sciences No. 127, Dordrecht, Matinus Nijhoff, pp. 517–552.

Lele, P. P. (1979) Safety and potential hazards in the current applications of ultrasound in Obstetrics and Gynecology, *Ultrasound in Med. and Biol.*, **5**, 307–320.

Lele, P. P. (1981) Letter: an annular-focus ultrasonic lens for production of uniform hyperthermia in cancer therapy, *Ultrasound in Med. and Biol.*, **7**, 191–193.

Lele, P. P. (1982) Local hyperthermia by ultrasound, in Nussbaum, G. H. (Ed.), *Hyperthermia. Medical Physics Monograph*, No. 8., New York, American Institute of Physics, pp. 393–440.

Lele, P. P. (1984) Ultrasound: Is it the modality of choice for controlled localized heating of deep tumors?, in Overgaard, J. (Ed.), *Hyperthermic Oncology, 1984*, Vol. 2, London, Taylor and Francis, pp. 129–154.

Lockwood, J. C. and Willette, J. G. (1973) High speed method for computing the exact solution for the pressure variations in the nearfield of a baffled piston, *J. Acoust. Soc. Amer.*, **53**, 735–741.

Martin, C. J. and Law, A. N. R. (1980) The use of thermistor probes to measure energy distributions in ultrasound *Ultrasonics*, **18**, 127–133.

Martin, C. J. and Law, A. N. R. (1983) Design of thermistor probes for measurements of ultrasound intensity and distribution, *Ultrasonics*, **21**, 85–90.

Mason, W. P. (1948) *Electromechanical Transducers and Wave Filters*, 2nd edn, New York, Van Nostrand, pp. 200–329.

Mason, W. P. (1950) *Piezoelectric Crystals and Their Application of Ultrasonics*, New York, Van Nostrand.

Meeks, S. W. and Ting, R. Y. (1983) Effects of static and dynamic stress on the piezoelectric and dielectric properties of PVF/SUB2, *J. Acoust. Soc. Am.*, **24**, 1681–1686.

Munro, P., Hill, R. P. and Hunt, J. W. (1982) The development of improved ultrasound heaters suitable for superficial tissue heating, *Med. Phys.*, **9**, 888–897.

NCRP Report 74 (1983) *Biological Effects of Ultrasound: Mechanisms and Clinical Implications*, Bethesda, Maryland, National Council on Radiation Protection and Measurements.

Oberhettinger, F. (1961) Transient solution of the "baffled piston" problems, *J. Res. Nat. Bur. Stand.*, **65B**, 1–6.

Ocheltree, K. B. and Frizzell, L. A. (1987) Determination of power deposition patterns for localized hyperthermia: a steady-state analysis, *Int. J. Hyperthermia*, **3**, 269–279.

Ocheltree, K. B. and Frizzell, L. A. (1988) Determination of power deposition patterns for localized hyperthermia: a transient analysis, *Int. J. Hyperthermia*, **4**, 281–296.

Penttinen, A. and Luukkala, M. (1976) The impulse response and pressure nearfield of a curved ultrasonic radiator, *J. Phys. D.*, **9**, 1547–1557.

Pohlman, R., Richter, R. and Parow, E. (1939) Uber die Austbreitand und Absorption des ultraschalls im menshleichen Gewebo und Seintherapeutisch Wirkung an Ischias Plexusneuralgie, *Dtsch Med Wochensche*, **65**, 251–256.

Pounds, D. W. and Britt, R. H. (1984) Single ultrasonic crystal techniques for generating uniform temperature distributions in Homogeneously perfused tissues, *IEEE Transactions on Sonics and Ultrasonics*, **31**, 482–490.

Sapareto, S. A., Hopwood, L. E., Dewey, W. C., Raju, M. R. and Gray, J. W. (1978) Effects of hyperthermia in survival and progression of Chinese hamster ovary cells, *Cancer Research*, **38**, 393–400.

Shimm, D. S., Hynynen, K. H., Anhalt, D. P., Roemer, R. B. and Cassady, J. R. (1988) Scanned focussed ultrasound hyperthermia: initial clinical results, *Int. J. Radiation Oncology Biol. Phys.*, **15**, 1203–1208.

Strutt, J. W. (1945) *Theory of Sound*, Vol. 2, Dover, New York.

Swartz, R. G. and Plummer, J. D. (1980) Monolithic silicon – PVF2 piezoelectric arrays for ultrasonic imaging, in Metherell, A. F. (Ed.), *Acoustic Imaging*, Vol. 8, New York, Plenum Press, pp. 69–95.

Swindell, W. (1985) A theoretical study of nonlinear effects with focussed ultrasound in tissues: an "acoustic Bragg peak", *Ultrasound in Medicine and Biology*, **11**, 121–130.

Umemura, S.-I. and Cain, C. A. (1987) Evaluation of a prototype sector-vortex array applicator, *IEEE Ultrasonics Symposium Proceedings*, **2**, 867–870.

Underwood, H. R., Burdette, E. C., Ocheltree, K. B. and Magin, R. L. (1987) A multi-element ultrasonic hyperthermia applicator with independent element control, *Int. J. Hyperthermia*, **3**, 257–267.

Wells, P. N. T. (1969) *Physical Principles of Ultrasonic Diagnosis*, London, Academic Press.

Wells, P. N. T. (1977) Wave fundamentals, *Biomedical Ultrasonics*, London, New York and San Francisco, pp. 1–44.

Woeber, K. (1949) Vorlaufige Erfahrugen mit Ultraschalltherapie bie Dermatosen, *Strahlentherapie*, **76**, 599–606.

Woeber, K. (1956) Biological basis and application in medicine, *Ultrasound in Biol. Med.*, **1**, 18.

Zemanek, J. (1971) Beam behavior within the nearfield of a vibrating piston, *J. Acous. Soc. Amer.*, **49**, 181–191.

15 Thermometry

T. C. Cetas

15.1. Introduction

A few basic points concerning thermometry can be made at the outset of this chapter. First, thermometry has been related to medicine since its beginning (Middleton, 1966). Second, the development of the science of thermometry and its contributions to medicine have depended on technical advances, and the converse is just as true. This latter point is, of course, true in all aspects of the development of scientific understanding and its engineering applications. In this chapter, some rather basic concepts of thermometry are discussed, followed by applications of these principles to the field of clinical hyperthermia. A useful source of information on thermometry, encompassing a broad spectrum from fundamental scale definition to practical problems, is the series *Temperature* (Schooley, 1982) which appears approximately each decade (only the latest edition is cited here). The fundamentals of thermometry are reviewed by Quinn and Compton (1975), Hudson (1980), Schooley (1986) and Swenson (1987). A number of reviews on thermometry related to hyperthermia have appeared in recent years (Cetas and Connor, 1978; Cetas *et al.*, 1978, 1980; Cetas, 1982a,b, 1984, 1985, 1987; Christensen, 1979, 1982, 1983; Fessenden *et al.*, 1984; Hand, 1985; Mangum and Furukawa, 1982; Martin, 1986; Samulski and Fessenden, 1990). This list is undoubtedly incomplete. Much of this chapter represents minor changes and updates to previously published reviews (Cetas, 1982b, 1987) which were, in turn, updates of earlier ones. The indulgence of the previous publishers and readers is appreciated.

15.2. Temperature scales

There are two distinct types of temperature scales. The first type is based on the theoretical thermodynamic concept of temperature. This was developed over several years of rigorous thermodynamic reasoning as man began to understand the nature of heat and temperature. The absolute temperature scale was established in terms of the energy and work involved when a system (heat engine) operated between two temperatures. When defined this way, temperatures are independent of material properties such as mercury or gas expansions. The fact that different materials do not expand proportionately caused perplexity to the early thermometer makers. An absolute zero exists; although it cannot be reached, it can be approached asymptotically. Ratios of temperature measured on the thermodynamic scale are more significant physically than addition of temperatures. When compared to the range of temperatures that has been produced by man, from a few microdegrees above zero kelvins in certain paramagnetic systems to millions of degrees in plasmas and thermonuclear explosions, the extremely close temperature regulation of the earth and the narrow range of temperatures compatible with mammalian survival are impressive indeed.

The second type of temperature scale is the practical or working scale. Until recently, the entire world, by agreement, referred all temperature measurements to the International Practical Temperature Scale of 1968 (IPTS–68) (1976). This scale was established by an international committee and was based upon careful thermodynamic experiments. It represented the state-of-the-art as an approximation to the Kelvin thermodynamic scale and was considered to be accurate to about 0.03 K over most of its range below 904 K. For comparison, the precision, but not necessarily the accuracy, of most fundamental thermometry standards work below the steam point is of the order of 0.1 mK.

The procedure of the IPTS-68 is to define the temperature of 11 natural fixed points between 13.81 and 1337.58 K such as the triple point of water (273.16 K=0.01°C), and to provide a protocol for interpolation and extrapolation. Between these fixed points a precisely constructed platinum resistance thermometer is used for interpolation. The interpolation function for the ideal platinum thermometer involves a polynomial with 21 terms carried to 16 significant figures. Thus, a given platinum resistance thermometer is calibrated at the fixed points to determine the corrections necessary to make it reproduce the scale of the ideal thermometer. This is a complicated routine, but it is quite accurate. In principle, the entire scale can be set up by following rules given in the text of the IPTS-68. Calibration of one platinum resistance thermometer against another is not necessary. In practice, only a few of the national standards laboratories in various countries have established the IPTS-68-defined fixed points and specified the scale in this way. The national laboratories calibrated many platinum resistance thermometers against the IPTS-68 and use these to calibrate the thermometer probes sent to them.

In late 1989, the 46-nation International Congress on Weights and Measures (CGPM) acting on advice of the Consultative Committee on

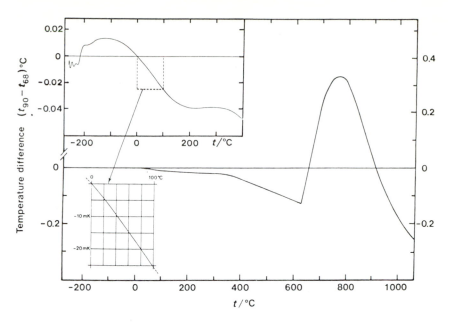

Figure 15.1 Difference between the new 1990 International Temperature Scale (ITS-90) and the 1968 scale (IPTS-68) it supersedes. The plot at the upper left shows the region below 400°C. The region from 0–100°C is further explained at the lower left. From Taylor (1989).

Thermometry (CCT) adopted a new practical scale ITS-90 (Taylor, 1989) which incorporates improvements in practical realization of the true thermodynamic scale over the last 20 years (Figure 15.1). In practice, in the context of hyperthermia research, the changes are insignificant but it is interesting to note that the difference between the 'old' IPTS–68 and the new ITS-90 scale is linear from 0.01 °C (the point which defines the unit of temperature, °C) to the boiling point of water which is now recognized to be at 99.975°C.

15.3. Thermometer probes

15.3.1. General principles

A thermometer is any instrument which measures temperature. Any temperature-dependent parameter can serve as the basis for a thermometer; the variety of thermometer types attests to this. Because of the large number of parameters which are temperature dependent, much engineering time is devoted to eliminating temperature sensitivities from instrumentation. Here, we are concerned with the converse problem — that of producing maximum dependence upon temperature, and eliminating any dependence upon extraneous or environmental parameters. It is not a trivial task to construct a stable, accurate and convenient instrument which measures only temperature. Consequently, we will discuss some of the sources of error of each type of thermometer.

Even if a thermometer behaves ideally, errors in measurement can arise from poor technique. If a few general concepts are kept in mind, however, these errors can usually be kept small, or at least their presence and magnitude can be noted. Probe thermometers sense their own temperature which is determined by establishing equilibrium with the medium to be measured. Consequently, the temperature sensed is an average over a finite volume which depends upon the presence of gradients, the thermal properties of the medium and those of the probes. Heat is exchanged by conduction, convection and radiation. Each must be considered when measuring temperatures.

Conduction is described by Fourier's Law which states that heat flow is proportional to the temperature gradient, the thermal conductivity of the medium and the cross-sectional area through which heat is flowing. Temperature measurements in the presence of steep gradients lose significance because the thermometer will sense an average value. Furthermore, thermal conduction down the thermometer leads or shaft may affect the temperature sensed and may change the gradients and temperature profiles drastically. If thermal equilibrium is desired, for example, when calibrating one thermometer against another, the system should be constructed so that the paths of heat flow do not cross the region where the thermometers are located and in which equilibrium is desired.

Convection can induce steep thermal gradients. For example, major blood vessels are not necessarily in thermal equilibrium with the surrounding tissue. Similarly, if a calibration bath is filled with viscous oil, proper mixing will not occur, and the temperature will vary significantly throughout the bath. Surface temperatures are also difficult to monitor because of radiant heat exchange as well as convective and conductive exchange. Probes placed on the surface will alter all three mechanisms and will sense not only the temperature of the altered surface, but also a temperature influenced by the radiative, convective and conductive exchange of the probe with the environment. Generally, radiometric techniques, such as thermography, are preferable for surface measurements if the surface has an emittance close to unity. Metals characteristically have very low emittance, but in some cases paint can be applied to raise the emittance. Tissues, on the other hand, have an emittance of the order of 0.98 (Steketee, 1973).

Thermometers can be characterized by several features. The choice of a particular type of thermometer for a given task will depend upon the relative merits of these parameters as well as personal preferences and availability of equipment:

Sensitivity. Some parameter, such as resistance, electromotive force or reflected light intensity, must change monotonically with temperature; and a means must be available for measuring this change with sufficient precision. In clinical applications, 0.1°C is usually adequate, although a resolution of 0.01°C is often helpful for determining the specific absorption rate (SAR) from the slope of the temperature versus time curve.

Insensitivity. The measured parameter and instrumentation must be insensitive to all other influences such as ambient temperature or the presence of moisture near the sensor. Satisfying this restriction often is more difficult than the first requirement.

Accuracy. The thermometer must reproduce absolute temperatures within the limits necessary for the task. In clinical hyperthermia, 0.2°C accuracy is sufficient, but certain biochemical studies inclusing cell survival studies with hyperthermia require an accuracy of at least 0.05°C if experimenters are to compare results properly.

Stability. Most thermometers drift with age, and so must be checked periodically and recalibrated when necessary. Examples are given later.

Response time. Thermometer readings must keep pace with the actual temperature fluctuations of interest. The response time is usually characterized by a time constant which is defined in the following way: A thermometer is subjected to a step change in temperature ΔT_0 by plunging the probe into a water bath. The indicated change in temperature ΔT will approach Δ_0 exponentially. The time constant is the time required for the difference between ΔT and ΔT_0 to reduce to $1/e$ or 37% of ΔT_0. Equilibrium, a residual difference within, say, 5% of ΔT_0, is assumed to be reached after an interval corresponding to about three time constants.

Passivity. The thermometer should not significantly perturb the medium to be measured, either through thermal conduction along the probe or through artifactual direct heating of the probe by an electromagnetic or ultrasonic therapeutic field. Neither should it cause a tissue reaction.

Size. The sensor should be small compared to the size of the object to be measured and small enough to resolve significant temperature gradients. Its heat capacity (or more properly, its thermal inertia which is the product of density, specific heat and thermal resistivity) should be sufficiently low so as not to produce a cold spot. The size of the probe is related intimately to the two preceding items and to the problem of durability.

Temperature indication. This refers to how temperatures are obtained from the thermometric parameter. For example, on a clinical thermometer, the height of a mercury column is read against a scale etched on a glass tube. Reading may be a little difficult, but neither expensive electronic equipment nor calibration tables are necessary. On the other hand, electronic thermometers are generally easier to read, to automate and to use for process control.

Interchangeability. High precision, accuracy and stability of the order of 0.1°C usually require individual calibration. By implication as well, one probe cannot be substituted for another and the same read-out mechanism still be used. With the possible exception of copper–constantan thermocouples, probes are not interchangeable at this level of accuracy.

These features are not all-inclusive nor entirely independent. Several common types of thermometers are discussed later in this chapter. An attempt is made to describe their features, drawbacks, instrumentation, and sources of error.

15.3.2. Mercury-in-glass thermometers

The mercury-in-glass thermometer is one of the oldest thermometers. It is a remarkable instrument in that it is essentially unchanged in form and accuracy from the time of Fahrenheit (ca. 1729). Frequently, the term thermometer is taken to mean this device only, although the word is more generally used here. Mercury-in-glass thermometers are quite inexpensive, ranging from a few dollars for a clinical thermometer to something less than one hundred dollars for the best standardized laboratory thermometer. If some care is used in the choice of a supplier, laboratory-quality thermometers which are accurate to about 0.1°C can be purchased for about 20 dollars. If calibrated individually, thermometers with 0.1°C divisions can be used to maintain a scale to approximately 0.02°C. Careful, well-trained technicians at national standards laboratories routinely and reproducibly read them to 0.005°C!

One additional feature of mercury-in-glass thermometers is that the mechanism for reading the temperature is etched on the glass stem. No expensive electronics are required for obtaining a reading. On the other hand, it is not convenient to automate or electronically monitor temperatures measured this way. Circuits have been designed for this propose, although electronic thermometers generally are used instead.

Deficiencies of mercury-in-glass thermometers are related mostly to their large mass. They cannot be used for point measurements in steep thermal gradients and have very long time constants. Finally, they are easily broken, which causes hazards from both broken glass and spilled mercury.

Several sources of error exist for mercury-in-glass thermometers. First, all thermometers should be checked upon receipt for a faulty scale, for example, the omission of a number or optical distortion along the capillary. Occasionally, the mercury column will become separated, especially during shipment. Carefully heating the thermometer to well above its range so that the mercury expands into the enlarged cavity at the top of the column and then cooling it slowly will usually correct the problem. Alternatively, the thermometer can be cooled in the neck of a liquid nitrogen flask so that all of the mercury is drawn into the bulb at the bottom. The thermometer may break, however, if it contacts the liquid or is cooled or warmed too quickly.

Glass and mercury do not expand proportionately, and glass exhibits some hysteresis in its thermal properties. Consequently, it is preferable in precision work to always work with increasing temperatures unless very long equilibrium times are possible. Gently tapping the thermometer before reading will help to relieve strains in the mercury column due to surface tension. It is preferable to use mercury-in-glass thermometers in a totally immersed fashion. If this is not possible, some corrections may be necessary because the stem and column are at temperatures different from the medium under study. Glass has a tendency to flow over a period of time, which will cause a change in the bulb volume and, consequently, a shift in the calibration of the thermometer. An occasional check of the reading of the thermometer at the ice point will reveal any changes that may have occurred. The shift at the ice point can be added to the entire calibration; a complete re-calibration is not necessary. If precise work with mercury-in-glass thermometers is planned, a copy of the NBS report by Wise and Soulen (1986) is extremely useful.

15.3.3. Platinum resistance thermometers

In pure ideal metals, at a temperature of absolute zero, electrons originating from the outer shells of the constituent atoms are free to travel about the well-organized lattice structure without hindrance. In real metals, electron flow is impeded by impurity atoms, dislocations in the lattice, strains and so forth, which leads to a finite, temperature-independent resistance. At temperatures well above absolute zero, the electrons also are scattered by collisions with vibrating atoms in the lattice structure. The higher the temperature becomes, the greater the vibrations become, and, consequently, the greater the resistance. A temperature-dependent resistance results, which is quite linear within the range of room temperature.

In 1887, H. L. Callender (Middleton, 1966) proposed the use of pure platinum wire as a means of constructing stable precision thermometers to maintain a standard temperature scale that can be compared with scales at various laboratories. The standard platinum resistance thermometer (SPRT) is specified as the interpolating instrument on the IPTS-68 (1976) for determining temperatures between the defined fixed points of 13.81 K and 903.89 K (630.74 °C). Platinum thermometers to be used as

SPRTs must meet strict standards of purity and must be strain free. The ratio of the resistance at 100°C to that at 0°C gives a measure of the purity because of the additive component to the resistance caused by the impurities. This ratio must be greater than 1.3925 for SPRTs. Standard laboratory-quality platinum resistance thermometers are quite expensive and must be handled with care to preserve their calibration. Riddle *et al.* (1972) have written an excellent monograph on the use of these instruments.

Most medical laboratories will find that the SPRT is more than they require. Nevertheless, many industrial-grade platinum resistance thermometers (RTDs) are available commercially. They will have typically a resistance of about 100 ohms at 0°C, compared to 25.5 ohms for the usual SPRT. The temperature coefficient is just under 0.4%°C. They exist in a variety of configurations, but cannot be made small enough practically to fit into hypodermic needles or fine catheters. They are stable to about 0.02°C if handled carefully (Schooley, 1986; Mangum and Furukawa, 1982); and because they are electronic devices, they can be automated easily for data acquisition and monitoring. The nearly linear variation of resistance with temperature simplifies the design of circuitry for direct readout in temperature units. Interchangeability is possible, but for precise work the probes must be calibrated individually.

Several sources of error exist for platinum resistance thermometers, although a little care will eliminate errors for all but the most-precise measurements. First of all, the probes are sensitive to shock. Strains cause resistance, thus shock can produce shifts in calibration. Heat can be conducted through the leads and affect the temperature of the sensing element, so the leads should be anchored to the body to be measured if possible or, if not, to another body at nearly the same temperature. Figure 15.2 gives a simplified measurement circuit which illustrates the basic

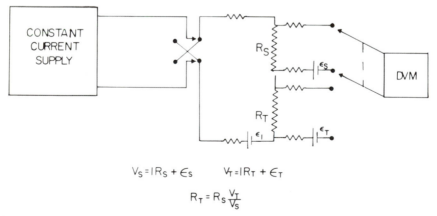

$$V_S = I R_S + \epsilon_S \qquad V_T = I R_T + \epsilon_T$$

$$R_T = R_S \frac{V_T}{V_S}$$

Figure 15.2 Circuit diagram illustrating the potentiometric method of resistance measurements and indicating some of the error sources. See text for discussion.

potentiometric method and illustrates several sources of measurement error. Current from a constant current supply is fed through a reversing switch to a series resistance circuit consisting of the standard resistor, R_s, and the resistance thermometers to be measured, R_T. A digital voltmeter is used to read the voltages across each resistor. An error can arise from the fact that the element will self-heat due to Joule losses (I^2R) from the measuring current I flowing through the element of resistance R. The magnitude of the temperature error will depend upon heat conduction from the thermometer element to the medium being measured which is just that produced by Joule losses and so is proportional to I^2. Measuring the resistance at two current levels permits the temperature error to be determined. Typically, a measuring current of 1 mA will produce insignificant self-heating. Thermal e.m.f's, as represented by the symbols ϵ_1, ϵ_2 in the figure can occur at junctions of leads or from strains in the lead wires if they pass through large temperature gradients. Their effects can be eliminated by reversing the current flow through the series circuit. The sign of the measured voltage will change, but the contribution from the contact e.m.f. will not, thus permitting its elimination by taking the average of the forward and reverse readings. Special care must be taken to avoid ground loops (spurious high-resistance current paths due to insufficient isolation). Otherwise the 'forward' current may be different from the 'reverse' current and will manifest itself as a large thermal e.m.f. with its magnitude proportional to the current.

Finally, the resistance of the thermometer leads can produce an error. For precise work, two leads are attached directly to each end of the thermometer sensor. This is referred to as the 'four-lead' technique. Resistances in the current leads are unimportant since the voltage is measured across the sensor only. Resistances in the potential leads are significant since a good voltmeter has a very high internal impedance. Only an infinitesimal current flows in those leads and no significant spurious potentials are added to that across the sensor. Separate leads are used for the current source and for measuring the potential (V). The resistance is the ratio $R = V/I$ (Ohm's law). For less-precise work, including most medical applications, the lead resistance is small compared to that of the element and usually does not change much since most of the leads remain at room temperature. Consequently, the two-lead configuration is quite acceptable and the lead resistance can be incorporated into the calibration of the probe.

A variety of circuits are used to read the resistance of a platinum thermometer. The most common for precision work are the DC potentiometric method and both DC and AC bridge methods. Bridge techniques usually involve some variation of the Wheatstone bridge, including the Mueller bridge for thermometers with four leads. Bridges using AC techniques have been developed which are convenient to use and which can be purchased with sensitivities adequate to the needs of many specific tasks. For biomedical applications, several companies produce instruments with linearization circuits such that the readings are in temperature units (°C or °F). The probes are not precisely similar, and the linearizations are not perfect, so individual calibrations are necessary to attain an accuracy of the order of 0.1°C.

15.3.4. Thermistors

Thermistors are also resistance devices. However, unlike platinum resistance thermometers, they are semiconductors and have different characteristics. The electrons in a semiconductor at low temperatures are bound to the atoms which make up the solid. In contrast to insulators, the energy required to excite the electrons into the conduction band and make them available to carry electrical current is of the order of thermal energies at ordinary temperatures (0.025 eV). The number of conduction electrons, and hence the conductivity σ, is determined by the Boltzman distribution function

$$\sigma = \sigma_0 \, e^{-\Delta/kT}, \tag{15.1}$$

where σ_0 and Δ are constants, k is the Boltzman constant and T is the absolute temperature in kelvins. The reciprocal of this expression gives the resistivity of the semiconducting medium. Thus, the conductivity increases exponentially with temperature, while the resistivity ($1/\sigma$) decreases exponentially. Taking the natural logarithm of the expression for resistivity and writing in terms of resistance R of a specific device, we obtain

$$\frac{1}{T} = A_0 + A_1 \ln R, \tag{15.2}$$

where A_0 and A_1 are constants. Finally, if we follow Steinhart and Hart (1968) and allow one additional term to correct for non-ideal semiconductors we have

$$\frac{1}{T} = A_0 + A_1 \ln R + A_3 (\ln R)^3. \tag{15.3}$$

Again, T is temperature in kelvins, R is resistance and A_0, A_1 and A_3 are constants to be determined from the calibration of the probe. This expression represents the temperature versus resistance characteristic for thermistors to better than 0.01°C from 0 to 55°C (Trolander *et al.*, 1972). An example is shown in Figure 15.3. A feature of this method is that only three constants are required to characterize the thermometer. Accurate temperatures can be computed directly by this expression from the

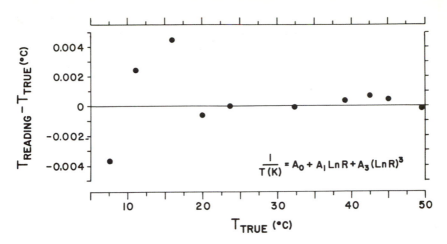

Figure 15.3 Differences between temperatures computed from the expression shown using measured resistance from the temperatures obtained from a laboratory standard calibrated thermistor. The coefficients were obtained from a least squares fit to the data.

measured resistance and displayed, tabulated or used in computer analysis.

Thermistors are constructed from a variety of sintered oxides such as magnesium oxide. A common form is a small bead of green material moulded about two fine platinum leads. The beads are fired and encapsulated in glass, and probes of various styles are constructed from them. They are made in a range of sizes from about 0.3 mm diameter, or less, to perhaps 2 mm. The fabrication of thermistors, however, is still much of an art. Their stability varies widely from probe to probe even if made from the same basic material. Glass-encapsulated beads tend to be more stable than discs and those encapsulated in epoxy. Thermistors have several advantages. First of all, they can be made quite small. We use commercial probes mounted in 25 AWG hypodermic needles or Teflon catheters. They have a high sensitivity of about

$$\frac{1}{R}\frac{dR}{dT} = -4\%/°C \qquad (15.4)$$

which is about an order of magnitude greater than that for platinum resistance resistors. They have high precision and, when calibrated, high accuracy with relatively modest instrumentation. The stability of thermistors varies, but if care is taken in selection, it can be very good (Wood *et al.*, 1978). The standard thermistor we use has not drifted in 13 years within our measurement uncertainty of $\pm 0.003°C$ at the ice point. Two identical needle probes with sequential serial numbers were purchased at the same time. One of them has not drifted, while the other has required recalibration several times to keep it accurate to within 0.05°C.

The most annoying deficiency of thermistors is that they are not really interchangeable. That is, they cannot be made to have the same R versus T characteristic to within 0.1°C. Schemes have been designed to keep them within about 0.3–0.5°C, especially if larger sensors are permissible and compound units can be used. Generally, a choice must be made between direct reading and small (needle) size. Electronic circuits to convert resistance measurements to direct reading temperature units are sensitive to the specific thermistor bead (see section on calibration for an example).

Many of the problems associated with producing interchangeable probes and circuits for direct reading in temperature units can be alleviated through use of modern microprocessor systems. The coefficients A_0, A_1, and A_3 of equation (15.3) above for a specific probe can be stored and the temperature computed from the specific resistance readings. Accuracy, small size, direct reading, and multiple probes all become possible at once.

Circuits for reading thermistors are similar in principle to those for platinum resistance thermometers but with different specific parameters, components and instruments. Generally, DC potentiometric methods or Wheatstone bridges are used, especially in precision work. We use the former in a 'four-lead' configuration for our standard thermometers. Probes used in the clinic or in biological experiments are calibrated as two-lead devices. Digital multimeters (DMM) with a resistance mode can be used, provided the measuring current can be kept low enough. Most

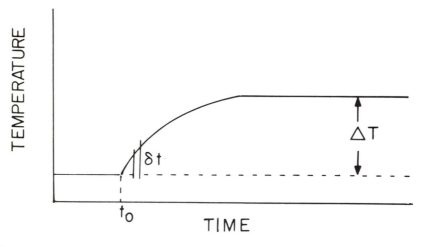

Figure 15.4 The self-heating error due to Joule losses (I^2R) in a resistance thermometer are illustrated. The thermometer is initially in equilibrium with the medium, but begins to warm above it when the reading current, 1, is switched to it at t_o. After several seconds the magnitude of the error stabilizes. For manual systems ΔT must be small compared to the desired level of accuracy; this limits the magnitude of the measuring current. For computerized systems, the thermometer resistance can be read in the interval, δt, before the probe has time to heat significantly.

DMMs use currents about an order of magnitude too large, causing significant self-heating on the order of 0.5–2°C depending upon the medium in which the probe is immersed (see Figure 15.4). In our calibration laboratory, we use a standard current of 10μA for a 1000 ohm thermistor. The self-heating error in a thermally conductive medium is less than 0.01°C; however, it is much greater in a thermally resistive medium such as air. Another way of circumventing the self-heating error is possible with computerized instrumentation and is also illustrated in the same figure. Fast data acquisition systems can read the resistance in a brief interval, δt, shortly after the reading current is switched to the resistance thermometer, and before the thermometer can warm significantly. Higher-reading currents thus are possible.

AC techniques have not been applied generally to thermistors. Circuits using AC bridge techniques, however, have been designed for germanium resistance thermometers used at temperatures below 30 K (Schooley, 1982; Mangum and Furukawa, 1982). Since the characteristics of those thermometers are similar to thermistors, the instrumentation would be suitable for use with thermistors.

Sources of error for thermistors are similar in principle to those for platinum resistance thermometers, although they tend to be less significant because of the higher sensitivity of thermistors. Self-heating

FOUR LEAD TECHNIQUE

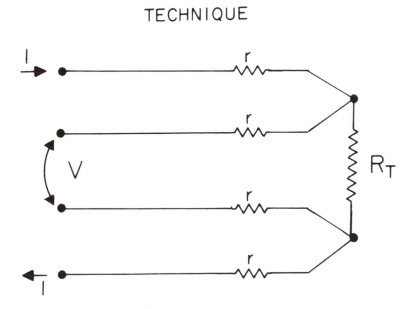

Figure 15.5 Illustration of the four-lead technique for measuring resistors, R. Lead resistances, r, are eliminated effectively from the measurement. See text for discussion.

problems encountered with thermistors are discussed above. Current reversal techniques should be used for precise thermometry with desired accuracies of better than 0.01 °C but are not usually necessary for clinical applications where 0.1°C is adequate.

Temperature measurements in the presence of the strong electro-magnetic fields required for inducing hyperthermia are complicated due to direct interactions of electrical thermometers with the field. While this is discussed in detail later, it is convenient to mention here the approaches by Bowman (1976) and by Larsen *et al.* (1979) to adapt thermistors for use in this application. Bowman used a small, very high-resistance thermistor and constructed leads from very high-resistance carbon-impregnated Teflon. They are connected in a four-lead configuration. The measuring current is carried by one pair of leads while the potential is measured across the other pair as in Figure 15.5. The high resistivity reduces the dipole currents induced by the electric field; close proximity of the leads minimizes the encompassed area of the leads and so minimizes mag-netically induced loop currents. The four-terminal configuration elimin-ates errors associated with lead resistances. Present thermometers are less than 1 mm in diameter. The electromagnetic losses that remain in this configuration are of the same order of magnitude as the tissue displaced by the probe. These thermometers, as single probes with an appropriate reading unit, are available from the Vitek Corporation. The BSD Corpor-ation markets a computerized data acquisition system based on this thermometer as part of its hyperthermia product line.

The nature of the leads also reduces errors from thermal conduction in steep thermal gradients. Nevertheless, as the manufacturer recommends, care must be taken to maintain very clean connectors without any oily or humid films that can lead to significant errors in this high-impedance system. Furthermore, it appears that stressing the leads can result in cali-bration drifts. We have found that this system performs very well in general although it is necessary before use to check the calibration at a temperature near that expected in use. We have noted on a few occasions that a probe will be accurate in a gallium cell (see section on calibrations later) but off by a degree or so near 42°C.

Larsen *et al.* (1979) developed related types of thermistor probes which employ microwave integrated circuit (MIC) techniques to plate miniature leads on sapphire needles. High-resistance extension leads connect to those on the needle. Again, four-lead techniques are used. The Narda Corporation has produced a thermometer system using a similar approach. The output of the analogue reading unit can be fed to a computerized data acquisition system. In this configuration, a resolution of 0.0001°C is possible which is very convenient for making SAR ($=C$ dT/dt, where C is the heat capacity of the medium) measurements in phantoms at low powers. The readout on the unit is only approximately calibrated in terms of temperature units (°C). A probe-specific calibration table must be generated and used if absolute temperatures are required. This unit is not really suitably configured for clinical use and is no longer available commercially.

15.3.5. Thermocouples

In electrical conductors, some of the electrons are free to move about, much as gas molecules do in air. If a temperature difference exists between the two free ends of the metal, electrons will have higher energies at the hotter end and tend to diffuse towards the cooler end. The excess charge resulting from this thermal diffusion of electrons produces a weak electrical potential difference E, the temperature coefficient of which is called the absolute thermoelectric force $dE/dT=\mu$, between the hot and cold ends. Since electrical leads must be connected to the two ends of the wire to measure this potential difference, and a similar effect will occur in the leads as in the original conductor, the voltmeter will measure the potential difference between the two metals (that is the leads and the original wire). Because the electronic structures are different for different metals (they have different concentrations of free electrons), the thermoelectric force also will be different.

A thermocouple thermometer is constructed by placing one junction of two metals at the point where the temperature T_X is to be measured and a second junction at a reference temperature T_R, such as in an ice point cell (see Figure 15.6). The potential difference is measured with a microvoltmeter. Frequently, for thermocouple readers used in clinical or industrial circumstances, the reference junction is simulated electrically and is incorporated into the reading instrument. Alternatively, the reference

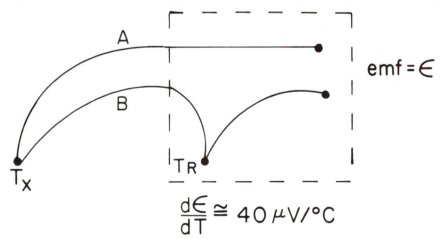

Figure 15.6 Illustration of a thermocouple circuit emphasizing the fact that it is a device for measuring temperature differences. One junction is placed at the unknown temperature T_X the other is at a known temperature, T_R.

junction can be bonded to a thermally isolated copper block, the temperature of which is monitored with a thermistor. This is the approach we use in our computerized data acquisition system and have been able to maintain an accuracy well within 0.1 °C. Because the thermoelectric force dE/dT is nearly constant near ambient and mammalian body temperatures, linearization circuits are relatively straightforward. Instruments which are capable of reading directly in temperature units are common.

Copper versus constantan (a copper–nickel alloy), also known as a type T thermocouple, is commonly used for biomedical purposes. It has a relatively high sensitivity of about 40 μV/°C and the materials are available at high quality and low cost. This permits the construction of probes which can be interchanged with one another to within about 0.1 °C without affecting the reading on the instrument. Probes are constructed by joining wires of pure copper with constantan by solder or a spot weld (Carnochan *et al.*, 1986). A cursory survey of supplier catalogues and technical brochures will reveal a variety of other thermocouple types and their characteristics. Nevertheless, all practical types have comparable thermoelectric powers at temperatures of interest here, and so require comparable instrumentation for reading (Hand, 1985).

Thermocouples can be made very small; probes in 29-gauge needles are common. Several workers (Cain and Welch, 1974; Priebe *et al.*, 1975; Battist *et al.*, 1969) have described techniques for constructing probes of the order of 10 μm diameter for special applications. Thermocouples are less sensitive than thermistors; and more sophisticated equipment and techniques are required to use them at resolutions greater than 0.1 °C. For thermistors, achieving 0.01 °C resolution is very easy if individual calibrations on each probe are accepted. Thermocouple systems are less suitable as secondary standards in a calibration facility.

Sources of error in thermocouple systems arise from anything that will add a spurious DC voltage to the reading network (Cetas, 1982a, b; Curley and Lele, 1984; Hand, 1985; Howard, 1972; Mossman *et al.*, 1982; Powell *et al.*, 1974). Contact potentials easily can exceed 10 μV unless care is taken. Thus, most commercial thermocouple systems maintain the same metal all the way back to the thermocouple reference junction; that is, connectors for type T thermocouples will have one terminal made of copper and the other of constantan. Another approach is to use ordinary DB-type connectors with gold-plated pins, as is illustrated in Figure 15.7. Spurious thermal potentials generated in the connector are very small for two reasons. First, by symmetry in the junction (for example, constantan to solder to gold to gold to solder to constantan), positive potentials on one side are offset by negative potentials on the other side. Second, the metals have relatively high thermal conductivity and are encased in thermally insulating plastic, so the temperature gradients across the junctions are small. In tests run in our laboratory, we could not distinguish any spurious

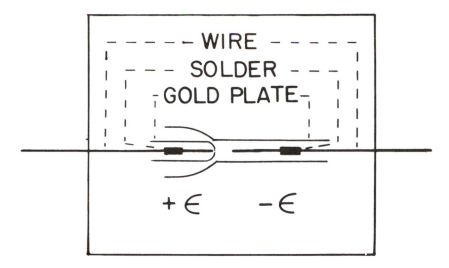

Figure 15.7 Schematic of the various dissimilar metal junctions at a connector. Each junction on the left side has its image on the right side, so the net thermal emf is nearly zero if no thermal gradient exists across the connector. The high thermal conductivity of the metal and the thermal insulation of the plastic housing help to insure uniform temperatures across the connector junction.

effects at the 0.01 °C level, even when the outside case of connector was shocked with a cold jet of compressed Freon.

Mechanical working of the wire and consequent induced strains will cause artifactual thermal voltages as will inhomogeneities in the alloying and in the purity of the wires (Mossman *et al.*, 1982). Typically, these effects can be kept below 0.1 °C for measurements in biomedical environments. Electronic biases and drifting in the reading instrumentation also appear as temperature shifts. This is one reason thermocouples are less satisfactory as thermometer secondary standards. Finally, since they are electrical conductors, they can be affected by electromagnetic fields in terms of noise, self-heating, and field perturbation. Some precautions can be taken to reduce problems, such as filtering the input to the electronic reading unit, inserting probes at normal incidence to electrical fields, electrically insulating the leads from current fields and conductors, and when possible, electromagnetically shielding the thermocouples. These measures are satisfactory in some cases and not at all in others and are discussed in a later section on measurements in intense electromagnetic fields.

A recent advancement in thermocouple thermometry in hyperthermia applications has been the development of multiple-junction thermocouples. They provide the great advantage of permitting measurements at multiple points but only requiring one probe insertion into a patient. Information is greatly increased with only a minor increase in trauma to patients from slightly larger probes. They exist in two forms. In one configuration, independent thermocouples are placed within the catheter. This has the advantage of straightforward applications of thermocouple principles without the introduction of new error sources. For an n-junction probe, $2n$ leads are required which implies a large overall size and increased errors from thermal conduction. The other configuration consists of a common lead of one metal and multiple leads of the second metal joined at increments to the common lead. Thus an n-junction probe requires only $n+1$ leads resulting in a smaller size and reduced thermal conduction errors when compared to the first configuration. Procedures have been developed in a number of laboratories for constructing these probes and they are available commercially from at least one source (Sensortek, Saddle Brook, New Jersey).

Andersen and his colleagues (1984) and Dickinson (1985) discuss an additional error that is associated with the second configuration and is illustrated in Figure 15.8. Suppose the junction at point 2 is poorly made, consisting of a solder bridge of a finite length which contacts the common lead at two points. When the thermocouple is placed in a steep temperature gradient, a small thermoelectric potential, $\delta\epsilon$, is produced. This appears in series with the thermoelectric potential measured across

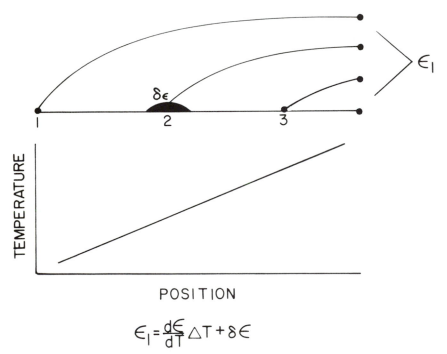

$$\epsilon_1 = \frac{d\epsilon}{dT}\Delta T + \delta\epsilon$$

Figure 15.8 Illustration of an error source in multiple junction thermocouples. If the junction at 2 is made poorly, say by being very large, a temperature gradient will result in a spurious thermal emf at 2 which expresses itself as an addition to the measured potential across the junction 1 as shown. Thus, problems with a proximal junction can affect readings of more distal junctions.

junction 1. Thus, an error in the temperature reading at junction 1 results from a problem with junction 2. This is really just a manifestation of the more general problem of heterogeneities and strains discussed by Mossman *et al.* (1982). It can be minimized by making very small junctions. Typically, 50 μm wire is used and great care is taken by most laboratories to avoid broad junctions from spreading solder or wide spot weld joints (Dickinson, 1985).

Another recent approach in thermocouple thermometry is to replace the copper in type T thermocouples (copper–constantan) with manganin, an alloy which has nearly the same thermoelectric potential against constantan as copper. Thus, intrumentation designed for type T probes can be used for the manganin–constantan probes with only minor calibration adjustments. The rationale for the substitution is three-fold (Hand, 1985). Manganin has a low thermal conductivity, 22 W/m/°C, comparable to constantan, 23 W/m/°C, which reduces errors from this source. Manganin and constantan have comparable electrical resistivities, 45 and 47×10^{-8} ohm·m, respectively, which leads to reduced errors from un-balanced currents in the leads included by electromagnetic heating fields (Chakraborty and Brezovich, 1982). Finally, manganin is much stronger than copper which means less breakage during assembly and use. The same advantages could be achieved by using other thermocouple types such as chromel–alumel (type K), chromel–constantan (type E) and iron–constantan (type J). However, alumel and iron are ferromagnetic which may cause difficulties in some applications such as heating with magnetic induction fields (Chakraborty and Brezovich, 1982).

15.3.6. Fibre optic thermometers

Over the last decade a new class of thermometers has been developed which uses light propagating through optical fibres as a means of trans-mitting information. A primary impetus has come from the necessity for determining temperatures in the presence of strong electromagnetic fields for inducing hyperthermia (Cetas, 1976, 1982a, b; Cetas and Connor, 1978; Christensen, 1979, 1983; Johnson and Guy, 1972). They also will be useful in other applications such as surgery where they intrinsically satisfy the most stringent requirements for electrical leakage. Eventually, they will be interchangeable, disposable and competitive with electronic thermo-meters although, currently, they remain expensive by comparison.

The instrumentation required to read each type of optical thermometer is peculiar to the given probe. A source is required which sends light to the sensor. Light returning from the sensor carries the temperature-related information through a second path to a photodetector where it is converted to an electronic signal which is manipulated to display the temperature. Most systems employ microprocessors so that the signal can be read in terms of temperature units and so that several probes can be multiplexed into one reading instrument. In some cases one fibre or fibre bundle is used to conduct the light to the sensor and another to carry the return signal. In others, a fibre optic directional coupler (or beamsplitter) is used to combine the two light paths into one fibre.

Several different types of sensors have been proposed, and new ones are added frequently. Hence any list is bound to be a strong reflection of the author's awareness of the state of the current literature and is likely to be incomplete. Nevertheless, I will list some of those that have been developed into commercial products. The first probe used a liquid crystal sensor. Several investigators worked on variations of this idea, but a group in Utah (Rozzell *et al.*, 1974) carried its development furthest and their device was marketed by the Ramal Corporation. A cholesteric liquid crystal undergoes a colour change with temperature within a certain finite range. If red light is incident on the crystal, the intensity of the reflected light will be a function of the sensor temperature. Unfortunately, liquid crystals do not appear to have the stability necessary for thermometry. They drift both in terms of colour play with temperature and in terms of the magnitude of the reflectance. Furthermore, they appear to show some hysteresis effects. This unit is no longer available.

One problem with many optical thermometers has been the drift in the wavelength of the emitted light with temperature of the gallium arsenide light-emitting diode. Christensen and his colleagues (Christensen, 1977; Vaguine *et al.*, 1984) exploited this effect by recognizing that its reciprocal, absorption of light, is also temperature dependent. They constructed a thermometer by placing a small chip of gallium arsenide on the end of the fibre bundle. They overcame the difficulty of light losses due to spurious scattering at the end of the fibre bundle by cutting the sensor in the shape of a prism such that light from the source fibres is offset by internal reflection in the sensor and directed into the photodetector fibres. The temperature characteristic of this probe is approximately linear over a broad range (of the order of 50°C) which is determined by the characteristics of the light source and the crystal. These probes range from 0.3 to 0.6 mm in diameter (Vaguine *et al.*, 1984). The unit is marketed by the Clinitherm Corporation and includes a built-in calibrator for the probe. Some users (Shrivastava *et al.*, 1988) experienced instabilities with the calibration of this instrument. Early versions had difficulties if the probe was near moisture. The use of closed-end catheters reduced this drift but did not eliminate it. Clinitherm literature indicates that improved versions of the probe in which these drifts are reduced have become available.

Wickersheim and Alves (1979) developed a fluorescent decay thermometer which was marketed by Luxtron as the Fluoroptic thermometer system (Model 2000). Light from a xenon flash tube excited a mixture of phosphors similar to those used in cathode ray tubes. The ratio of two wavelengths of fluorescent light is used as the temperature-sensitive parameter. Thus, the effects of bending losses in the optical fibre leads are reduced but a relatively complicated optical network is required to process the signals. Resolution is of the order of 0.1 °C and stability within a few tenths of a degree. An industrial version of this unit is available with a lower resolution and a wider temperature range.

Recently, Luxtron introduced a new type of fluorescent decay thermometer (Model 3000) based upon the technologies of Samulski and Shrivastava (1980) and of Scholes and Small (1980). In this system, the rate of decay of a fluorescing crystal is the temperature-dependent parameter which can be determined by measuring the time for the emitted light to fall from one intensity level to a lower level. The indicated temperature value is not related to the analogue intensity level but rather to a time interval. Problems with changes in bending losses, in connector stability and repeatability and in drifts in the light source are all eliminated. This unit appears to be a stable and repeatable system as substantiated by an evaluation (Samulski and Lee, 1986). The Model 3000 is also available from several hyperthermia system manufacturers.

A significant characteristic of fibre optic thermometers is that they are insensitive to the strong electromagnetic heating fields used in hyperthermia. They also have several other attractive features. Most are very small; 1mm diameter now is considered large. Optical fibres have relatively low thermal conductivity, so the thermometers cause little thermal perturbation and measure the temperature of a very small region near the sensor. A consequence of their small size is that they have very fast time constants. The time constants increase somewhat by using protective Teflon jackets, but these can be removed in special circumstances.

15.4. Calibration laboratory

Accuracy to 0.1 °C cannot be guaranteed in off-the-shelf thermometers without individual calibration. Small probe thermometers mounted either in flexible catheters or in hypodermic needles are not direct reading, interchangeable and stable at this level of accuracy. Thus, a calibration facility accurate to 0.02 °C should be part of every laboratory or clinic in which temperatures are a significant parameter. When secondary, working thermometers are calibrated against the standard thermometer, uncertainties still should be well below the levels of biological interest. Furthermore, calibration drifting in the secondary probes can be detected before it becomes significant. Figure 15.9 illustrates typical reading accuracies for three different types of direct-reading analogue systems. $T_{reading}$ is the value indicated by the thermometer, while T_{true} is the value maintained on the standard. For thermometers and thermocouples the shape of the curve depends on the electronic

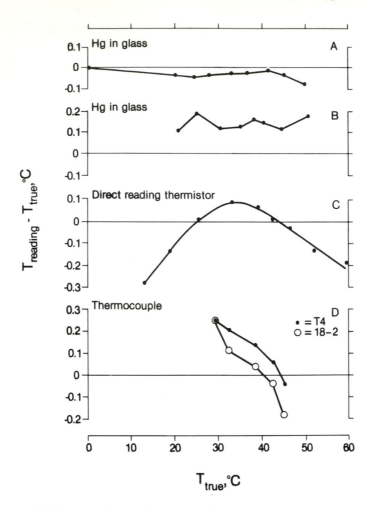

Figure 15.9 Calibration data on four typical direct reading analogue thermometer systems. A: Hg-in-glass with good accuracy. B: Hg-in-glass which shows a zero shift of ±0.15°C which can be corrected by recalibration in an ice point cell. C: Direct reading thermistor system. Other 'interchangeable' probes read by the same unit showed the same shape of calibration curve, but were shifted vertically on the scale. D: Two thermocouple probes read by the same reader (Cetas, Connor and Manning, 1980).

reader; the absolute values of temperature depend on the specific compound probe (two thermistor beads in one probe).

With the advent of computerized data acquisition system, or even instrumentation with microprocessors incorporated, it is possible to dramatically improve the precision and accuracy of a readout in temperature units. For example, Figure 15.3 shows the results of a calibration of a thermistor probe. The calibration coefficients for the Steinhart and Hart equation, as discussed earlier, can be stored in the computer and accurate temp-

Thermocouple Calibration

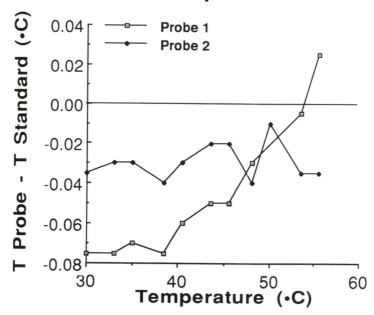

Figure 15.10 Calibration data on two typical thermocouples. The emfs were measured with a computerized data acquisition system and temperatures were computed by an analytical algorithm stored in the computer from the measured emfs.

eratures computed and displayed automatically from the measured resistances. Similarly, Figure 15.10 shows typical calibration data for two copper–constantan thermocouples read by a Hewlett-Packard (HP) 3497A data acquisition system. In this case, the thermocouples were referenced to a thermally lagged (but not temperature-controlled) copper plate, the temperature of which was monitored by four stable thermistors. The internal HP algorithm was used to convert thermoelectric potentials to temperatures. We have tested a wide range of thermocouples, both commercial and constructed in our laboratory from thermocouple-grade materials, and have found them accurate, and interchangeable, to about 0.1 °C. The particular data in Figure 15.10 are typical and were selected by chance from over 80 plots.

Two approaches to a calibration facility are possible. The first, and most common, is to maintain a standard thermometer which has a calibration that can be referred back to the ITS-90. Some type of temperature-controlled medium, such as a water bath or a copper or aluminium block, is used to maintained the unknown and standard thermometers at the same temperature for comparing readings. The other approach is to use established cells of natural fixed point temperatures, such as the ice point, 0°C, or gallium point, 29.764°C, and observe the thermometer reading at these

Table 15.1 Intercomparison data (9/85) on five standard-quality thermistor thermometers in the hyperthermia laboratory at the University of Arizona. Thermometer 113 is the designated laboratory standard. The dates and source of the primary calibration on each is given in the table footnotes.

T(113)	T(194)	113–194 (millidegrees)	T(442)	113–442 (millidegrees)	T(459)	113–459 (millidegrees)	T(6322)	113–6322 (millidegrees)
16.356	16.356	0	16.352	4	16.314	42	16.370	−14
20.590	20.589	1	20.586	4	20.585	5	20.595	−5
25.218	25.218	0	25.214	4	25.214	4	25.220	−2
30.088	30.088	0	30.085	3	30.084	4	30.087	1
34.932	34.933	−1	34.930	2	34.930	2	34.932	0
40.251	40.252	−1	40.249	2	40.250	1	40.253	−2
45.258	45.260	−2	45.257	1	45.258	0	45.263	−5
50.162	50.163	−1	50.162	0	50.164	−2	50.174	−12
54.619	54.619	0	54.617	2	54.617	2	54.626	−7

Thermistor 113 Calibrated by Thermometrics 6/76.
Thermistor 6322 Calibrated with T113 11/7/84.
Thermistor 194 Calibrated with T113 6/12/78.
Thermistor 442 Calibrated by Thermometrics 9/81.
Thermistor 459 Calibrated by Thermometrics 9/81.

points. Each of these approaches is discussed below.

Precision mercury-in-glass thermometers are available from a number of scientific suppliers and can be read to this accuracy, especially if a reading lens is used. Wise and Soulen (1986) of the National Institute of Standards and Technology (NIST, formerly National Bureau of Standards, NBS), have prepared a very useful manual on the calibration and use of these thermometers. In addition, NIST has developed special thermometers (SRM 933 and 934) for their Standard Reference Materials programme which have precise calibrations at four specific temperatures: 0, 25, 30 and 37°C.

We use a standard thermistor thermometer to maintain our scale. The calibration of this thermometer can be referenced back to NIST and is recorded with millidegree precision. Over the 13 years we have had it in our laboratory, we have not detected drifts at the ice point within our measurement uncertainty (≈ 3 m°C). Table 15.1 shows the result of inter-comparisons of five stable standard-quality thermistor probes in our laboratory. The dates and sources of the original calibrations for each are given as well. In addition to these data, further confirmation of the accuracy of our laboratory standards is the fact that stable thermistors calibrated against these standards are accurate in a gallium cell to within 5 mK. A second observation is that very accurate calibrations can be made with straightforward laboratory instrumentation. A constant current source of 10–$100\mu A$ and a 5 1/2 digit voltmeter with a 0.1 V scale is all that is required for this level of precision. Four-lead techniques with current reversals as discussed earlier are used for standard thermometer intercomparisons.

Thermistors have the advantage of high sensitivity over a selected range, and can be calibrated with five or so accurate data points. Industrial platinum resistance thermometers are more linear, but require somewhat more sophisticated equipment to attain the same sensitivity. The ideal primary thermometer is the SPRT, which is the defined interpolation instrument for the IPTS–68. Unfortunately, these and the instruments required to use them are expensive. To summarize, electronic thermometry is easier to use, especially to automate and record, than that based on mercury in glass. It is more expensive, however, since auxiliary instrumentation is necessary to obtain a reading.

A water bath with a good temperature controller and high fluid circulation is the most common apparatus used for thermometer calibration. Since thermal gradients and fluctuations can exist, care must be taken that the primary and secondary thermometers are at the same temperature when the calibration point is recorded. These effects are especially noticeable for thermocouples mounted in hypodermic needles. In order to average out spatial and temporal variations of the order of 0.2°C in our bath, we (Cetas and Connor, 1978) constructed an aluminium calibration block inside a can which is placed in the bath (Figure 15.11). The main heat

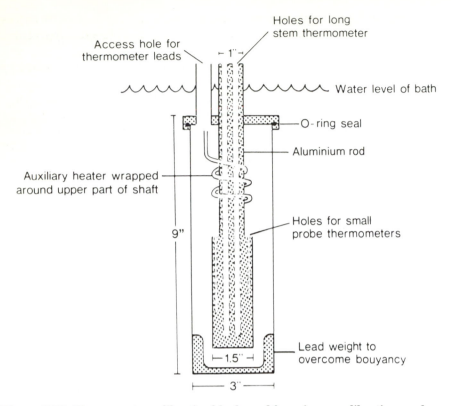

Figure 15.11 Thermometer calibration block used for primary calibrations and standard thermometer intercomparisons (Cetas and Connor, 1978).

flow is along a central aluminium rod which is the main thermal contact with the bath. When the bath stabilizes, the block comes to the average temperature of the bath, and no significant gradients can exist, so the thermometers mounted in the block are all in equilibrium at the same temperature. More time is required for the block to attain thermal equilibrium for each calibration point than when the bath is used alone, so calibration runs take longer. A booster heater on the block speeds the process somewhat. If only small thermometers need calibration, a block such as this can be packed in polyurethane insulation, and an electronic controller with a resistance heater can be used to maintain the temperature. It is important to consider the heat flow paths that could result in temperature gradients. Automated calibration systems are common in many laboratories and in commercial hyperthermia systems. Calibration accuracies of the order of 0.01 °C are possible if adequate time is allowed for each data point. If the data are taken too quickly, thermal gradients will remain in the calibration block and errors of 0.1–0.5°C or greater are likely. Automation will bring a calibration block near the set temperature quickly, but the intrinsic thermal time constant (of the order of minutes) of the system will govern the ultimate speed with which accurate data can be acquired. Usually, it is necessary to let the block stabilize for up to three time constants.

Occasionally, calibration systems are designed which use a continuous warming (ramp) technique. Because the heater is not uniformly distributed throughout the block, heat is flowing and by Fourier's law, gradients must exist within the calibration block. Ramp techniques are probably better with liquid baths than with solid blocks but, even then, errors can result due to a thermal lag between the sensor and jacket of the probe. An illustration of this problem is given in the later section on sources of error in clinical measurements. It follows that this form of calibration must be undertaken only with caution.

Two other precepts of thermometry are that no thermometer is necessarily stable and that none is indestructible. Hence, three standard-quality thermometers must be maintained to determine which one may have drifted, and all thermometers must be checked regularly, especially the standards. We have found that for various reasons, including operator carelessness, thermometer element deterioration and several unknown factors, thermometers have a finite lifetime. In our laboratory, the average lifetime is of the order of several months, except for our well-guarded standards.

Thermometric fixed points also can be used for calibration. Indeed, this is the basic philosophy behind the world standard of thermometry, the IPTS-68 and its recent update ITS-90. Table 15.2 shows several fixed points that are of interest to studies requiring precise thermometry between 0 and 100 °C. The most readily available is the ice point. It is easy to set up and, if done properly, will reproduce much better than 0.01 °C. The technique is to crush ice from distilled water, pack it into a clean vacuum flask (Dewar), and add just enough distilled water to provide thermal contact. Any excess water which causes the ice mass to float should be siphoned off and replaced with crushed ice. The purist, of course,

Table 15.2 Fixed point temperatures (ITS.90) of use for calibration and verification of thermometers. See text for citations.

Fixed point	Temperature (°C)
H_2O (f.p.)	0.00
H_2O (t.p.)	0.01 (exact)
Ga (m.p.)	29.7646
$Na_2SO_4 \bullet 10H_2O$ (s.h.p.)	32.367
Rb (m.p.)	39.29
$KF \bullet 2H_2O$ (s.h.p.)	41.411
$Na_2HPO_4 \bullet 7H_2O$ (s.h.p.)	48.209
Succinonitrile (t.p.)	58.0642
H_2O (b.p.)	99.975

f.p., freezing point; t.p., triple point; m.p., melting point; s.h.p., salt hydration point; b.p., boiling point.

will instead use a triple-point cell (0.01°C=273.16K), which provides better reproducibility and is basic to the definition of the unit of temperature. These are more expensive to set up and maintain, however. The melting point of very pure (99.9999%) gallium has been shown to be very useful as a reference temperature for calibration (Mangum and Thornton, 1977; 1979; Thornton, 1977; Sostman, 1977a, b). Complete cells ready for use can be obtained from the NIST and from the Yellow Springs Instruments Company. Similarly, rubidium and succinonitrile (SCN) cells can be obtained from the NIST office of Standard Reference Materials. According to the papers by Mangum and his colleagues (Mangum, 1983a, b; Mangum and El-Sabban, 1986), SCN is the better of these two. Three points are sufficient to determine the three coefficients in the Steinhart and Hart equation noted earlier, and Mangum suggests the fixed points of ice, gallium and SCN listed in Table 15.2. Thermocouples are more linear and the three points are adequate for them also.

We use a gallium cell to check the stability of our thermometers on a regular basis. The cell is placed in a water bath at 28°C and is allowed to equilibrate. The temperature of the water bath is raised to about 30°C, or about 0.3°C above the melting point at 29.772°C or 29.764 (ITS-90). After about 30 min, the melting plateau is established and the temperature of the cell remains stable for several hours as illustrated in Figure 15.12. Thermometers are placed in the re-entrant well of the cell, one after another, and their readings are compared with the published value of the

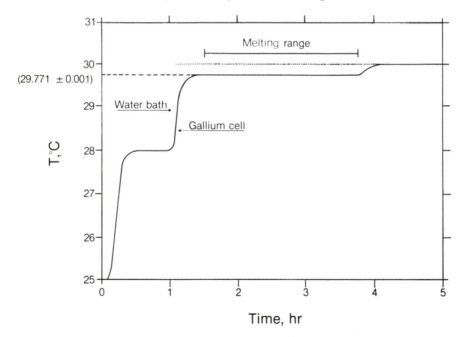

Figure 15.12 Illustration of the melting characteristics of a Ga melting point cell.

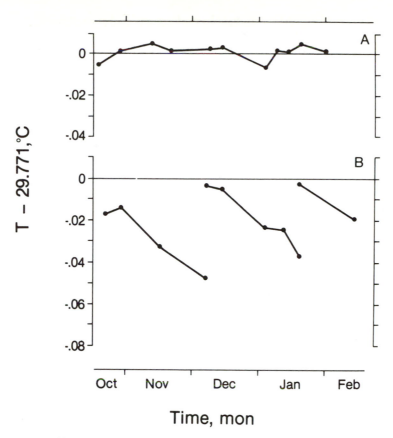

Figure 15.13 Characteristics of two calibrated thermistor probes over several months when checked in a Ga cell. The probe labelled B was recalibrated in early December and mid-January and finally failed in February (Cetas and Connor, 1978).

melting point. Figure 15.13 shows gallium-fixed point data taken over several months on two thermistor probes which were purchased from the same supplier at the same time. The one probe remained very stable while the other drifted off, was recalibrated and drifted off again. This occurred several times until it finally failed.

A note of caution: we found that after about 18 months, the melting temperature of a prototype gallium cell we were using became unstable and somewhat erratic. We first noticed it by observing that all newly calibrated thermistors began to show a calibration bias when compared to the gallium melting point. Intercomparisons of several stable thermistors at several temperatures indicated that the melting temperature of the cell was drifting. The shape of the melting curve at the beginning and end of the melt was consistent with an impurity in the cell (Thornton, 1977). We then obtained another gallium cell and were able to confirm the fact that the selected thermistors were stable and that the prototype gallium cell

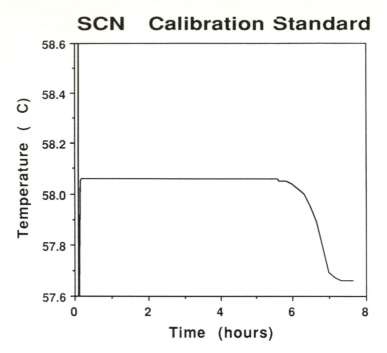

Figure 15.14 **Illustration of the freezing characteristic of a SCN cell.**

was not. For perspective, however, this erratic behaviour was still only of the order of 0.01 °C, which is insignificant in most biomedical situations.

An SCN cell (Mangum and El-Sabban, 1986) is used somewhat differently from a gallium cell in that it is more stable upon freezing as opposed to melting. Figure 15.14 shows the time–temperature characteristic for the SCN cell which we are using. The cell is first melted completely and then the cell is placed in a bath at 57.62 °C. Just prior to beginning thermometer calibrations a cold rod is slipped into the re-entrant tube to start a freeze of the cell about the tube. The rest of the procedure is similar to the use of a gallium cell. An electronic controller analogous to that marketed by the Yellow Springs Instruments Company for the gallium cell has been constructed in our laboratory to make the procedure more convenient for routine thermometer verifications (D. Anhalt, unpublished work).

Magin and coworkers (1981) devised a technique for using the hydration–dehydration point of certain salts in solution as temperature fixed points (see Table 15.2). In simple terms, fresh reagent-grade chemicals are obtained from the local chemistry stores and are heated a few degrees above the salt hydration transition temperature. The salt is poured into a vacuum flask and is stirred mechanically. A small-diameter glass tube, sealed at one end and filled with thin mineral or silicone oil, is inserted into the salt solution and is used as a re-entrant well to protect the thermometers from the caustic salts. The thermometers are placed in the well and are calibrated as with gallium cells. New salts are purchased for each calibration to avoid contamination.

In conclusion, calibration facilities can be set up without unreasonable capital cost, especially when compared to the overall cost of a medical or clinical programme which requires accurate ($\pm 0.1\,°C$) thermometry. The most significant expense is perhaps that of maintaining an appropriately skilled, careful technician who can calibrate the thermometers initially and periodically check them for drift. A fully automated, microprocessor-controlled calibration unit would be helpful to laboratories that do not wish to invest time in training an individual to develop and maintain a calibration facility. Alternatively, three fixed point cells (ice, gallium, SCN) are simple to use and yield accurate calibrations.

15.5. Quality assurance

Quality assurance routines are necessary in clinical hyperthermia programmes. While many of the concepts have been mentioned already, it is useful to outline the basic approach to ensuring the quality of the thermometry systems. First, laboratory-standard thermometers and a calibration system are necessary and this has been outlined in the above section. Second, many hyperthermia data-acquisition systems have internal standards and it is necessary to verify the accuracy of the entire system including both the sensor and the readout. That BSD uses standard-quality thermistors (Thermometrics model S-10) suggests that the probes will be stable and accurate as is shown in Table 15.1. Nevertheless, the reading accuracy of the system must be verified periodically by checks against external resistance and temperature standards. The third step is to calibrate the working clinical thermometers, usually according to the manufacturer's protocol. This, then, provides for calibrated probes that are ready for use.

The next important step is to establish procedures for routinely checking thermometry systems prior to each clinical use. The most convenient check is to insert the probe into a patient and then judge if the reading is sensible. The reliability of this approach is not always adequate since temperatures in patients can vary by a few degrees, especially in superficial locations and each patient's basal temperature will differ from other patients' and with time. Another easy test is to use a gallium cell. We had a great deal of confidence in this procedure until on three separate occasions we found that certain clinical probes read correctly to within a few tenths of a degree at $29\,°C$, but were off by as much as $2\,°C$ near $42\,°C$! (see Figure 15.15). Thus, it is necessary to have an independent check of each thermometer near the working temperature immediately prior to use. As a consequence, we now use both gallium and SCN cells for this purpose. Verifications are straightforward if blind-ended catheters are

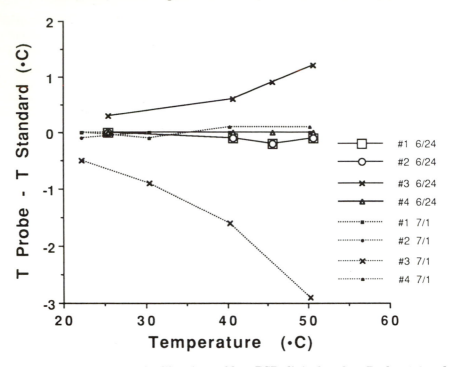

Figure 15.15 Verification of calibrations of four BSD clinical probes. Probes 1, 2 and 4 were stable to within 0.1°C, the accuracy of the data acquisition system. Probe 3 drifted off one direction one week and the other direction the next week. All four probes were considered acceptable for clinical use when checked in a Ga cell at 29.77°C.

used since the probes need not be sterile. In other cases it is more difficult due to the need of maintaining sterility while verifying accuracy. At present, we are only using sterile thermocouples in open catheters and have found that they tend to fail catastrophically rather than by calibration shifts.

In addition to verification of temperature accuracy, other routine checks should be made. All systems should be checked for basic functionality and each thermometer probe inspected prior to use. With high-resistance systems, such as that by BSD, dirt, moisture, poor contacts and static electrical buildups and discharges can cause difficulties and system failures. On occasion, we have experienced some difficulty of the order of 0.5°C or more with basic probe calibrations. These difficulties are mentioned here as concrete examples of the necessity of routine verifications for any clinical system. Care must be taken to avoid placing confidence even in systems that initially appear to warrant it. Shrivastava *et al.* (1988) have evaluated several thermometry systems as part of their hyperthermia physics quality assurance programme sponsored by the US National Cancer Institute. In another report, Shrivastava *et al.* (1989) published guidelines recommending the frequency at which calibration checks should be made.

15.6. Source of error in clinical measurements

Errors can affect the accuracy of temperature measurements in the clinic independently of those associated with thermometer calibrations. The

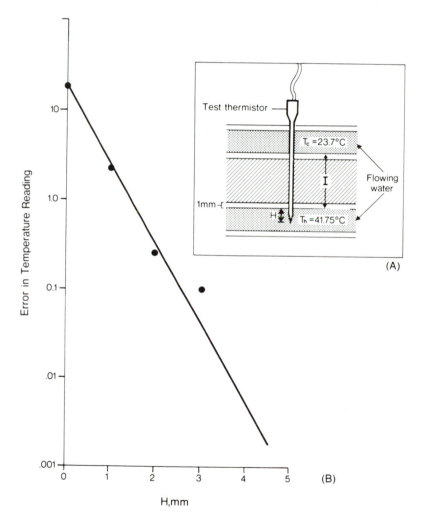

Figure 15.16 Evaluation of thermal conduction errors for a needle mounted thermistor probe. The insert gives the experimental arrangement (Cetas and Connor, 1978).

three sources are related to thermal conduction, to response times and to intense heating fields. Each of these is discussed in turn.

15.6.1. Thermal conduction errors

The leads and supporting structure of a thermometer probe affect the reading of the sensor by providing an extra path for heat transfer from the sensor. Insulating jackets required for sensor protection and biocompatibility reduce the thermal contact of the bead of the tissue. The insert in Figure 15.16 (Cetas and Connor, 1978) illustrates one experimental method of demonstrating the effect. A needle thermistor was inserted through a plastic tube or running water at 23.7°C through an insulator and then into another plastic tube filled with 41.75°C flowing water. The distance between the tubes was held constant at 2.2 cm, resulting in a gradient of 8.2°C/cm. The tip was inserted into the second tube until it registered the correct water temperature. The data are plotted in Figure 15.16 and show that for probe insertions greater than 3 mm, the errors will be less than 0.1°C. The opposite test also was performed. The insertion depth was held at 4 mm and the distance between the tubes was reduced until an 0.1°C error was observed at 7 mm spacing. Finally, the cold-water tube was removed and the needle was inserted directly into the warmer-water tube (from air) until an accurate temperature was recorded, again between 3 and 4 mm insertion depth. The difference between this test and the last is that the thermal contact of the needles to the air is much poorer that that of flowing water.

Figure 15.17 shows a similar experiment reported by Fessenden *et al.* (1984) in which a three-junction thermocouple is compared to a Luxtron fibre optic probe. The gradient in this experiment is 30°C/cm which is perhaps somewhat high for a clinical situation except very near to an interstitial heat source. Note that significant errors are present for the thermocouple until it is immersed 5–7 mm into the uniform, thermally heat-sunk medium (flowing water). On the other hand, the very low thermal conductivity of the optical fibres in the Luxtron probe result in only minimal conduction errors. Samulski *et al.* (1985) have carried out a numerical analysis of this problem and describe various probe and catheter combinations in terms of a characteristic length parameter. Dickinson (1985) evaluated this effect experimentally and theoretically for copper–constantan and manganin–constantan multiple-junction probes.

It is to be noted that the thermal contact between the probe and its catheter is very important. If possible, close-fitting or liquid-filled catheters are helpful in reducing errors from thermal conduction. In the steep thermal gradients, often 10°C/cm or greater, frequently experienced in localized hyperthermia, the measured gradients must be recognized as underestimates.

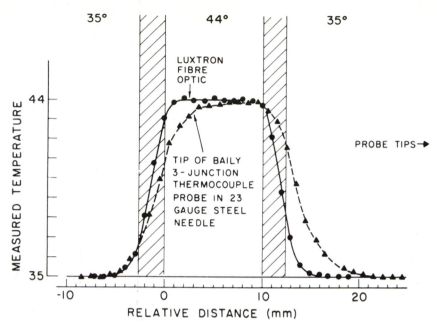

Figure 15.17 Evaluation of thermal conduction errors for a needle mounted thermocouple and an optical thermometer (Fessenden *et al.*, 1984)

To summarize, errors due to thermal conduction can be minimized by using thermometers with leads of high thermal resistance. The thermometers designed for use in intense electromagnetic heating fields (the optical probe as well as the BSD/Vitek units) also satisfy this requirement very well. Good thermal contact of the sensor element to the tissue must be maintained. Coupling through air or thick, insulating jackets should be minimized. Small sensors and probes average temperatures over a small volume rather than a large one and so give better spatial resolution (less thermal smearing). When possible, the thermometer should not pass through a steep temperature gradient. Thus, for invasive thermometry, avoid a measurement where just the tip pierces the skin.

15.6.2. Response time errors

Errors due to the response time of the thermometer probes share many similarities to those related to thermal conduction. The three primary situations in which the time constant of the thermometer can affect the readings are in measuring the temperature in a rapidly warming environment, in determining the temperature transient, dT/dt, for use in determining SAR values, and when analysing power-off and power-on thermal transients to determine artifacts from heating fields. Waterman (1985) has looked at the first effect in a systematic manner as is shown in Figure

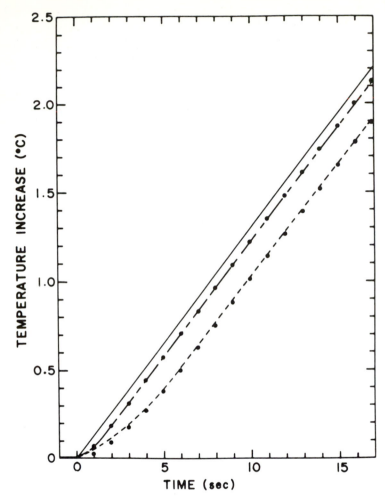

Figure 15.18 Response of a Vitek thermistor thermometer probe following the transition from a constant temperature to a temperature that increases at a constant rate, $dT/dt=7.8C/min$. The true temperature is shown by the solid line (—). The measured response of the bare probe (— —) and of the probe inserted in a closed catheter (– – –) are shown by dashed lines. The closed circles (•) show the response calculated by Waterman for both configurations (Waterman, 1985).

15.18 which is taken from his paper. First, note that because of the time required for the heat to conduct through the catheters and jackets to the sensor, the thermometer signal lags behind the actual temperature of the medium. This is also a good illustration of the gradients and associated errors with calibration systems which use continuous warming (ramp) techniques.

Waterman also points out that the slope of the temperature versus time curve approaches that of the real increase after a brief time lag and so

dT/dt and hence SAR can be determined accurately. However, the initial data recorded within about three time constants (a few seconds) of the probes must be ignored. If blood flow or sharp thermal gradients lead to non-linearities in dT/dt shortly after the power is switched on, then selecting the proper slope for calculating SAR is difficult.

Uncertainties in determining temperatures and heating field artifacts by extrapolation after power-off events can be analysed in a similar manner as the first two since the phenomena are the same. This type of problem is illuminated in the next section.

15.6.3. Errors in intense electromagnetic fields

The nature and significance of the perturbing effect of thermometers in intense electromagnetic fields, has been demonstrated graphically with a thermographic camera by Johnson and Guy (1972) and by others (Cetas, 1976; Chan *et al.*, 1988a). Conversely, some have claimed that their data were only minimally affected by the presence of electromagnetic fields. The discrepancy lies in the fact that many factors influence the magnitude of the perturbation. These include the field strengths, the orientation and position of the thermometer leads with respect to field orientation and phase, the degree of shielding including that due to tissue, the geometry of the subject and of the radiator and the heat sinking of the probe to the medium.

Several techniques can be used to reduce the magnitude of these effects. Electromagnetic interference (EMI) can be prevented by shielding and grounding the instruments. Special care should be taken to avoid ground loops, including those involving capacitive coupling. Radio-frequency filters in the sensing circuitry prevent noise from entering the reading instrument. Chakraborty and Brezovich (1982) and Carnochan *et al.* (1986) describe resonant filters and Vojnovic and Joiner (1985) describe wide-band filters that can be placed in the thermocouple leads.

In the thermocouple data acquisition system which we assembled, a sensor interface box (SIB) serves as the main junction point for thermo-couples and other instruments into the computer. It is connected by a shielded cable to the computer-controlled scanner. Inside the SIB is a heavy copper plate which serves as the thermocouple reference. Mounted in the lower side of the plate are four thermistors for determining the plate (thermocouple reference) temperature. Radio-frequency gasket material between the plate and its mounting isolates the upper chamber of the SIB which is not radio-frequency tight from the lower which is radio-frequency tight. The gasket also thermally lags the copper plate from the chassis. Commercial radio-frequency filters (MuRata Erie, part No. 9910-381-6004) screwed into the plate, one for each thermocouple lead, provide termin-ations for the thermocouples and block radio-frequency noise from passing through to the measuring instrumentation. This leaves only the other two sources of interactions of the probes with the electromagnetic field — direct thermocouple sensor heating and perturbation of the heating fields by the presence of the thermocouples.

Chakraborty and Brezovich (1982) discuss in detail thermocouple errors when used in 13.56 MHz electromagnetic fields. Apart from the EMI discussed above, they list three sources of induced temperature changes in the junction of the thermocouple: (a) junction current heating in which the unequal resistances of the thermocouple leads cause a net current flow from the higher- to lower-resistance wire across the junction, (b) heating in the surrounding resistive material (tissue) and (c) eddy current heating of the thermocouple wires in the oscillating magnetic field. They conclude that if care is taken, the errors will be less than 0.1 °C at 13.56 MHz. Dunscombe and his colleagues (Dunscombe *et al.*, 1986, 1988; Constable *et al.*, 1987), have attempted to quantify the magnitude of the temperature errors both experimentally and theoretically. Their analysis is informative in terms of improving our understanding of the nature of some of the errors from self-heating. Part is due to heating of the sensor in its sheath and part is due to some of that excess power absorption being conducted out to warm the media or tissue. The latter, then, is additional tissue heating that arises from the presence of the thermometer probe. Finally, as they note, this effect is distinct from that discussed by Chakraborty and Brezovich, and others (e.g. Johnson and Guy, 1972; Cetas, 1976), for which the presence of a metallic probe perturbs the electromagnetic field and hence modifies tissue heating. The final conclusions to be drawn from this series of papers by Dunscombe and co-workers is that corrections are dubious. It is really only acceptable to use methods for which the artifacts are demonstrably small to insignificant.

Chan and his colleagues (Chan *et al.*, 1988a, b) also have used thermography to demonstrate errors from probes implanted in electromagnetic fields generated by microwave applicators from commercial superficial hyperthermia systems. First they note that the probes cause perturbations in the field when a component of the electromagnetic field is parallel to the probe and the perturbation is dependent upon the extent to which the probe is inserted under the applicator. Then they go on to show that a similar effect can arise from the presence of the dielectric catheter. This is a consequence of replacing lossy media with the lossless catheter, resulting in an electromagnetic cold spot.

Astrahan *et al.* (1988) considered a somewhat different problem, that of the accuracy of thermometer probes embedded in microwave interstitial microwave antennas. They show that this configuration is not particularly good for accurate thermometry since heating will occur in the antennas due to reflections from the inevitable electromagnetic mismatch of fixed-design antennas inside catheters inserted into a variety of tissues with varying properties. The effects of the electromagnetic mismatch are enhanced by the variation of thermal heat sinking through the catheter walls. They suggest the caveat that this configuration is useful for dynamically controlling the relative power to a number of antennas but do not present evidence to demonstrate it.

We performed several studies with localized, single-aperture microwave sources at 915 MHz and 400 MHz, with a regional heating system at 80 MHz and a magnetic induction system at 100 mHz used for heating ferromagnetic seeds implanted in tumours. Our results are consistent with those of Chakraborty and Brezovich (1982). In particular, with care, thermocouples can be used with bolus-coupled aperture microwave systems and with the 100 kHz magnetic induction system. Systems which produce extensive radiative fields introduce more difficulties in thermometry.

An experiment with a single microwave applicator operating at 915 MHz will illustrate some of the ways in which spurious heating is manifested in thermocouple thermometry as well as means of reducing the

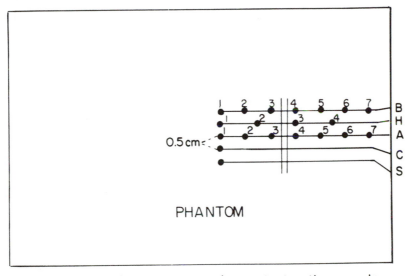

B	Multiple sensor manganin-constantan thermocouple
H	Multiple sensor Clinitherm Optical thermometer
A	Multiple sensor copper-constantan thermocouple
C	Single sensor Clinitherm Optical thermometer
S	Single sensor thermocouple

Figure 15.19 View of the top surface of the experimental phantom configuration for evaluating the artifacts induced in thermocouple probes during electromagnetic heating. The various probes, as indicated in the legend, were laid into shallow grooves in the muscle phantom surface, and began to emerge from the phantom about 3 cm from the tips as indicated by the double vertical lines. Microwave sources operating at 915 MHz were centred approximately over probe A, sensor 1, and were either in direct contact through a water bolus or were spaced 5 mm above the surface.

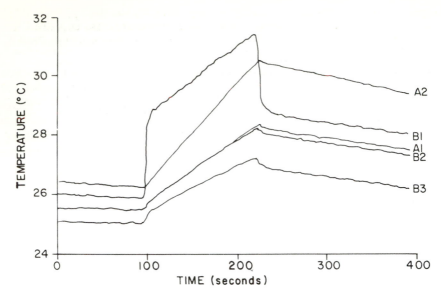

Figure 15.20 Temperature versus time curves on five of the sensors for Experiment I discussed in the text. The experimental configuration is shown in Figure 19 and the results tabulated in Table 3.

effects to manageable proportions. Figure 15.19 shows the experimental arrangement. Various thermocouple and optical thermometer probes, as indicated in the figure, were laid into the surface of muscle equivalent phantom materials. The tips of the thermometer probes were in good thermal contact with the phantom, but began to come away from the surface about 3 cm back from the tips. For the multiple-junction probes, this meant that sensors 1 to 3 were in good thermal contact while 5 through 7 were not. The A probe is a copper–constantan thermocouple supplied by Sensortek, but with a DB-type connector to interface with our SIB. The B probe is manganin–constantan, has a very fine silica sheath in close contact with the thermocouple wires and has its first sensor very near the tip. Two experiments are illustrated here. Experiment I used a water-bolused applicator, centred near the tip of probe A. Experiment II used an air-coupled applicator, similarly centred, but positioned 5 mm above the surface. The field in both cases was polarized parallel to the thermocouple leads. Figure 15.20 shows the resulting temperature versus time curves for five of the sensors when about 50 W forward power was applied. Note that for sensor A1 a small artifact is discernible, while for B1 a very large artifact is obvious. From the warming slope of these curves, the local SAR was calculated from SAR$=C\mathrm{d}T/\mathrm{d}t$ where C is the phantom specific heat, T is temperature and t is time. The magnitude of the electric field, E, was computed from $E=(\rho\mathrm{SAR}/\sigma)^{\frac{1}{2}}$ where ρ is the phantom density taken to be 1000 kg/m^3 and σ is the phantom electrical conductivity of 1.27

Table 15.3 Results of tests to determine the magnitude and effective parameters of electromagnetically induced artifacts in thermocouple probes. The experimental configuration is shown in Figure 15.17. S1 is a single sensor thermocouple from Sensortek; A1 ... A7 are sensors (tip is A1) on a copper–constantan multijunction probe; B1 ... B7 are sensors (junction tip is B1) on a manganin–constantan multijunction probe. The two experiments are described in the text. Δ T is the measured artifact (see Figure 15.18); SAR is the specific absorption rate in the phantom determined from the slope of the heating curve; E is the magnitude of the electric field computed from the SAR; Δ TN is the artifact (Δ T) normalized to an SAR of 100 W/kg.

Probe	ΔT (°C)	SAR (W/kg)	E (V/m)	ΔTN (°C)
Experiment I: bolus coupling				
S1	0.08	104.00	286.16	0.08
A1	0.16	80.00	250.00	0.20
A2	0.04	125.00	313.73	0.03
A3	0.11	46.00	190.32	0.24
A4	0.11	20.00	125.49	0.55
A5	0.17	7.80	78.37	2.18
A6	0.44	7.90	78.87	5.57
A7	0.28	5.60	66.40	5.00
B1	2.70	89.00	264.72	3.03
B2	0.12	76.00	244.63	0.16
B3	0.28	56.00	209.99	0.50
B4	0.25	30.00	153.69	0.83
B5	0.50	11.00	93.07	4.55
B6	0.60	7.90	78.87	7.59
B7	0.60	7.90	78.87	7.59
Experiment II: Air coupling				
S1	0.63	73.00	239.75	0.86
A1	0.55	53.00	204.28	1.04
A2	0.31	65.00	226.23	0.48
A3	0.55	40.00	177.47	1.38
A4	0.91	29.00	151.11	3.14
A5	1.30	27.00	145.81	4.81
A6	2.68	33.00	161.20	8.12
A7	1.65	27.00	145.81	6.11
B1	4.72	50.00	198.42	9.44
B2	0.31	53.00	204.28	0.58
B3	0.63	36.00	168.36	1.75
B4	0.79	21.00	128.59	3.76
B5	2.50	15.00	108.68	16.67
B6	2.68	21.00	128.59	12.76
B7	4.50	21.00	128.59	21.43

S/m. The results are tabulated in Table 15.3. The columns labelled ΔTN represent expected artifacts by scaling the measured ΔT value for the observed SAR values to an SAR of 100 W/kg.

Several observations can be made here. First, for experiment I, the bolus-coupled case, the power absorbed by the thermocouples is very small, and is effectively absorbed into the tissues. The net temperature rise (artifact) is of the order of 0.1 °C, even with the leads parallel to the electric field. This is evident for sensors S1 and A1 to A4. Second, the sensor near the tip, A1, has a larger artifact than A2, or even for S1 which is further back from its tip than is A1. The B1 sensor which is very close (≈1mm) to its tip has a large artifact. A third observation from this experiment is that the sensors A5 to A7 and B5 to B7 are not heat sunk as well and the artifact is larger, even though the field intensity (SAR or E) is less. This is even more evident from the scaled artifacts, ΔTN.

Experiment II differed from experiment I by using an air-coupled applicator. The forward power was the same, 50 W, but the reflected power was 16 W, compared to 6 W for experiment I, yielding a net power of 34 W compared to 44 W. The main point to be made here is that the scattered fields are higher, much more of the thermocouple leads are exposed to them and the resulting artifacts are four to eight times larger. In fact, none of these sensors have artifacts that are less than 0.1°C.

Several other analyses and experiments were run. The temperature rise indicated by the thermocouples (including B1) that resulted from the heat pulse, measured with power off, was the same as that measured with the optical thermometer adjacent to it. That is, if the artifact is subtracted, the thermocouple measurements were correct. This also supports the statement that the net power absorbed by the thermocouples was small. Thermographic evaluations were made of the heating, but in the cases studied here, the spurious heating of the phantom was too small to be resolved, either spatially or thermally by thermography. The thermocouple sensors themselves were far more sensitive to these small effects. If the probes were mounted orthogonal to the field, no artifact was observed. Finally, similar studies with the same configuration performed at 400 MHz gave similar results.

To summarize this section then, thermometry can be performed with thermocouples with accuracy under certain conditions. The instrumentation should be filtered, guarded and shielded. The probes must be electrically insulated from the medium; even the capacitive impedance should be high. Thick jacketing on the probe helps this, but it impairs the thermal contact. The sensor must be removed from the extreme tip of the probe. When possible, the probe should be placed perpendicular to the electric fields and the leads should be twisted to reduce magnetic induction currents. The leads should be shielded by tissues when possible and stray fields kept to a minimum. Bolus-coupled applicators help here as well as in other ways. The probes and leads should be small to keep the total absorbed power small and they should be in good thermal contact with absorbing masses to dissipate extraneous power.

These solutions are not entirely satisfactory for a number of reasons. First, they are frequently inadequate, as has been demonstrated, so care must be taken. Shielding a sensor may reduce EMI, but it also may produce a reflection or shunting of the field. Hot spots caused by the presence of the probe may occur somewhat removed from the sensor and not be detected by a switching transient. Finally, switching off the field for a few seconds will result in a fast temperature drop if the probe is in a region of high blood flow or if especially steep temperature gradients are present as is the case with interstitial heating. It is not always obvious whether an abrupt temperature change when power is switched off is due to these effects or to artifacts associated with the probe. Alternatively, minimally perturbing probes can be used with little concern for electromagnetic effects, although in some cases other concerns such as calibration accuracy, noise, stability, convenience or cost may be overriding.

Comments with regard to thermocouples should not be generalized to include thermistors. The high-resistance bead at the end of the leads electrically differs from a low-resistance thermocouple junction. The goal here was to determine the conditions for which thermocouples can be used and to indicate the tests necessary to validate any proposed method using electrical probes. A few brief tests in our laboratory suggest that thermocouples are easier and more reliable than thermistors for the purposes discussed here. However, detailed comments are inappropriate since these were not systematic tests.

15.6.4. Errors in intense ultrasound fields

Errors in the presence of strong ultrasonic fields have been understood in general by practitioners of the discipline for a number of years (Fry and Fry, 1954a, b; Lele, 1980; Hynynen *et al.*, 1983; Wells, 1977), but only recently have the problems been discussed and appreciated by others in the field of hyperthermia. The errors are associated with self-heating of the probe, including, perhaps, adjacent tissues, and with perturbation of the ultrasonic field by the probe. Hynynen and Edwards (1989) in an excellent paper on thermometry in ultrasonic fields discuss these effects in detail. This paper should be studied by anyone using ultrasound to induce hyperthermia. The errors are of the same order of magnitude as in electromagnetically induced hyperthermia. Self-heating arises from two principal effects, preferential absorption of ultrasound by plastic coatings and heating from viscous flow effects between media of differing mechanical properties such as wire and soft tissues. Figure 15.21, taken from Fessenden *et al.* (1984), illustrates dramatically the effect of the preferential absorption of ultrasound in a Teflon catheter and compares these with the minimal perturbation in a thermocouple encased in a metal needle. The ultrasonic energy absorption coefficient (in dB/cm) of polyethylene is approximately five times greater than that of soft tissues (Wells, 1977). Of course, thermal conduction errors are worse in the needles so this solution is not particularly good either. Kuhn and Christensen (1986) have measured the artifacts that result with a number of different sheathing materials and displayed them as a histogram (Figure 15.22). Of particular interest is the silica tubing which has a small

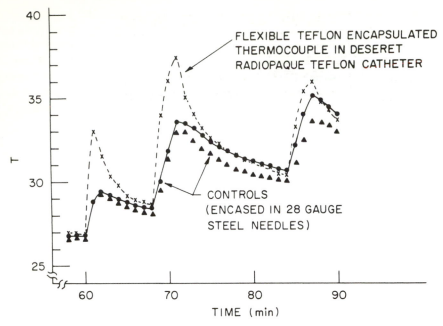

Figure 15.21 Examples of temperature artifacts from Teflon encapsulated thermo-couples in Teflon catheters compared to that from needle encapsulated thermo-couples when subjected to ultrasonic heating field (Fessenden *et al.*, 1984).

residual artifact of the order of 1°C. We have found this approach to be effective in our laboratory as well, although we have trouble with the tube breaking when it is tightly filled with thermocouple leads. Hynynen and Edwards (1989) describe a probe (Figure 15.23), on which short sections of stainless steel reduces the viscous heating dramatically. Sectioning the stainless-steel rod recovers the flexibility of the probe and reduces errors from thermal conduction down the shaft.

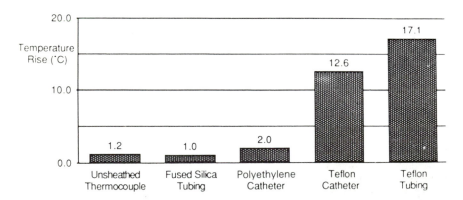

Figure 15.22 Histogram illustrating the magnitude of the ultrasonic induced temp-erature artifact for several jacketing materials (Kuhn and Christensen, 1986).

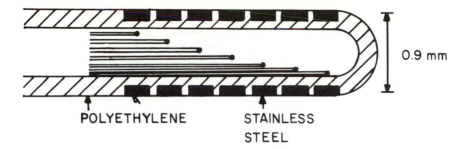

POLYETHYLENE   STAINLESS
                    STEEL

Figure 15.23 Diagrams of the thermocouple probe showing stainless steel around polyethylene probes (Hynynen and Edwards, 1989).

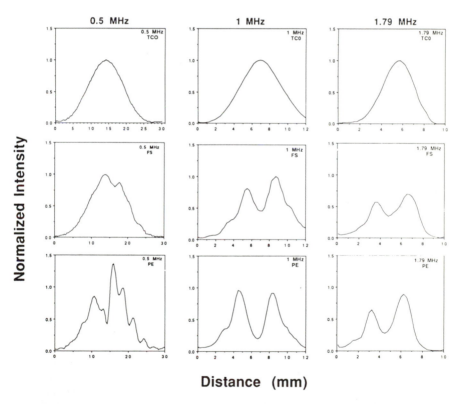

Distance (mm)

Figure 15.24 The normalized (to the peak value of the undisturbed field) intensity profile across the acoustical focus of three different transducers for the undisturbed case (top), with 0.7 mm diameter fused silica (middle), and with 0.96 mm polyethylene (bottom) seven sensor thermocouple probes 1 mm in from the scan. Results with three different frequencies are presented in the graph (Hynynen and Edwards, 1989)

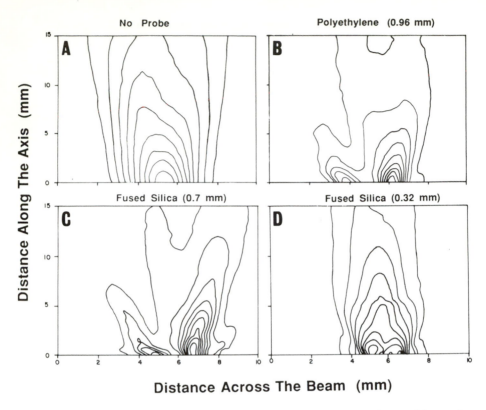

Figure 15.25 The normalized iso-intensity contour (contour interval 10% of the peak value) plot of the axial ultrasonic field from a transducer beyond the acoustical focus when tubes of different material were placed at the acoustical focus 1 mm in front of the scan. (a) Undisturbed field, (b) polyethylene 0.9 mm o.d., (c) fused silica 0.7 mm o.d., and (d) fused silica 0.32 mm o.d. (Hynynen and Edwards, 1989).

A second artifact is due to viscous heating resulting from the relative motions of two media of differing ultrasonic properties (Fry and Fry, 1954a, b). Hynynen *et al.* (1983) have shown that this error is proportional to the ultrasonic intensity which is consistent with the theory of Fry and Fry (1954a). Since it is highly localized at the interface, it decays rapidly, within 200 ms for bare thermocouples. Coatings often increase this effect due to increased shear viscosity in addition to having greater absorption coefficients as discussed above. Finally, these effects are orientation dependent and hence are difficult to calibrate out. The only means of avoiding this error is to switch off the field and estimate the artifact from the temperature transient as was discussed for the electromagnetic case. With scanned focused ultrasonic systems, it is possible to use scanning patterns which do not strike the probe directly or, alternatively, to switch off the field as the beam passes the sensor.

The remaining artifact is that of the distortion of the ultrasonic field by

Optimal strategies for superficial bulky malignancies

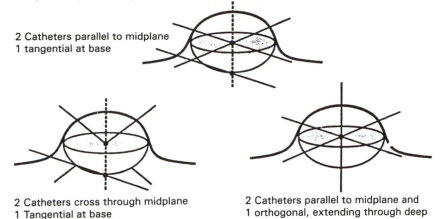

2 Catheters parallel to midplane
1 tangential at base

2 Catheters cross through midplane
1 Tangential at base

2 Catheters parallel to midplane and
1 orthogonal, extending through deep
base along the midaxis

Note: Catheters extend beyond margin of tumour when possible,
thereby allowing for measurement of normal tissue temperature

Less Desirable Catheter Placement Strategy

1. Catheters that do not intersect at
   midplane in terms of depth

2. Catheters that intersect away from
   central axis

Unacceptable Strategies

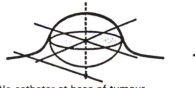

1. No catheter at base of tumour
   or catheter misses base of tumour

2. Catheters that do not intersect

Figure 15.26 Strategies for probe placement suggested by Dewhirst, *et al.*, 1990.

the presence of a thermometer probe with dimensions comparable to the
wavelength of the ultrasound beam. Figure 15.24 and 15.25 (Hynynen and
Edwards, 1989) illustrate this very well. For low-frequency beams and
small probes, the error is insignificant, but shadowing and diffraction
effects become important as the ratio of probe diameter to $\lambda^{1/2}$, where λ is

the wavelength, becomes larger. The effect is seen from another perspective in Figure 15.25 where the shadowing behind the beam is apparent and appears similar to waves near a pier. The authors indicate that for probe diameters less than $0.2\lambda^{1/2}$, with all parameters in millimeters, the errors are detectable but are probably insignificant. In this matter, as in all others in this chapter, the original papers should be consulted for a complete understanding.

The most difficult aspect of thermometry or thermal dosimetry in ultrasonic fields is common to all forms of heating by propagating energy waves. Heterogeneous tissues and the interfaces between them cause reflections, refractions and scattering of the waves. Consequently, regions of high energy absorption (hot spots) or low field intensity resulting from shadow or wave interference are very difficult to predict. Bone–soft tissue interfaces and air–soft tissue interfaces are particularly troublesome with ultrasound. Reflections from bone will cause hot spots in the soft tissue just in front of the bone. Conversely, if the coupling to the bone becomes very good, then the high absorption of ultrasound by bone will cause the bone to heat preferentially. Measurements of the temperature at single points with a probe will not necessarily imply anything at all about the temperature of a point of a few millimeters away.

15.7. *Application to the clinic*

The principles of thermometry have been discussed in the proceeding but the direct application in the clinic raises several more questions that should be addressed. The primary goal of thermometry is to determine the temperature distributions that occur during the course of the treatment and from these infer the magnitude and spatial distribution of the thermal dose that occurred. This leads to three major questions: How can accurate values of temperature be measured at various points? Where are these points located? How can this information be reduced to meaningful values for predicting responses? The first question has been the emphasis of this chapter. The other two are currently major areas of investigation.

Physical or geometric dosimetry is concerned specifically with the issue of determining where to place the thermometer probes and then independently locating them in three-dimensional co-ordinate space with respect to the tumour, the normal anatomy and the heating applicator. Samulski and Fessenden (1990) discuss in some detail methods for locating the thermometer probes. For deep pelvic lesions treated with regional devices, bony anatomy, such as the tips of the femurs, can be located by stereoscopic radiographs along with the thermometer catheters (using radiomarkers if necessary). A reconstruction algorithm such as that developed by Siddon and Barth (1987) can be used to determine relative co-ordinates. These then serve to relate and verify computed-tomography (CT) images used to guide probe placements and determine the location of tumours.

For deep-seated lesions treated with scanned focused ultrasound, K. Hynynen *et al.* (personal communication) have constructed a plastic device that is mounted in a specific foam moulding which supports the patient (Alpha-Cradle, Smithers Medical products, Inc., Hudson, Ohio). The rods, with variable spacing similar to those on a BRW stereotactic frame for CT imaging (Brown Roberts Wells CT Stereotactic System, Radionics, Burlington, Mass.) show up on the CT scan during probe placement. The co-ordinates of the plastic rods can be correlated to the co-ordinate system of the scanned focused system. The thermocouple sensors also can be located by pulsing the focused beam while scanning it along a raster pattern. When the focus of the beam hits the sensor, a sharp temperature elevation is detected (Hynynen *et al.*, 1987). Now, this sensor location can be used to correlate the CT co-ordinates to the treatment co-ordinate system. Thus, they have excellent localization directly in the treatment machine computer. Oleson and Samulski (1989) describe the use of the same material for use with superficial ultrasonic systems. They also use the foam to fix a template that holds the ultrasonic applicator on the patient. This aids in reproducibility in treatment set-ups.

A BRW stereotactic positioning device has been used by Lulu *et al.* (1989) to position precisely arrays of interstitial catheters into brain tumours for combined brachytherapy and hyperthermia induced with ferromagnetic implants (Stea *et al.*, 1990). That group uses a modification of the BRW unit to orient a pair of carefully registered templates for drilling and for catheter placement. Following stereotactic treatment planning and the surgical placement of the catheters, a box with fiducial marking points is attached to the frame and stereoscopic simulator films are taken to verify relative co-ordinate positions of the implants and the thermometer catheters. Registration with respect to the tumour is possible since the BRW frame remains in place throughout the procedure until this time.

The precise localization of thermometer sensor positions for superficial hyperthermia treatments remains a tedious task. Carefully placed clear plastic sheets placed over the tumour to be heated can be used to establish the direction of the thermometer track in relation to the surface of the patient. The orientation of the heating applicator can be noted as well. Depth of the sensor from the skin remains an estimate based upon the angle of insertion and so forth. Because of this uncertainty, future trials in the USA carried out under the Radiation Therapy Oncology Group (RTOG) will require CT guidance and verification for all thermometer probes placed deeper than 1.5 cm (Dewhirst *et al.*, 1990). The obvious advantage is that information is available for relating the probes and the tumours. The obvious disadvantage is the necessity of scheduling yet another service in order to carry out the treatments. The same document

gives strict rules for probe placement in order to standardize documentation until improved methods of determining position information are
developed and implemented, Figure 15.26. Methods of accurately determining the geometrical co-ordinates of probes, applicators and tumours in
an efficient manner suitable for a routine clinical environment is one of
the greater challenges facing the implementation of good thermal
dosimetry in hyperthermia. Several years ago, everyone gave up trying to
read temperatures manually from a few probes and began to use computerized data acquisition systems. An analogous improvement for physical
dosimetry must be devised and implemented generally.

The final question to be addressed relates to what is to be made of the
dosimetric information. Minimum sampled temperatures give a crude
indication of temperature distribution in that the probability of
measuring a low temperature is related to the likelihood of inadequate
heating. These have been shown to be an indicator of prognosis (Dewhirst
and Sim, 1984). However, Corry *et al.* (1988) demonstrated that as more
probes are placed, the likelihood of sampling cool temperatures increases
and so the projected prognosis should fall. Thus the fraction of measured
tumour temperatures greater than some index temperature ultimately
will not be a good parameter even though it has been useful in several
recent studies (Corry *et al.*, 1988; Kapp *et al.*, 1988; Shimm *et al.*, 1988; Stea
et al., 1990). Oleson *et al.* (1989) discuss the concept that the outer shell of a
tumour, which is the most difficult to heat, contains the largest volume of
tumour. Hence, they suggest that the temperatures of the peripheral
portions of the tumours are most critical. These are quantified by the
temperature and its gradient at the edge of the tumour. An important
point in this paper is that a practical method of reducing clinical data for
the interim is proposed. Practical, if tedious, data acquisition also is the
goal of the RTOG document (Dewhirst *et al.*, 1990) discussed above until
more refined approaches are available routinely.

The 'more refined' approach is quite clear conceptually, but in practice it
still remains elusive due to the indeterminate nature of the blood flow
patterns: both of diffuse perfusion and of discrete thermally significant
vessels. The general approach is to determine the power deposition
pattern, also referred to as the SAR, in a phantom for a given heating
applicator. From measurements of temperature transients at the location
of the implanted thermometer probes, the SAR can be determined at the
probe locations (Roemer *et al.*, 1985) and adjustments to the phantom SAR
pattern can be made. This information then is inserted into a proper
thermal model with assumed levels of blood perfusion. That is, perfusion is
used as an adjustable parameter (Roemer and Cetas, 1984). The actual
level of blood flow can be inferred by combining information from the
model and from the measured steady-state temperatures again at the
measurement points. For 'hot source' interstitial techniques such as
inductively heated ferromagnetic implants, this approach is quite

straightforward. Unfortunately, it is still not clear how to interpolate any of this information between the measurement points since blood flow tends to change abruptly rather than in a continuous fashion and temperatures are highly dependent upon the level of perfusion cooling.

Thus, we are still left with only coarse indications of the thermal dose that is delivered to each point within the tumour. This imprecision is reflected clearly in the discussions of clinical trials of hyperthermia. Comparisons between radiation alone and combined hyperthermia and radiation carried out at a single institution where the methodology, good or bad, is relatively consistent have shown major benefits for this therapy in a large number of trials (Overgaard, 1989). However, a recent multiple-institutional trial (Perez *et al.*, 1989) did not yield that same benefit because the level of quality control was not sufficient to compensate for the lack of detailed, geometrically reliable thermal dosimetry. Analysis from the subset of the data where adequate heating could be inferred suggested the same result as seen in single-institution trials. Dosimetry research still has a large task before it.

15.8. Summary

The necessity of thermometry in conducting clinical trials of hyperthermia is self-evident. The basic principles and underlying assumptions of the science have been presented as a guide to good technique. Thermometry is not particularly difficult, but some care is necessary to avoid errors. It is quite easy to obtain precise results which are inaccurate; the apparent sophistication of computers is deceptive.

Various common thermometers and a few not so common ones have been described. Techniques for establishing a thermometer calibration facility are discussed and routine quality assurance methods are suggested. Finally, a few comments have been given regarding the practical issues of sampling temperatures in the clinic and inferring what these may mean.

References

Andersen, J. B., Baun, A., Harmark, K., Heinzl, L., Rasmark, P. and Overgaard, J. (1984) A hyperthermia system using a new type of inductive applicator, *IEEE Transactions on Biomedical Engineering*, **31**, 21–27.

Astrahan, M. A., Luxton, G., Sapozink, M. D. and Petrovich, Z. (1988) The accuracy of temperature measurement from within an interstitial microwave antenna, *International Journal of Hyperthermia*, **4**, 593–607.

Battist, L., Goldner, F. and Todreas, N. (1969) Construction of a fine wire thermocouple capable of repeated insertions into and more accurate positioning within a controlled environmental chamber, *Medical and Biological Engineering*, **7**, 445–448.

Bowman, R. (1976) A probe for measuring temperature in radio-frequency heated material, *IEEE Transactions on Microwave Theory and Techniques*, **24**, 43–45.

Cain, C. P. and Welch, A. J. (1974) Thin-film temperature sensor for biological measurements, *IEEE Transactions on Biomedical Engineering*, **21**, 421–423.

Carnochan, P., Dickinson, R. J. and Joiner, M. C. (1986) The practical use of thermocouples for temperature measurement in clinical hyperthermia, *International Journal of Hyperthermia*, **2**, 1–19.

Cetas, T. C. (1976) Temperature measurement in microwave diathermy fields: Principles and probes, in Robinson, J. E. and Wizenberg, M. J. (Eds) *Proceedings of the International Symposium on Cancer Therapy by Hyperthermia, Drugs and Radiation*, Bethesda, American College of Radiology, pp. 193–203.

Cetas, T. C. (1982a) Invasive thermometry, in Nussbaum, G. H. (Ed.), *Physical Aspects of Hyperthermia*, New York, American Institute of Physics, pp. 231–265.

Cetas, T. C. (1982b) Thermometry, in Lehman, J. F. (Ed.), *Therapeutic Heat and Cold*, 3rd edn, Baltimore, Williams and Wilkins, pp. 35–69.

Cetas, T. C. (1984) Will thermometric tomography become practical for hyperthermia monitoring? *Cancer Research*, **44**, (Suppl), 4805s–4808s.

Cetas, T. C. (1985) Thermometry and thermal dosimetry, in Overgaard, J. (Ed.), *Hyperthermic Oncology 1984*, Vol. 2, London, Taylor and Francis, pp. 91–112.

Cetas, T. C. (1987) Thermometry, in Field, S. B. and Franconi, C. (Eds), *Physics and Technology of Hyperthermia*, Dordrecht, Martinus Nijhoff, pp. 470–508.

Cetas, T. C. and Connor, W. G. (1978) Thermometry considerations in localized hyperthermia, *Medical Physics*, **5**, 79–91.

Cetas, T. C., Connor, W. G. and Boone, M. L. M. (1978) Thermal dosimetry: some biophysical considerations, in Streffer, C. *et al.* (Eds), *Cancer Therapy by Hyperthermia and Radiation*, Munich, Urban and Schwarzenberg, pp. 3–12.

Cetas, T. C., Connor, W. G. and Manning, M. R. (1980) Monitoring of tissue temperature during hyperthermia, *Annals of the New York Academy of Science*, **335**, 281–297.

Chakraborty, D. P. and Brezovich, I. A. (1982) Error sources affecting thermocouple thermometry in RF electromagnetic fields, *Journal of Microwave Power*, **17**, 17–28.

Chan, K. W., Chan, C. K., McDougall, J. A. and Luk, K. H. (1988a) Changes in heating patterns due to perturbations by thermometer probes at 915 and 434 MHz, *International Journal of Hyperthermia*, **4**, 447–456.

Chan, K. W., Chan, C. K., McDougall, J. A. and Luk, K. H. (1988b) Perturbations due to the use of catheters with non-perturbing probes, *International Journal of Hyperthermia*, **4**, 699–702.

Christensen, D. A. (1977) A new non-perturbing temperature probe using semiconductor band edge shift, *Journal of Bioengineering*, **1**, 541–545.

Christensen, D. A. (1979) Thermal dosimetry and temperature measurements, *Cancer Research*, **39**, 2325–2327.

Christensen, D. A. (1982) Current techniques for noninvasive thermometry, in Nussbaum, G. H. (Ed.), *Physical Aspects of Hyperthermia*, New York, American Institute of Physics, pp. 266–279.

Christensen, D. A. (1983) Thermometry and thermography, in Storm, F. K. (Ed.), *Hyperthermia in Cancer Therapy*, Boston, G. K. Hall, pp. 223–232.

Constable, R. T., Dunscombe, P. and Tsoukutos, A. (1987) Perturbation of the temperature distribution in microwave irradiated tissue due to the presence of metallic thermometers, *Medical Physics*, **14**, 385–388.

Corry, P. M., Jabboury, K., Kong, J. S., Armour, E. P., McCraw, F. J. and LeDuc, T. (1988) Evaluation of equipment for hyperthermia treatment of cancer, *International Journal of Hyperthermia*, **4**, 53–74.

Curley, M. G. and Lele, P. P. (1984) Some potential errors in measurement of temperature *in vivo* during hyperthermia by ultrasound and electromagnetic energy, in Overgaard, J. (Ed.), *Hyperthermic Oncology 1984*, Vol. 1, London, Taylor and Francis, pp. 561–564.

Dewhirst, M. W. and Sim, D. A. (1984) The utility of thermal dose as a predictor of tumour and normal tissue responses to combined radiation and hyperthermia, *Cancer Research*, **44**, (Suppl.), 4772s–4780s.

Dewhirst, M. W., Phillips, T. L., Samulski, T. V., Stauffer, P., Shrivastava, P., Paliwal, B., Pajak, T., Gillim, M., Sapozink, M., Myerson, R., Waterman, F. M., Sapareto, S. A., Corry, P., Cetas, T. C., Leeper, D. B., Fessenden, P., Kapp, D., Oleson, J. R. and Emami, B. (1990) RTOG quality assurance guidelines for clinical trials using hyperthermia, *International Journal of Radiation Oncology, Biology and Physics*, in press.

Dickinson, R. J. (1985) Thermal conduction errors of manganin–constantan thermocouple arrays, *Physics in Medicine and Biology*, **30**, 445–453.

Dunscombe, P. B. and McLellan, J. (1986) Heat production in microwave-irradiated thermocouples, *Medical Physics*, **13**, 457–461.

Dunscombe, P. B., Constable, R. T. and McLellan, J. (1988) Minimizing the self-heating artefacts due to microwave irradiation of thermocouple, *International Journal of Hyperthermia*, **4**, 437–445.

Fessenden, P., Samulski, T. V. and Lee, E. R. (1984) Direct temperature measurement, *Cancer Research*, **44**, (Suppl.), 4799s–4804s.

Fry, W. J. and Fry, R. B. (1954a) Determination of absolute sound levels and acoustic absorption coefficients by thermocouple probes — theory, *Journal of the Acoustical Society of America*, **26**, 294–310.

Fry, W. J. and Fry, R. B. (1954b), Determination of absolute sound levels and acoustic absorption coefficients by thermocouple probes — experiment, *Journal of Acoustical Society of America*, **26**, 311–317.

Hand, J. W. (1985) Thermometry in hyperthermia, in Overgaard, J. (Ed.), *Hyperthermia Oncology 1984*, Vol. 2, London, Taylor and Francis, pp. 299–308.

Howard, J. L. (1972) Error accumulation in thermocouple thermometry, in Plumb, H. H. (Ed.), *Temperature IV*, Pittsburgh, Instrument Society of America, pp. 2017–2029.

Hudson, R. P. (1980) Measurement of temperature, *Review of Scientific Instruments*, **51**, 871–881.

Hynynen, K. and Edwards, D. K. (1989) Temperature meaurements during ultrasound hyperthermia, *Medical Physics*, **16**(4), 618–626.

Hynynen, K., Martin, C. J., Watmough, D. J. and Mallard, J. R. (1983) Errors in temperature measurement by thermocouple probes during ultrasound induced hyperthermia, *British Journal of Radiology*, **56**, 969–970.

Hynynen, K., Roemer, R. B., Anhalt, D., Johnson, C., Xu, Z. X., Swindell, W. and Cetas, T. C. (1987) A scanned focused, multiple transducer ultrasonic system for localized hyperthermia treatments, *International Journal of Hyperthermia*, **3**, 21–35.

International Practical Temperature Scale of 1968 (1969) *Metrologia*, **5**, 35–44; Preston-Thomas, H. (1976) Amended edition of 1975, *Metrologia*, **12**, 7–17.

Johnson, C. C. and Guy, A. W. (1972) non-ionizing electromagnetic wave effects in biological materials and systems, *Proceedings of the IEEE*, **60**, 692–718.

Kapp, D. S., Fessenden, P., Samulski, T. V., Bagshaw, M. A., Cox, R. S., Lee, E. R., Lohrbach, A. W., Meyer, J. L. and Prionas, S. D. (1988) Stanford University Institutional Report, Phase I evaluation of equipment for hyperthermic treatment of cancer, *International Journal of Hyperthermia*, **4**, 75–115.

Kuhn, P. K. and Christensen, D. A. (1986) Influence of temperature probe sheathing materials during ultrasonic heating, *IEEE Transactions on Biomedical Engineering*, **33**, 536–538.

Larsen, L. E., Moore, R. A., Jacobi, J. H., Halgas, F. A. and Brown, P. V. (1979) A microwave compatible MIC temperature electrode for use in biological dielectrics, *IEEE Transactions on Microwave Theory and Techniques*, **27**, 673–679.

Lele, P. P. (1980) Induction of deep, local hyperthermia by ultrasound and electromagnetic fields, *Radiation and Environmental Biophysics*, **17**, 205–217.

Lulu, B., Lutz, W., Stea, B. and Cetas, T. C. (1989) Treatment planning of template-guided stereotaxic brain implants, *International Journal of Radiation Oncology, Biology and Physics*, in press.

Magin, R. L., Mangum, B. W., Statler, J. A. and Thornton, D. D. (1981) Transition temperatures of the hydrates of Na_2SO_4, Na_2HPO_4 and KF as fixed points in biomedical thermometry, *Journal of Research of the NBS*, **86**, 181–192.

Mangum, B. W. (1983a) SRM 1969, Rubidium triple-point standard: a temperature reference standard near 39.30°C, *NBS Special Publication 260–87*, Washington DC, US Government Printing Office.

Mangum, B. W. (1983b) Triple point succinonitrile and its use in the calibration of thermistor thermometer, *Review of Scientific Instruments*, **54**, 1687–1692.

Mangum, B. W. and El-Sabban, S. (1986) SRM 1970, succinonitrile triple-point standard a temperature reference standard near 58.08°C, *NBS Special Publication 260–101*, Washington, DC, US Government Printing Office.

Mangum, B. W. and Furukawa, G. T. (1982) Report on the sixth international symposium of temperature, *Journal of Research of the NBS*, **87**, 387–406.

Mangum, B. W. and Thornton, D. D. (Eds) (1977) The gallium melting-point standard, *NBS Special Publication 481*, Washington, DC, US Government Printing Office.

Martin, C. J. (1986) Temperature measurement in tissues by invasive and non-invasive techniques, in Watmough, D. J. and Ross, W. M. (Eds) *Hyperthermia*, Glasgow, Blackie, pp. 154–179.

Middleton, W. E. K. (1966) *A History of the Thermometer*, Baltimore, Johns Hopkins Press.

Mossman, C. A., Horton, J. A. and Anderson, R. L. (1982) Testing of thermocouples for inhomogeneities: a review of theory with examples, in Schooley, J. F. (Ed.), *Temperature: Its Measurement and Control in Sciences and Industry*, Vol. 5, New York, American Institute of Physics, pp. 923–929.

Oleson, J. R. and Samulski, T. V. (1989) Improved coupling of ultrasound hyperthermia applicators to patients, *International Journal of Radiation Oncology, Biology and Physics*, **16**, 609–612.

Oleson, J. R., Dewhirst, M. W., Harrelson, J. M., Leopold, K. A., Samulski, T. V, and Tso, C. Y. (1989) Tumor temperature distributions predict hyperthermia effect, *International Journal of Radiation Oncology, Biology and Physics*, **16**, 559–570.

Overgaard, J. (1989) The current and potential role of hyperthermia in radiotherapy, *International Journal of Radiation Oncology, Biology and Physics*, **16**, 535–549.

Perez, C. A., Gillespie, B., Pajak, T., Hornback, N. B., Emami, B. and Rubin, P. (1989) Quality assurance problems in clinical hyperthermia and their impact on therapeutic outcome: a report by the Radiation Therapy Oncology Group, *International Journal of Radiation Oncology, Biology and Physics*, **16**, 551–558.

Powell, R. L., Hall, W. J., Hyink, C. H., Sparks, L. L., Burns, G. W., Scroger, M. G. and Plumb, H. H. (1974) Thermocouple reference tables based on the IPTS-68. *NBS Monograph 125*, Washington DC, US Government Printing Office.

Priebe, L. A., Cain, C. P. and Welch, A. J. (1975) Temperature rise required for production of minimal lesions in macaca mulatta retina, *American Journal of Opthalmology*, **79**, 405–413.

Quinn, T. J. and Compton, J. P. (1975) The foundations of thermometry, *Reports on Progress in Physics*, **38**, 151–239.

Riddle, S. L., Furukawa, G. T. and Plumb, H. H. (1972) Platinum resistance thermometry, *NBSG Monograph 126*, Washington, DC, US Government Printing Office.

Roemer, R. B. and Cetas, T. C. (1984) Applications of bioheat transfer simulations in hyperthermia, *Cancer Research*, **44**, (Suppl.), 4788s–4798s.

Roemer, R. B., Fletcher, A. M. and Cetas, T. C. (1985) Obtaining local SAR and blood perfusion data from temperature sensors: steady state and transient techniques compared, *International Journal of Radiation Oncology, Biology and Physics*, **11**, 1539–1550.

Rozzell, T. C., Johnson, C. C., Durney, C. H., Lords, J. L. and Olsen, R. G. (1974) A non-perturbing temperature sensor for measurements in electromagnetic fields, *Journal of Microwave Power*, **9**, 241–249.

Samulski, T. V. and Fessenden, P. (1990) Thermometry in therapeutic hyperthermia, in Gautherie, M. (Ed.), *Clinical Thermology*, subseries *Therapy*, Vol. 3, Munich, Springer-Verlag, in press.

Samulski, T. V. and Lee, E. R. (1986) Premedical evaluation of Luxtron 3000, in Kondraske, G. V. and Robinson, C. J. (Eds), *Proceedings of 8th Annual Conference IEEE-EMB*, New York, IEEE, pp. 1493–1495.

Samulski, T. and Shrivastava, P. N. (1980) Photoluminescent thermometer probes: temperature measurements in microwave fields, *Science*, **208**, 193–194.

Samulski, T. V., Lyons, B. E. and Britt, R. H. (1985) Temperature measurements in high thermal gradients: II analysis of conduction effects, *International Journal of Radiation Oncology, Biology and Physics*, **11**, 963–971.

Schooley, J. F. (Ed.) (1982) *Temperature: Its Measurement and Control in Science and Industry*, Vol. V, New York, American Institute of Physics.

Schooley, J. F. (1986) *Thermometry*, Boca Raton, CRC Press.

Shimm, D. S., Hynynen, K. H., Anhalt, D. P., Roemer, R. B. and Cassady, J. R. (1988) Scanned focused ultrasound hyperthermia: initial clinical results, *International Journal of Radiation Oncology, Biology and Physics*, **15**, 1203–1208.

Sholes, R. R. and Small, J. G. (1980) Fluorescent decay thermometer with Biological applications, *Review of Scientific Instruments*, **51**, 882–884.

Shrivastava, P. N., Saylor, T. K., Matloubieh, A. Y. and Paliwahl, B. R. (1988) Hyperthermia thermometry evaluation: criteria and guidelines, *International Journal of Radiation Oncology, Biology and Physics*, **14**, 327–335.

Shrivastava, P. N., Luk, K., Oleson, J., Dewhirst, M., Pajak, T., Paliwahl, B., Perez, C., Sapareto, S., Saylor, T. and Steeves, R. (1989) Hyperthermia quality assurance guidelines, *International Journal of Radiation Oncology, Biology and Physics*, **16**, 571–587.

Siddon, R. L. and Barth, N. H. (1987) Stereotaxic localization of intercranial targets, *International Journal of Radiation Oncology, Biology and Physics*, **13**, 1241–1246.

Sostman, H. E. (1977a) Melting point of gallium as a temperature calibration standard, *Review of Scientific Instruments*, **48**, 127–130.

Sostman, H. E. (1977b) The gallium melting-point standard: its role in the manufacture and quality control of electronic thermometers for the clinical laboratory, *Clinical Chemistry*, **23**, 725–739.

Stea, B., Cetas, T. C., Cassady, J. R., Guthkelch, A. N., Iacono, R., Lulu, B., Lutz, W., Obbens, E., Rossman, K., Seeger, J., Shetter, A. and Shimm, D. S. (1990) Interstitial Thermoradiotherapy in brain tumours: preliminary results of a phase I clinical trial, *International Journal Radiation Oncology, Biology and Physics*, submitted.

Steinhardt, J. S. and Hart, S. R. (1968) Calibration curves for thermistores, *Deep Sea Research*, **15**, 497–503.

Steketee, J. (1973) Spectral emissivity of skin and pericardium, *Physics in Medicine and Biology*, **18**, 686–694.

Swenson, C. A. (1987) Thermometry, in Meyers, R. A. (Ed.), *Encyclopedia of Physical Science and Technology*, Vol. 13, New York, Academic Press, pp. 830–849.

Taylor, B. N. (1989) New measurement standards for 1990, *Physics Today*, **42**, 23–26.

Thornton, D. D. (1977) The gallium melting-point standard: a determination of the liquid solid equilibrium temperature of pure gallium or the international practice temperature scale of 1968, *Clinical Chemistry*, **23**, 719–724.

Trolander, H. W., Case, D. A. and Harruff, R. W. (1972) Reproducibility, stability and linearization of thermistor resistance thermometers, in Plumb, H. H. (Ed.), *Temperature*, Vol. IV, Pittsburgh, Instrument Society of America, pp. 997–1009.

Vaguine, V. A., Christensen, D. A., Lindley, J. H. and Walston, T. E. (1984) Multiple sensor optical thermometry for application in clinical hyperthermia, *IEEE Transactions on Biomedical Engineering*, **31**, 168–172.

Vojnovic, B. and Joiner, M. C. (1985) A multiple RF heating system for experimental hyperthermia in small animals, *International Journal of Hyperthermia*, **1**, 287–298.

Waterman, F. M. (1985) The response of thermometer probes inserted into catheters, *Medical Physics*, **12**, 368–372.

Wells, P. N. T. (1977) *Biomedical Ultrasonics*, New York, Academic Press.

Wickersheim, K. A. and Alves, R. B. (1979) Recent advances in optical temperature measurement, *Industrial Research and Developments*, **21**, 82–89.

Wise, J. A. and Soulen Jr, R. J. (1986) Thermometer calibration: a model for state calibration laboratories, *NBS Monograph 174*, Washington, DC, US Government Printing Office.

Wood, S. D., Mangum, B. W., Filliben, J. J. and Tillett, S. B. (1978) An investigation of the stability of thermistors, *Journal of Research NBS*, **83**, 247–263.

16 Thermal models: principles and implementation

J. J. W. Lagendijk

16.1. Introduction

This chapter addresses heat transfer phenomena in tissues. It begins with the fundamentals of general heat transfer and ends with the theory of heat transfer in vascularized tissues. Since bioheat transfer theory is currently somewhat controversial, particular attention is drawn to the points of controversy.

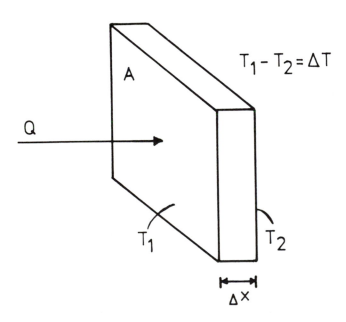

Figure 16.1 Heat transport through a plate.

16.2. General theory of heat transfer

16.2.1. One-dimensional heat transfer by conduction

16.2.1(a) Fourier's law
Heat flow (Q) through a plate (Figure 16.1) is proportional to plate thickness Δx, plate surface area A and to the temperature difference ΔT across the plate. Introducing k, the thermal conductivity coefficient (units: W/m/°C) of the material of which the plate is composed, we obtain:

$$Q = -k\,A\,\frac{\Delta T}{\Delta x}\,\text{(W)}. \tag{16.1}$$

The minus sign has been added to make the heat flow in the direction of decreasing temperature. If we consider the heat flow per unit surface (heat flux Q'') for a plate of infinitesimal thickness dx, equation (16.1) becomes

$$Q'' = -k\,\frac{\mathrm{d}T}{\mathrm{d}x}\,\text{(W/m}^2\text{)}, \tag{16.2}$$

which is the well-known Fourier law of heat conduction in solids. This equation, together with appropriate boundary conditions and absorbed power distribution (SAR), describes the steady-state temperature distribution in solid materials.

16.2.1(b) Heat conduction through a plane wall

As an example of the use of the Fourier law we consider the one-dimensional heat transport through an infinite plane wall of layers of materials having different thicknesses and thermal conductivities (Figure 16.2). We assume that there is good thermal contact between the layers. Under steady-state conditions no heat storage takes place and so the heat flux must be constant over all layers. This fact, together with equation (16.2), gives Q'', the heat flux through the wall, as

$$Q'' = \text{constant} = k_1\frac{\Delta T_1}{\Delta x_1} = k_2\frac{\Delta T_2}{\Delta x_2} = k_3\frac{\Delta T_3}{\Delta x_3}\,\text{(W/m}^2\text{)}. \tag{16.3}$$

If Q'' is known, the overall temperature difference across the wall is

$$\Delta T_{\text{total}} = Q''\left[\frac{(x_2-x_1)}{k_1} + \frac{(x_3-x_2)}{k_2} + \frac{x_4-x_3)}{k_3}\right]\text{(°C)}. \tag{16.4}$$

The temperature distribution in the wall is given in Figure 16.2.

The terms $(x_i-x_{i-1})/k_i$ can be considered as heat resistances. Equation (16.4) shows, as in the electrical equivalent (Ohm's law), that the 'tension'

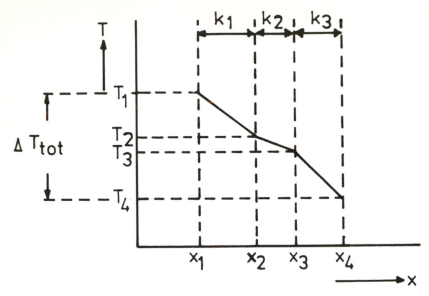

Figure 16.2 Steady-state temperature distribution in an infinite plane wall of layers having different thicknesses and thermal conductivities. Layer 2 has the highest thermal conductivity.

ΔT equals the 'current' Q'' multiplied by the sum (U^{-1}) of the different heat resistances $\Delta x/k$. U^{-1} is the total heat resistance. U is the total heat transfer coefficient describing the heat transport over the plate:

$$U = \frac{Q''}{\Delta T} \ (\text{W/m}^2/^\circ\text{C}). \tag{16.5}$$

It can be determined experimentally to describe the total heat flux through (un)known complex structures.

16.2.1(c) Heat conduction in the wall of a cylindrical tube

As an introduction to heat transfer around blood vessels, we consider the heat flow through the wall of a cylindrical tube (Figure 16.3). The heat flow per unit length through the surfaces $r=R_1$ and $r=R_2$ and through all surfaces $r=$constant in between, must be equal for steady-state conditions. Using equation (16.2) and polar co-ordinates we have:

$$Q' = -k \frac{\mathrm{d}T}{\mathrm{d}r} 2\pi r = \text{constant (W/m)}. \tag{16.6}$$

The temperature distribution in the wall follows from the solution of equation (16.6) with the boundary conditions shown in Figure 16.3.

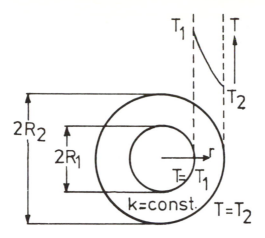

Figure 16.3 Steady-state heat transport in the wall of a cylindrical tube.

$$\frac{T(r)-T_2}{T_1-T_2} = \frac{\ln(r/R_2)}{\ln(R_1/R_2)} \tag{16.7}$$

Thus the heat flow per unit length of tube is

$$Q'=2\pi k \frac{(T_1-T_2)}{\ln(R_2/R_1)} \text{ (W/m)}. \tag{16.8}$$

In this case the temperature in the wall varies logarithmically, in contrast to the linear temperature variations in plane walls (equation (16.4)).

16.2.2. Heat transfer coefficients

Equation (16.5) defines the total heat transfer coefficient U, which describes the heat transfer through (un)known structures provided that the boundary temperatures are known. A common situation is the one in which a liquid or a gas is in thermal contact with a solid wall. The heat transfer from the gas or liquid to the wall can also be described by a heat transfer coefficient h:

$$Q''=h(<T>-T_w)\text{(W/m}^2), \tag{16.9}$$

where $<T>$ is the mean temperature of the gas or liquid at some distance from the wall and T_w is the temperature at the wall's surface. The value of h, which depends greatly on the properties of the liquid or gas and the presence of convection, can be determined experimentally. Table 16.1 lists values of h for some specific cases.

Table 16.1 Values of heat transfer coefficient between gases and liquids and a solid wall.

Medium	Heat transfer coefficient (W/m²/°C)
Gas (free convection)	5– 15
Gas (forced convection)	10– 100
Liquid (free convection)	50– 1000
Liquid (forced convection)	500– 3000 (water 5×as high)
Vapour condensation	1000– 4000 (water 5×as high)
Liquid boiling	1000–20000 (water 3×as high)

If we neglect the influence of blood perfusion on the thermal conductivity for the moment (note that this is not allowed for normal tissues such as muscle and skin), then the heat transfer coefficient for a 1 cm thick fat layer is only $h=k_{fat}/0.01=25$ W/m²/°C (see Section 16.2.1(b)), explaining the good insulating properties of fat layers.

If no forced cooling is used, the heat transfer between skin and the surroundings is described by the sum of the heat transfer coefficients describing the air cooling ($h=5$–15 W/m²/°C) and infra-red cooling. The infra-red heat flow is described by

$$Q'' = e\,\sigma\,T^4_{skin} - a\,\sigma\,T^4_{air}\ (W/m^2),\qquad(16.10)$$

where σ, the Stefan–Boltzmann constant, is 5.67×10^{-8} W/m²/K⁴ and e and a are, respectively, the emissivity and the absorptivity of the surface — both equal and about unity for tissues. The temperatures in equation (16.10) must be expressed in the Kelvin absolute temperature scale. The difference of the fourth power of the absolute temperatures can be simplified in the clinical temperature range of 20–50°C to

$$Q'' = 4\,\sigma\,T^3_{air}(T_{skin} - T_{air}) = h\,(T_{skin} - T_{air})\,(W/m^2),\qquad(16.11)$$

with

$$h = 4\,\sigma\,T^3_{air}\ (W/m^2/°C).\qquad(16.12)$$

At normal room temperatures $h \approx 6$ W/m²/°C. The low values of the heat transfer coefficients for infra-red cooling and non-forced air cooling show that a patient with dry skin has a small capacity for cooling. On the other hand, the heat transfer coefficient for liquid evaporation is high, indicating the importance of sweating in cooling the human body.

16.2.3. The general three-dimensional heat conduction equation

16.2.3(a) General heat conduction equation

The general three-dimensional heat conduction equation describing the spatial and time-dependent temperature distribution in inhomogeneous media is given by

$$\rho\, C \frac{\mathrm{d}T}{\mathrm{d}t} = \mathrm{div}(k\, \mathrm{grad}(T)) + P, \qquad (16.13)$$

where ρ and C are the density and the specific heat of the medium and $P(\mathrm{W/m^3})$ is the absorbed power density. Equation (16.13) follows from heat balance within a small volume element and from the Fourier law (equation (16.2)). Equation (16.13) may be approximated by a forward finite difference operator (e.g. Croft and Lilley, 1977)*.

In the one-dimensional case (see Figure 16.4) we have

$$\frac{\mathrm{d}x\, \rho\, C(T'-T)}{\mathrm{d}t} = k\, \frac{[T(x-\mathrm{d}x)-T(x)]}{\mathrm{d}x} + k\, \frac{[T(x+\mathrm{d}x)-T(x)]}{\mathrm{d}x} + \mathrm{d}x\, P. \qquad (16.14)$$

Equation (16.14) which describes the heat balance applied on element x plays an important role in thermal modelling. Two- and three-dimensional

---

*In a one-dimensional homogeneous medium we have

$$\rho\, C \frac{dT}{dt} = k\, \frac{\mathrm{d}^2T}{\mathrm{d}t^2} + P. \qquad (16.\mathrm{i})$$

The forward Taylor's expansion of a function $f(x+\mathrm{d}x)$, as $\mathrm{d}x$ tends to zero, gives

$$f(x+\mathrm{d}x)=f(x)+\mathrm{d}x\, \frac{\mathrm{d}f(x)}{\mathrm{d}x} + \frac{\mathrm{d}x^2}{2}\, \frac{\mathrm{d}^2f(x)}{\mathrm{d}x^2}+\dots . \qquad (16.\mathrm{ii})$$

The backward Taylor's expansion gives

$$f(x-\mathrm{d}x) = f(x) - \mathrm{d}x\, \frac{\mathrm{d}f(x)}{\mathrm{d}x} + \frac{\mathrm{d}x^2}{2}\, \frac{\mathrm{d}^2f(x)}{\mathrm{d}x^2} - \dots . \qquad (16.\mathrm{iii})$$

Neglecting terms of $\mathrm{d}x^2$ and smaller gives

$$\text{forward: } \frac{\mathrm{d}f}{\mathrm{d}x} = \frac{[f(x+\mathrm{d}x)-f(x)]}{\mathrm{d}x} \qquad (16.\mathrm{iv})$$

and

$$\text{backwards: } \frac{\mathrm{d}f}{\mathrm{d}x} = \frac{[f(x)-f(x-\mathrm{d}x)]}{\mathrm{d}x}. \qquad (16.\mathrm{v})$$

Adding equations (16.ii) and (16.iii) we get the central finite difference approximation to the second derivative of $f(x)$ in x:

$$\frac{\mathrm{d}^2f(x)}{\mathrm{d}x^2} = \frac{[f(x+\mathrm{d}x)-2\,f(x)+f(x+\mathrm{d}x)]}{\mathrm{d}x^2}. \qquad (16.\mathrm{vi})$$

Applying equation (16.vi) on the conductivity term of equation (16.i) and equation (16.iv) on the $\mathrm{d}T/\mathrm{d}t$ term of equation (16.i) gives

$$\frac{\rho\, C(T'-T)}{\mathrm{d}t} = \frac{k[T(x-\mathrm{d}x)-2T(x)+T(x+\mathrm{d}x)]}{\mathrm{d}x^2} + P. \qquad (16.\mathrm{vii})$$

With T' the new temperature of the volume element at time $t+\mathrm{d}t$. Rewriting equation (16.vii) gives equation (16.14).

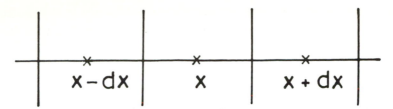

Figure 16.4 One-dimensional heat transfer.

models to solve equation (16.13) based on equation (16.14) or on finite element techniques are readily available. Applications of these models to examples relevant to hyperthermia are demonstrated by Paulsen *et al.* (1985).

The general heat conduction equation (16.13) consists of a heat storage term, a conductivity term $K=\mathrm{div}(k\ \mathrm{grad}(T))$ which describes how well heat is transported, and a heat production term P which describes the amount of heat produced in each volume element by the heating system. P can result from electromagnetic, ultrasonic or chemical (metabolic heat production) sources. Metabolic heat sources are usually small with respect to P (approximately 1 W/kg). However, differences in metabolic heat production can cause significant temperature differences in the human body. For example, liver core temperature is about 38 °C, 1 °C above the average core temperature. This may explain why liver damage is seen sometimes in total-body hyperthermia. Negative values for P (i.e. a heatsink term) can be useful in describing some heat transfer phenomena (see Section 16.4).

For steady-state conditions, $\mathrm{d}T/\mathrm{d}t=0$ and equation (16.13) reduces to

$$\mathrm{div}(k\ \mathrm{grad}(T))+P=0. \tag{16.15}$$

In the case of a homogeneous medium without heat production, equation (16.15) reduces to the Laplace equation:

$$\frac{\mathrm{d}^2T}{\mathrm{d}x^2} + \frac{\mathrm{d}^2T}{\mathrm{d}y^2} + \frac{\mathrm{d}^2T}{\mathrm{d}z^2}=0. \tag{16.16}$$

16.2.3(b) Microwave heating of a uniform half space

We consider an analytical solution to microwave heating of an infinite tissue layer, thickness L (m) in which the absorbed power distribution is

$$P=P(0)\,\mathrm{e}^{-2x/d}\,(\mathrm{W/m^3}), \tag{16.17}$$

where $P(0)$ is the microwave intensity at the surface and d is the penetration depth (i.e. the distance over which the field is reduced by a factor e).

The surface is cooled with water at a temperature T_w. A heat transfer coefficient h describes the heat transfer between the surface and the water cooling. The steady-state temperature distribution is described by

$$k \frac{d^2 T}{dx^2} + P = 0. \tag{16.18}$$

Solving equation (16.18) by integration and assuming that the temperature at a depth $L=0.05$ m is equal to the body core temperature T_{tis} gives

$$kT = P(0) \frac{d^2}{4} e^{-2x/d} + C_1 x + C_2, \tag{16.19}$$

where

$$C_1 = -h(T_w - T(0)) - P(0) \, d/2 \tag{16.20}$$

$$C_2 = k \, T(0) + P(0) \, d^2/4 \tag{16.21}$$

$$T(0) = \frac{P(0)(d/2)^2 (e^{-2L/d} - 1) + LhT_w + kT_{tis} + LP(0)(d/2)}{(k + L \, h)}. \tag{16.22}$$

An example calculation is shown in Figure 16.5

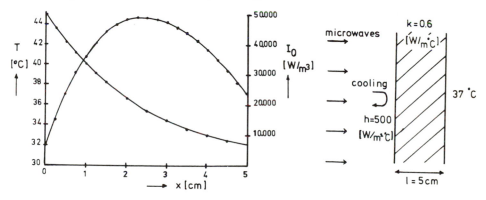

Figure 16.5 Temperature distribution in an infinite uniform half-space with properties: $h=500$ W/m²/°C, $T_{water}=30$°C, $T_{tis}=37$°C, $d=3$ cm, $k=0.6$ W/m²/°C and $L=5$ cm.

16.2.4. Heat penetration theory

16.2.4(a) Heat penetration depth

To evaluate the time-dependent behaviour of the general heat conduction equation (16.13) we first analyse the heat penetration into a uniform

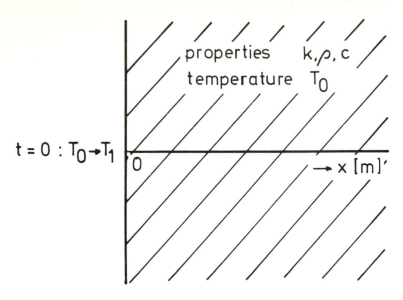

Figure 16.6 Heat penetration into a uniform infinite half space; model geometry.

infinite half space (Figure 16.6). The initial temperature of the medium is
T_0. At time $t=0$ the wall temperature is changed stepwise to $T=T_1$. We can
analyse this problem in one dimension by defining the x-co-ordinate into
the medium. Equation (16.13) becomes

$$\rho C \frac{\mathrm{d}T}{\mathrm{d}t} = k \frac{\mathrm{d}^2 T}{\mathrm{d}x^2} \rightarrow \frac{\mathrm{d}T}{\mathrm{d}t} = \frac{k}{\rho C} \frac{\mathrm{d}^2 T}{\mathrm{d}x^2} = a \frac{\mathrm{d}^2 T}{\mathrm{d}x^2}, \tag{16.23}$$

where a is the thermal diffusivity of the medium. Boundary conditions are
$x=0$, $T=T_1$; $x=\infty$, $T=T_0$ with the initial condition $t=0$, $T=T_0$ for all x.
Dimensional analysis indicates that equation (16.34) can be solved by
introducing two dimensionless parameters:

$$f = \frac{T-T_0}{T_1-T_0} \text{ and } g = \frac{x}{\sqrt{a t}}. \tag{16.24}$$

Rewriting equation (16.23) as

$$\frac{\mathrm{d}^2 f}{\mathrm{d}g^2} + \frac{g}{2} \frac{\mathrm{d}f}{\mathrm{d}g} \, 0, \tag{16.25}$$

where $f=0$ for $g=\infty$ and $f=1$ for $g=0$. The exact solution for equation
(16.23) becomes

$$1 - \frac{T-T_0}{T_1-T_0} = \frac{T_1-T}{T_1-T_0} = \frac{2}{\sqrt{\pi}} \int_0^{x/2\sqrt{at}} e^{-y^2} \, \mathrm{d}y. \tag{16.26}$$

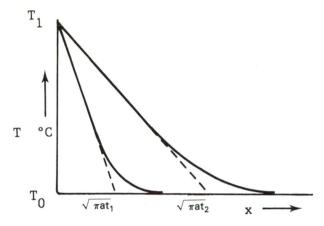

Figure 16.7 Heat penetration by conduction in a uniform half space.

The right-hand side of equation (16.26) is called the 'error function' and is described in most mathematical handbooks. The most important result is that the tangent to the temperature distribution at $x=0$ crosses $T=T_0$ at $x=\sqrt{(\pi\ a\ t)}$ (Figure 16.7). The distance $x=\sqrt{(\pi\ a\ t)}$, known as the heat penetration depth, gives the point in the medium at which the temperature markedly differs from T_0. The amount of heat passing through the surface $x=0$ is

$$Q'' = \frac{k(T_1-T_0)}{\sqrt{(\pi\ a\ t)}}\ (\text{W/m}^2). \qquad (16.27)$$

The rate of heat transfer and therefore the rate at which the temperature equilibrates in the medium is determined by the thermal diffusivity $a=k/\rho\ C$.

As indicated above and in Section 16.2.1(c), analytical solutions of the temperature distribution in media are cumbersome and often completely impossible. Numerical techniques based on equation (16.22) which readily solve this and most other heat transfer problems are available (Croft and Lilley, 1977).

16.2.4(b) Heating of objects by conduction

If the heat penetration depth $x=\sqrt{(\pi a t)}$ is large compared to the dimensions of the object considered (see Figure 16.8), the heat penetration theory described above fails. After $t=t_2$ the situation is characterized by zero heat flow at location $x=0$. Although an analytical solution is complicated, the numerical analysis of the problem is simple. The results of calculations of the rate of heating of different objects are of interest. Figure 16.9 shows the time it takes to heat or cool down an object with

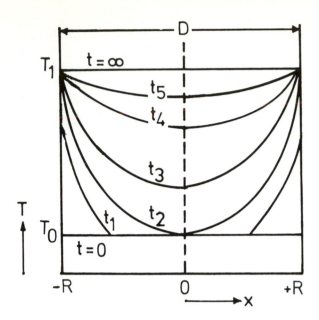

Figure 16.8 Heating of a cylyndrical object of radius *R* by conduction.

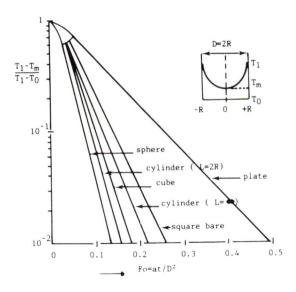

Figure 16.9 Heating of a body by conduction. Temperature in the middle of the body as a function of time.

dimensions D. Figures 16.7 and 16.9 give us insight into how rapidly measurements of SAR distributions in phantoms must be performed to avoid the temperature distribution from being changed by the effects of conduction. For example, Figure 16.9 shows that after plunging an egg in boiling water, it takes at least 4 min for the central temperature to rise above 60°C. This explains why the yolk is soft in most boiled eggs.

16.2.5. Heat transfer between a wall and flow

The heat transfer between plane walls and flow has already been described in Section 16.2.2. Of far greater importance is the heat transfer between solid media and flow in tubes (vessels). Man contains well over 10^6 m of blood vessels which are extremely important in tissue heat transport.

The heat transfer coefficient between the wall of the tube and the fluid in the tube is often described in terms of the difference between the temperature of the inner wall (diameter D) and the mean fluid temperature $<T>$. This mean temperature is defined such that $<T>$ is proportional to the amount of heat transported by the fluid,

$$<T> <v> \frac{\pi D^2}{4} = \int_0^{D/2} 2 \pi r v(r) T(r) \, dr, \qquad (16.28)$$

and is the temperature which we would measure if we were to collect the fluid in the tube and mix it. $<T>$ is called the 'mixing cup' temperature. $<v>$ is the mean flow velocity and $v(r)$ the flow velocity at radius r.

Dimensional analysis indicates that the heat transfer in tubes with laminar flow is described by the Nusselt number $Nu=hD/k$ and the Greatz number $Gz=(ax)/(<v>D^2)$. If a tube enters a medium at a different temperature, as with the transient heating of solid objects (Figure 16.8), there is a certain length x from the beginning of the tube where we have a sort of 'penetration' theory. Beyond this length ($G\bar{z}<0.1$) the flow is thermally developed and the value of the heat transfer coefficient between the fluid and the wall becomes constant.

During the 'thermal entry length'

$$h(x)= \frac{1.08 \, k}{D} \frac{(ax)^{-1/3}}{<v> D^2} \text{ (W/m}^2\text{) for } Gz <0.1, \qquad (16.29)$$

whilst for the thermally fully developed flow

$$h=3.66 \frac{k}{D} \text{ (W/m}^2\text{) or } Nu=3.66. \qquad (16.30)$$

It is assumed that the flow is already laminar at $x=0$ and that the fluid's viscosity and density do not depend on temperature. The description of turbulent flow falls beyond the content of this chapter.

Knowing $h(x)$ and the heat transport in the medium around the tubes, as described in Section 16.2.1.(c), we can calculate the heating of liquids in laminar flow tubes and the temperature around tubes in solid materials.

16.2.6. Heat exchangers

The simplest form of a heat exchanger consists of two concentric tubes (Figure 16.10). If the flow direction in the tubes is equal we refer to a parallel flow heat exchanger; if the directions of flow are opposed, we have

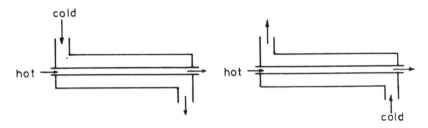

Figure 16.10 Concentric tube heat exchangers.

a counterflow heat exchanger. The total amount of heat exchange between the tubes is

$$Q = U A (T_c - T_h) \text{ (W)}, \tag{16.31}$$

where A is the exchange surface and U is the total heat transfer coefficient between the two tubes. Subscripts c and h are used to indicate cold and hot fluid, respectively. In the following, the subscripts i and o indicate the inlet and outlet conditions, respectively. The overall heat balance of the heat exchanger gives

$$M_c C_c (T_{c,o} - T_{c,i}) = M_h C_h (T_{h,i} - T_{h,o}). \tag{16.32}$$

M_c and M_h are the mass flow rates of the fluids. Figure 16.11 shows the relative temperature variations through heat exchangers for both flow situations when $M_c C_c = M_h C_h$.

Provided that the heat exchanger is long, $T_{h,o} = (T_c + T_h)/2$ for parallel flow and $T_{h,o} \to T_{c,i}$ for counterflow. If there is no heat loss to the surroundings, $T_h - T_c$ is constant at each location in the counterflow heat exchanger since all heat loss from the hot fluid enters the cold tube.

We shall see in Section 16.3.3 that counterflow heat exchange is important in describing the heat transfer around artery–vein vessel pairs in heated tissues.

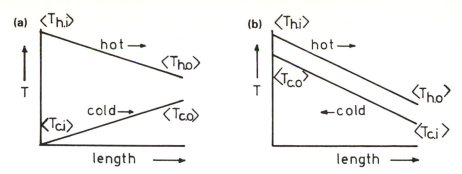

Figure 16.11 Temperature variations in (a) parallel flow and (b) counter-flow heat exchangers.

16.3. Heat transfer in vascularized tissues

16.3.1. Introduction

In terms of heat transfer, tissues differ from solid media mainly by the presence of blood vessels. To indicate the importance of blood flow in tissue heat transfer, I shall present first two typical results from clinical hyperthermia practice.

Result 1. In our clinical treatment of inoperable breast tumours using a system with two 434 MHz applicators, the net amount of power from each applicator during steady-state treatment is 55 ± 19 W (Lagendijk *et al.*, 1988). The amount of power required to obtain therapeutic steady-state temperatures in breast phantoms simulating the same treatment is 4 ± 1 W. The breast phantoms were modelled according to the average patient geometry. The well-known 'Super Stuff' muscle phantom (Guy, 1971; Chou *et al.*, 1984) is used to simulate muscle, gland, tumour and skin. The fatty tissues have been simulated using a fat–bone phantom (Lagendijk and Nilsson, 1985).

This large difference in the powers needed to maintain elevated temperatures in phantoms and in patients clearly indicates the extreme importance of blood perfusion in tissue heat transfer. Thermal models which do not take into account blood flow correctly are of no value for hyperthermia research.

Result 2. Figure 16.12 shows the transient temperature response in a heated breast tumour. Sensor 6, which was in the middle of the tumour, could not be raised above 40°C while the remaining part of the heated volume was at 44°C. Further analysis of this case showed that a large vessel crossed the tumour volume. It was not possible to heat the blood in the vessel and the tissue directly around the vessel.

This example clearly illustrates the local cooling caused by large thermally significant vessels, a situation analysed further in Section 16.3.2.

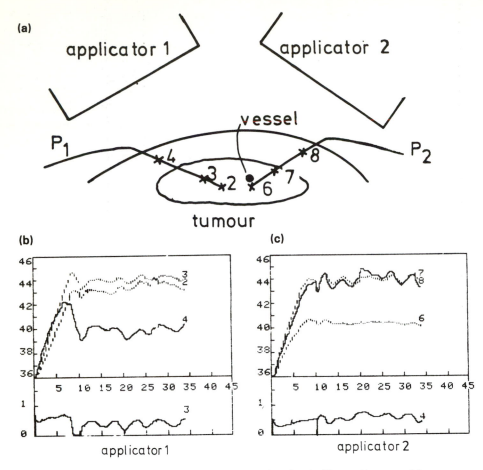

Figure 16.12 Transient temperature response in a heated breast tumour. (a)
Treatment geometry. (b) Probe beneath the left applicator. (c) Probe beneath the
right applicator.

16.3.2. Discrete blood vessels

The bases of the description of the influence of blood flow in tissue heat
transfer are: (i) the influence of single discrete blood vessels; (ii) the
influence of artery–vein vessel pairs; (iii) the influence of randomly
oriented small vessels. We shall first consider a semi-analytical solution
for the single-vessel system (Lagendijk, 1982) to illustrate the physical
principles involved.

We must consider both heating of the blood in the vessel and the temperature distribution in the surrounding tissue. Our model, outlined in
Figure 16.13, consists of a cylinder of tissue of radius R_2 around a vessel of
radius R_1. The temperature distribution in the tissue and the temperature

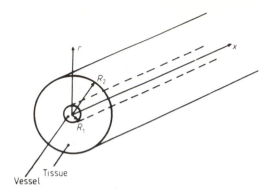

Figure 16.13 Model geometry. Blood vessel radius R_1 in a tissue cylinder radius R_2, x-co-ordinate along vessel axis.

of the blood are calculated as functions of the position x along the vessel. The flow in the vessel is assumed to be laminar. The temperature of the blood at each location x is described by the 'mixing cup' temperature as defined in equation (16.28). The heat flux through the vessel wall at a point x is

$$Q'' = h(x)\,(T_{\text{wall}}(x) - <T_{\text{blood}}(x)>)\ (\text{W/m}^2). \tag{16.33}$$

Since there is no discontinuity in the temperature distribution along the vessel in the clinical case, the heat transfer coefficient $h(x)$ is given by equation (16.30) with $D=2R_1$. Assuming that (i) there is no axial heat flow through the tissue, (ii) there is no absorption of energy from the heating system in the tissue around the vessel and (iii) the temperature of the tissue at distance R_2 from axis of the vessel is equal to T_{tumour}, we can calculate the heat flow per unit length of the vessel through the tissue towards the vessel [equation (16.8)]. The temperature at a distance R_2 must be kept at a fixed value. This can be the tumour temperature which is kept constant by a heating system (microwave, ultrasound, etc.). This is the easiest way to take into account the heating system. R_2 must be chosen in such a way that the energy absorbed within the cylinder with radius R_2 is greater than, but on the same order of magnitude as, the energy flowing through the vessel wall of the blood. The results will be essentially the same if we consider a real three-dimensional system including a realistic absorbed power distribution in tissue (Lagendijk *et al.*, 1984). To calculate the temperature distribution, the blood vessel is divided along its axis into small elements Δx, where Δx is taken to be so small that the heat flux is constant over it. If we now consider the heat flow through the vessel wall at a point x over a length Δx we find, using equation (16.33), that

$$Q(x) = 2\,\pi\,R_1\,h\,\Delta x\,(T_{\text{wall}}(x) - <T_{\text{blood}}(x)>)\ (\text{W}). \tag{16.34}$$

Using equation (16.8), the heat flow through the tissue over the same length is given by

$$Q(x) = \frac{2 \pi k \Delta x}{\ln(R_1/R_2)} (T_{\text{tumour}} - T_{\text{wall}}(x)) \text{ (W)}. \tag{16.35}$$

The heat flow through the vessel wall will heat the blood. The heating of the blood at a point x is given by

$$Q(x) = \pi R_1^2 <v> C_{\text{blood}} \rho (<T_{\text{blood}}(x + \Delta x)> - <T_{\text{blood}}(x)>) \text{ (W)} \tag{16.36}$$

The heat flows given in equations (16.34), (16.35) and (16.36) are equal in the steady-state case. If both $<T_{\text{blood}}(x)>$ and T_{tumour} are known, $<T_{\text{blood}}(x + \Delta x)>$ and $T_{\text{wall}}(x)$ can be calculated. The temperature gradient at the point x in tissue is given by equation (16.7). Since the temperature of the blood at the vessel entry point is usually known, for arteries usually 37 °C, we can calculate the heating of the blood and the temperature distribution around the vessel by a step-by-step computer calculation. To give an impression of the results, the axial temperature distribution for some vessels is given in Figure 16.14. The two-dimensional plot of the temperature distribution in a plane through the axis of vessel 2 (Figure 16.14) along the x axis is given in Figure 16.15. Figure 16.16 shows the temperature distribution for vessel 2 for the value $x = 3$ cm.

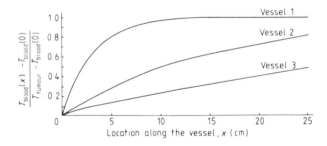

Figure 16.14 The ratio between the difference of the temperature of the blood at point x and the initial temperature and the difference of the tumour temperature and the initial temperature is plotted as a function of location x along the vessels. R_2 is 1 cm. Vessel dimensions: vessel 1, $<v> = 1.0$ cm/s, $R_1 = 0.5$ mm; vessel 2, $<v> = 1.5$ cm/s, $R_1 = 1.0$ mm; vessel 3, $<v> = 2.0$ cm/s, $R_1 = 1.5$ mm.

If we introduce the variable $u = x(<v>R_1^2)$ and plot temperature rise versus u (rather than x), we obtain a more general result describing the temperature rise in all laminar flow vessels entering a heated tissue area, independent of vessel size and flow velocity (Figure 16.17). The most important parameter we get from Figure 16.17 is the 'thermal equilibrium length', i.e. the length over which the temperature difference between the blood and the tissue will be reduced by a factor e. We find that the thermal

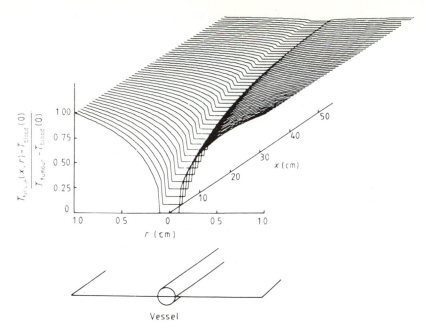

Figure 16.15 Two-dimensional plot of the temperature distribution in a plane through the axis of vessel 2 along the x axis. Note the difference in scale between the x axis and the r axis.

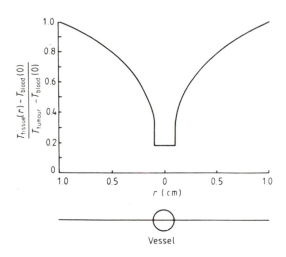

Figure 16.16 Tissue temperature distribution around vessel 2 at point $x=3$ cm.

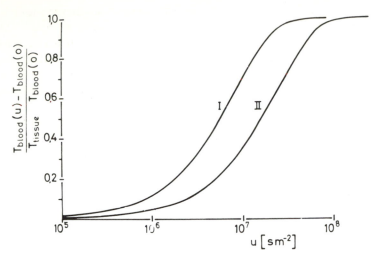

Figure 16.17 General plot of the relative heating of laminar flow vessels entering a 'muscle' tissue volume (*I*) and a 'fat' volume (II) with temperature T_{tissue}. $u=x/(<v>R_1^2)$; $x=0$ at the location the vessel enters the heated area.

equilibrium length for vessels in 'muscle' tissue is given by $X_{eq}=9\times10^6 <v>R_1^2$ and in fatty tissue by $X_{eq}=25\times10^6 <v>R_1^2$. Table 16.2 gives diameter, blood flow velocity and computed thermal equilibrium length for some specific vessels. Due to the dearth of physiological data on the human vascular system, data on the vascular system of a 13 kg dog (Mall, 1888; Green, 1950) are used in Table 16.2.

Table 16.2 Thermal equilibrium length X_{eq} for single vessels in high water content tissues. Data of the vascular system of a 13 kg dog (Mall, 1888; Green, 1950).

Vessel	Diameter (mm)	Number	Length (cm)	Velocity (cm/s)	X_{eq} (cm)
Aorta	10	1	40	50	1.12×10^4
Large arteries	3	40	20	13	2.7×10^2
Main branches	1	600	10	8	18
Secondary branches	0.6	1.8×10^3	4	8	6.5
Tertiary branches	0.14	7.6×10^4	1.4	3.4	1.5×10^{-1}
Terminal arteries	0.05	10^6	0.1	2	1.1×10^{-2}
Terminal branches	0.03	1.3×10^7	0.15	0.4	8.9×10^{-4}
Arterioles	0.02	4×10^7	0.2	0.3	2.9×10^{-4}
Capillaries	0.008	1.2×10^9	0.1	0.07	1.0×10^{-5}
Venules	0.03	8×10^7	0.2	0.07	1.4×10^{-4}
Terminal branches	0.07	1.3×10^7	0.15	0.07	8.9×10^{-4}
Terminal veins	0.13	10^6	0.1	0.3	1.1×10^{-2}
Tertiary veins	0.28	7.6×10^4	1.4	0.8	1.4×10^{-1}
Secondary veins	1.5	1.8×10^3	4	1.3	6.7
Main veins	2.4	600	10	1.5	19
Large veins	6	40	20	3.6	2.9×10^2
Vena cava	12.5	1	40	33	1.17×10^4

It should be noted that the calculated thermal equilibrium lengths in Table 16.2 are valid for vessels entering a tissue volume with solid tissue properties (without blood flow). Normally, vessels will have an interaction which modifies the thermal equilibrium lengths of the individual vessels. As will be shown in section 16.3.3, vessel–vessel(s) interactions reduce the thermal equilibrium length (Weinbaum and Jiji, 1985). Table 16.2, therefore, gives the maximum lengths possible.

Table 16.2 shows that: (i) small vessels such as capillaries are thermally insignificant; their thermal equilibrium length is very small and they are always at local tissue temperature; (ii) the large arteries are always at body core temperature and therefore produce cold tracks through the heated volume (Figures 16.15 and 16.16); (iii) because of their number and equilibrium lengths, the main, secondary and tertiary branches and veins are very important in local hyperthermia.

16.3.3. Artery-vein vessel pairs

Every atlas of human anatomy shows that vessels run in counterflow artery–vein vessel pairs up to the level of arterioles, venules and capillaries. Of course, there are exceptions such as the superficial veins in the extremities, but these do not make the above statement any less important.

The heat transfer around a single vessel is characterized by the vessel's thermal equilibrium length, which is defined by properties of the vessel such as diameter and flow, and the thermal properties of the surrounding tissue (Section 16.3.2). If we assume that the interaction between the vessels is much more important (low heat resistance between the vessels) than the interaction between the vessels and the surrounding tissue (Weinbaum and Jiji, 1985), we have the heat exchange theory as described in Section 16.2.6 which suggests that a constant ΔT exists between the arterial and venous blood because of the equal mass flow in the vessels. In reality there is an interaction with the surrounding tissue which greatly complicates the situation. We shall first consider the case of a cold artery entering a heated tissue volume (the thermal equilibrium definition of Section 16.3.2) but now with a hot vein leaving the heated area. The vein and the artery are in counterflow and have a distance L (Figure 16.18). The cold artery entering the heated volume gains heat not only from the surrounding tissue but also from the hot vein. The thermal equilibrium length of the artery will be reduced, to an extent that depends not only on the flow and diameter of the artery but also on the heat resistance between the two vessels. This heat resistance depends mainly on the distance between the two vessels and can be calculated in a way analogous to that described in Section 16.2.1(b). The hot vein–cold artery counterflow vessel pair is of great clinical importance as indicated later in Section 16.5.

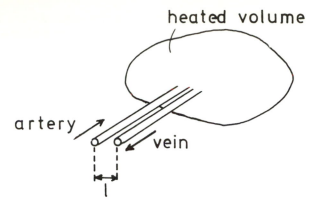

Figure 16.18 Cold artery entering hot tissue volume, hot vein leaving the volume.

Vessels transport heat through tissue, cooling down hot areas and heating cold areas, and so smooth the temperature distribution. Normal body thermal housekeeping is based on such heat transport by blood mass flow. Vessels transport heat in the direction of flow. Provided that the vessels are small and can be described collectively, this heat transport results in

$$Q'' = C\,U\,\mathrm{grad}\,(T) \qquad\qquad (16.37)$$

as described by Wulff (1974). Here U is the mean perfusion velocity.

In most tissues, vessels run in counterflow vessel pairs, resulting in a net $U=0$. However, counterflow vessel pairs contribute greatly to tissue heat transfer as shown in some model calculations. Figure 16.19 shows the

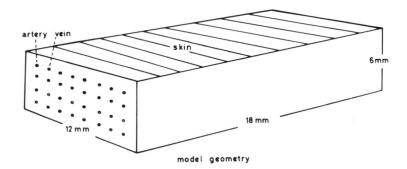

Figure 16.19 Model geometry. The 'muscle' phantom block $36 \times 18 \times 12$ mm³, has a thermal conductivity of 0.6 W/m/°C and a specific heat of 4.2 J/kg/°C. The block contains 32 vessels each with a diameter of 0.48 mm and a flow velocity of 1.5 cm/s. The initial block temperature is 37°C. At time $t=0$ the temperature of the left side changes stepwise to 44°C. The right side is kept at 37°C, the other sides are thermally isolated. The vessel properties and the mutual vessel distances are chosen because of model limitations and not according to clinical geometry.

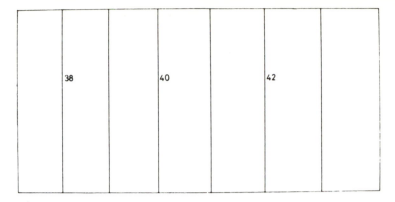

Figure 16.20 Isotherm plot of the steady-state temperature distribution in the mid-plane parallel to the 'skin' of the tissue block without vessels.

model geometry. The rate of heating in this tissue block has been analysed for two cases: (i) without vessels and (ii) with 32 vessels (16 vessel pairs). Figure 16.20 shows the steady-state temperature distribution in the block without vessels while Figure 16.21 shows the transient temperature distribution along the central axis through the model. Slow heating from left to right is shown starting from the initial tissue temperature of 37°C, with the rate of heating dependent on the intrinsic thermal diffusivity of the tissue as described in Section 16.2.4(a).

Figure 16.22 shows the steady-state temperature distribution in the tissue block perfused with 16 arteries (flow direction from left to right, entering the tissue block at 37°C) and 16 veins (right to left, entering the tissue block at 44°C). The diameter of each vessel is 0.48 mm and the flow

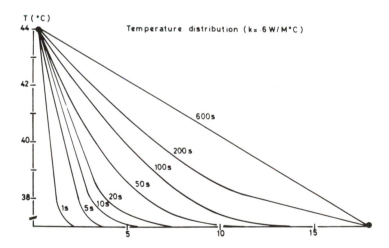

Figure 16.21 The case shown in Figure 16.20. Transient heating of the tissue block along the central axis through the mid-plane of the block.

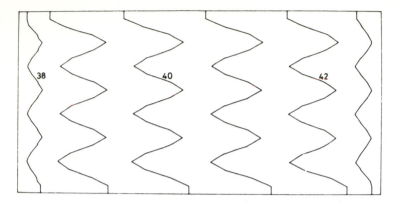

Figure 16.22 Isotherm plot of the steady-state temperature distribution in the mid-plane parallel to the 'skin' of the tissue block with vessels as described in Figure 16.19.

velocity in each is $<v> = 1.5$ cm/s. The separation between the vessel pairs is 1.5 mm. The temperature uniformity is disturbed slightly by the presence of the vessels. Figure 16.23 shows the transient temperature distribution for the same case. Due to the greatly enhanced heat transfer by blood mass flow, the stationary distribution is reached in 60 s. This can be compared with 10 min for the case without blood flow (Figure 16.21). If we simulate the presence of vessels by altering the thermal conductivity of the tissue in a block without vessels, we find that the heat transfer due to this group of vessels can be described by an effective thermal conductivity, k_{eff}, equal to 7.2 W/m/°C (Figure 16.24).

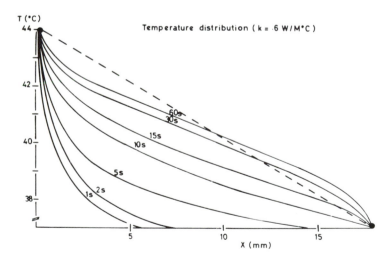

Figure 16.23 Mean transient temperature distribution in the mid-plane of the tissue block with vessels as described in Figure 16.19.

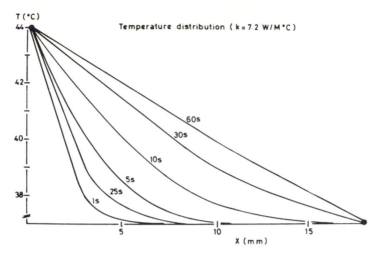

Figure 16.24 Transient temperature distribution as Figure 16.20 but now with an effective thermal conductivity of 7.2 W/m/°C.

The temperature distribution shown in Figure 16.22 is slightly disturbed by the limited size of the model and the boundary conditions used. By describing the discrete vessels in terms of an effective thermal conductivity, the temperature distribution is smoothed which implies that, dependent on vessel type, details of the temperature distribution are lost if the vessels are described collectively.

From the above numerical demonstration we can conclude that (i) provided the thermal equilibrium lengths are small enough and (ii) the arteries and veins run in counterflow, and consequently the net blood flow velocity U is zero, we can describe these small vessels collectively by an enhanced thermal conductivity. The calculation of the effective thermal conductivity related to the capillaries, venules and arterioles has been presented previously by Chen and Holmes (1980). Our calculations also suggest a blood flow heat transfer description by an enhanced thermal conductivity for the larger vessels, at least up to and including the tertiary branches and veins (Table 16.2). Extensive calculations to determine the effective thermal conductivity for all relevant vessel pairs and to find the limits of this collective description for medium-sized vessels have yet to be made. Present numerical calculations indicate effective thermal conductivity values, describing all small vessels up to and including the secondary branches and veins, of 5–10 times the intrinsic thermal conductivity of the tissue. This is in reasonable agreement with the power levels required in clinical practice (Section 16.3.1) and theoretical estimations by Weinbaum and Jiji (1985). In Sections 16.3.4 and 16.3.5 further evaluation of the use and limitations of the effective thermal conductivity model will be given.

16.3.4. Heat transfer in vessel networks

So far, the single-vessel system, two-vessel counter flow, and a possible collective description of small and medium-sized vessels by an effective thermal conductivity have been decribed. In this section the important effects on tissue heat transfer related to vessel branching and complete vessel networks are evaluated.

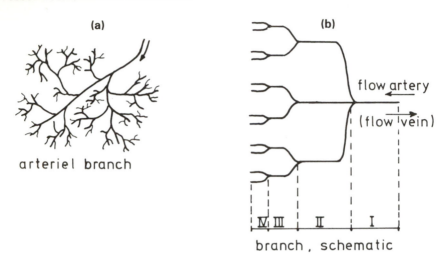

Figure 16.25 Arterial branching from large vessels I to capillaries IV: (a) tissue geometry, (b) schematic geometry.

We first consider the arterial branch entering a heated area (Figure 16.25). According to Table 16.2, the blood vessel in section I of Figure 16.25 has a thermal equilibrium length which is too long to attain the local tissue temperature. Beyond the branches I→II the blood in the smaller vessels with shorter equilibrium lengths will tend to reach local tissue temperature with increasing speed. After branches II→III the blood will reach the local tissue temperature. It is clear that because of the continuous branching towards smaller thermal equilibrium lengths, the blood in the arterial branch is inevitably forced to reach the local tissue temperature.

If we now consider the venous vessel branch we have an essentially different situation. Blood starts in section IV with the tissue temperature, at branch IV→III blood collects in the larger section III vessels which have larger thermal equilibrium lengths. For branches III→II and II→I we have an identical situation; blood from vessels with relatively small equilibrium lengths gathers in vessels with larger equilibrium lengths. If the thermal equilibrium length is longer than the actual length of the vessels in each section, heat can escape the system in the section I vessel by the stepwise collection of the blood in larger vessels. If the thermal equilibrium lengths are small compared to the actual length of the vessels in each section, the blood temperature will follow the tissue temperature and almost no heat will escape the system by the venous section I vessel. The two cases described above present one of the bottlenecks in the conventional 'bioheat transfer' theory (Pennes, 1948) (Section 16.4) in which all venous blood leaves the heated tissue area at the local tissue temperature. In this theory the thermal equilibrium length of the venous vessels is infinite in all vessel sections except section IV. To heat the incoming arteries we need an amount of heat

$$B = w\, C(T-37)\,(\mathrm{Wm}^{-3}). \tag{16.38}$$

where w is the volumetric perfusion rate and C is the specific heat of the blood. If this amount of heat escapes the volume by the venous blood flow, this results in a heat sink factor equal to $-B$ for the tissue.

From Table 16.2 we see that the actual vessel length is much smaller than the thermal equilibrium length until the secondary branches are reached when the actual length becomes comparable with the equilibrium length. However, if we remember that the thermal equilibrium lengths in Table 16.2 are reduced by the counterflow heat exchange between the arteries and the veins and the enhanced heat transfer towards the vessels by the effective thermal conductivity caused by the small vessels, it is expected that even in the secondary veins heat will not escape the system. The heat lost by the veins is transferred to the tissue and incoming arteries, so diminishing the heat sink factor in the conventional theory.

The clinical example discussed in Section 16.3.1 showed that a large amount of heat, related to local perfusion, is lost in the tissue during local hyperthermia. The effective thermal conductivity describing the influence of the counterflow vessel pairs is non-isotropic. The thermal conductivity is only enhanced in the flow direction of both vessels (Figure 16.26). However, anatomical studies indicate that vessels smaller than the main branches and veins are oriented randomly, producing a highly isotropic effective thermal conductivity in tissue. Thus heat loss is not only by the main venous vessels leaving the heated area but also by an increased heat conduction caused by the whole venous–artery network in the periphery of the heated volume.

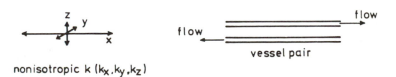

Figure 16.26 Non-isotropic effective thermal conductivity due to counterflow vessel pairs.

It is clear that the effective thermal conductivity model depends to a large extent on the vessel network properties (which will be different for different tissues and organs) and on size, flow velocity and mutual distance of the vessels.

16.3.5. Bioheat transfer thermal modelling

In the previous sections the basis for a bioheat transfer theory has been presented. In this section a general description of this theory suitable for thermal modelling purposes is formulated.

To describe the heat transfer in thermal modelling we divide the heat transfer due to the blood flow into three phases, according to the actual physiological length relative to the thermal equilibrium length of the vessels. Phase I consists of large vessels, both arteries and veins, with a thermal equilibrium length much longer than the actual physiological length. Phase II consists of vessels with a thermal equilibrium length of the order of actual vessel length and phase III consists of small vessels with an equilibrium length small in comparison with vessel length.

Phase I represents the large vessels such as the aorta, the large arteries, the main branches and their venous counterparts. These vessels must be considered individually in thermal models and they are the main cause of inhomogeneous temperature distributions measured in clinical practice. Numerical models which can describe these discrete vessels are being developed (Lagendijk *et al.*, 1984; Mooibroek and Lagendijk, 1987). If these vessels run through an area which is relatively small in comparison with their thermal equilibrium length, they can be replaced in the model by heat sink lines at body core temperature. This simplifies the computer calculations.

Phase III represents the very small vessels including not only the capillaries but also the arterioles, terminal branches and terminal arteries and their venous counterparts (Table 16.2). No temperature difference exists between these vessels and the surrounding tissue temperature. Their orientation is highly isotropic and their relatively small influence on tissue heat transfer can be taken into account by introducing a slightly enhanced effective thermal conductivity (Chen and Holmes, 1980).

This effective thermal conductivity, which is spatially variable and mostly temperature dependent, can easily be taken into account in numerical thermal models.

The heat transfer process in phase II falls between the collective description of phase III and the individual description of phase I. Phase II vessels are too small and too numerous to be taken into account individually in thermal models. The models would become too complex while the spatial resolution would require such a large number of discrete elements that the capacities of normal present computer facilities would be exceeded.

Furthermore, the anatomical vessel network data required as input to the model are not available. As shown in Section 16.3.3, vessels up to and including tertiary branches and veins can be described collectively by a greatly enhanced effective thermal conductivity. If we assume that the thermal equilibrium length of vessels is reduced by the counterflow heat exchange and the increased effective thermal conductivity due to the small surrounding vessels (even secondary branches and veins) can be described collectively with a greatly enhanced thermal conductivity, the distribution is smoothed.

Increasing the fine structure of the model, i.e. shifting the boundary between discrete description and collective description towards the smaller vessels, makes the model more complex and requires a larger amount of physiological blood flow data. Shifting the boundary towards the larger phase I vessels simplifies the model and decreases the amount of anatomical data necessary but leads to a loss of information on the actual small-scale temperature distribution in tissue.

Due to the vessel network properties as described in Section 16.3.4, we have to distinguish two situations:

(a) The whole organ (tissue volume with own vessel supply) is heated. In this case, because of the heated venous blood, there is an optimal reduction of the thermal equilibrium lengths allowing a collective description of most vessels to be made. Only the large vessels must be taken into account discretely. In this situation the temperature uniformity is optimal. The practical model situation is outlined in Figure 16.27.

(b) Only a small tissue volume is heated. We now have enlarged thermal

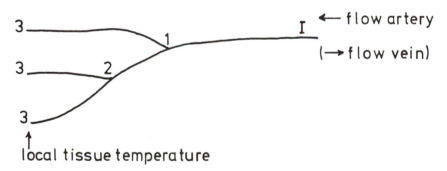

Figure 16.27 Discrete vessel model. Practical example. Arteries: vessel 1 starts with the body core temperature; at branch 1 the blood is divided over two vessels with blood starting temperature equal to the outlet temperature of vessel 1; at branch 2 we have the same situation; at location 3 the vessels terminate. Veins: vessels start at location 3 with the local tissue temperature; at branches 2 and 1 the vessels gather and continue with the resulting mixing cup temperature (Mooibroek and Lagendijk, 1987). The whole tissue around these discrete vessels has a high effective thermal conductivity (which can be spatially dependent) to stimulate the influence of all small vessels.

equilibrium lengths due to cooling from both sides. This increases the complexity of the calculations of the temperature. The vessels to be included discretely in the model become so numerous and small that (i) the numerical model cannot handle them and (ii) the vessel anatomy is not available. When the volume to be described becomes very small the model situation improves again. It should be stressed that the uniformity of the temperature distribution in clinical applications is very poor when enlarged thermal equilibrium lengths are present.

Summarizing, we have: Depending on the size of the heated volume we can describe that heat transfer contribution of all small vessels by an enhanced effective thermal conductivity (expected values 5–10 times the intrinsic thermal conductivity of the tissue) according to equation (16.13) while we have to describe the large discrete vessels separately.

The effective thermal conductivity and the discrete vessel description are inseparably related. The errors made by using only the effective thermal conductivity and failing to take the discrete large vessels into account will be large, particularly when the heated volume decreases in size. We must realize that the lowest tumour temperatures are of utmost clinical relevance in hyperthermia and that these temperatures are directly related to the large vessels.

If the vessels in the volume considered are reasonably isotropic, an isotropic effective thermal conductivity can be used. Otherwise the thermal conductivity must be broken into factors $k(x)$, $k(y)$, $k(z)$ which complicates the solution of equation (16.13).

The resolution of the model depends on the presence of vessels with significant thermal equilibrium lengths. According to Table 16.2. we can expect about one significant vessel every square centimeter. Taking into account the winding paths of vessels, this implies that a real three-dimensional resolution of about 1 mm is required. This means that 5×10^5 nodes are needed to describe a relatively small $10 \times 10 \times 5$ cm^3 volume. This spatial resolution and the necessary anatomical and vascular details lead to a memory requirement around 16 Mb. By choosing fast algorithms, the model is in the range of present 32 bit computer workstations. Cheap parallel processing boards using transputers may speed up calculations. The memory requirement may be lowered by varying the resolution within one model, depending on the presence of vessels, but this greatly increases the complexity of the model.

We need the following input to the thermal models:

(i) The SAR distribution calculated by a SAR model (see Chapter 12). Computed tomography and (particularly) magnetic resonance imaging can provide details of the three-dimensional anatomy. To check and support the calculations we can measure the SAR at the thermometry sensor locations during actual treatments using the first pulse (thermal decay) technique or by using PPT (pulsed power technique) algorithms (Roemer *et al.*, 1985; Knol and Lagendijk, 1989; Lagendijk *et al.*, 1988).

(ii) The vascular anatomy and flow velocities of the thermally significant vessels in the target volume. For limited locations Digital Subtraction Angiography (DSA) can provide excellent information of vascular properties. Magnetic resonance imaging angiography is a fast growing field with great potential, particularly for superficially located regions in which the use of surface coils can provide information of arterial and venous vessels with excellent spatial resolution. Algorithms for the reconstruction of the three-dimensional vascular network are under development.

(iii) The perfusion of the heated volume.

(iv) Transient/temperature responses of blood flow.

(v) The anatomy (thermal properties). Due to the relatively small contribution of intrinsic conduction to total tissue heat transfer, this item is less important than the blood flow items.

16.4. *The conventional bioheat transfer equation*

In general, the temperature distribution in vascularized tissues is described by the conventional 'bioheat transfer' equation (Pennes, 1948). In this theory, the influence of blood is described by the addition of a heat sink term B to equation (16.13):

$$\rho\, C \frac{\mathrm{d}T}{\mathrm{d}t} = \mathrm{div}\,(k\,\mathrm{grad}\,(T)) + P - B. \tag{16.39}$$

P is the total heat produced in each volume element by both the external heating system and the metabolic heat production. B is the amount of heat withdrawn (heat sink) from each tissue volume by the blood flow and is represented by

$$B = w\, C\,(T-37)\ (\mathrm{W/m^3}), \tag{16.40}$$

where w is the volumetric perfusion rate and C is the specific heat of the blood.

The basic assumption of this theory is that the blood enters the local tissue volume at the arterial temperature (usually $37\,^{\circ}\mathrm{C}$) and leaves this volume at the local tissue temperature. Conventional 'bioheat transfer' theory neglects the following points: (1) heat transport related to the mass transport of blood, (2) the actual temperature of the blood entering the local tissue volume, (3) the individual cooling of discrete large vessels and (4) last, but certainly not least, the fundamental importance of the entire venous vessel network by assuming an infinite thermal equilibrium length for all venous vessels.

The reason for this conventional theory remaining the basis of most thermal modelling for over 40 years lies in its mathematical simplicity and in its intelligent form. It fails completely in its description of heat transfer processes in vascularized tissues. The combination of the heat production term (P) and the heat sink term (B), which are both spatially variable, together with the fact that the solid tissue heat transfer contribution by conduction is relatively unimportant (see Section 16.3.1), makes it possible to fit almost any measured temperature distribution. However, this fit is just a mathematical one, the values of B found with this fitting process being of no prospective value for hyperthermia treatment planning systems.

A major difference between the conventional theory and the theory as described in this chapter is the amount of heat flowing between two tissue volume elements. In the conventional theory this heat flow is only determined by the instrinsic thermal conductivity of the tissue which is very small. This makes the thermal interaction between tissue volume elements negligible in comparison to the amount of heat withdrawn by the heat sink factor. Since the thermal conductivity is small, the thermal diffusivity is small too, making the thermal response of the system slow. In the new theory the thermal conductivity is greatly enhanced resulting in a much greater heat transfer between volume elements and a greatly enhanced thermal diffusivity. This results in a system with a much faster thermal response and increased interaction between volume elements. *In vivo* tests are being designed at our institute to distinguish between the theories.

16.5. Local hyperthermia

At present we do not have the numerical models to calculate three-dimensional temperature distributions in vascularized tissues. However, it is of course possible to draw some general conclusions from tissue heat transfer theory and to obtain guidelines to optimize clinical hyperthermia treatments.

(i)   In localized hyperthermia the heated areas vary in size from a few cm^3 to about 1000 cm^3. According to Table 16.2, the blood in the large vessels passing the tumour area will usually fail to reach tumour temperature. The predicted size of the underdosed area and the depth of the temperature minimum in the tumour tissue depend on the thermal behaviour of the surrounding tissue. Since the heat transfer coefficient from the (inside) vessel wall to the blood is fixed (equation (16.30)) (and consequently the thermal resistance is fixed too), the

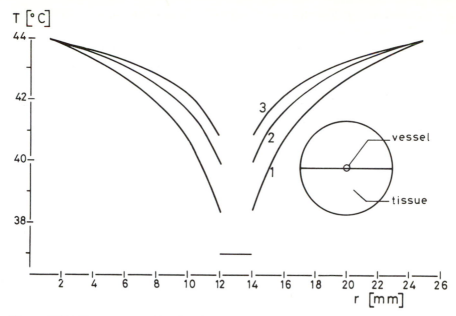

Figure 16.28 Temperature distribution around vessels. Case 1, *K*=0.6, *B*=0; case 2, *K*=1.8, *B*=0; case 3, *K*=3.0, *B*=0.

minimum temperature of the tissue directly surrounding the vessel depends on the heat resistance of the surrounding tissue. If the heat resistance of the surrounding tissue is small (high k_{eff}) most of the temperature gradient will be inside the vessel itself (Figure 16.28). If the vessel has a relatively thick vessel wall (e.g. as in the case of arteries), the minimum temperature in the surrounding tissue will be higher than the temperature of the inside vessel wall because of the steep temperature gradient close to the vessel. With a low k_{eff} in the surrounding tissue, most of the temperature gradient will be inside the surrounding tissue. As a consequence of the large vessel cooling, parts of the surrounding tumour will fail to reach the desired temperature and thus will be underdosed. If large vessels are present, especially with tumour vessel infiltration definitive radiotherapy must be given.

(ii) We must realise that regions of tissue with low perfusion are characterized by low thermal conductivity and, for a given absorbed power distribution, by relatively high local temperatures in the steady-state temperature distribution compared to those produced in well-perfused tissue regions (Strang and Patterson, 1980). The clinical application of hyperthermia allows the determination of data related to local tissue blood flow by analysing the relation between the amount of power absorbed in the tissue and the resulting tissue temperature. Recent analyses of this 'blood flow' determination are presented by Roemer *et al.* (1985), Lagendijk *et al.* (1988) and Samulski *et al.* (1987).

(iii) The effective thermal conductivity of normal tissue will be high because of the high perfusion. In almost all local hyperthermia treatments we have superficial tumours with relatively low perfusion and a heating system which has a small aperture field and a penetration depth which is usually barely adequate. This situation, with small penetration depth and a transition from low thermal conductivity in tumour to high thermal conductivity in normal tissue, will inevitably result in subtherapeutic temperatures in peripheral regions of the tumour. As an example, consider the temperature distribution in a large spherical tumour surrounded by normal tissue. Regardless of the thermal theory used, we see that if we have a uniform SAR in the tumour and zero SAR in the normal tissue then the temperature gradient is in the tumour periphery and not in the healthy tissue, with the result that parts of the tumour are inadequately heated (Figure 16.29). This underdosing of the peripheral region is independent of tumour size.

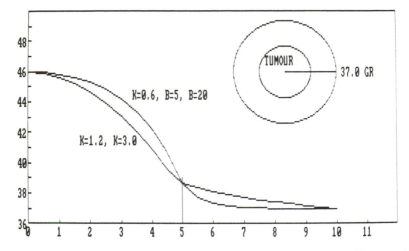

Figure 16.29 Temperature underdosage in tumour periphery. Tumour diameter 10 cm. Case 1: conventional bioheat transfer theory tumour, $K=0.6$, $B=5$; normal tissue, $K=0.6$, $B=20$. Case 2: K_{eff} theory, tumour $K_{eff}=1.2$, $B=0$; normal tissue, $K_{eff}=3.0$, $B=0$.

(iv) The non-uniformity of the intratumoral temperature can be considerably reduced by increasing the heated volume. It is my strong opinion that in all hyperthermia centres, including our own, the heating methods used produce heated volumes which are much too small compared with the tumour volume. Heating only the tumour and not heating a reasonable margin around the tumour results in a poor and undesirable temperature uniformity. Our clinical practice has shown that recurrences after hyperthermia treatment are almost always located in the peripheral zone of the original tumour volume or along the large supplying vessels. Systems able to heat large volumes for the treatment of superficial tumours are not yet available. For deep-body heating we must wonder whether the present trend to go to small volume heating using focused ultrasound or interstitial techniques instead of regional heating methods is a good development. I will not deny the quality of interstitial or ultrasound hyperthermia heating methods. If the heated volume is small and the tumour is small too, if there are no large vessels present or if vessels can be blocked, then these heating systems can produce excellent hyperthermia treatments because of their flexible power control. However in deep-body hyperthermia we usually deal with relatively large tumour volumes in inhomogeneous tissue areas. In these cases a mild systemic heating combined with loco-regional hyperthermia offers physically the best thermal dose distributions.

16.6. Conclusions

Due to the complexity of tissue blood flow, the lack of information on vessel network properties of individual patients and the considerable difficulties in designing these computational models, the actual thermal dose distributions applied in most clinical treatments will stay unknown for several years. However, knowledge of the thermal behaviour of phantoms and tissues is essential to the development of hyperthermia physics. Current thermal modelling can provide valuable information for treatment optimization. In particular, knowledge of the thermal behaviour of large vessels in combination with limited anatomical and vascular information of the target volume can at least indicate where thermal underdosing may be expected and the means to avoid this.

References

Chen, M. M. and Holmes, K. R. (1980) Micro-vascular contributions in tissue heat transfer, *Annals of the New York Academy of Science*, **335**, 137–151.

Chou, C. K., Chen, G. W., Guy, A. W. and Luk, K. H. (1984) Formulas for preparing phantom muscle tissues at various radiofrequencies, *Bioelectromagnetics*, **5**, 335–441.

Croft, D. R. and Lilley, D. G. (1977) *Heat Transfer Calculations Using Finite Difference Equations*, London, Applied Science Publishers.

Green, H. D. (1950) Circulatory system, *Medical Physics II*, Chicago, Year Book Publishers.

Guy, A. W. (1971) Analyses of electromagnetic fields induced in biological tissues by thermographic studies in equivalent phantoms, *IEEE Transactions on Microwave Theory and Techniques*, **19**, 205–214.

Knol, R. G. F. and Lagendijk, J. J. W. (1989) Hyperthermia treatment control and quality evaluation using a pulsed power control technique (PPCT) in Sugahara, T. and Saito, M. (Eds), *Hyperthermic Oncology 1988*, Vol. 1, London, Taylor and Francis, pp. 748–750.

Lagendijk, J. J. W. (1982) The influence of bloodflow in large vessels on the temperature
distribution in hyperthermia, *Physics in Medicine and Biology*, **27**, 17–23.

Lagendijk, J. J. W. and Mooibroek, J. (1986) Hyperthermia treatment planning, *Recent
Results in Cancer Research*, **101**, 18–35

Lagendijk, J. J. W. and Nilsson, P. (1985) Hyperthermia dough; a fat and bone equivalent
phantom to test microwave/radiofrequency hyperthermia heating systems, *Physics in
Medicine and Biology*, **30**, 709–712.

Lagendijk, J. J. W., Schellekens, M., Schipper, J. and van der Linden, P. M. (1984) A three-
dimensional description of heating patterns in vascularized tissues during hyperthermic
treatment, *Physics in Medicine and Biology*, **29**, 495–507.

Lagendijk, J. J. W., Hofman, P. and Schipper, J. (1988) Perfusion analyses in advanved breast
carcinoma during hyperthermia, *International Journal of Hyperthermia*, **4**, 479–495.

Mall, F. (1888) Die Blut und Lymphwege im Dunndarm des Hundes. Koniglich Sachsische
Gesellschaft der Wissenschaft, *Abhandelungen der Mathematischphysischen Classe*, **14**.

Mooibroek, J. and Lagendijk, J. J. W. (1987) Computer implementation of vessel bending and
branching in a 3-D hyperthermia thermal model, *International Journal of Hyperthermia*, **3**, 582.

Paulsen, K. D., Strohbehn, J. W. and Lynch, D. R. (1985) Comparative theoretical performance for two types of regional hyperthermia systems, *International Journal of Radiation Oncology, Biology and Physics*, **11**, 1659–1671.

Pennes, H. H. (1948) Analysis of tissue and arterial blood temperatures in the resting human
forearm, *Journal of Applied Physiology*, **1**, 93–122.

Roemer, R. B., Fletcher, A. M. and Cetas, T. C. (1985) Obtaining local SAR and blood perfusion
data from temperature measurements: steady state and transient techniques compared,
International Journal of Radiation Oncology, Biology and Physics, **11**, 1539–1550.

Samulski, T. V., Fessenden, P., Valdagni, R. and Kapp, D. S. (1987) Correlations of thermal
washout rate, steady state temperatures, and tissue type in deep seated recurrent or
metastatic tumors, *International Journal of Radiation Oncology, Biology and Physics*,
13, 907–916.

Strang, R. and Patterson, J. (1980) The role of thermal conductivity in hyperthermia, *International Journal of Radiation Oncology, Biology and Physics*, **6**, 729–735.

Weinbaum, S. and Jiji, L. M. (1985) A new simplified bioheat equation for the effect of blood
flow on local average tissue temperature, *Journal of Biomechanical Engineering*, **107**,
131–139.

Wulff, W. (1974) The energy conservation equation for living tissues, *IEEE Transactions on
Biomedical Engineering*, **21**, 494–495.

17 Quality assurance in hyperthermia

J. W. Hand

17.1. Introduction

The experimental use of hyperthermia in human cancer therapy has become popular in recent years in many institutes worldwide. The general experience to date is that 'routine' treatment of superficial tumours is feasible but non-invasive treatment of tumours deep in the body remains problematic (Sugahara and Saito, 1989). Nevertheless, despite limitations in heating techniques, the results of many studies in which hyperthermia has been given as an adjuvant to conventional therapies suggest that the combined treatment may be advantageous compared with the conventional treatment alone, as discussed in the chapters on clinical topics in this book. There is currently a need to assess the efficacy of hyperthermia by phase III clinical trials. For the near future, trials will, in general, be restricted to treatments of superficial tumours. However, a positive outcome of such trials will not only benefit treatment of superficial tumours but is also highly likely to increase incentive and support for developing and improving our ability to deliver and monitor hyperthermic treatments of deep-seated tumours.

Experience has shown that factors such as tumour size and location, radiation dose and the general condition of the patient as well as other factors inherent to the hyperthermic therapy such as an appropriate match of equipment and tumour and the minimum temperature monitored in tumour together with the period at this temperature can be associated with outcome of treatment (Arcangeli *et al.*, 1987; Luk *et al.*, 1984; Perez *et al.*, 1986; van der Zee *et al.*, 1986). Recently, a preliminary analysis of a randomized trial in which 300 patients with superficial measurable tumours received fractionated radiotherapy, either alone or with hyperthermia, highlighted the need to apply strict eligibility criteria in the selection of patients and tumours (Perez *et al.*, 1989). Only those patients with lesions for which it is feasible both to heat the tumour volume with the equipment available and to monitor the resulting temperature distribution should be considered for phase III studies. It is important that each treatment in such studies is fully documented and that the characteristics of all equipment used are known and documented.

The purpose of this chapter is to consider technical aspects of quality assurance. Clinical aspects and details of protocols are discussed elsewhere in this book.

17.2. Treatment planning

Treatment planning in hyperthermia is in its infancy. One reason for this is that clinical hyperthermia, as we know it today, is a relatively new modality. Furthermore, the incentive to develop sophisticated treatment planning systems has been small until very recently. This was because most of the clinical hyperthermia systems which were in use were relatively simple (often the only degree of control available was to increase or decrease the power delivered to a single applicator) and so even if such routines had been available, there would have been little impact on the quality of clinical treatments. This situation is changing as 'second generation' applicators with more degrees of freedom are being developed. However, the greatest problem for hyperthermic treatment planning is that it is conceptually very difficult.

In Chapter 12, Paulsen discusses four areas of hyperthermic dosimetry described as comparative, prospective, concurrent and retrospective, concepts initially proposed by Roemer and Cetas (1984). In the first area a direct comparison between heating devices under the same conditions can be made. Standardized models or phantoms which contain only essential anatomical and physiological characteristics of the 'average' patient and salient features of the device(s) can be used for this purpose. In contrast, the second area refers to the prediction of temperature distributions in a specific patient using specific heating devices. With this information available, a 'hyperthermic treatment plan' could be drawn up to optimize the forthcoming treatment. Concurrent dosimetry refers to the real (or quasi-real) time prediction of temperature distribution for a specific treatment using data obtained during that treatment and should be of use in maintaining optimum power delivery through the course of the treatment. In the fourth area, predictions of the temperature field achieved are made from measured data following completion of a particular treatment. These predictions should be of use when trying to correlate the clinical outcomes with details of the treatment actually administered.

It is clear from Chapters 12 and 16 that major problems must be solved before prospective, concurrent and even retrospective dosimetry become routine practices. Not only must the three-dimensional problem of power deposition in real patients due to real applicators be solved but a realistic and manageable thermal model must be identified. The first of these problems is likely to be tractable given sufficient effort to developing numerical models and availability of increased computing power. However, a more realistic yet practical solution to the second than the bioheat transfer equation (see Chapter 16) currently appears more remote (Strohbehn *et al.*, 1989). On the other hand, the requirements for comparative dosimetry appear less stringent and the information gleaned from such studies, even within the restrictions of current capabilities, may be useful in drawing up criteria for quality assurance guidelines and elementary treatment plans. Suggestions as to how such comparisons might be made include the acceptable power band (APB) approach (Roemer and Cetas, 1984) and the Hyperthermia Equipment Performance (HEP) rating (Paulsen *et al.*, 1984). In the APB approach the range of absorbed power values which give acceptable temperature distributions is determined. An acceptable temperature distribution is one in which the minimum temperature in the tumour exceeds an agreed value, the maximum temperature in the tumour is less than an agreed value and the maximum temperature in normal tissue is below an agreed value. The HEP rating is based on the percentage of tumour volume which reaches a specific temperature (e.g. 42 or 43°C), subject to temperatures in tumour and normal tissues being less than agreed values. As explained in Chapter 16 (Section 16.5), a relatively high temperature must be accepted in surrounding normal tissue if therapeutic temperatures in the peripheral regions of the tumour and acceptable temperature uniformity are to be achieved.

A consensus has yet to be formed regarding the structure of one or more standard models and phantoms. Examples of simple phantoms that have been used in several studies concerning the evaluation of SAR distri-

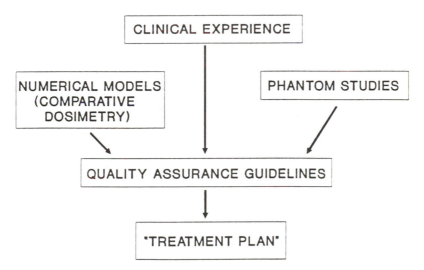

Figure 17.1 Empirical approach to quality assurance and treatment planning.

butions from applicators are those described by Kantor (1981) and Allen *et al.* (1988). Ideally, the models for numerical simulations and the phantoms for experimental studies should simulate the same 'average' patient.

In view of our current limited capabilities, a pragmatic approach to treatment planning and quality assurance guidelines, as outlined in Figure 17.1, must be adopted. Here, information gained from numerical and experimental evaluations of heating devices, together with the considerable practical experience gained over several years by physicians and physicists or engineers actively involved in clinical hyperthermia programmes is collated to form a consensus of informed opinion regarding procedures which must be followed if a high 'quality' of treatment is to be achieved. For the reasons alluded to earlier, much of the work in hyperthermia quality assurance to date has concentrated on localized treatments of superficial tumours.

17.3. *Choice of applicator*

To answer the question 'How do we select an applicator for treatment of a specific tumour?' we must first define some parameters which describe the characteristics of the applicators at out disposal. Ideally, we would like to predict the volume of tissue over which an applicator could induce therapeutic temperatures. Currently, the absence of a clinically realistic thermal model usually prevents us from reaching this goal. One characteristic which has been used is the *effective volume* (EV), defined as the volume in which specific absorption rate (SAR)* is equal to or greater than an agreed percentage of the maximum SAR. Ideally, this should be determined in three dimensions, but since the necessary measurements can be very time consuming, two important related parameters are often considered. These are the *effective field size* (EFS) and the *penetration depth* (PD).

For non-invasive electromagnetic devices used for localized superficial treatments, EFS has been defined in a plane at a particular depth (usually 10 mm) in a planar muscle-like phantom in terms of the area enclosed by the contour drawn at an agreed percentage of the maximum SAR. The level of 50% peak SAR has often been chosen (arbitrarily) to define EFS (Hand *et al.*, 1989a; Shrivastava *et al.*, 1989) although arguments for a more conservative estimation of the useful field (70–80% peak SAR) have been put forward (P. Fessenden, personal communication; Chapter 14).

---

*SAR is the time derivative of the incremental energy (dW) absorbed or dissipated in an incremental mass (dm) contained in a volume element (dV) of a given density (ρ) and may be expressed as (National Council on Radiation Protection and Measurements, 1981)

$$\mathrm{SAR} = -\frac{\mathrm{d}}{\mathrm{d}t}\left[\frac{\mathrm{d}W}{\mathrm{d}m}\right] = \frac{\mathrm{d}}{\mathrm{d}t}\left[\frac{\mathrm{d}W}{\rho\,\mathrm{d}V}\right].$$

Likewise, PD has been defined as the distance below the position of maximum SAR in the plane used to define EFS at which SAR is reduced to 50% of that maximum. Similar definitions of EFS and PD may be used when characterizing stationary ultrasound transducers. The concept of EV may also be used to characterize the SAR distribution due to microwave interstitial antennas. This is useful for making comparisons between arrays with different interantenna spacings, insertion depths and modes of operation (i.e. coherent or incoherent arrays). It should be stressed that the temperature distribution achieved during clinical use of interstitial techniques is very dependent upon heat transport mechanisms within the implanted tissue (Strohbehn and Mechling, 1986). The use of SAR alone to characterize techniques such as radio-frequency localized currents and hot sources is of little help in predicting performance in patients. This caution is generally applicable but it is particularly true for interstitial techniques where high SAR gradients are always present.

There is little variation in PD for most electromagnetic applicators used for localized, superficial treatments (it is approximately 10 mm) although radiative applicators with large apertures operating at relatively low frequencies and large capacitive electrodes may achieve a modest increase over this value. PD values for ultrasound transducers are more dependent upon driving frequency. In the case of interstitial techniques, the spacing between sources (electrodes, antennas, hot sources) should be 1–2 cm. The EV for microwave antennas can be dependent upon insertion depth (when this is shallow) as well as interantenna spacing, driving frequency and mode of operation.

When selecting an applicator for use in a specific localized superficial treatment, the absolute minimum requirement which must be satisfied is that the EV (or EFS and PD) covers the tumour volume. It should be borne in mind that we are usually attempting to heat tumours located in well-perfused normal tissues and that many of the applicators used produce greater SAR beneath their central region than under their edges. This combination can easily lead to inadequate hyperthermia in the peripheral regions of the tumour. It is therefore good practice to choose an applicator with an EV (or EFS) which provides a margin of approximately 5 cm beyond the lateral dimensions of the tumour. When using an interstitial technique, antennas or sources should be implanted approximately 1 cm beyond tumour margins. By including some normal tissue in the field, the temperature distribution may be improved since more of the surrounding vasculature will be heated (see Chapter 16).

An attraction of using SAR as a parameter for defining field size is that it is clearly defined and is easily measured. However, it must be stressed that since there is no clear correlation between SAR and the resulting temperature distribution, caution is necessary when relating EVs, EFSs and PDs measured under simple, idealized conditions to clinical performance. It is particularly important to assess the effects of heat transfer mechansims within the tissue when using an interstitial technique. General clinical experience has shown that many electromagnetic applicators for superficial hyperthermia can raise tissue temperatures to about 42–43 °C at depths up to approximately 2 cm (Sapozink *et al.*, 1988). For stationary, plane ultrasound transducers operating at a frequency of about 1 MHz, the typical useful treatment depth is 3 cm (Kapp *et al.*, 1988). If significant thickness of underlying fat is present, these values are usually increased. For interstitial techniques, the distance between sources should be 1–2 cm, depending upon technique and the perfusion within the tissue. Even these figures require careful interpretations since temperature distributions are very dependent upon tissue site, geometry and perfusion. Ultimately, the choice of applicator is based on the practical experience of the clinical and physics/engineering staff.

17.4. Characterization of heating devices

17.4.1. Electromagnetic devices

SAR distributions produced by applicators should be measured using standard phantoms. The simplest approach appropriate to radiative electromagnetic applicators used for localized superficial treatments and microwave interstitial antennas is to use a plane homogeneous phantom which simulates the dielectric properties of muscle tissue. In the case of external applicators the phantom should be at least 10 cm thick to ensure that fields reflected from its bottom surface undergo sufficient attenuation for them not to perturb measurements in more superficial regions. It should also extend at least 5 cm beyond the edges of the applicator and its bolus to avoid modifications in the SAR distribution due to the presence of boundaries. The type and geometry of the bolus used with an applicator can modify the SAR distribution significantly. EFS and PD should therefore be determined in the presence of a bolus similar to that used in treatments. In the case of invasive antennas the dimensions of the phantom should be sufficiently large that a boundary of at least 5 cm of phantom exists around the implanted array.

Although intercomparisons of applicators based on measurements obtained from such simple phantoms are relatively easy to make, the assessment of the performance of an applicator in a particular treatment based on these data is difficult. For example, the presence and thickness of a plane layer of fat-like phantom can have considerable influence on the SAR distribution beneath an applicator, as illustrated in the example discussed by Chou in Chapter 18. It is prudent therefore to characterize applicators using several phantoms with fat layers of varying thickness. Indeed, realistic phantoms which simulate the geometry of the treatment set-up and relevant anatomical features should be used to determine SAR distributions whenever possible. This is particularly important when using radio-frequency capacitive electrodes since SAR is strongly dependent upon the separation and orientation of the electrodes as well as the electrical conductivity of the tissues.

Several phantom materials which simulate the dielectric properties of tissues with high or low water content have been developed and some of these have been recommended for use in quality assurance procedures (see Table 17.1). A summary of other materials which have been developed to simulate specific tissues (e.g. brain and lung) is given by Hand (1990). Descriptions of the manufacture of these phantom materials to gether with details of their density and specific heat and the temperature at which their dielectric properties were determined can be found in most of the references cited.

Table 17.1. **Phantom materials useful in characterizing electromagnetic devices. Those in bold type are cited in existing quality assurance procedures** ([1]Kikuchi *et al.*, 1986; [2]Hand *et al.*, 1989a; [3]Shrivastava *et al.*, 1989).

Tissue simulated	Frequency range (MHz)	Type	Reference
Muscle[1]	**1–40**	**Gel**	**Kato and Ishida (1987)**
Muscle	13.56–40.68	Gel	Bini *et al.* (1984)
Muscle	**13·56–100**	**Gel**	**Chou *et al.* (1984)**
Muscle	100–900	Gel	Hartsgrove *et al.* (1987)
Muscle[2,3]	**200–2450**	**Gel**	**Chou *et al.* (1984)**
Muscle	750–2450	Gel	Andreuccetti *et al.* (1988)
Muscle	450	Liquid	Hand *et al.* (1989b)
Muscle	915–2450	Liquid	Kopecky (1980)
Muscle	2450	Liquid	Gajda *et al.* (1979)
Fat–bone	27.12	Gel	Bini *et al.* (1984)
Fat–bone	451	Dough	Lagendijk and Nilsson (1985)
Bone	100–900	Liquid	Hartsgrove *et al.* (1987)
Fat–bone	100–900	Castable solid	Hartsgrove *et al.* (1987)
Fat–bone[3]	**915–2450**	**Castable solid**	**Guy (1971)**
Fat	2450	Liquid	Gajda *et al.* (1979)

A common method of measuring SAR distributions is to determine the rate of change of temperature (dT/dt) at known positions in a solid or gel phantom following a brief period of heating at high power. SAR may be inferred from

$$\text{SAR} = \frac{dT}{dt} c \,(\text{W/kg}),$$

where $c(\text{J/kg/}^\circ\text{C})$ is specific heat of the phantom material. With this method, measurements should be within a short period (no more than 60 s) of the power being turned on to minimize artifacts caused by thermal conduction within the phantom. In all cases the relationship between temperature and time should be reasonably linear. If temperature sensors are mechanically scanned through the phantom to obtain the SAR distribution, temperatures should be allowed to return to steady state following each power pulse to avoid temperature-related changes in the dielectric properties of the phantom. Such measurements are often time consuming and for convenience should be made using an automated system (Wong *et al.*, 1985; Jones *et al.*, 1988; Hand *et al.*, 1988). A quicker but more costly approach is to use an infra-red thermographic camera to obtain two-dimensional distributions in predetermined planes in a phantom which is constructed so that it can be opened quickly to reveal these particular cross-sections (Guy, 1971; Cetas, 1983). Care should be taken that the interfaces between the planes do not perturb the SAR distribution (Kantor and Cetas, 1977; Lehmann *et al.*, 1983; Schaubert, 1984). This is particularly important at lower frequencies (e.g. 27 MHz). Although the use of liquid crystal sheets can provide a rapid and low-cost qualitative assessment of heating patterns, this method is not recommended for quality assurance purposes. When using thermographic methods, care should be taken to avoid artifacts due to cooling of films or surfaces.

Periodic measurements of the total power output and efficiency of applicators should also be made. These parameters may be readily determined using a calorimetric method (Anhalt and Hynynen, 1988; Lagendijk and Schipper, 1986). The calorimeter consists of a near-adiabatic enclosure filled with a lossy liquid such as saline and with dimensions sufficiently large to ensure that all of the energy from the applicator is absorbed. The liquid is carefully stirred and the temperature increase resulting from exposure to the energized applicator is measured. If the clinical arrangement of applicator bolus is simulated in the calorimetric measurement, a calibration factor relating the power reading displayed at the generator to the power delivered from the applicator during treatment (for a given arrangement of cables and components between generator and applicator) can be esimated. To determine the efficiency, the net power delivered to the input of the applicator must be measured using forward and reflected power meters. It should be realized that the efficiency of most electromagnetic applicators is low, on the order of 50%. The SAR normalized to net power delivered at the applicator (W/kg per W net power) is also a useful parameter in applicator characterization. According to Bassen and Coakley (1981), therapeutic applicators should be capable of producing an SAR of 235 W/kg in the muscle region of a fat–muscle phantom.

17.4.2. Ultrasound devices

Characterization of ultrasound devices requires measurement of the power output from the transducer, the spatial peak intensity in the ultrasound field and the intensity distribution. Both power output and spatial peak intensity should be measured as functions of applied radio-frequency power (Hynynen, 1989). The intensity distribution may be determined from the pressure distribution using a non-directional hydrophone [for example, one based on the piezoelectric polyvinylidene fluoride (Shotton *et al.*, 1980; Preston *et al.*, 1983)] or from thermal transients using a small thermocouple or thermistor probe coated with an absorbing material such as glue, silicone rubber or polyurethane varnish (Martin and Law, 1983). Although a complete three-dimensional measurement is desirable, it is acceptable to determine the fields of superficial heating devices in the axial plane together with measurements in planes perpendicular to the axis at one or two distances from the transducer. For focused transducers, measurements should be obtained in the focal plane.

The total acoustic power output from the transducer can be obtained using either a radiation force measurement technique (Chapter 14; Robinson, 1984) or a calorimeter containing an absorbing fluid (Hynynen, 1990). Such measurements should be performed over the complete range of radio-frequency power applied to the transducer.

Degassed water may be used as a general phantom material in ultrasound studies since the acoustic impedance and the speed of sound in this medium are similar to those of soft tissues. Field profiles measured in

Table 17.2 Phantom materials useful for characterizing ultrasound devices.

Tissue simulated	Ingredients	Speed of sound (m/s)	α_0 (dB/cm/MHz^n)	n
Various soft tissues[1]	Gelatin/Graphite powder/n-Propanol	1520–1650	$\approx$0.2–1.5	$\approx$1
Various soft tissues[2]	Agar/n-Propanol/ Water/Graphite powder	1498– > 1600	0.04–1.4	$\approx$1
Glandular tissue for breast[3]	Gelatin plus: 34% (v/v) olive oil, 3% (v/v) n-propanol, 0.11 mg/cm^3 scatterers	1518.8	0.80	1.01
Fat[3]	Gelatin plus: 25% (v/v) olive oil, 25% (v/v) kerosene, 2.2% (v/v) n-propanol, 0.065 mg/cm^3 scatterers	1458.5	0.42	1.02
Liver[3]	Gelatin plus: 13.6% (v/v) castor oil, 20.4% (v/v) olive oil, 8.8% (v/v) n-propanol	1539.4	0.51	1.08
Various tissues[4]	Gelatine–alginate gel plus scatterers	1519	0.12–0.5	$\approx$1

[1]Madsen *et al.* (1978) (values at 22–25 °C).
[2]Burlew *et al.* (1980) (values at 22 °C).
[3]Madsen *et al.* (1982) (values at 22 °C).
[4]Bush and Hill (1983) (at 20–22 °C).

water must be adjusted mathematically to account for the attenuation which would be present in tissue. An alternative procedure for determining the SAR distribution is to use a phantom which simulates the acoustic properties of tissue and into which is embedded an array of thermocouple sensors (Edmonds *et al.*, 1985; Bassen *et al.*, 1985). Details of several tissue-mimicking materials are listed in Table 17.2.

17.5. Thermometry

It is current practice to use invasive thermometry in hyperthermic treatments. Conventional probes (using either thermistors or thermocouples as sensors), electromagnetic non-perturbing probes (using some temperature-dependent optical property of the sensor) or electromagnetically minimally perturbing probes (e.g. thermistor with high-resistance leads) are usually passed through plastic catheters which have been implanted into the tissue. Although this procedure is widely adopted, caution is called for since the presence of plastic catheters can perturb heating fields and give rise to artifacts (Chan *et al.*, 1988; Hynynen and Edwards, 1989).

17.5.1. General requirements and calibration

The general requirements of thermometers used in hyperthermic treatments are: (i) minimal trauma and risk to the patient; (ii) an accuracy of ±0.2°C or better over the temperature range 37–46°C; (iii) a stability of ±0.2°C or better for the duration of a treatment; (iv) a resolution of ±0.1°C or better; (v) a response time of a few seconds.

To be compatible with requirement (i), sensors or probes should be as small as possible. This also helps minimize artifacts which are the result of field perturbations or probe heating or which may occur when probes are used in the presence of temperature gradients, as discussed later in this section. To satisfy requirements (ii) and (iii), all thermometers used in treatments must be calibrated against a working (tertiary) standard (i.e. a reference thermometer or reference temperature maintained locally which is traceable to a national secondary standard) which is accurate to approximately ±0.02°C. Suitable working standards are precision mercury-in-glass thermometers or highly stable thermistors. Three working standards should be maintained and compared periodically to test for drift. Calibration should be carried out in an isothermal medium. This may be a well-stirred water bath (baths in which the temperature uniformity and stability are better than ±0.02°C and ±0.02°C per minute are readily available), avoiding regions near the walls and surface where temperature gradients are present, or a copper or aluminium block. In many cases, probes are not interchangeable at the required level of

Table 17.3. Fixed point temperatures relevant to thermometry in hyperthermia.

Substance	Fixed point	Temperature (°C)	Uncertainty associated with maintaining standard (°C)	Reference
H_2O	Triple point	0.01	±0.0002	1
Diphenyl ether (phenoxybenzene)	Triple point	26.869	±0.002	1
Gallium	Melting point	29.772	±0.002	2
$Na_2SO_4 \cdot 10H_2O$	Hydrate transition	32.373	±0.002	3
Ethyl carbonate (1,3-dioxolan-2-1)	Triple point	36.325	±0.002	1
$Na_2HPO_4 \cdot 7H_2O$	Hydrate transition	48.222	±0.007	2

[1]National Physical Laboratory, Division of Quantum Metrology, Teddington TW11 0LW, UK.
[2]Mangum and Thornton (1977).
[3]Magin *et al.* (1981).

accuracy. The frequency at which calibrations need to be carried out is dependent upon the type of probe used. Further details of calibration procedures are discussed by Cetas in Chapter 14. Table 17.3 lists some fixed point temperatures which can be useful for calibrating thermometers for clinical hyperthermia. In general, requirements (iv) and (v) are readily met by the various invasive probes used in clinical hyperthermia although the response time of a probe within a catheter or loose-fitting sheath can be significantly greater than that associated with the bare probe (Waterman, 1985).

17.5.2. General comments on the use of probes

Multisensor probes should be used whenever possible since they provide more thermometric data than single-sensor probes (e.g. temperature gradients along the probe) without additional trauma to the patient. Probes with up to 10 thermocouple sensors at, say, 5 or 10 mm spacing and typically 0.5–1 mm in diameter are easily made. Artifacts due to thermal conduction along thermocouple probes may be reduced by using manganin–constantan rather than copper–constantan. The resistivities of the leads in the former pair are also better balanced leading to a reduction in direct heating of the junction in radio-frequency fields. Possible artifacts which may arise due to conduction when making temperature measurements in high thermal gradients using any type of probe should be investigated by testing in a temperature step-function (Lyons *et al.*, 1985; Samulski *et al.*, 1985). In addition, there are other artifacts associated with the use of multi-sensor thermocouple probes in these conditions which should be evaluated (Bach Andersen *et al.*, 1984; Dickinson, 1985; Fessenden *et al.*, 1984). Most manufacturers of electromagnetically non-perturbing probes currently offer a multisensor probe with typically four sensors per probe, although probes with up to seven sensor probes are being evaluated as future products. In the absence of multisensor probes a thermal maping technique in which single-sensor probes are moved within their catheters should be used (Gibbs, 1983; Engler *et al.*, 1987). With these techniques, care is needed in defining the minimum time for which the sensors are stationary since the response time of a sensor or probe will be increased by the presence of the catheter or tubing through which it is scanned. A three- or four-fold increase over the response time of the bare sensor can be expected (Waterman, 1985).

Probes intended for use in ultrasound-induced hyperthermia should be small in diameter to keep field perturbations low. Hynynen and Edwards (1989) have shown that the presence of probes having a diameter greater than 0.6 mm for systems operating at 1 MHz (the size reduces as a function of $\sqrt{\lambda}$, where λ is the wavelength in mm) can induce significant distortions in the field. Although artifacts are small if bare thermocouples are used, it is more practical to use sheathed probes in clinical studies. The use of a stainless-steel sheath (or catheter) introduces relatively small artifacts arising from viscous heating and absorption of ultrasound energy although thermal conduction along the probe is likely to be increased and rigid probes can prove to be a source of discomfort for the patient. The absorption of ultrasound energy by plastic sheaths or catheters is well documented (Hynynen *et al.*, 1983; Fessenden *et al.*, 1984; Hand and Dickinson, 1984; Kuhn and Christensen, 1986; Waterman *et al.*, 1989). However, thermocouple probes with a sheath of thin-walled polyethylene tubing placed approximately parallel with the direction of propagation of the beam are acceptable if the ultrasound field is turned off for a few seconds before measurements are made (Hynynen and Edwards, 1989). In the case of scanned focused ultrasound systems, the power should be turned off while a sensor is within the focal volume. The use of Teflon sheathing should be avoided and plastic catheters should be withdrawn from the tissue during ultrasound hyperthermia.

Although an accuracy of $\pm 0.2\,°C$ is attainable under calibration conditions by most thermometers used in clinical hyperthermia (Shrivastava *et al.*, 1988), this figure is likely to be degraded in the conditions prevalent during clinical use. To minimize additional inaccuracies, conventional thermocouple and thermistor probes should be oriented perpendicular to the direction of the electromagnetic field whenever possible. When used in the presence of ultrasound, thermocouple probes sheathed with thin-walled polyethylene tubing should be oriented parallel with the direction of propagation of the beam. Heating fields should be switched off typically 2–3 s before temperature measurements are made with these conventional sensors. An estimate of the change in tissue temperature over this period in which the power is off should be made. The rate of decrease in tumour temperature following cessation of power is on the order of 0.1 °C/s. To minimize errors arising from the finite time required to interrogate multiple temperature sensors within acceptable bounds, data acquisition systems used with conventional thermometry should be capable of scanning all sensors within 1 s. Temperature sensors should be scanned at typically 10–20 s intervals.

17.6. Treatment practices

17.6.1. Minimum number of temperature sensors

As many temperature sensors as possible, compatible with patient tolerance and clinical safety, should be used in every hyperthermic treatment. However, a minimum criterion relating to the number of implanted sensors, which can be site specific, should always be satisfied. For example, for breast tumours, a minimum requirement of two sensors located in the tumour at diametrically opposed positions and within 1 cm of the periphery, a third sensor implanted within 1 cm of the deepest part of the tumour and a sensor on the skin (or tumour surface) is proposed (Hand *et al.*, 1989a; Medical Research Council, 1989). The insertion of several probes into neck nodes is probably less acceptable and so a slightly less-stringent minimum requirement is proposed, namely one implanted probe and one sensor on the skin. Ideally the probe should have at least three sensors, two of which should be located in peripheral regions of the tumour, but a mechanically scanned probe with a single sensor is acceptable. For melanomas or chest wall recurrences of carcinoma of the breast, the minimum number of sensors may be determined by the area of the lesion. For example, in addition to a sensor on the surface, at least two sensors should be implanted in tumours with an area not greater than 16 cm². For larger tumours, an extra sensor should be implanted for each additional 16 cm² of area. Particular care should be taken to monitor temperatures in poorly perfused regions and in areas with scar tissue. The size and location of the tumour and the locations of the temperature sensors should be estimated using radiographic or ultrasound diagnostic techniques whenever possible.

It is useful to estimate SARs at the locations of the temperature sensors by measuring dT/dt following an initial brief period during which power is on or by measuring the initial rate of decrease in temperature when power is turned off (Roemer *et al.*, 1985).

There is a need to locate the position of the implanted temperature sensors with respect to the tumour volume. Ideally this information should be obtained from computed tomography or ultrasound images.

Anatomical landmarks can be used as fiducial points (Samulski and Fessenden, 1990). If a scanned focused ultrasound machine is used the temperature sensors may be located with respect to the co-ordinate system of the machine by performing a raster scan and detecting the transient response of the sensors as the focus moves over their locations (Hynynen *et al.*, 1987). In the case of interstitial treatments, catheters for both sources and thermometer probes can be implanted under stereotactic guidance (Onik *et al.*, 1987).

17.6.2. Positioning of applicator(s)

For local treatments applicators and bolus or coupling bags should be positioned so as to optimize the temperature distribution produced (i.e. minimizing the presence of hot regions in normal tissue and cold regions within tumour). In the case of ultrasound applicators, it is essential that good acoustic contact between applicator and tissues is maintained. This may be achieved using acoustic gel or degassed water. The geometrical set up should also be compatible with good impedance matching (minimal reflected power) and, in the case of electromagnetic applicators, with acceptably low leakage fields. Details of radio-frequency safety standards are discussed in detail by Chou in Chapter 18.

If possible, the position and orientation of an applicator with respect to the patient should be described in terms of a suitable co-ordinate system (Shrivastava *et al.*, 1989).

17.6.3. Duration of treatment(s)

Hyperthermic treatments are often prescribed in terms of a target temperature (typically 42–43°C) and their duration (typically 45–60 min). In some protocols (European Society for Hyperthermia Oncology, 1986; Medical Research Council, 1989) the target temperature is the minimum temperature (43°C) to be achieved at the locations of all temperature sensors within the tumour volume. For these protocols duration of treatment has been identified (Hand *et al.*, 1989a) as the period beginning either 10 min after commencement of heating or from the time at which all temperature sensors located in the tumour record 43°C or higher, if this occurs within 10 min of commencement of heating. If any break occurs during treatment, the period during which the power is turned off should be added to the overall treatment time to a maximum of 20 min. These guidelines limit the time allowed for a typical treatment to a maximum of 90 min. Experience has shown that an acceptable rate of increase in temperature during the initial phase of treatment is approximately 1°C per minute.

17.7. Documentation

Adequate documentation of the characteristics of all equipment used during clinical treatments and details of each treatment delivered must be made. In this section a summary of details which must be recorded is given. The design of forms to contain this information is likely to be protocol specific.

17.7.1. Equipment characteristics

Crucial details which must be recorded include:

Applicators: type (manufacturer, model number, etc.) and operating frequency, overall dimensions, type of bolus used, dimensions of bolus, effective volume (or effective field size and penetration depth).

Thermometry: type of thermometer (manufacturer, model number, etc.), type (thermocouple, thermistor, etc.) and number of probes, number of sensors per probe, accuracy and stability, response time, details of calibration procedure, date of calibration.

This information should be updated as necessary. It must be made available for external assessment as part of a quality assurance programme. In multi-institutional studies, the co-ordinating centre must be provided with technical details of all equipment used in the study at each institute. This information is an aid in assessing whether individual institutes meet quality assurance guidelines pertinent to a particular study.

17.7.2. Treatment details

The following details should form part of the record of each series of hyperthermic treatments delivered to a particular tumour. In addition, the record must contain details of the radiotherapy or other adjunct therapy and other clinical information.

- Patient's name, date of birth, sex.
- A list of criteria for eligibility, all of which must be satisfied before a patient can be considered for a randomized trial involving hyperthermia. An affirmative answer to the question '*Is hyperthermia treatment feasible for this tumour?*' must be one of these criteria.
- Treatment protocol number and treatment allocated.
- Date of randomization.
- Anatomic site of lesion.
- Histologic type.
- Tumour classification and staging.

- Previous treatment.
- Tumour measurements (length, width, depth) and method of measuring (e.g. ultrasound, computed tomography, magnetic resonance imaging, calipers, photography, etc.).

The crucial details of *each* hyperthermia treatment must be documented. These must include:

- The hyperthermia system used.
- The applicator(s) and bolus used.
- The thermometry systems used.
- Tumour measurements and method of measuring (ultrasound, computed tomography, magnetic resonance imaging, calipers, photography, etc.).
- The locations of the temperature sensors (e.g. tumour, normal tissue, skin).
- The relative positions of the applicator (and bolus), thermometry probes and tumour. If possible, these should be given in terms of convenient co-ordinate system. A photograph (including a length scale) and a diagram of the set-up should be included in the record.
- The name(s) of the operator(s).
- The date of treatment.
- The time of treatment relative to adjuvant therapy.
- The treatment number (e.g. within a series of fractions).
- The duration of treatment.
- The occurence and reason for any breaks in treatment (e.g. readjustment of applicators, patient discomfort, etc.).
- The occurence and nature of any pain suffered by the patient.
- If the treatment is terminated prematurely, the reason for this must be recorded (e.g. patient pain, inability to maintain position, technical malfunction, etc.).
- The complete time–temperature record (temperatures should be monitored at least at 30 s intervals) at the location of each temperature sensor.
- When thermal mapping is carried out a scan should be recorded immediately before power-on and then at least at 10 min intervals during the quasi-plateau phase of the time–temperature relationship.
- A condensed form of the time–temperature record can be used on the summary form relating to each treatment (e.g. temperatures at 5 min intervals, maxima and minima in temperatures recorded in tumour and normal tissue).

Further details of reporting hyperthermic treatments have been discussed by Shrivastava *et al.* (1989) and Hand *et al.* (1989a). Sapareto and Corry (1989) present a comprehensive data file for this purpose in a proposed standard format.

Acknowledgements

Many of the concepts presented in this chapter have been the subject of extensive discussions with colleagues involved in ESHO and MRC hyperthermia trials. I am also grateful to Dr P. Fessenden, Dr K. Hynynen and Dr J. J. W. Lagendijk for helpful suggestions.

References

Allen, S., Kantor, G., Bassen, H. and Ruggera, P. (1988) CDRH RF phantom for hyperthermia systems evaluations, *International Journal of Hyperthermia*, **4** (1), 17–23.

Andreuccetti, D., Bini, M., Ignesti, A., Olmi, R., Rubino, N. and Vanni, R. (1988) Use of polyacrylamide as a tissue equivalent material in the microwave range, *IEEE Transactions on Biomedical Engineering*, **35** (4), 275–277.

Anhalt, D. and Hynynen, K. (1988) The efficiency of clinical microwave applicators measured by a calorimetric technique, *Medical Physics*, **15** (6), 919–921.

Arcangeli, G., Cividalli, G., Lovisolo, G. and Mauro, F. (1987) The combination of heat and radiation in cancer treatment, in Field, S. B. and Franconi, C. (Eds), *Physics and Technology of Hyperthermia*, Dordrecht, Martinus Nijhoff, pp. 574–585.

Bach Andersen, J., Baun, A., Harmark, K., Heinzl, L., Rasmark, P. and Overgaard, J. (1984) A hyperthermia system using a new type of inductive applicator, *IEEE Transactions on Biomedical Engineering*, **31**, 21–27.

Bassen, H. I. and Coakley, R. F. (1981) United States radiation safety and regulatory considerations for radiofrequency hyperthermia systems, *Journal of Microwave Power*, **16**, 215–226.

Bassen, H., Allen, S., Herman, B., Kantor, G. and Robinson, R. (1985) Quality assurance of RF and ultrasound cancer hyperthermia systems, *Proceedings of the Seventh Annual Conference of the Engineering in Medicine and Biology Society*, New York, IEEE, pp. 346–351.

Bini, M. G., Ignesti, A., Millanta, L., Olmi, R., Rubino, N. and Vanni, R. (1984) The polyacrylamide as a phantom material for electromagnetic hyperthermia studies, *IEEE Transactions on Biomedical Engineering*, **31** (3), 317–322.

Burlew, M. M., Madsen, E. L., Zagzebski, J. A., Benjavic, R. A. and Sum, S. W. (1980) A new ultrasound tissue-equivalent material, *Radiology*, **134**, 517–520.

Bush, N. L. and Hill, C. R. (1983) Gelatine–alginate complex gel: a new acoustically tissue–equivalent material, *Ultrasound in Medicine and Biology*, **9** (5), 479–484.

Cetas, T. C. (1983) Physical models (phantoms) in thermal dosimetry, in Storm, S. K. (Ed.), *Hyperthermia in Cancer Therapy*, Boston, G. K. Hall, pp. 257–278.

Chan, K. W., Chou, C. K., McDougall, J. A. and Luk, K. H. (1988) Perturbations due to the use of catheters with non-perturbing thermometry probes, *International Journal of Hyperthermia*, **4** (6), 699–702.

Chou, C. K., Chen, G. W., Guy, A. W. and Luk, K. H. (1984) Formulas for preparing phantom muscle tissue at various radiofrequecies, *Bioelectromagnetics*, **5** (4), 435–441.

Dickinson, R. J. (1985) Thermal conduction errors of manganin–constantan thermocouple arrays, *Physics in Medicine and Biology*, **30**, 445–453.

Edmonds, P. D., Ross, W. C., Lee, E. R. and Fessenden, P. (1985) Spatial distributions of heating by ultrasound transducers in clinical use, indicated in a tissue-equivalent phantom, *Proceedings of 1985 IEEE Ultrasonics symposium*, New York, IEEE, pp. 908–912.

Engler, M. J., Dewhirst, M. W., Winget, J. M. and Oleson, J. R. (1987) Automated temperature scanning for hyperthermia treatment monitoring, *International Journal of Radiation Oncology, Biology and Physics*, **13**, 1377–1382.

European Society for Hyperthermic Oncology (1986) *European Society for Hyperthermic Oncology Protocols*, 3rd ed, Study coordinator: J. Overgaard, Institute of Cancer Research, DK-8000 Aarhus C.

Fessenden, P., Samulski, T.V. and Lee, E.R. (1984) Direct temperature measurement, *Cancer Research*, **44** (Suppl.), 4799s–4804s.

Gajda, G., Stuchly, M.A. and Stuchly, S.S. (1979) Mapping of the near field pattern in simulated biological tissues, *Electronics Letters*, **15** (2), 120–121.

Gibbs, F.A. (1983) 'Thermal mapping' in experimental cancer treatment with hyperthermia: description and use of a semiautomatic system, *International Journal of Radiation Oncology, Biology and Physics*, **9**, 1057–1063.

Guy, A.W. (1971) Analysis of electromagnetic fields induced in biological tissues by thermographic studies on equivalent phantom models, *IEEE Transactions on Microwave Theory and Techniques*, **19** (2), 205–214.

Hand, J.W. (1990) Biophysics and technology of electromagnetic heating, in Gautherie, M. (Ed.), *Clinical Thermology (Therapy)*, Vol. 2, Methods of External hyperthermia heating, Munchen, Springer-Verlag, pp. 1–59.

Hand, J.W. and Dickinson, R.J. (1984) Linear thermocouple arrays for *in vivo* observation of ultrasonic hyperthermia fields, *British Journal of Radiology*, **57**, 656.

Hand, J.W., Prior, M.V., Johnson, R.H., Forse, G.R. and Roberts, T. (1988) Effective field size measurements and quality assurance for clinical applications of localized hyperthermia. *Medical Applications of Microwaves (IEE digest 1988/60)*, London, Institution of Electrical Engineers, pp. 6/1–6/4.

Hand, J.W., Lagendijk, J.J.W., Bach Andersen, J. and Bolomey, J.C. (1989a) Quality assurance guidelines for ESHO protocols, *International Journal of Hyperthermia*, **5** (4), 421–428.

Hand, J.W., Paulsen, K.D., Lumori, M.L.D., Gopal, M.K., Cetas, T.C. and Alkhairi, S. (1989b) Microwave array applicators for superficial hyperthermia, in Sugahara, T. and Saito, M. (Eds), *Hyperthermic Oncology*, Vol. 1, London, Taylor and Francis, pp. 827–828.

Hartsgrove, G., Kraszewski, A. and Suriec, A. (1987) Simulated biological materials for electromagnetic radiation absorption studies, *Bioelectromagnetics*, **8** (1), 29–36.

Hynynen, K. (1989) Biophysics and technology of ultrasound hyperthermia, in Gautherie, M. (Ed.) *Clinical Thermology (Therapy)*, Vol. 2, Methods of external hyperthermia heating, Munchen, Springer-Verlag, pp. 61–115.

Hynynen, K. and Edwards, D.K. (1989) Temperature measurements during ultrasound hyperthermia, *Medical Physics*, **16** (4), 618–626.

Hynynen, K., Martin, C.J., Watmough, D.J. and Mallard, J.R. (1983) Errors in temperature measurement by thermocouple probes during ultrasound induced hyperthermia, *British Journal of Radiology*, **56**, 969–970.

Hynynen, K., Roemer, R., Anhalt, D., Johnson, C., Xu, Z.X., Swindell, W. and Cetas, T.C. (1987) A scanned, focussed, multiple, transducer system for localized hyperthermia treatments, *International Journal of Hyperthermia*, **3**, 21–35.

Jones, K.M., Mechling, J.A., Trembly, B.S. and Strohbehn, J.W. (1988) SAR distributions for 915 MHz interstitial microwave antennas used in hyperthermia for cancer therapy, *IEEE Transactions on Biomedical Engineering*, **35** (10), 851–857.

Kantor, G. (1981) Evaluation and survey of microwave and radiofrequency applicators, *Journal of Microwave Power*, **16** (2), 135–150.

Kantor, G. and Cetas, T.C. (1977) A comparative heating pattern study of direct contact applicators in microwave diathermy, *Radio Science*, **12**, 111–120.

Kapp, D.S., Fessenden, P., Samulski, T.V., Bagshaw, M.A., Cox, R.S., Lee, E.R., Lohrbach, A.W., Meyer, J.L. and Prionas, S.D. (1988) Stanford University institutional report. Phase I evaluation of equipment for hyperthermic treatment of cancer, *International Journal of Hyperthermia*, **4** (1), 75–115.

Kato, H. and Ishida, T. (1987) Development of an agar phantom adaptable for simulation of various tissues in the range 5–40 MHz, *Physics in Medicine and Biology*, **32** (2), 221–226.

Kikuchi, M., Egawa, S., Onoyama, Y., Kato, H., Kanai, H., Saito, M., Tsukiyama, I., Hiraoka, M., Mizushina, S. and Yamashita, T. (1986) *Guide for Use of Hyperthermia Devices (1): Capacitive Heating (Tentative)*, Document produced by Quality Assurance Committee of the Japanese Hyperthermia Association, Tokyo.

Kopecky, W.J. (1980) Using liquid dielectrics to obtain spatial thermal distributions, *Medical Physics*, **7** (5), 566–570.

Kuhn, P.K. and Christensen, D.A. (1986) Influence of temperature probe sheathing materials during ultrasonic heating, *IEEE Transactions on Biomedical Engineering*, **33** (5), 536–538.

Lagendijk, J. J. W. and Nilsson, P. (1985) Hyperthermia dough: a fat and bone equivalent phantom to test microwave/radiofrequency hyperthermia heating systems, *Physics in Medicine and Biology*, **30** (7), 709–712.

Lagendijk, J. J. W. and Schipper, J. (1986) Clinical hyperthermia systems, in Hand, J. W. and James, J. R. (Eds), *Physical Techniques in Clinical Hyperthermia*, Letchworth, Research Studies Press, pp. 452–505.

Lehmann, J. F., McDougall, J. A., Guy, A. W., Chou, C. K., Esselman, P. C. and Warren, C. G. (1983) Electrical discontinuity of tissue substitute models at 27.12 MHz, *Bioelectromagnetics*, **4**, 257–265.

Luk, K. H., Pajak, T. F., Perez, C. A., Johnson, R. J., Connor, N. and Dobbins, T. (1984) Prognostic factors for tumour response after hyperthermia and radiation, in Overgaard, J. (Ed.), *Hyperthermic Oncology 1984*, Vol. 1, London, Taylor and Francis, pp. 353–356.

Lyons, B. E., Samulski, T. V. and Britt, R. H. (1985) Temperature measurements in high thermal gradients: I. The effects of conduction, *International Journal of Radiation Oncology, Biology and Physics*, **11**, 951–962.

Madsen, E. L., Zagzebski, J. A., Banjavie, R. A. and Jutila, R. E. (1978) Tissue mimicking materials for ultrasound phantoms, *Medical Physics*, **5** (5), 391–394.

Madsen, E. L., Zagzebski, J. A. and Frank, G. R. (1982) Oil-in-gelatin dispersions for use as ultrasonically tissue-mimicking materials, *Ultrasound in Medicine and Biology*, **8** (3), 277–287.

Magin, R. L., Mangum, B. W., Statler, J. A. and Thornton, D. D. (1981) Transition temperatures of the hydrates of NA_2SO_4, Na_2HPO_4 and KF as fixed points in biomedical thermometry, *Journal of Research of the NBS*, **86** (2), 181–192.

Mangum, B. W. and Thornton, D. D. (Eds) (1977) The gallium melting point standard. 1982, Triple point of gallium as a temperature fixed point, *NBS Special Publication 481*, Washington, DC, US Government Printing Office.

Martin, C. J. and Law, A. N. R. (1983) Design of thermistor probes for measurement of ultrasound distributions, *Ultrasonics*, **21**, 85–90.

Medical Research Council (1989) *M.R.C. Phase III Trial (January 1989)*, Trial coordinator: C. C. Vernon, Hammersmith Hospital, London.

National Council on Radiation Protection and Measurements (1981) *NCRP Report No. 67: Radiofrequency Electromagnetic Fields — Properties, Quantities and Units, Biophysical Interaction and Measurement*, Washington, DC, National Council on Radiation Protection and Measurements, p. 15.

Onik, G., Cosman, E., Wells, T., Goldberg, H. I., Moss, A., Kane, R., Moore, S., Stauffer, P., Costello, P. and Hoddick, W. (1987) CT body sterotaxic system for placement of needle arrays, *International Journal of Radiation Oncology, Biology and Physics*, **13**, 121–128.

Paulsen, K. D., Strohbehn, J. W. and Lynch, D. R. (1984) Theoretical temperature distributions produced by an annular phased array type system in CT-based patient models, *Radiation Research*, **100**, 536–552.

Perez, C. A., Kuske, R. R., Emami, B. and Fineberg, B. (1986) Irradiation alone or combined with hyperthermia in the treatment of recurrent carcinoma of the breast in the chest wall: a nonrandomized comparison, *International Journal of Hyperthermia*, **4** (2), 179–188.

Perez, C. A., Gillespie, B., Pajak, T., Hornback, N. B., Emami, B. and Rubin, P. (1989) Quality assurance problems in clinical hyperthermia and their impact on therapeutic outcome: a report by the Radiation Therapy Oncology Group, *International Journal of Radiation Oncology, Biology and Physics*, **16** (3), 551–558.

Preston, R. C., Bacon, D. R., Livett, A. J. and Rajendran, K. (1983) PVDF membrane hydrophone performance properties and their relevance to the measurement of the acoustic output of medical ultrasonic equipment, *Journal of Physics E: Scientific Instruments*, **16**, 786–796.

Robinson, R. A. (1984) Performance evaluation of a digital readout hyperthermia range ultrasonic wattmeter, *IEEE Transactions on Sonics and Ultrasonics*, **31** (5), 467–472.

Roemer, R. B. and Cetas, T. C. (1984) Applications of bioheat transfer simulations in hyperthermia, *Cancer Research*, **44**, 4788s–4798s.

Roemer, R. B., Fletcher, A. M. and Cetas, T. C. (1985) Obtaining local SAR and blood perfusion data from temperature measurements: steady state and transient techniques compared, *International Journal of Radiation Oncology, Biology and Physics*, **11**, 1539–1550.

Samulski, T. V. and Fessenden, P. (1990) Thermometry in therapeutic hyperthermia, in Gautherie, M. (Ed.) *Clinical Thermology (Therapy)*, Vol. 3, Munchen, Springer-Verlag, in press.

Samulski, T. V., Lyons, B. E. and Britt, R. H. (1985) Temperature measurements in high thermal gradients: II. Analysis conduction effects, *International Journal of Radiation Oncology, Biology and Physics*, 11, 963–971.

Sapareto, S. A. and Corry, P. M. (1989) A proposed standard data file format for hyperthermia treatments, *International Journal of Radiation Oncology, Biology and Physics*, 16, 613–627.

Sapozink, M. D., Cetas, T., Corry, P. M., Egger, M. J., Fessenden, P. *et al.* (1988) Introduction to hyperthermia device evaluation, *International Journal of Hyperthermia*, 4 (1), 1–15.

Schaubert, D. H. (1984) Electromagnetic heating of tissue-equivalent phantoms with thin, insulating partitions, *Bioelectromagnetics*, 18, 123–126.

Shotton, K. C., Bacon, D. R. and Quilliam, R. M., (1980) A PVDF hydrophone for operation in the range 0.5 MHz to 15 MHz, *Ultrasonics*, 18, 123–126.

Shrivastava, P. N., Saylor, T. K., Matloubieh, A. Y. and Paliwal, B. R. (1988) Hyperthermia thermometry evaluation: criteria and guidelines, *International Journal of Radiation Oncology, Biology and Physics*, 14, 327–335.

Shrivastava, P., Luk, K., Oleson, J., Dewhirst, M., Pajak, T., Paliwal, B.,Perez, C. Sapareto, S., Saylor, T. and Steeves, R., (1989) Hyperthermia quality assurance guidelines, *International Journal of Radiation Oncology, Biology and Physics*, 16, 571–587.

Strohbehn, J. W. and Mechling, J. A. (1986) Interstitial techniques for clinical hyperthermia, in Hand, J. W. and James, J. R. (Eds) *Physical Techniques for Clinical Hyperthermia*, Letchworth, Research Studies Press, pp. 210–287.

Strohbehn, J. W., Paulsen, K. D. and Lynch, D. (1989) Numerical modelling of SAR and temperature in 1992: where we might be, in Sugahara, T. and Saito, M. (Eds), *Hyperthermic Oncology 1988*, Vol. 2, London, Taylor and Francis, pp. 770–773.

Sugahara, T. and Saito, M. (Eds) (1989) *Hyperthermic Oncology 1988*, Vols. 1 and 2, London, Taylor and Francis.

van der Zee, J., van Putten, W. L. J., van den Berg, A. P., van Rhoon, C. G., Wike-Hooley, J. L., Broekmeyer-Reurink, M. P. and Reinhold, H. S. (1986) Retrospective analysis of the response of tumours in patients treated with a combination of radiotherapy and hyperthermia, *International Journal of Hyperthermia*, 2 (4), 337–349.

Waterman, F. M. (1985) The response of thermometer probes inserted into catheters, *Medical Physics*, 12 (3), 368–379.

Watermann, F. M., Nerlinger, R. E. and Leeper, J. B. (1990) Catheter induced temperature artifacts in ultrasound hyperthermia, *International Journal of Hyperthermia*, 6, 371–381.

Wong, T. Z., Strohbehn, J. W., Smith, K. F. and Douple, E. B. (1985) Automated measurement of power deposition patterns from interstitial microwave antennas used in hyperthermia, Kuklinski, W. S. and Ohley, W. J. (Eds), *Proceedings of the 11th Annual Northeast Bioengineering Conference (IEEE 85CH2203-8)*, New York, IEE, pp. 58–61.

18 Safety considerations for clinical hyperthermia

C. K. Chou

18.1. Introduction

Diathermy, the heating of body parts by radio-frequency (RF) electro-magnetic (EM) energy (shortwaves and microwaves) or by sonic energy (ultrasound), has been shown to increase extensibility of collagenous tissues, to decrease joint stiffness, to relieve pain, and to resolve muscle spasms. It has been widely used in physical therapy for more than 40 years. Hyperthermia is similar to diathermy. Both are used to elevate tissue temperature. The main difference between them lies in the nature of the target tissues. Indigenous tissues and organs are treated by diathermy to ameliorate pain and to enhance function. In contrast, neo-plasms are the principal targets of hyperthermia. When considering the beneficial effects of hyperthermia, one must consider also the potential for hazards that application of this modality may portend, not only for the patient but also for the operating personnel. Much of the guidance for safe and effective application of diathermy is applicable to hyperthermia (Kloth *et al.*, 1984). Both guidance for and contraindications of therapeutic heat have been described in detail by Lehmann (1990).

When diathermy and hyperthermia fail as therapies, the outcomes are decidedly different. Failure of diathermy means persistence of pain or dys-function, both of which may readily yield to another modality of treatment. But the failure of hyperthermia in combination with other therapy to control a malignant tumour can have grave consequences, since it can be fatal. Effective heating, therefore, is the most important clinical consideration in the application of hyperthermia. Because heating patterns vary with many factors, it is critical to have characterized patterns for every treatment condition. The previous chapter on quality assurance has already dealt with this subject. A few examples are illus-trated here to emphasize the importance of dosimetry.

In this chapter, safety of patients and operating personnel, and proper use of equipment, are considered. Because hyperthermia, whether local or regional, may require shortwave, microwave or ultrasound energy to elevate a tumour's temperature, the biological response to EM fields and ultrasound is the first topic to be summarized. Ultrasound hyperthermia does not pose a problem of stray radiations as is the case with shortwaves and microwaves. Accordingly, EM fields are addressed in greater detail. Electrical safety, and standards and guidelines for applying EM fields are discussed. It is concluded that hyperthermia is a highly complicated therapy, and that it should be administered only by trained personnel if effectiveness and safe use are to be assured.

18.2. Biological effects of electromagnetic waves*

There are more than 7000 reports in the biomedical literature on EM waves. Both salutary and deleterious properties are cited in the reports, and many of the reported findings are controversial and lacking in independent confirmation. A comprehensive review of the literature is not attempted in this chapter, but the interested reader is directed to the extensive offerings in the following publications: Tyler (1975), Baranski and Czerski (1976), Adair (1983). Osepchuk (1983), Polk and Postow (1986), National Council on Radiation Protection and Measurements (NCRP, 1986, 1988), Heynick, (1987),the *Journal of Bioelectromagnetics*, and the *Journal of Microwave Power and Electromagnetic Energy*.

Because so many reports are speculative and unsubstantiated, critical reviews and analysis of the literature have been necessary. To minimize bias, collaborative reviews by committees of experts have been performed. The US Environmental Protection Agency completed a comprehensive review and evaluation of the available scientific information on the biological effects of RF EM radiations. The report was published in the mid-1980s (Elder and Cahill, 1984), and it was updated by Elder in 1987. The NCRP published a report in 1986 entitled 'Biological effects and exposure criteria for radiofrequency electromagnetic fields'. My summary of biological effects borrows heavily from these reviews, and the interested reader is urged to consult them for additional information and extensive citations of the literature.

18.2.1. Quantification of biological effects

Experiments that ethically cannot be performed on human beings are performed on laboratory animals. The experimental data are not always directly applicable to man, and extrapolation is therefore fraught with

---

*This section is condensed from the paper 'Biological effects of electromagnetic waves' previously written for the NATO Hyperthermia Summer School held at Urbino, Italy, in 1986 (Chou, 1987a.)

complications. Various biological differences, such as species, age, sex and body mass must be considered. In research involving non-ionizing radiations, not only must one consider differences in biological end-points, one must deal also with the difficult problem of physical scaling. Further complicating the task of extrapolation is a problem of inequality: Approaches and concepts used in ionizing radiation often have been incorrectly applied to non-ionizing radiations (Chou and Guy, 1985).

The absorption and distribution of EM energy in the body are determined by many factors: dielectric properties of tissues, tissue geometry and mass, the body's orientation in the field, the field's polarization and frequency, configuration of the source, exposure duration, field intensity and finally, the many variables in the exposure environment, including temperature, humidity and air velocity. In sum, these factors can alter the quantity of energy coupled to a body or its thermal response to radiation by several orders of magnitude.

Dosimetry is critical to the science and the art of hyperthermia. Prime among dosimetric measures for the experimenter is the specific absorption rate (SAR) (Justesen and King, 1970; Justesen, 1975). The SAR is a dose rate. Specifically, it is the time rate at which EM energy is coupled to a body of standardized mass, i.e. in units of watts per kilogram (W/kg). The SAR is derived, in turn, from specific absorption (SA), the energy dose in units of joules per kilogram (J/kg). Both SA and SAR are defined in NCRP Report 67 (1981). The SAR has been widely adopted by laboratory investigators of biological effects of EM fields. In the hyperthermia clinic, the SAR is important, but it is secondary to the variable of temperature. For the clinician, it is more convenient to use heating rate (°C/min) (Chou, 1990).

In the experimental study of hazardous and beneficial effects of exposure to EM fields, the power density of the incident field, and the resulting SARs and temperature increases in targeted tissues, must be determined. Only when the invariances among these three variables are firmly established can there emerge a clear understanding of how EM fields interact with tissues, i.e. whether an effect is microscopic or macroscopic, and whether an effect is by nature thermal or athermal — or merely an artifact of experimental or judgemental error. Only by taking into account body mass and dimensions of experimental animals and by gaining accurate dosimetries *in vivo*, both electrical and thermal, can results of different investigators with differing species and end-points be related to humans.

18.2.2. Epidemiological studies

Soviet and other Eastern European authors have frequently described a set of symptoms expressed by workers after prolonged employment in presumed proximity to microwaves or other RF EM fields. These symptoms, which variously have been labelled 'neurasthenia' and 'microwave sickness', are based on subjective complaints that include headache, sleep disturbance, motor weakness, decrease of sexual drive, impotence, pain in the chest and, in general, poorly defined feelings of malaise (Baranski and Czerski, 1976). Also described are changes in cardio-vascular function, which include bradycardia, hyper- or hypotension and changes in cardiac conduction. These reports are difficult to analyse because of a lack of detailed information on parameters of exposure.

Several epidemiological studies have failed to reveal evidence that occu-pational exposure to EM fields can affect health or mortality. Yet other studies have yielded debatable evidence of adverse effects on health. However, evidence of adversity has not been found among the larger, better-designed studies such as those of Robinette *et al.* (1980) on 40 000 US Naval personnel and of Lilienfeld *et al.* (1978) which involved nearly 5000 employees and dependents in the Moscow embassy and more than 2500 employees and dependents in 'control' embassies.

Two reports by physiotherapists who use generators of RF EM fields in their work contain potentially significant findings of heart disease, primarily ischaemic, in male adults, and poor pregnancy outcomes in female therapists (Hamburger *et al.*, 1983; Kallen *et al.*, 1982). In neither study, unfortunately, were the authors able to provide adequate inform-ation on radiation levels in the working environment.

Some workers have attributed ocular defects to microwave exposure, but no findings have been advanced that can support a conclusion that low-level, chronic exposure to microwaves can induce cataracts in human beings (Elder and Cahill, 1984).

The question of microwaves as a cause or promoter of cancer has also been raised. The results of Lilienfeld *et al.* (1978) did not show any statis-tically significant association between microwave radiation and inci-dences of mortality or morbidity from any illness, including cancer. In contrast, Milham (1982) and Coleman *et al.* (1983) have reported increased mortality and morbidity ratios for leukaemia among workers with presumed exposure to electric and magnetic fields, but Heynick (1987) suspects the integrity of Milham's methodology.

The currently available epidemiological data on human exposures to EM fields suffer a common, if understandable, failing: Data on stengths of fields to which individual human beings have been exposed are virtually nil, as are durations of individual exposures. Personal dosimeters of the sort that measure particulate radiations do not exist for the non-ionizing radiations. The comparisons of morbidity and mortality that exist for exposed and non-exposed populations are based solely on groups. That is, the general environment of the work place, not the person or the individual, has provided the index of exposure. In addition, the environ-ments in which the populations have been studied have usually been con-taminated by noise, dust and toxic agents. Thus, the data on EM exposures of human beings are confounded and cannot be used to formulate general population or occupational limits on exposure to EM fields. To construct such limits, the committees charged with development of exposure standards have had to rely on the results of experimental studies, as described below.

18.2.3. Experimental studies

18.2.3(a) Cellular and subcellular effects

In vitro studies have been conducted on enzyme and DNA preparations, cell membranes, phospholipid vesicles, isolated mitochondria and neurons, red blood cells and bacteria. Well-defined dosimetry is required to minimize artifacts from inhomogeneous elevations of temperature in such preparations. No consistent cellular and subcellular effects can be attributed to EM-specific interactions for 1–3.2 GHz microwave exposures at SARs that have ranged from 5 W/kg to 375 kW/kg. Some effects have been seen at high SARs, but these can be attributed to significant elevations of temperature.

Exposure of isolated neurons from *Aplysia* sp., a marine gastropod, caused changes in the spontaneous firing rate of pacemaker neurons. The threshold SAR of the effect was as low as 1 W/kg. Several similar studies of electrical properties of the cell also have yielded reports of various effects. Because all living cells have gradients of electrical potentials across cell membranes, the effect of microwaves on the firing rate can be significant. However, these effects are reversible.

18.2.3(b) Haematological and immunological effects

Conflicting results on haematological effects have been reported. The experiments include *in vivo*, partial, or whole-body exposures of mice, rats, Chinese hamsters, guinea-pigs, rabbits and dogs to EM waves ranging from 26 MHz to 24 000 MHz at SARs of 1–70 W/kg, and *in vitro* exposure of human blood to 2450 MHz at 120–2000 W/kg. Both stimulatory and inhibitory effects have been reported. In general, the effects are similar to those associated with stress. This effect is true for high-intensity exposures. At lower intensities, when no detectable elevations of temperature have been observed, the possibility still exists that the animal is compensating by thermoregulation for an added thermal burden.

Both stimulation and suppression of the immune system by EM fields have been reported. In most cases, these effects can be attributed to an acute stress response. It is well known that stress alters physiological systems that regulate immunological function. At present, there is no convincing evidence of a direct, athermal effect of EM exposure on the immune system.

18.2.3(c) Reproductive effects

Teratogenic studies have been performed in which bird eggs and gravid mice and rats have been exposed to microwaves at discrete frequencies from 915 MHz to 6 GHz. In general, no adverse effects at low SARs under normal environmental conditions have been reported. At high power levels resulting in whole-body hyperthermia, EM radiation is unequivocally teratogenic. Exposure of rodents during gestation to EM fields at levels lower than those that result in death or malformations is associated with smaller foetal mass. In addition, gestational exposure to EM radiation may cause functional changes later in the animal's life.

Concerns for teratogenesis notwithstanding, an interesting application of microwaves to ease labour pain was reported by Daels (1976) of Belgium. The pelvic area of women in labour was irradiated by intense fields as a manoeuver to relax uterine muscles, minimize pain and facilitate delivery. No evidence of injury was manifest in a 1-year follow-up of 2000 children delivered by this method. However, the only exposure in this application was at time of parturition.

Regarding reproductive efficiency, only irradiation at extremely high SARs has been found to affect the female reproductive pattern. In contrast, moderate elevations of testicular temperature above the norm can cause reversible sterility. Exposure of mice and rats to 1.7–3.0 GHz microwaves has produced morphological and functional changes in these animals. The effects are dependent on the exposure intensity and duration. Sterility is seen at thresholds of temperature near 41–42°C. This effect has been applied in China for birth control in several hundred male subjects (Jiang *et al.*, 1987). Sperm counts were kept below fertility levels by a 20 min exposure per day of the testes to high-intensity micowave fields on two consecutive days per month. However, whether repeated elevations of testicular temperature might produce untoward effects in the recipients of microwave heating — or genetic defects in offspring later conceived — are unanswered questions. Because of concerns aroused by these questions, the study by Jiang and co-workers has been suspended.

18.2.3(d) Nervous system effects

Because there are myriad electrical activities in the nervous system, this system has been thought by many scientists to be the most sensitive to EM radiation. Many studies of the nervous system and behaviour (see below) have been performed. Morphological observations have revealed that EM radiation can cause changes in the central nervous system of experimental animals following acute or chronic exposures at SARs of 2 W/kg or higher.

Electrophysiological recordings from the brains and spinal cords of animals have been made during and following EM exposure. Johnson and Guy (1972) showed a threshold of 2.55 W/kg for decreased latency of evoked thalamic responses in cats. High-intensity exposure of the rat's head can stun the animal in a manner akin to electroconvulsive shock. The stun threshold of absorbed energy was near 28 kJ/kg, which resulted in an 8°C maximal temperature rise in the animals' brains (Guy and Chou, 1982). The effect on monosynaptic responses of the cat's spinal cord was shown to be similar to that induced by conventional heating. Chou and Guy (1978) also demonstrated, so long as the tissue temperature was held constant, that SARs as high as 2.2 kW/kg produced no change in compound action potentials of peripheral nerves.

Pharmacological studies have indicated that low-level microwave irradiation (i.e. at an SAR as low as 0.1 W/kg), especially when pulse modulated, may elicit a drug-specific interaction in the nervous system (for a review see Lai *et al.*, 1987). Effects on neurotransmitters were seen only at SARs higher than 8 W/kg. Whether there was an increase or decrease of the concentration level depended on the specific neurotransmitter.

18.2.3(e) Behavioural effects

Behavioural effects of microwave radiation have been extensively studied. There are reports from the USSR and other Eastern European countries indicating effects at power densities below 1 mW/cm². Properly measured, behaviour reflects the functioning of the intact organism and the status of the nervous system and many other systems functioning as a whole.

Naturalistic behaviour, i.e. spontaneously or naturally occurring behaviour, has been studied during exposures at various microwave frequencies and intensities. All effects reported in the Western literature showed thresholds of SAR near or above 1 W/kg. A Soviet group (Rudnev *et al.*, 1978) reported effects on rats at lower SARs near 0.1 W/kg. The effects included reductions in food intake, in time on a treadmill and on an inclined rod, and in exploratory activity. Sensitivity to a mild electric shock first increased and later decreased. Some effects persisted as long as 90 days after exposure. Despite attempts at duplication, the Soviet data generally have not been confirmed by Western investigators.

Decreases in ongoing operant behaviour of rats were observed during exposures to a 1280 MHz field at an SAR of 2.5 W/kg, and cessation of the behaviour was observed at an SAR of 10 W/kg in rats exposed at 500 MHz. King *et al.* (1971) reported that the threshold for detection of microwaves was as low as 0.6 W/kg in rats exposed to a 2450 MHz microwave field.

From all the reported behavioural effects, it is concluded that behaviour can be disrupted with an energy input at a rate that approximates one-quarter to one-half of the resting metabolic rate of the animal. The altered behaviours are, in general, reversible.

18.2.3(f) Special senses

To answer the question whether microwave exposure can cause cataracts in the human eye, experiments involving rabbits and monkeys have been conducted. The threshold intensity for cataract formation in rabbits was reported to be 150 mW/cm^2 (SAR=138 W/kg) for 2450 MHz near-field exposures. The threshold for multiple exposures was only slightly lower than that causing cataract after a single exposure. When tested over a frequency range of 0.8–107 GHz, radiation at frequencies between 1 and 10 GHz was more effective in causing cataracts. The mechanism of cataract formation in the rabbit is believed to be due to temperature elevation above 41°C in the posterior lens. Kramar *et al.* (1975) showed that cataracts were not produced in hypothermic rabbits exposed to a normally cataractogenic dose of microwaves. When monkeys were exposed to 2450 MHz microwaves at power densities to 500 mW/cm^2, no cataracts were observed during an observation period of 13 months. The exposures at 500 mW/cm^2 were sufficient to produce burns of tissue over the nasal bridge. McAfee *et al.* (1979) studied the effects of chronic, daily, 'voluntary' exposures of the faces of monkeys (*Macaca mulatta*) to pulsed 9.3 GHz microwaves at 150 mW/cm^2 for 3 months. In aspirating fruit juice from a straw, the animals activated the microwave field. No cataracts or corneal lesions were observed during a 1-year follow-up.

When the head of a human with normal hearing is exposed to pulsed microwaves, a sound is heard that is described as a click, buzz, chirp or a knocking sensation. The sound appears to originate from within the head. Individuals with middle-ear impairment but with intact bone conduction can hear the microwave pulses. The threshold energy density per pulse for the auditory sensation is very low (2–40 μJ/cm^2). The maximal rise in temperature of the exposed tissue is on the order of 10^{-5} to 10^{-6}°C per individual microwave pulse at the energy density threshold. Behavioural, physiological, psychophysical and physical studies of the microwave hearing effect have revealed that thermal expansion is the mechanism. The absorption of microwave energy produces non-uniform heating of the exposed head, which, through thermo-elastic expansion, launches a wave of pressure through bony tissues to the cochlea, in which it is detected. After excitation of the auditory nerve in the high-frequency portion of the cochlea, transmission of the microwave-induced neural response follows the same auditory pathway as do all acoustically induced responses, i.e. through the brainstem and thalamus to the auditory cortex (Chou *et al.*, 1982).

Exposure of the human body to microwaves can cause heating that is detectable by the temperature receptors in the skin. Justesen *et al.* (1982) exposed the ventral surface of the forearm of normal men and women to microwaves and then to infra-red radiation. The total energy absorbed at the threshold of detection was near 9.5 mJ/cm^2 for 2450 MHz microwaves and 1.7 mJ/cm^2 for the infra-red radiation. The difference is easily explained because infra-red has limited depth of penetration and stimulated the cutaneous thermoreceptors, which are situated superficially in the skin, much more efficiently than the more deeply penetrating microwaves. It follows that the cutaneous perception of radiant fields is not a reliable warning stimulus of potentially harmful levels of RF EM radiations at shortwave frequencies (30–300 MHz) and at microwave frequencies below 5 GHz.

18.2.3(g) Endocrine, physiological and biochemical effects

The endocrine system in combination with the nervous system is a major regulatory system in the body. Thyroid function and corticosteroid levels have been measured to study the effects of microwaves on the endocrine system. Effects on thyroid function have been reported at SARs as low as 2.1 W/kg, but negative results were also reported for SARs as high as 25 W/kg. The duration of exposure and the SAR are both important in producing the effect on thyroxine levels. Increase of corticosterone levels has been reported at SARs as low as 10 W/kg but not at 6.25 W/kg, and adrenal responses have been reported at levels of 4.6 W/kg but not at 3 W/kg.

Serum chemistry changes that have been reported are those that would be expected from heat stress. Altered oxygen consumption was reported at 10.4 W/kg for mice, 6.5 W/kg for rats and 0.9 W/kg for squirrel monkeys. Growth and development were not affected by microwave exposures at SARs below 2.8 W/kg in rats exposed pre- and post-natally, or postnatally only.

The results of microwave exposure on the cardiovascular system indicate that whole-body exposure of sufficient intensity to produce heating also produces an increase in heart rate similar to that from conventional heating. Bradycardia observed on isolated heart preparations and in whole animals during microwave exposure has also been reported, but it is not explainable as a thermal effect. It is highly probable that, if an electrode of high electrical conductivity was used for recording, the artifactual electrical stimulation of the myocardium was responsible for the bradycardia.

18.2.3(h) Modulation effects

The biological effects of modulated EM radiation are intriguing. Experimental results from four laboratories have shown that EM fields, sinusoidally modulated at sub-ELF (extremely low frequency) levels, especially at 16 Hz, cause central nervous system changes in various *in vitro* preparations and in live animals. The SARs were less than 0.05 W/kg. Because the effect is dependent on modulation and not on power density, it cannot be based on a simple thermal mechanism. Other than the nervous tissues, studies have also been carried out on T lymphocytes and pancreatic tissue. The physiological significance of these field-induced effects is not known. The response to specific frequencies and intensities is unusual and at present is unexplained.

18.2.3(i) Genetic and mutagenic effects

Experiments designed to examine the genetic consequences of exposure to EM radiation have been conducted on a variety of preparations, including isolated DNA, prokaryotic and eukaryotic cells and the intact animal. Radiation-induced effects in biochemical properties of DNA, chromosomal structures, induction of mutations and reproductive capabilities have been studied. Evidence of mutagenesis was observed only with highly intense fields that resulted in substantial thermal loading.

18.2.3(j) Longevity

In studies by different investigators, mice have been exposed to microwaves at frequencies from 800 to 9270 MHz, then observed for longevity. Most studies revealed slight increases in life span, some significant, others lacking in statistical reliability. A study conducted at the University of Washington showed a small but statistically unreliable enhancement of life span of 100 rats (versus 100 controls) that were exposed to a 2450 MHz field at 0.2–0.4 W/kg, 21 hours per day for 25 months (Guy *et al.*, 1985).

18.2.3(k) Carcinogenesis

Preskorn *et al.* (1978) reported that the development of tumours following injection of sarcoma cells into 16-day-old mice was significantly delayed if the mice had been exposed to 2450 MHz microwaves in utero on days 11, 12, 13 and 14 of gestation (20 min/day; SAR of 35 W/kg). However, the total tumour counts were eventually the same in both exposed and control groups as the animals reached an advanced age. In contrast, Szmigielski *et al.* (1982) demonstrated that repeated exposure of mice to 2450 MHz microwaves at 5 or 15 mW/cm^2 (SARs of 2–3 or 6–8 W/kg) 2 h per day, 6 days per week, for various periods to 10 months, accelerated the appearance of spontaneous mammary cancer in C3H/HeA mice, and accelerated skin cancer in male Balb/c mice treated with 3,4-benzopyrene during or after microwave exposure. Whether these effects are a direct, field-specific action of the microwaves on sensitive cells, a stress reaction or a generalized inhibition of the immune response, is not clear. Adey (1988) has suggested that signal amplification of weak, applied fields at the level of the cell membrane may be a factor in field-dependent promotion of cancer.

The University of Washington study showed that 85 of 200 animals had neoplastic lesions; 45 had been exposed to microwaves and 40 were controls. Statistical evaluation revealed no significant difference between the total counts of neoplasia. The neoplasms were then identified as primary or metastatic, benign or malignant. When all primary malignancies were pooled, the number was higher in the exposed animals (18 of 100) than in controls (five of 100), a statistical difference that is highly significant (P=0.006). The collapsing of data on malignant tumours without regard for tissue type and origin can be useful in detecting statistical trends. The finding of a near four-fold increase of primary malignancies in the exposed animals is provocative, although the findings must be considered in the light of the other biological data, especially those on longevity, which actually slightly favoured the controls. It is possible that the difference (18 compared to five) is spurious — not real — but prudence dictates the need to suspend judgement on the end-point of malignancies, and the need is manifest for independent evaluation of cancer induction or promotion in large-scale studies akin to that of Guy *et al.* (1985).

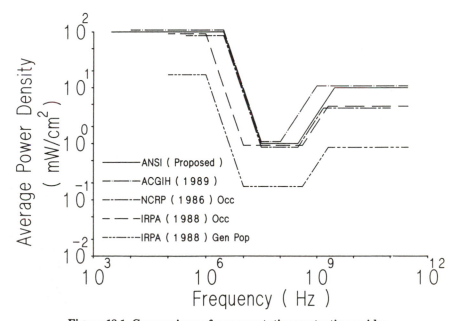

Figure 18.1 Comparison of representative protection guides.

18.3. RF safety standards

Figure 18.1 shows exposure limits that have been proposed by the American National Standard Institute (ANSI, 1990), the National Council on Radiation Protection and Measurements (NCRP, 1986), the American Conference of Governmental Industrial Hygienists (ACGIH, 1989), and the International Radiation Protection Association (IRPA, 1988). For clarity, overlapping curves are slightly shifted. Although the concerns for stray radiation are primarily to protect operating personnel, the general population standard of IRPA is also shown for comparison. At low frequencies (<100 MHz), near-field exposures are expressed in terms of magnitudes (root mean square) of electric and magnetic field strengths (Table 18.1). Shock and burn problems in the 3 kHz to 3 MHz range are also addressed in the proposed ANSI limits. For details on rationales, exclusions and special conditions the texts of original standards should be consulted.

Table 18.1 Proposed 1990 ANSI RF protection guide for controlled environments.

Part A. Electromagnetic fields[1].

Frequency f (MHz)	Electric field (V/m)	Magnetic field (A/m)	Equivalent power density (mW/cm²)	Averaging time (min)
0.003 – 0.1	614	163		6
0.1 – 3.0	614	16.3/f		6
3 – 30	1842/f	16.3/f		6
30 – 100	61.4	16.3/f		6
100 – 300	61.4	0.163	1.0	6
300 – 3000			f/300	6
3000 – 15 000			10	6
15 000 – 300 000			10	$616\,000/f^{1.2}$

[1]The electric and magnetic field strengths are the value obtained by spatially averaging values over a volume equivalent to the human body.

Part B. Induced RF Currents[2]

Frequency f (MHz)	Maximum current (mA)		
	Through both feet	Through each foot	Contact
0.003–0.1	2000f	1000f	1000f
0.1–100	200	100	100

[2]It should be noted that the currents given above may not adequately protect against startle reactions caused by transient discharges when contacting an energized object.

Exposure standards of the UK, proposed by the National Radiological Protection Board (NRPB, 1986) and the FRG, proposed by the Deutsche Institut für Normung (DIN, 1984), are shown in Tables 18.2 and 18.3 (Hand, 1990). The principal considerations in exposure limitation of the FRG standards are direct effects on nerve and muscle cells. The threshold values of current densities for the induction of the effects on nerve, muscle,

Table 18.2 UK RF safety standards 1986: NRPB occupational exposure.

Frequency f (MHz)	Electric field (V/m)	Magnetic field (A/m)	Power density (mW/cm²)
0.3 – 10	600/f	5.0/f	
10 – 30	60	5.0/f	
30 – 100	60	0.16	1
100 – 500	$6f^{1/2}$	$0.016f^{1/2}$	f/100
500 – 3×10⁵	135	0.36	5

brain, heart and single cells depend on the frequency and duration of applied stimulus. These neurological factors are the basis of limits on fields below 30 kHz, the effects of which are mainly on excitable membranes. Excessive heating is the limiting factor for fields at frequencies above 30 kHz. Many other countries also have published safety guidelines, which should be observed by practitioners of hyperthermia in those particular countries. If not available, the IRPA standard is an alternative. Protective measures listed in the document should be followed.

Table 18.3 FRG RF Standards 1984: DIN 57 848 Teil 2.

Frequency f (MHz)	Electric field (V/m)	Magnetic field (A/m)	Power density (mW/cm²)
0.010–0.03	1500	500	
0.030–2	1500	7.5/f	
2–30	3000/f	7.5/f	
30–3×10³	100	0.25	2.5
3×10³–12×10³	$100(f/3000)^{1/2}$	$0.25(f/3000)^{1/2}$	2.5(f/3000)
12×10³–3×10⁶	200	0.5	10

In 1982, ANSI C95 subcommittee IV conducted a review of the literature that, in total, included preparations and species that ranged from quail embryos to intact primates, and frequencies that ranged from 0.147 to 9.6 GHz. In this review, the subcommittee took into consideration the resonant frequencies of human bodies, smallest infant to tallest adult (30–300 MHz), in devising frequency-dependent limits on human exposure. The 1982 ANSI standard replaced the frequency-independent 1974 standard. In addition, the ANSI subcommittee took the following steps:

1. Determination of whole-body SARs at which acute adverse biological effects occur in animals. It was consensually agreed, after an intensive review of the literature, that the threshold level of harm — as determined by evidence of behavioural incapacitation — lies near 4 W/kg.

2. Application of a safety factor of 10 for human beings, which reduced that maximal permissible, whole-body averaged SAR to 0.4 W/kg for any period of time in excess of 6 min.
3. Derivation of theoretical 'worst-case' power densities of incident radiation that would result in a whole-body SAR of 0.4 W/kg.
4. Compilation of the resulting, frequency-dependent guides on limits of exposure.

Nationally and internationally, other regulatory organizations have adopted the SAR in particular and dosimetry in general in formulating their own standards. The 1982 ANSI standard is now under revision. Although the NCRP and IRPA have adopted two-tier standards, i.e. general-population limits at one-fifth those of the occupational standards, the ANSI subcommittee debated the scientific rationale of this approach.

None of the standards shown in Figure 18.1 applies to the purposeful exposure of patients by or under the direction of medical practitioners. Therefore, the patients are exempt from these regulations. However, hyperthermia treatment personnel will have to conform to the occupational limits. As research on biological effects of EM exposure progresses, the regulatory guidelines will periodically undergo revision to provide the most enlightened protection of operating personnel.

18.4. Biological effects of ultrasound

18.4.1. Human studies

Most human exposures to ultrasound occur as a result of medical applications, such as diagnostics, physical therapy, dental cleaning, surgery, hyperthermia and renal litholysis (Ziskin, 1988). Ultrasound has been used in physical therapy at power densities circa 3000 mW/cm^2 for more than 40 years. According to a large survey involving millions of treatments, a very small fraction of patients have experienced temporary pain or other adverse symptoms during or after treatment. There has been no recorded instance of significant harm to a patient attributable to an ultrasound treatment that was applied according to recommended practice. Epidemiological studies of diagnostic ultrasound at power densities below 100 mW/cm^2 have been performed but have not revealed evidence of ill effects (NCRP Report 74, 1983).

18.4.2. Animal studies

Interest in the effects of ultrasound on biological tissues, organs and intact animals has centred on the role of the thermal events in the production of irreversible structural changes and determination of dose relationship at threshold levels (NCRP, 1983). Many experiments have involved exposure of gravid mice or chicken embryos. The end-points have been foetal mass, postpartum mortality, and/or developmental abnormalities. Other studies have assayed for focal lesions, hind limb paralysis, effects on gonads, blood flow stasis, wound healing, reduction of tumours, DNA synthesis and immunological changes.

Although many of the above reported effects were due to elevations of tissue temperature, athermal effects from radiation force and cavitation (stable or transient cavitation in liquid media due to high intensity acoustic pressure) also have been observed. Power densities have varied from 120 mW/cm^2 in teratological studies to 20 kW/cm^2, an extremely high density at which the sonic field can rapidly produce focal lesions. With respect to archival reports of studies of experimental animals, the following statement on *in vivo*, mammalian biological effects of ultrasound was affirmed in 1982 by the American Institute of Ultrasound in Medicine (AIUM) BioEffects Committee (1983):

> In the low megahertz frequency range there have been no independently confirmed significant biological effects in mammalian tissues exposed to intensities below 100 mW/cm^2. Furthermore, for ultrasonic exposure times less than 500 seconds and greater than one second. such effects have not been demonstrated, even at higher intensities, when the product of intensity and exposure time is less than 50 J/cm^2.

The AIUM statement will be modified, if necessary, as new data are reported in the literature. This 100 mW/cm^2 level is simply a generalization based on animal observations. The AIUM does not claim that this level is safe. For diagnostic ultrasound, it is reasonable to keep the level low to avoid any biological effect. On the other hand, when there is obvious medical benefit, higher intensities will be necessary.

18.5. Ultrasound exposure criteria

Hill and ter Haar (1982), of the World Health Organization's Regional Office for Europe, have addressed the question of exposure limits on ultrasound diathermy: 'There does not seem to be any justification for using exposure levels above 3 W/cm^2'. Although there is a federal performance standard for ultrasound equipment in the USA, limits on exposure levels have not been set.

The AIUM and the National Electrical Manufacturer's Association published a safety standard for diagnostic ultrasound equipment in 1981 (AIUM–NEMA). There is no specification of hazardous energy or power densities. Instead, one finds the following statement on risk assessment and exposure criteria:

> The use of any active system on a human being involves risk, however low.

Although no definitive risk assessment can be made for diagnostic ultrasound, it must be presumed that such risk may exist. Therefore, the use of the minimal, practical, acoustic intensities and exposure durations is prudent. Given our imprecise knowledge of any potential hazard to human beings based upon current experimental and epidemiological studies, it is premature, and potentially regressive, to recommend quantitative guidelines.

Representatives of the World Health Organization's Regional Office for Europe also chose not to recommend an exposure limit in diagnostic ultrasound:

It appears at present that, with suitable existing equipment and techniques, many diagnostic procedures can be carried out entirely satisfactorily under conditions where the patient is exposed to a relatively low beam intensity such as 10 mW/cm^2 temporal–spatial average, or less (Hill and ter Haar, 1982).

The Environmental Health Directorate of Canada issued a guidance in 1981 stating that setting an upper limit on acoustic intensity for diagnostic ultrasound devices '...does not seem necessary from the data available at present. [However], ... manufacturers should make every endeavour to design and construct equipment so that it functions at spatial and temporal averaged intensities of less than 100 mW/cm^2' (EHD, 1981).

The NCRP Report 74 (1983) recommends the following exposure criteria:

1. Manufacturers and users should be guided by the following general principle: In a diagnostic examination, intensities, dwell times, and total exposure time should be no longer than are required to obtain the relevant clinical information.
2. Routine ultrasound examination of the human fetus should not be performed under exposure conditions where a significant temperature rise might be expected. (An increase of 1 °C in intra-uterine temperature is the limit.)
3. The establishment of a complete system of optimal exposure parameters for balancing benefit against risk should be accepted as a long-range goal, at least for those situations where it is found that there is a reasonable expectation of significant risk. Such a system would have to distinguish between different kinds of equipment and different applications and would allow for new technological and medical developments and for clinical judgement in individual cases.

18.6. Electrical safety

All hospitals should have standard procedures for inspecting electrical safety. The International Electrotechnical Commission has issued guidelines (IEC 601–1), and there are a few non-federal regulations available in the USA (Federal Register, 1987). Requirements for line cords, AC plugs, wire gauge and composition, ground-line resistance, and leakage current are specified in these regulations.

Leakage current is in general specified for three classifications of medical equipment:

1. Risk 3 equipment (in electrically conductive contact with patients having a foreign conductive pathway to the heart or to the great vessels such as pacing leads or catheters): 10 μA from any or all patient leads to ground, 100 μA from any point on equipment case to ground, 10 μA from any or all patient leads to ground through a 120 kΩ resistor.
2. Risk 2 equipment (intentionally placed in electrically conductive contact with patients not having foreign conductive pathways to the heart or to the great vessels): 50 μA from any or all patient leads to ground or 100 μA from any point on equipment to ground.
3. Risk 1 equipment (unlike risk 3 or 2 equipment, risk 1 equipment only makes incidental contact with a patient): 500 μA from any point on equipment case to ground.

The IEC 601–1 requires that the maximal leakage current be less than 10 μA. The Underwriters Laboratories publication (UL-544, 1988) on standards for safety of medical and dental equipment also specifies a 10 μA limit for isolated patient connection. Hyperthermia systems are considered risk 3 equipment because of their invasive thermometry. A thermometer's wires or interstitial needles or antenna are treated as conductive leads, and the applicators are defined as part of the chassis of equipment. If the leakage current is higher than allowed, a transformer should be installed to isolate the patient from the system. Problems like this should have been taken care of by the manufacturers or maintenance personnel and should not be a concern for the clinical staff.

18.7. Patient safety

As mentioned in Section 18.1, hyperthermia in many aspects is similar to diathermy. However, there are also differences between them. In diathermy, the existence of a malignant tumour is one of the contraindications because of concerns that subtherapeutic temperatures may accelerate tumour growth, and that increased blood flow may promote metastases (Lehmann, 1990). In hyperthermia, the tumour is the target of treatment. Therefore, some of the concerns for diathermy are not applicable to hyperthermia. Nevertheless, most of the contraindications to diathermy discussed by Lehmann are useful in consideration of hyperthermia and patient safety. Adverse effects of hyperthermia are very minor when compared with those attendant to radiation and chemotherapy. Burns and pain are the primary problems, and ulceration is commonly observed during regression of a thermally devitalized tumour. Vomiting and nausea are seen on rare occasion (Hahn, 1982).

18.7.1. Patients: either cautious treatment or no treatment

Diathermy treatments of an anaesthetized area of the body or of a patient with obtunded sensorium are generally contraindicated. The reason for this precaution is that dosimetry is not exact and, therefore, one has to rely on the ability of the patient to perceive pain as a warning that a threshold of thermal injury may be exceeded. In hyperthermia, treatments of insensitive areas are sometimes obligatory as, for example, in recurrence of breast cancer after mastectomy, or in anaesthetized patients requiring interstitial heating by RF electrodes. Special precautions must be taken that include careful monitoring of temperature and cooling of the skin.

Pregnant women should not undergo hyperthermia of the pelvis because of the possibility of thermal damage to the fetus. This precaution is especially important during the first trimester, when conception may not be known to the patient, and when the fetus is most sensitive to elevations of temperature. Although it is possible that the amniotic fluid selectively absorbs RF or microwaves energy, there is no evidence at this time that the uterine contents in pregnancy are selectively heated. However, in the absence of clear evidence to the contrary, and with the knowledge that a temperature of 38.9 °C (Smith *et al.*, 1978) is damaging to the human fetus, pregnancy should be considered a contraindication to hyperthermia.

Patients with metallic implants should be treated with caution. Thermocouples should be used with great care when monitoring temperatures during microwave-induced hyperthermia. In addition to electrical interference, perturbation of fields can modify heating patterns in the tumour (Chan *et al.*, 1988). If an appreciable amount of energy can reach the site of a metallic implant, hyperthermia should not be applied because it is difficult to predict the temperatures achieved near implants. Depending on the size and orientation of the implant, intense hot spots may form in unexpected areas. Even though it is theoretically possible to reduce selective heating of a metal implant by orienting it appropriately in an EM field, it is impossible to predict field orientation in tissues. On the other hand, if metal implants are located at sites not reached by an appreciable quantity of EM energy, their use is acceptable. Examples of metal implants include artificial joints, surgical clips, metal wires used in fracture repair, cardiac pacemakers, intrauterine devices and dental braces. Metallic objects exterior to the body, such as earrings, necklaces, wrist watches or bracelets and wire-framed glasses, should be removed from the treatment area. In regional heating by EM applicators, because of the intensity of stray radiations, metallic jewellery should be removed even if it is not within the treated area. These considerations also apply to local ultrasound hyperthermia because ultrasound reflects from a metal surface much as it does from bone.

Patients with cardiac problems should not be subjected to whole-body or regional hyperthermia. The large thermal input to the body causes increased cardiac output, which can be a severe stress to the cardiovascular system. Even for patients with strong hearts, monitoring of vital signs is highly recommended.

In general, children have not been treated with hyperthermia although there have been exceptions such as those treated for retinoblastoma and for advanced sarcomas. If necessary, the application should be done in such a way that a significant rise in temperature does not occur in the growth zones of their bones. Bone covered with minimal soft tissue may, under certain circumstances, be overheated significantly.

18.7.2. Critical tissues

Hyperthermia should not be applied to the eyes unless it is intended for ophthamological treatment because of the limited ability of the avascular lens to dissipate heat and the susceptibility of the eye to adverse effects. As mentioned before, high-density exposure of rabbit's eyes produces cataracts. It should be noted that a cataractogenic dosage is painful and in animals anaesthesia is required. This may indicate in conscious man that a cataractogenic dose cannot be applied without pain, which in turn would produce an immediate withdrawal from the exposure. However, although more than 40 years of diathermy practice has yielded no clear evidence that cataracts have been produced in man, exposure of the eye should be avoided in hyperthermia, except when this organ is the target.

Contact lenses should be removed before hyperthermia is induced in the vicinity of the eye. There may be a concentration of current, which could caused hot spots in the region of the ciliary body (Scott, 1956). It can also be assumed that contact lenses may interfere with evaporative cooling of the cornea. In general, caution should be observed when irradiating areas near the eyes.

Ultrasound should not be applied to fluid media in the body because cavitation is readily produced at therapeutic power levels and may damage tissues. This precaution applies to exposures of the eye and of the gravid uterus at hyperthermic intensities. In the latter situation, of course, heating of the foetus would also be detrimental. However, it should be recognized that because of the excellent focusing properties of ultrasound, it is quite possible to treat certain body parts without affecting the ocular or reproductive organs.

When the head of a human being is exposed to 915 MHz plane waves, energy absorption is focused on the centre of the brain near the hypothalamus, which controls thermoregulation (Johnson and Guy, 1972).

Theoretical calculations show a peak SAR of 0.458 W/kg in a head of 14 cm diameter when exposed at a power density of 1 mW/cm^2. This peak SAR value increases to 1.1 W/kg for a small head 7 cm in diameter. Behavioural and neurophysiological effects in human beings and other mammals have been reported, as described in a previous section. From these considerations, exposure of the skull to high-intensity microwaves should be avoided unless a careful dosimetric analysis is performed, to ensure absence of hot spots. Interstitial microwave applicators have been used to treat brain tumours because better control of temperature can be realized *in situ* than that realized with externally applied fields (Salcman and Samaras, 1983; Lyons *et al.*, 1984).

The heating of tissues with an inadequate vascular supply is contraindicated in diathermy because elevations of temperature will increase metabolic demands without associated vascular compensation. The result may be an ischaemic necrosis. Also, heating of hemorrhagic diatheses by diathermy is contraindicated, because bleeding is enhanced by the increase of vascularity. However, this problem is resolved in hyperthermia. An inadequate vascular supply to a tumour results in higher tumour temperatures, which is desirable. Hyperthermia can stop cancer bleeding because high temperatures can cause collapse of blood vessels.

The testes are sensitive to elevations of temperature and, therefore, exposure should be avoided (VanDemark and Free, 1971). The testicles are external to the body, and they are easily exposed to stray radiations during therapeutic application. In contrast to the testicles, the ovaries are covered by a thick layer of soft tissue and are difficult to expose to any significant amount of microwave radiation and heating.

Because of the high conductivity of beads of perspiration, they may be selectively heated by microwaves. The same selectivity has been used to heat mouse bladder tumours — by injecting high-salinity fluid into the bladder (Chou *et al.*, 1987a). Cooling the skin with air or a bolus of water can minimize accumulation of perspiration beads.

Application of ultrasound over open wounds should be avoided. Because of the required acoustical coupling, a gel is needed between the body and the ultrasound applicator. Actually, it is best to avoid contact with open wounds by any direct contact applicator, which can irritate the wound. Treating soft tissues that are thinly spread over bony areas with microwaves or ultrasound can also cause localized hot spots in the tissue–bone interface. Extra care must be taken with this condition. Ultrasound therapy should not be applied to the heart, in accord with recommendations in the NCRP Report 74 (1983).

18.7.3. Treatment recommendations

Adequate patient sensitivity to pain and temperatures in the exposed region must be assured prior to hyperthermia treatment. This assurance includes patient willingness to indicate when pain is experienced. Skin sensation should be tested over the region or area to be treated, and if found insensitive, hyperthermia should be used with extreme caution. The operator should be present throughout the treatment and should be alert to signs of patient distress. If, during a treatment, a patient experiences any pain in the involved area, the applicator should be adjusted, the temperature of a water bolus or of air used in cooling should be lowered, and/or the power level should be decreased to a level that eliminates pain. To reduce the probability of thermal injury, the therapist should encourage the patient to report any sensations of discomfort or pain.

18.7.4. Dosimetry

To prescribe a treatment, the desired heating pattern must cover the targeted tumour's volume. The instantaneous heating pattern is determined by the SAR and the specific heat of tissue. Over a long period of time, thermal diffusion and blood flow will alter the heating pattern into a time-varying temperature pattern. This is especially true when power levels are under feedback control. The rate of absorption of EM energy (i.e. the SAR) by tissues is a function of many factors. Among them are: dielectric properties, size, geometry, and depth of tissues; size and shape of applicator; as well as spacing or coupling between tissue and applicator. Therefore, for effective treatment in cancer hyperthermia, measurements of energy absorption in tissues are required to apply the EM energy for tissue heating.

Tissue-simulating models have been fabricated to simulate the dielectric properties and the geometry of tissues (Chou *et al.*, 1984). Through thermographic or other temperature-mapping techniques, SAR or heating patterns can be obtained in these models (Guy, 1971; Chou *et al.*, 1986; Guy *et al.*, 1986). When inhomogeneous tissues are involved, heating patterns instead of SAR patterns are more convenient to use. Three examples are shown below to illustrate the complex variations of the heating pattern. It should be pointed out that some characteristics can be generalized from one applicator to the other. However, in most situations, they are unique for a particular combination of physical parameters. Therefore, a case-by-case analysis is necessary to ensure that a heating pattern is acceptable.

18.7.4(a) Patient and applicator spacing

Figure 18.2 shows the heating pattern along the sagittal plane of a model consisting of fat 2 cm thick and muscle 10 cm deep exposed at 434 MHz by a TAG-MED applicator (TCA 434-1). The pattern on the left was for the applicator in direct contact with the fat surface. The two hot spots were

TAG MED   TCA 434−1   434 MHz   SLAB

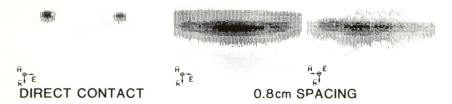

DIRECT CONTACT   0.8cm SPACING

Figure 18.2 Heating pattern in bisected plane showing hot spots at the edges of applicator with no spacing between applicator and model; with a spacing of 0·8 cm, uniformity of heating in muscle is improved.

produced in the fat with only slight heating of the muscle. When 0.8 cm spacing was used between the applicator and the model, the heating was greatly improved in the muscle region for the vector of electric fields either parallel or perpendicular to the interface (Chou *et al.*, 1986). Although quantitative data are not shown in this figure, our data show that fat heating was six times higher in the direct contact case as compared to that with the 0.8 cm spacing. This finding indicates that if the applicator is in

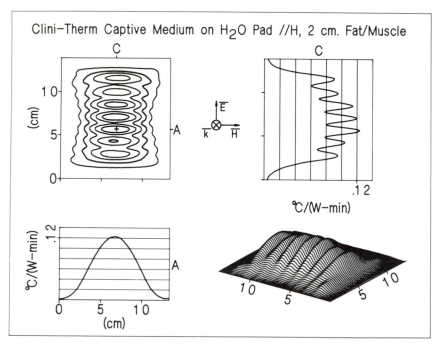

Figure 18.3 Perturbation of heating patterns on a fatty surface, which was produced by flow of water in a direction perpendicular to the electric field vector. When the direction of flow was parallel to the electric field vector, the perturbation was not observed.

direct contact with the patient, not only may the hot spots cause burns in the patient, but also little heating will occur in the muscle (or tumour).

18.7.4(b) Water bolus orientation

Figure 18.3 shows the heating pattern on the fat surface of a model exposed at 915 MHz by a Clini-Therm M applicator with a water bolus between the model and the model applicator (Chou *et al.*, 1990b). The water bolus was constructed with parallel tubes looping in a plastic bag. When the tubes were oriented perpendicular to the electric field, perturbations of the heating pattern appeared, as shown. After a 90° rotation, which made the water flow parallel to the electric field, there was minimal perturbation.

18.7.4(c) Fat thickness

Scott *et al.* (1986) reported unexpected blisters on a patient treated with a BSD MA151 applicator operated at 660 MHz. The blisters were within the 6 cm diameter of the circular aperture but outside the rectangular wave-

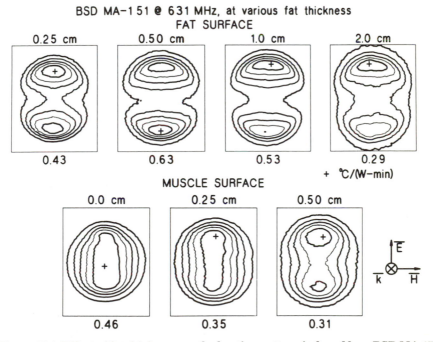

Figure 18.4 Effect of fat thickness on the heating pattern induced by a BSD MA-151 applicator. [Numbers above each heating pattern are fat thickness in cm and the numbers below are maximum rates of heating in °C/(W-min).]

guide. A study was conducted to investigate the cause of this problem, because studies of a model based on homogeneous muscle showed good central heating. Figure 18.4 shows the effect of fat thickness on the heating patterns at 631 MHz. A fat layer 0.25 cm thick can alter the central heating to a bipolar pattern. The variations at 915 MHz are different from those at 631 MHz (Chou *et al.*, 1990a).

The examples cited above show that detailed dosimetry is necessary for clinical applications that require different combinations of physical parameters. Some variables are more sensitive than others, depending on the design of each applicator. Details on the studies of models can be found in Chou (1987b, 1988). It is ultimately important to the physician and hyperthermia treatment team (i.e. physicist, engineer, nurse, technologist) that the proper functions of hyperthermia equipment be fully understood for the sake of the patient and for operator safety. When using EM heating, the external applicators are critical components which are in proximity to or actually make contact with the patient, and they can be the determining factors of effective and safe treatment. Before an applicator is used, detailed measurements should be made of its EM energy deposition for each condition.

In addition to the desired heating pattern, it is necessary to have a powerful generator and applicators that are capable of handling the power levels that are needed for effective heating. As Lehmann noted in his chapter on shortwave diathermy:

> The greatest present limitation of the equipment is the lack of power and, in many cases, its inability to produce peak temperatures any place except the subcutaneous fat. Thus, many of the commercially available pieces of equipment are safe but ineffective (Lehmann, 1990).

18.8. Personnel safety

Stray radiation during treatment may expose the therapist and normal tissues and organs of the patient to high densities of non-ionizing EM energy. Because of the impedance mismatch between air and tissue, stray radiation with ultrasound is negligible. It is essential that personnel using shortwave or microwave fields in cancer therapy are familiar with their biological effects and with guidelines for safe use; only by gaining this familiarity can they protect their patients and themselves from unnecessary exposure.

In the clinical application of any potent therapy, there is always a trade-off between benefit and risk, whereas in occupational exposure no hazards are acceptable. Two approaches to the development of an exposure standard can be used:

1. *Performance or emission standard.* The output and stray radiations of equipment are controlled in such a fashion that inadvertant exposure of the patient's or the therapist's sensitive organs cannot occur. This approach has been used for microwave ovens.
2. *Exposure standard.* Safeguards must be taken to ensure that neither the patient's nor therapist's sensitive organs are exposed at levels exceeding current limits of exposure. This type of standard in the medical setting must rely heavily on the intelligent use of the equipment by trained personnel who have adequate information about the performance of the machine. Control can be exercised to allow the use of such equipment only by physicians or therapists. In either case, it is necessary to identify safe exposure levels, as described in previous sections.

In their paper on 'United States safety and regulatory considerations and procedures for radiofrequency hyperthermia systems', Bassen and Coakley (1982) described many US government regulations such as pre-market approval (PMA), investigational device exemption (IDE), and the Institutional Review Board (IRB), of which all new investigators should be aware. The authors also describe a proposed microwave diathermy performance standard and safety protocols for microwave hyperthermia. Although this proposal was not promulgated because of cutbacks in US government programmes, users should be aware of the four concepts it addressed:

1. The leakage of microwave radiation is limited to 10 mW/cm^2 at a distance of 5 cm from any point on the surface of the diathermy system, including the applicator.
2. Various controls, indicators, RF-power stabilizers and controls, and safety locks are required. Net levels of power, not forward-power levels, should be monitored.
3. Effectiveness of heating must be quantified through tests involving each diathermy generator/applicator combination, on one or more of the three standard models developed by the Center for Device and Radiological Health. Production of a peak SAR of 235 W/kg (equivalent to a heating rate of 4 °C per min) in a standard model is required.
4. Informational requirements are developed to provide the operator of a microwave diathermy system with sufficient information to treat a patient with maximal heating effectiveness and minimal leakage radiation.

Lehmann (1982) has commented that the use of performance standards is not practical because no readily available applicator could fulfil the proposed leakage requirement of 5 or 10 mW/cm^2, 5 cm from the applicator's edge. Even applicators designed to fit perfectly the large and small cylindrical models and the plane-layered tissue-substitute models, as proposed by the US Food and Drug Administration, would be unable to fulfil the same criteria in human application to the corresponding parts of the anatomy, the arm, the leg and the back. It has been demonstrated that leakage occurs primarily in the direction in which direct contact is lost. In a linear polarized applicator, leakage is always greater in the direction of the electric field (Lehmann *et al.*, 1979). Knowledgeable use of such equipment would direct the stray radiation away from the patient's and therapist's sensitive organs. Of course, the standards that decree who may apply such equipment must also require that the physician and therapist be equipped with information on the stray radiation patterns produced by the various applicators as applied to the proposed models such that adequate judgement can be applied when prescribing treatment.

Correct placement of an applicator will not only couple energy more efficiently to the patient but also will reduce stray radiation. For example, the electric field should be oriented along the long axis of a tumour if it is elliptical in shape or is situated in an arm or leg. Direct contact applicators or water-bolus-coupled applicators are designed for better coupling and less stray radiation. They should be applied according to their design, otherwise hot spots may form at the edges of the applicator, as shown in Figure 18.2. Stray radiation can be minimized further by placing absorbing materials such as saline bags or commercial, lossy, absorbing foam around the treated area. Shielding with conductive material to reflect the EM energy can alter the heating pattern in the tumour. Beam shaping such as that used in ionizing radiations cannot be applied to hyperthermia applicators. At a further distance, a metallic, woven fabric used as a curtain or protective suit may be useful to further reduce stray radiations (Chou *et al.*, 1987b).

The magnitude of stray radiations should be monitored at the start of every treatment. If the patient's or the applicator's position is changed significantly during treatment, the measurement should be repeated. There are many commercial survey meters available. It is important to use one with enough sensitivity and range (at least down to 0.1 mW/cm² and up to 150 mW/cm²), with adequate accuracy ± 3 dB), and a fast response time (less than 1 s) (Bassen and Coakley, 1982). The safety standards as shown in Figure 18.1. should be referred to as guides.

18.9. Equipment safety

The most commonly used frequencies for microwave hyperthermia are 434, 915 and 2450 MHz, and for shortwave hyperthermia, 13.65 and 27.12 MHz. The RF local current field method uses 500 kHz, and the ferromagnetic seed technique operates at 100 kHz. Among these frequencies, 434 MHz and 100 kHz are not in the industrial, medical and scientific (ISM) bands as designated by the US Federal Communications

Commission. An electrically shielded room is necessary for frequencies outside ISM bands to minimize interference with sensitive communication networks. A screen window on the shielded room will allow operators to observe the patient during treatment instead of being inside the shielded room. Equipment such as the video monitor inside the room should be shielded to minimize interference. Unused line cords should be unplugged to avoid unwanted coupling with stray radiations. In operating high-power ISM-band hyperthermia equipment, interference with nearby medical instrumentation such as electroencephalogram or electroradiogram machines, blood pressure monitor, and intravenous drip pump, is possible. Interference with cardiac pacemakers was a problem many years ago, but newer versions are immune. Disturbance of television and radio reception is commonly observed.

Because field strengths diminish as distance from a source increases, usually the interference or disturbance can be eliminated or minimized by increasing the distance of the hyperthermia source and other affected machines. Shielding with metallic foil or screen is an alternative. Interference with thermocouples is a common problem. Low-pass filtering of the signal has proven to be useful. However, it is necessary to ensure that the metallic thermocouples or thermistors do not perturb the heating pattern in tissues.

18.10. Conclusions

Hyperthermia is still an investigational method for cancer treatment. Refinements to improve treatment continue. Although beneficial effects of shortwave, microwave and ultrasound hyperthermia have been observed, it is equally important to make the treatment safe for the patients and for the operators. In this chapter, various safety aspects have been considered. It is impossible to cover all aspects in detail. As time passes, other important safety considerations may evolve.

Hyperthermia is a complicated technique and should be applied only by individuals well trained in use of this modality. They should be knowledgeable in indications and contraindications, and in the recommended procedures pertinent to the application. They should also have a basic understanding of the underlying physical principles and a knowledge of potentially harmful effects. Safety guidelines should be observed. Certification of hyperthermia operators is a necessary step to ensure that qualified personnel perform effective and safe treatments (Luk, 1988). Equipment should be routinely checked at a frequency recommended by the quality assurance committee for proper calibration, performance and electrical safety (Shrivastava *et al.*, 1989). Careful dosimetry should be performed on all exposure geometries and contingencies prior to treatment to assure the best treatment conditions for the patient. Since hyperthermia in combination with high energy radiotherapy cannot be repeated after the tumour receives a maximal dosage of ionizing radiation, the physician must therefore try to reach the critical tumour temperatures in optimal conjunction with radiotherapy. Anaesthetizing or tranquilizing the patient may be necessary for an effective treatment. Accurate thermometry is particularly important in all phases of clinical hyperthermia, especially when the patient is anaesthetized. Compared with acute problems such as burns and pain, a successful treatment can result in permanent eradication of a malignant tumour. Thus, the benefit of the treatment outweighs minor risks. If there is no choice, it would be more beneficial for the patient to have an effective treatment with a few blisters rather than a safe but ineffective treatment. It is easy to treat the burns, but not the cancer.

Acknowledgements

This work was supported in part by National Cancer Institute Grant CA 33752. The author appreciates the generous help of his former Chairman , Professor Justus F. Lehmann, Department of Rehabilitation Medicine, University of Washington, in Seattle, and Dr Don Justesen of the University of Kansas School of Medicine and the Kansas City VA Medical Center.

References

ACGIH (1989) *Threshold Limit Values (TLV) and Biological Exposure Indices for 1988–1989*, American Conference of Governmental Industrial Hygienists.

Adair, E. R. (Ed.) (1983) *Microwaves and Thermoregulation*, New York, Academic Press, pp. 1–490.

Adey, W. R. (1988) Electromagnetic fields, cell membrane amplification and cancer promotion, *Nonionizing Electromagnetic Radiations and Ultrasound*, NCRP Proceedings No. 8, Bethesda, Maryland, National Council of Radiation Protection and Measurements, pp. 80–110.

AIUM BioEffects Committee (1983) *Safety Considerations for Diagnostic Ultrasound*, Washington, DC, American Institute of Ultrasound in Medicine.

AIUM–NEMA (1981) AIUM–NEMA safety standard for diagnostic ultrasound equipment, *AIUM–NEMA Publication UL 1–1981*, Washington, DC, National Electrical Manufacturers Association.

ANSI (1982) *American National Standard Safety Levels with Respect to Human Exposure to Radiofrequency Electromagnetic Fields, 300 kHz to 100 GHz*, New York, IEEE, pp. 1–24.

ANSI (1990) *American National Standard Proposed Safety Levels with Respect to Human Exposure to Radiofrequency Electromagnetic Fields, 3 kHz to 300 GHz*, proposed.

Baranski, S. and Czerski, P. (1976) *Biological Effects of Microwaves*, Stroudsburg, Pennsylvania, Dowden, Hutchinson and Ross, pp. 1–234.

Bassen, H. I. and Coakley Jr, R. F. (1982) United States Radiation safety and regulatory considerations and procedures for radiofrequency hyperthermia systems, in Nussbaum, G. H. (Ed.), *Physical Aspects of Hyperthermia*, New York, American Institute of Physics, pp. 372–392.

Chan, K. W., Chou, C. K., McDougall, J. A. and Luk, K. H. (1988) Changes in heating patterns due to perturbations by thermometer probes at 915 and 434 MHz, *International Journal of Hyperthermia*, 4(4), 447–456.

Chou, C. K. (1987a) Biological effects of electromagnetic waves, in Field, S. B. and Franconi, C. (Eds) *Physics and Technology of Hyperthermia*, Dordrecht, Martinus-Nijhoff, pp. 319–353.

Chou, C. K. (1987b) Phantoms for electromagnetic heating studies, in Field, S. B. and Franconi, C. (Eds), *Physics and Technology of Hyperthermia*, Dordrecht, Martinus-Nijhoff, pp. 294–318.

Chou, C. K. (1988) Electromagnetic energy deposition patterns in phantoms, in Paliwal, B., Hetzel, F. and Dewhirst, M. (Eds) AAPM Monograph 16, New York, American Institute of Physics, pp. 132–151.

Chou, C. K. (1990) Use of Heating Rate (HR) and Specific Absorption Rate (SAR) in the Hyperthermia Clinic, *International Journal of Hyperthermia*, 6(2), 367–370.

Chou, C. K. and Guy, A. W. (1978) Effects of electromagnetic fields on isolated nerve and muscle preparations, *IEEE Transactions on Microwave Theory and Techniques*, 26, 141–147.

Chou, C. K. and Guy, A. W. (1985) Research on nonionizing radiation: Physical aspects in extrapolating infrahuman data to man, in Monahan, J. C. and D'Andrea, J. A. (Ed.), *Behavioural Effects of Microwave Radiation Absorption*, HHS Publication FDA 85–8238, Rockville, Maryland Food and Drug Administation, pp. 135–149.

Chou, C. K., Guy, A. W. and Galambos, R. (1982) Auditory perception of radiofrequency electromagnetic fields, *Journal of the Acoustic Society of America*, 71(6), 1321–1334.

Chou, C. K., Chen, G. W., Guy, A. W. and Luk, K. H. (1984) Formulas for preparing phantom muscle tissue at various radio frequencies, *Bioelectromagnetics*, 5(4), 435–441.

Chou, C. K., Guy, A. W., McDougall, J. A., Dong, A. and Luk, K. H. (1986) Thermographically determined SAR patterns of 434 MHz applicators, *Medical Physics*, 13(3), 385–390.

Chou, C. K., See, W. A., Luk, K. H. and Chapman, W. H. (1987a) Microwave hyperthermia system for treating bladder carcinoma in rats, *Endocurietherapy and Hyperthermia Oncology*, 3, 147–152.

Chou, C. K., Guy, A. W. and McDougall, J. A. (1987b) Shielding effectiveness of improved microwave-protective suit, *IEEE Transactions on Microwave Theory and Techniques*, 35(11), 995–1001.

Chou, C. K., McDougall, J. A., Chan, K. W., Luk, K. H., (1990a) Effects of fat thickness on heating patterns of the microwave applicator MA–151 at 631 and 915 MHz, *International Journal of Radiation Oncology Biology and Physics* (in press).

Chou, C. K., McDougall, J. A., Chan, K. W., and Luk, K. H., (1990b) Evaluation of captive bolus applicators, *Medical Physics* (in press).

Coleman, M., Bell, J. and Skeet, R. (1983) Leukemia incidence in electrical workers, *Lancet*, I, 982–983.

Daels, J. (1976) Microwave heating of the uterine wall during parturition, *Journal of Microwave Power*, 11, 166–168.

DIN (1984) *Hazards by Electromagnetic Fields: Protection of Persons in the Frequency Range 10 kHz to 3000 GHz*, DIN 57 848 Teil 2, Berlin, Deutsche Elektrotechnische Kommision im DIN und VDE (DKE).

EHD (1981) *Safety Code 23 Guidelines for the Safe Use of Ultrasound. Part I. Medical and Paramedical Applications*, Report 80-EHD-59, Ottawa, Environmental Health Directorate, Health Protection Branch.

Elder, J. A. and Cahill, D. F. (1984) *Biological Effects of Radiofrequency Radiation*, Report EPA-600/8-83-026F, Research Triangle Park, North Carolina, Health Effects Research Laboratory, Office of Research and Development, USEPA.

Federal Register (1987) *Safety of electrical Medical Devices — Withdraw of Notice of Intent*, Docket number 80N-0005, 52(199), 38276-38279.

Guy, A. W. (1971) Analyses of electromagnetic fields induced in biological tissues by thermographic studies on equivalent phantom models, *IEEE Transactions on Microwave Theory and Techniques*, 19, 205–214.

Guy, A. W. and Chou, C. K. (1982) Effects of high-intensity microwave pulse exposure on rat brain, *Radio Science*, 17(5S), 169s–178s.

Guy, A. W., Chou, C. K., Kunz, L. L., Crowly, J. and Krupp, J. H. (1985) Effects of long-term low-level radiofrequency radiation exposure on rats, Summary, *USAF Report SAM-TR-85-64*, **9**, 1–23.

Guy, A. W., Chou, C. K. and Luk, K. H. (1986) 915 MHz phased-array system for treating tumours in cylindrical structures, *IEEE Transactions on Microwave Theory and Techniques*, **34**(5), 502–507.

Hahn, G. M. (1982) *Hyperthermia and Cancer*, New York, Plenum Press.

Hamburger, S., Logue, J. N. and Silverman, P. M. (1983) Occupational Exposure to non-ionizing radiation and an association with heart disease, an exploratory study, *Journal of Chronic Disease*, **36**(1), 791–802.

Hand, J. W. (1990) Biophysics and technology of electromagnetic heating, in Gautherie, M. (Ed.) *Methods of External Hyperthermic Heating*, Munich, Springer-Verlag, pp. 1–59.

Heynick, L. N. (1987) Critique of the literature on bioeffects of RF radiation, a comprehensive review pertinent to Air Force operations, *USAF Report SAM-TR-87-3*, Texas, Brooks Air Force Base, pp. 1–692.

Hill, C. R. and ter Harr, G. (1982) Ultrasound, in Suess, M. (Ed.), *Nonionizing Radiation Protection*, Copenhagen, World Health Organization Regional Office for Europe, Chapter 6.

IEC 601-1 (1977) *Specification for Safety of Electrical Equipment Used in Medical Practice*, Part1: General requirements, Geneva, International Electrotechnical Commission.

IRPA (1988) Guidelines on limits of exposure to radiofrequency electromagnetic fields in the frequency range from 100 kHz to 300 GHz, *Health Physics*, **54**(1), 115–123.

Jiang, H. B., Chou, C. K., Guy, A. W. and Luk, K. H. (1987) Heating patterns in male reproductive organs exposed to high-intensity microwaves (in Chinese), *Reproduction and Contraception*, **7**(4), 38–42.

Johnson, C. C. and Guy, A. W. (1972) Nonionizing electromagnetic wave effects in biological materials and systems, *Proceedings of the IEEE*, **60**, 692–718.

Justesen, D. R. (1975) Toward a prescriptive grammar for the radiobiology of non-ionizing electromagnetic radiations: Quantities, definitions, and units of absorbed electromagnetic energy, *Journal of Microwave Power*, **10**, 333–356.

Justesen, D. R. and King, N. W. (1970) Behavioral effects of low-level microwave irradiation in the closed-space situation, in Cleary, S. F. (Ed.), *Symposium on Biological Effects and Health Implications of Microwave Radiation*, BRH/DBE 70–2, Rockville, Maryland, USPHS, pp. 154–179.

Justesen, D. R., Adair, E. R., Stevens, J. C. and Bruce-Wolfe, V. (1982) A comparative study of human sensory thresholds: 2450-MHz microwaves vs far-infrared radiation, *Bioelectromagnetics*, **3**(1), 117–125.

Kallen, B., Malmquist, G. and Moritz, U. (1982) Delivery outcome among physiotherapists in Sweden: is nonionizing radiation a fetal hazard? *Archives Environmental Health*, **37**(2), 81–85.

King, N. W., Justesen, D. R. and Clarke, R. L. (1971) Behavioral sensitivity to microwave irradiation, *Science*, **172**, 398–401.

Kloth, L., Morrison, M. A. and Ferguson, B. H. (1984) *Therapeutic Microwave and Shortwave Diathermy — A Review of Thermal Effectiveness, Safe Use, and State of the Art, 1984*, HHS Publication FDA, 85–8237, US Department of Health and Human Services, pp. 1–37.

Kramar, P. O., Emery, A. F., Guy, A. W. and Lin, J. C. (1975) The ocular effects of microwaves on hypothermic rabbits: a study of microwave cataractogenic mechanisms, *Annals of New York Academy of Science*, **247**, 155–165.

Kramar, P., Harris, C., Emery, A. F. and Guy, A. W. (1978) Acute microwave irradiation and cataract formation in rabbits and monkeys, *Journal of Microwave Power*, **13**, 239–249.

Lai, H., Horita, A., Chou, C. K. and Guy, A. W. (1987) A review of microwave irradiation and actions of psychoactive drugs, *IEEE Engineering in Medicine and Biology Magazine*, **6**(1), 31–36.

Lehmann, J. F. (Ed.) (1982) *Therapeutic Heat and Cold*, 3rd edn, Baltimore, Williams and Wilkins, pp. 1–641.

Lehmann, J. F. (Ed.) (1990) *Therapeutic Heat and Cold*, 4th edn, Baltimore, Williams and Wilkins, pp. 1–725.

Lehmann, J. F., Stonebridge, J. B. and Guy, A. W. (1979) A comparison of patterns of stray radiation from therapeutic microwave applicators measured near tissue substitute models and human subjects, *Radio Science*, **14**(5S), 271s–283s.

Lilienfeld, A. M., Tonascia, J., Tonascia, S., Libauer, C. A. and Cauthen, G. M. (1978) *Foreign Service Health Status Study — Evaluation of Health Status of Foreign Service and Other Employees from Selected Eastern European Posts, Final Report*, Contract No. 6025–619073, NTIS PB–288163, Washington, DC, Department of State.

Luk, K. H. (1988) Training and certification issues in hyperthermia, in Paliwal, B., Hetzel, F. and Dewhirst, M. (Eds), *Biological, Physical and Clinical Aspects of Hyperthermia*, AAPM Monograph 16, New York, American Institute of Physics, pp. 476–483.

Lyons, B. E., Britt, R. H. and Strohbehn, J. W. (1984) Localized hyperthermia in the treatment of malignant tumors using interstitial microwave antenna array, *IEEE Transactions on Biomedical Engineering*, **31**(1), 53–62.

McAfee, R. D., Longacre, Jr, A., Bishop, R. R., Elder, S. T., May, J. G., Holland, M. G. and Gordon, R. (1979) Absence of ocular pathology after repeated exposure of unaneasthetized monkeys to 9.3-GHz microwaves, *Journal of Microwave Power*, **14**(1), 41–44.

Milham Jr, S. (1982) Mortality from leukemia in workers exposed to electric and magnetic fields, *New England Journal of Medicine*, **207**, 249.

NCRP (1981) *Radiofrequency electromagnetic fields: Properties, Quantities and Units, Biophysical Interaction, and Measurements*, Report 67, Bethesda, Maryland National Council of Radiation Protection and Measurements, pp. 1–134.

NCRP (1983) *Biological Effects of Ultrasound: Mechanisms and Clinical Implications*, Report 74, Bethesda, Maryland, National Council of Radiation Protection and Measurements, pp. 1–266.

NCRP (1986) *Biological Effects of Exposure Criteria for Radiofrequency Electromagnetic Fields*, Report 86, Bethesda, Maryland, National Council of Radiation Protection and Measurements, pp. 1–382.

NCRP (1988) *Nonionizing Electromagnetic Radiations and Ultrasound*, Proceedings No. 8, Bethesda, Maryland, National Council of Radiation Protection and Measurements, pp. 1–384.

NRPB (1986) *Advice on the Protection of Workers and Members of the Public from the Possible Hazards of Electric and Magnetic Fields with Frequencies Below 300 GHz: a Consultative Document*, Didcot, UK, National Radiological Protection Board.

Osepchuk, J. M. (Ed.) (1983) *Biological Effects of Electromagnetic Radiation*, New York, IEEE, pp. 1–593.

Polk, C. and Postow, E. (Eds) (1986) *CRC Handbook of Biological Effects of Electromagnetic Fields*, Boca Raton, CRC Press, pp. 1–503.

Preskorn, S. H., Edwards, W. D. and Justesen, D. R. (1978) Retarded tumour growth and greater longevity in mice after fetal irradiation by 2450-MHz microwaves, *Journal of Surgical Oncology*, **10**, 483–492.

Robinette, C. D., Silverman, C. and Jablon, S. (1980) Effects upon health of occupational exposure to microwave radiation (radar), *American Journal of Epidemiology*, **112**, 39–53.

Rudnev, M., Bokina, A., Eksler, N. and Navakatikyan, M. (1978) The use of evoked potential and behavioral measures in the assessment of environmental insult, in Otto, D. A. (Ed.), *Multidisciplinary Perspectives in Event-related Brain Potential Research*, EPA-600/9-77-043, US EPA, Research Triangle Park, North Carolina, pp. 444–447.

Salcman, M. and Samaras, G. M. (1983) Interstitial microwave hyperthermia for brain tumors, Results of a phase-I clinical trial, *Journal of Neurooncology*, **1**(3), 225–236.

Scott, B. O. (1956) Effect of contact lenses on shortwave field distribution, *British Journal of Ophthalmology*, **40**, 696–697.

Scott, R., Chou, C. K., McCumber, M., McDougall, J. A. and Luk, K. H. (1986) Complications resulting from spurious fields produced by a microwave applicator used for hyperthermia, *International Journal of Radiation Oncology, Biology and Physics*, **12**, 1883–1886.

Shrivastava, P., Luk, K., Oleson, J., Dewhirst, M., Pajak, T., Paliwal, B., Perez, C., Sapareto, S., Saylor, T., Steeves, R. (1989) Hyperthermia Quality assurance guidelines, *International Journal of Radiation Oncology, Biology and Physics*, **16**, 571–587.

Smith, D. W., Clarren, S. K. and Harvey, M. A. S. (1978) Hyperthermia as a possible teratologic agent, *Journal of Pediatrics*, **92**, 878–883.

Szmigielski, S., Szudzinski, A., Pietraszek, A., Bielec, M., Janiak, M. and Wrembel, J. K. (1982) Accelerated development of spontaneous and benzopyrene-induced skin cancer in mice exposed to 2450–MHz microwave radiation, *Bioelectromagnetics*, **3**(2), 179–191.

Tyler, P. E., (Ed.) (1975) Biological effects of nonionizing radiation, *Annals of the New York Academy of Science*, **247**, 1–545.

UL-544 (1988) *Standard for Safety: Medical and Dental Equipment*, Chicago, Underwriters Laboratories, Inc, pp. 30–33.

VanDemark, N. L. and Free, M. J. (1971) Temperature effects, in Johnson, A. D., Gomes, W. R. and VanDemark, N. L. (Eds), The Testes, New York, Academic Press, **3**, 223–312.

Ziskin, M. C. (1988) Ultrasound: medical applications and medical exposures, *Nonionizing Radiation and Ultrasound*, Proceedings No. 8, Bethesda, Maryland, National Council of Radiation Protection and Measurements, pp. 308–328.

Index